Genetic Modification: Methods and Techniques

Genetic Modification: Methods and Techniques

Editor: David Rhodes

CALLISTO REFERENCE

www.callistoreference.com

Callisto Reference,
118-35 Queens Blvd., Suite 400,
Forest Hills, NY 11375, USA

Visit us on the World Wide Web at:
www.callistoreference.com

ISBN: 978-1-63239-860-4 (Hardback)

The publisher's policy is to use permanent paper from mills that operate a sustainable forestry policy. Furthermore, the publisher ensures that the text paper and cover boards used have met acceptable environmental accreditation standards.

Trademark Notice: Registered trademark of products or corporate names are used only for explanation and identification without intent to infringe.

Printed in the United States of America.

Cataloging-in-publication Data

Genetic modification : methods and techniques / edited by David Rhodes.
    p. cm.
Includes bibliographical references and index.
ISBN 978-1-63239-860-4
1. Genetic engineering. 2. Biotechnology. 3. Genetically modified foods. I. Rhodes, David.
QH442 .G46 2017
660.65--dc23

Table of Contents

Preface

This book provides comprehensive insights into the field of genetic modification. It elucidates the concepts and innovative models around prospective developments with respect to this subject. Genetic modification is the application of biotechnology to alter the genomes in order to enhance the characteristics of an alike organism. The main techniques used under this branch are gene transfer, molecular cloning, DNA synthesizing, gene targeting, etc. The aim of this field is to produce genetically modified organisms (GMOs), genetically modified food and most recently genetically modified crops. The objective of this text is to give a general view of the different areas of genetic modifications and its applications. Most of the topics introduced in the book cover new techniques and the applications of this field. Researchers and students in this subject will be greatly assisted by the text. In this book, using case studies and examples, constant effort has been made to make the understanding of the difficult concepts of genetic modification as easy as informative as possible, for the readers.

This book has been a concerted effort by a group of academicians, researchers and scientists, who have contributed their research works for the realization of the book. This book has materialized in the wake of emerging advancements and innovations in this field. Therefore, the need of the hour was to compile all the required researches and disseminate the knowledge to a broad spectrum of people comprising of students, researchers and specialists of the field.

At the end of the preface, I would like to thank the authors for their brilliant chapters and the publisher for guiding us all-through the making of the book till its final stage. Also, I would like to thank my family for providing the support and encouragement throughout my academic career and research projects.

Editor

Expression of Intracellular Interferon-Alpha Confers Antiviral Properties in Transfected Bovine Fetal Fibroblasts and Does Not Affect the Full Development of SCNT Embryos

Dawei Yu[1,3], Shoufeng Zhang[2], Weihua Du[1]*, Jinxia Zhang[2], Zongxing Fan[1], Haisheng Hao[1], Yan Liu[1], Xueming Zhao[1], Tong Qin[1], Huabin Zhu[1]

1 Embryo Biotechnology and Reproduction Laboratory, Institute of Animal Science, Chinese Academy of Agricultural Sciences, Beijing, China, **2** Institute of Military Veterinary, Academy of Military Medical Science, Changchun, China, **3** State Key Laboratories of Agrobiotechnology, College of Biological Science, China Agricultural University, Beijing, China

Abstract

Foot-and-mouth disease, one of the most significant diseases of dairy herds, has substantial effects on farm economics, and currently, disease control measures are limited. In this study, we constructed a vector with a human *interferon-α* (*hIFN-α*) (without secretory signal sequence) gene cassette containing the immediate early promoter of human cytomegalovirus. Stably transfected bovine fetal fibroblasts were obtained by G418 selection, and *hIFN-α* transgenic embryos were produced by somatic cell nuclear transfer (SCNT). Forty-six transgenic embryos were transplanted into surrogate cows, and five cows (10.9%) became pregnant. Two male cloned calves were born. Expression of hIFN-α was detected in transfected bovine fetal fibroblasts, transgenic SCNT embryos, and different tissues from a transgenic SCNT calf at two days old. In transfected bovine fetal fibroblasts, expression of intracellular IFN-α induced resistance to vesicular stomatitis virus infection, increased apoptosis, and induced the expression of double-stranded RNA-activated protein kinase gene (*PKR*) and the 2'-5'-oligoadenylate synthetase gene (2'-5' OAS), which are *IFN*-inducible genes with antiviral activity. Analysis by qRT-PCR showed that the mRNA expression levels of *PKR*, 2'-5' OAS, and *P53* were significantly increased in wild-type bovine fetal fibroblasts stimulated with extracellular recombinant human IFN-α-2b, showing that intracellular IFN-α induces biological functions similar to extracellular IFN-α. In conclusion, expression of intracellular hIFN-α conferred antiviral properties in transfected bovine fetal fibroblasts and did not significantly affect the full development of SCNT embryos. Thus, *IFN-α* transgenic technology may provide a revolutionary way to achieve elite breeding of livestock.

Editor: Glenn Francis Browning, The University of Melbourne, Australia

Funding: This work was partially supported by grants from the Agricultural Science and Technology Innovation Program (ASTIP-IAS06 to WHD), State Major Project of Transgenic (2009ZX08007-007B to WHD). The funders had no role in study design, data collection and analysis, decision to publish, or preparation of the manuscript.

Competing Interests: The authors have declared that no competing interests exist.

* Email: dwh@iascaas.net.cn

Introduction

Transgenic technology enables the introduction of exogenous genes into animal genomes and provides a revolutionary way to achieve elite breeding of livestock. The main applications of transgenic technology in livestock breeding include improving their disease resistance, carcass composition, lactational performance, wool production, growth rate, and reproductive performance, as well as reducing their environmental impact [1].

Our efforts have focused on developing foot-and-mouth disease (FMD) resistance in dairy cattle using transgenic somatic cell nuclear transfer (SCNT) technology. FMD is a highly contagious vesicular disease of cloven-hoofed animals [2]. Outbreaks of FMD can have severe economic and social consequences that result in the loss of billions of dollars ($US) in direct and indirect costs, as well as the slaughter of millions of animals [3]. Current vaccines and disease-control measures to eliminate FMD have many drawbacks [4]. Several new strategies, such as RNA interference

[5,6], have been developed to control FMD, but few reports have detailed transgenic livestock strategies [1].

Evidence suggests that expression of exogenous IFN-α in livestock confers resistance to FMDV infection [2,7]. Interferons (IFNs) are widely expressed cytokines that have potent antiviral and growth-inhibitory effects; they are the first line of defense against virus infections [8,9]. However, several reports indicate that side effects are associated with over-expression of secreted IFN-α in animal models, such as disrupted spermatogenesis in male transgenic mice [10,11].

In this study, cloned transgenic cattle containing IFN-α were generated to produce FMDV-resistant cattle. A secretory signal sequence of IFN-α was deleted to verify whether intracellular expression of IFN-α has side effects in transgenic cattle. We hypothesized that IFN-α without the secretory signal sequence would elicit the same biologic response as the secreted counterpart, but would not be secreted in transgenic SCNT embryos, would not trigger a signal transduction pathway in transgenic SCNT embryos and between pre-implant transgenic SCNT embryos and endometrial cells [12], and would have reduced toxicity to neighboring tissues [13].

To produce transgenic SCNT cattle that can resist FMDV, we constructed a vector with a human *IFN-α* gene cassette containing the immediate early promoter of human cytomegalovirus (HCMV). Transfected bovine fetal fibroblasts and transgenic SCNT embryos were obtained. After embryo transfer, transgenic cattle containing IFN-α were born at full term.

Materials and Methods

Animals

Cows were from the farm of the Chinese Academy of Agricultural Sciences, Beijing, China. They were allowed access to feed and water *ad libitum* under normal conditions. Experiments were performed according to the Regulations for the Administration of Affairs Concerning Experimental Animals (Ministry of Science and Technology, China, revised in June 2004) and approved by the Institutional Animal Care and Use Committee Chinese Academy of Agricultural Sciences, Beijing, China (Permit Number: IAS10012).

Study design

In this study, a vector with a human *IFN-α* (*hIFN-α*) (without secretory signal sequence) gene cassette containing the immediate early promoter of HCMV was constructed. Stably transfected bovine fetal fibroblasts were obtained by G418 selection, and expression of intracellular hIFN-α, resistance to vesicular stomatitis virus infection, apoptosis, and expression of the genes *PKR*, *2'-5' OAS* and *P53* were investigated.

Transgenic embryos were produced by SCNT. Transgenic SCNT embryos were transplanted into surrogate cows. Expression of intracellular hIFN-α was detected in the transgenic SCNT embryos and different tissues from a transgenic SCNT calf.

Construction of plasmid vector

A new plasmid, pIRESneo-*IFN*-bCP-*LFCIN B-EGFP*, was constructed. pIRES1-neo (GenBank Acc. No. U89673; Cat.no. 6060, Clontech Laboratories, Palo Alto, CA, USA) was used as the basal plasmid for vector construction. The *hIFN-α* gene (GenBank Acc. no. X66186.1) was cloned into the vector at the *Eco*RV site. A mammary-gland-tissue-specific expression sequence, including the *lactoferricin B* (*LFCIN B*) gene (GenBank Acc. no. L08604) and goat β-casein promoter sequence (GenBank Acc. no. AY311384) [14,15] was subcloned into the vector at *Bst*1107I; this can direct the expression of a foreign gene in the lactating mammary gland. The HCMV promoter sequence and *EGFP* gene were cloned into the vector at *Xho*I, forming pIRESneo-*IFN*-bCP-*LFCIN B-EGFP* (Fig. S1). The plasmid pEGFP-N1 (Clontech) was used as the control vector.

Transfection of vector into bovine fetal fibroblasts

Bovine fetal fibroblasts were obtained from a fetus of a Holstein cow slaughtered 40 days after artificial insemination. Fibroblasts were isolated from a skin biopsy and cultured in DMEM (Gibco, Life Technology, Carlsbad, CA, USA) supplemented with 10% fetal bovine serum (Gibco, Life Technology) in a 37.5°C humidified incubator with 5% CO_2.

Bovine fetal fibroblasts were transfected with the vector using an Amaxa Nucleofector (Lonza, Cologne, Germany), according to the manufacturer's protocol [16]. Forty-eight hours after transfection the cells were split 1:16 and neomycin (Sigma, St. Louis,

MO, USA) was added to a final concentration of 500 μg/mL. After 2 weeks, resistant colonies were obtained and stably transfected cells were established by expansion culture. Integration of *hIFN-α* DNA was verified by polymerase chain reaction (PCR) using a forward primer binding to the HCMV sequence (5'-GAA GTT GTT CGT GCT GAA AT-3') and a reverse primer binding to the *hIFN-α* sequence (5'-GAG AAA GGC AAA GTG GAT GT-3').

Vesicular stomatitis virus infection

To harvest enough cells for vesicular stomatitis virus infection, the stably transfected bovine fetal fibroblasts were expanded for 10 passages in DMEM supplemented with 10% FBS (Gibco, Life Technology) and 300 ng neomycin/mL (Sigma). They were then plated at a density of 2.0×10^5 cells/well in six-well, flat-bottomed plates. After incubation for 24 h under standard conditions, fibroblasts were infected with vesicular stomatitis virus (VSV) (provided by Professor Gang Li [Chinese Academy of Agricultural Sciences, Beijing, China]) at a multiplicity of infection (MOI) of 10 for 2 h. After culture in normal medium (without virus) for a further 72 h, culture medium was collected and centrifuged at 1,000 rpm/min for 5 min to remove cell debris. Cell viability was estimated using an apoptosis assay. Serial dilutions of each supernatant were added to non-transfected bovine fibroblasts in 96-well plates, and 24 h later the cytopathic effect (CPE) was observed using TE-2000 phase-contrast microscopy (Nikon, Tokyo, Japan). VSV in the supernatants was titrated according to the method of Kaerber [12].

Somatic cell nuclear transfer

Transgenic SCNT embryos were produced using an improved method that combines micro-extrusion and zona-free fusion. The donor cells in the transfected group were the bovine fetal fibroblasts transfected stably with vector pIRESneo-*IFN*-bCP-*LFCIN B-EGFP* and cultured in DMEM supplemented with 10% FBS (Gibco, Life Technology) and 300 ng neomycin/mL (Sigma) for 8–10 passages. The donor cells in the control group were non-transfected bovine fetal fibroblasts and cultured in DMEM supplemented with 10% FBS (Gibco, Life Technology) for the same number of passages as above. First, matured oocytes were enucleated by micromanipulation; next, two zona-free enucleated oocytes were fused with single donor cell [17]; finally, reconstructed embryos were activated and cultured using the well-of-the-well (WOW) system [18]. On day 7, SCNT blastocysts cultured *in vitro* were transferred nonsurgically into Holstein heifers (15–18 months of age) at 6.5–7 d after estrus on a local farm (Beijing, China). A single embryo was transferred in bovine embryo transfer medium (Agtech, Manhattan, KS, USA) into a recipient uterine lumen ipsilateral to the corpus luteum. Recipient synchronization was achieved using a standard double-injection prostaglandin (Lutalyze; Pharmacia & Upjohn, Kalamazoo, MI, USA) treatment. Transgenic embryos were transferred into 46 Holstein heifers; non-transgenic embryos were transferred into 77 Holstein heifers. Pregnancies were diagnosed by rectal palpation between days 55 and 60.

Real-time RT-PCR analyses

Quantitative real-time PCR (qRT-PCR) was used to detect the expression levels of the *hIFN-α*, *LFCIN B*, *PKR*, *2'-5' OAS*, and *P53* genes in transfected bovine fetal fibroblasts cultured for 10 passages.

For real-time RT-PCR analyses of cells, total RNA was extracted using TRIzol (Invitrogen, Life Technology), RNA purity was determined by measuring absorbance at 260 and 280 nm

(A260/280), and RNA was reverse-transcribed using SUPER-SCRIPT III (Invitrogen, Life Technology) as previously described [19]. The 7500 Fast Real-Time PCR System (Applied Biosystems, Foster City, CA, USA) was used to determine the level of cDNA using Taqman probe assays (Invitrogen, Life Technology) for the bovine $2'-5'$ OAS, PKR, P53, and GAPDH mRNAs in transfected cells. The program consisted of pre-denaturing at $95°C$ for 10 min; 40 cycles of denaturing at $95°C$ for 10 sec, annealing at $60°C$ for 30 sec and extension at $70°C$ for 45 sec. Data were normalized to the expression level of GAPDH in each sample. The primers and probe sequences are shown in Table 1. The Ct value was defined as the number of PCR cycles in which the fluorescence signal exceeded the detection threshold value. The $2^{-\Delta\Delta Ct}$ value, where $\Delta Ct = Ct_{Gene} - Ct_{GAPDH}$ and $\Delta\Delta Ct = \Delta Ct_{treated} - \Delta Ct_{control}$, was calculated to represent the relative mRNA expression of bovine $2'-5'$ OAS, PKR, P53 and GAPDH mRNA in antivirus analysis of transfected cells, non-transfected cells and wild type cells. SYBR green I (Invitrogen, Life Technology) was used for the hIFN-α and LFCIN B expression analysis. The ratio $= (Etarget)^{\Delta CTtarget\ (control\ -\ sample)}/(Eref)^{\Delta CTref\ (control\ -\ sample)}$ was calculated to represent the relative mRNA expression of hIFN-α and LFCIN B in transfected bovine fetal fibroblasts, non-transfected cells, wild type cells and transgenic SCNT embryos [20].

For real-time RT-PCR analyses of embryos, cDNAs of single embryos were synthesized with Cell to cDNA II kits (Ambion, Austin, TX, USA), and quantitative PCR was performed as previously described [21].

Western blotting analysis

Tissues (liver, lung, spleen, heart, ear skin, muscle, stomach, intestine, colon, kidney) from a transgenic SCNT calf and transfected bovine fetal fibroblasts expanded for 10 passages were extracted in lysis buffer (1% NP-40, 150 mM NaCl, 5 mM EDTA, 50 mM NaF, 20 mM Tris–HCl, pH 7.4, 2 mM Na3VO4, 1 mM PMSF, and protease inhibitor cocktail). Proteins were quantified using a BCA Protein Assay kit. For western blot assays, equal amounts of protein (20 µg) were subjected to 12% SDS–PAGE and electrophoretically transferred to nitrocellulose (NC) membranes (0.45 µm, Millipore, Billerica, MA, USA). The non-specific sites on each blot were blocked with 3% BSA diluted in TBS with 0.05% Tween 20 (TBST) for 90 min. Proteins were detected using mouse monoclonal [9D3] antibody against IFNα-2b (Abcam, Cambridge, UK), purified mouse monoclonal antibody against LTF (Abgent, San Diego, CA, USA), or mouse monoclonal antibody against β-actin (Santa Cruz Biotechnology, Dallas, TX, USA) as primary antibody and horseradish peroxidase (HRP)-conjugated goat anti-mouse Ig G (Jackson ImmunoResearch, West Grove, PA, USA) as secondary antibody. Proteins were detected using an enhanced chemiluminescence (ECL) reagent (Millipore).

Apoptosis assay

FITC-conjugated annexin V and propidium iodide (PI) staining was performed using a kit (Neobioscience, Shanghai, China). Briefly, after infection with VSV for 2 h and culture for a further 72 h in normal medium, the transfected bovine fetal fibroblasts were resuspended in binding buffer, washed twice for 3 min with PBS, and annexin V-FITC and PI were added. Following a 10 min incubation in the dark, samples were subjected to flow cytometric analysis with a FACScalibur flow cytometer (Becton Dickinson, San Jose, CA, USA) using Cell-Quest software [22]. Mechanically damaged cells were located in the upper left quadrant, late apoptotic or necrotic cells in the upper right quadrant, dual negative and normal cells in the lower left quadrant, and early apoptotic cells in the lower right quadrant of the flow cytometric dot plot. The total apoptotic rate was calculated as the late apoptotic rate + early apoptotic rate.

Immunofluorescence and confocal microscopy

For staining of hIFN-α, transgenic SCNT blastocysts (generated using stably transfected bovine fetal fibroblasts as donors, followed by SCNT) (n = 10) cultured for 7 days in vitro were fixed and permeabilized in 2% paraformaldehyde and 0.5% Triton X-100 in PBS (pH 7.4) for at least 30 min at room temperature. Blastocysts were blocked in 0.1% Triton X-100 and 2% BSA-supplemented PBS for 1 h and incubated overnight at $4°C$ with 1:200 V450-conjugated mouse anti-human IFN-α-2b (Becton Dickinson). After three washes in PBS, blastocysts were mounted on glass slides and examined with a Confocal Laser-Scanning Microscope (C-1, Nikon) under the Violet laser Em Max at 450 nm.

Table 1. Primer and probe sequences.

Gene	Primers sequences (5'-3')	Size (bp)	Probe sequences (5'-3')
hIFN-α	F: GAT GCG TCG TAT CTC TCT GTT C	103	
	R: CGG GAT GGT TTC AGC TTT CT		
LFCIN B	F:TTG GAC TGT GTC TGG CTT TC	86	
	R: CCT CCT CAC ACA GGT GAT AGA		
PKR	F: AGT GCT GCG TGT GGT GAT GT	321	AAG AAA CTC TCC AGC AGT GAG TAC
	R: TGG AGA CAC GGA AGA GCT GTT		
2'-5' OAS	F: AAA GAG GAA GCG GCG TGC	171	TCA GCT TCG TGC TCA GGT C
	R: CTC CTC AGG TTC TCG CAC TCT		
P53	F: AAC ACC AGC TCC TCT CCA CAG	115	AAG AAG AAA CCA CTG GAT GGA
	R: CCA AGG CAT CAT TCA GCT CTC		
GAPDH	F: AAG GCC ATC ACC ATC TTC CA	113	AGC GAG ATC CTG CCA ACA T
	R: CCA GCC TTC TCC ATG GTA GTG		

Note: hIFN-α: human interferon-α; LFCIN B: Lactoferricin B; PKR: double-stranded RNA-activated protein kinase; 2'-5' OAS: 2'-5'-oligoadenylate synthetase.

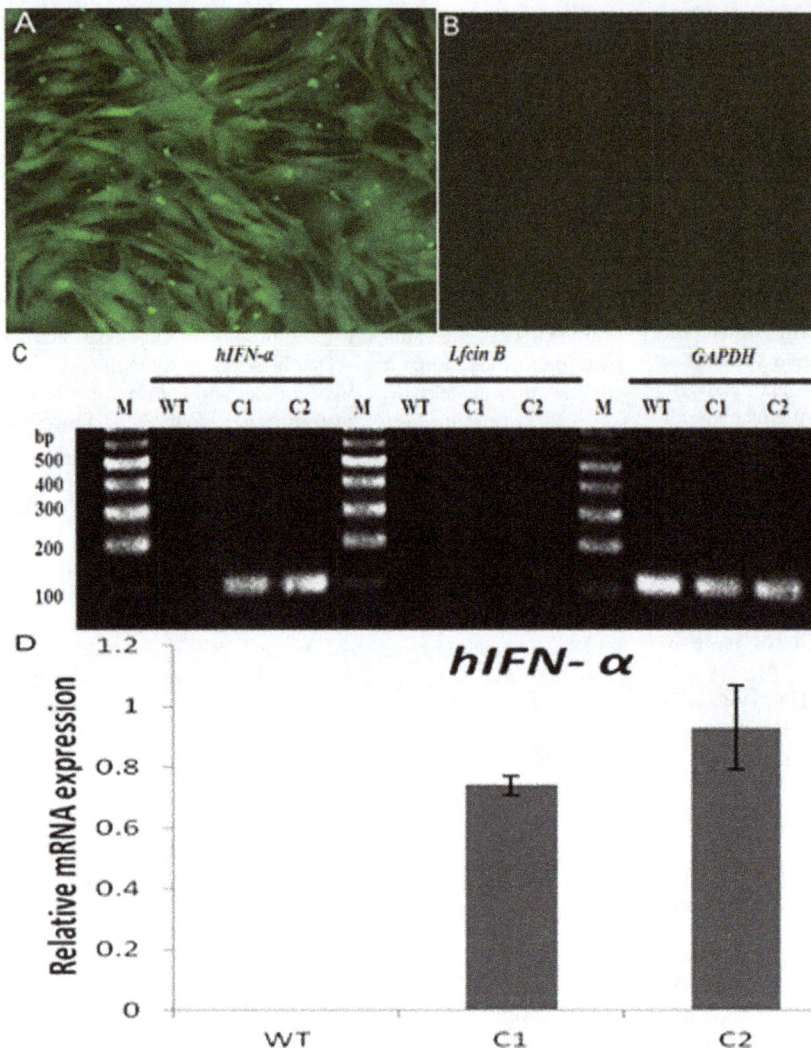

Figure 1. *hIFN-α* and *LFCIN B* **gene expression in transfected bovine fetal fibroblasts.** GFP was expressed in transfected bovine fetal fibroblasts (×200) (A), but not in wild-type bovine fetal fibroblasts (×200) (B). Green: green fluorescent protein (GFP). (C, D) *hIFN-α* and *LFCIN B* gene expression in transfected bovine fetal fibroblasts was analyzed using real-time RT-PCR, with results normalized to expression levels of *GAPDH*. WT, Wild-type bovine fetal fibroblasts; C1, transfected bovine fetal fibroblast clone 1; C2, transfected bovine fetal fibroblast clone 2. Relative mRNA expression levels are shown as means ± standard deviation (SD) from three independent experiments.

Statistical analysis

Data in each experiment were expressed as means ± standard deviation (SD) from three independent experiments. Statistical analyses were performed using one-way analysis of variance (ANOVA) with Duncan's test for post hoc analysis using SAS version 8 (SAS Institute Inc., Cary, NC, USA). $P<0.05$ was considered significant.

Results

Expression of hIFN-α in transfected bovine fetal fibroblasts

The vector pIRESneo-*IFN*-bCP-*LFCIN B-EGFP* (Fig. S1) was constructed and used to transfect early-passage bovine fetal fibroblasts. Stable G418 resistant clones were derived. Under fluorescence microscopy, GFP expression was detected in stably transfected bovine fetal fibroblasts (Fig. 1A), but not in wild-type bovine fetal fibroblasts (Fig. 1B). RT-PCR analysis showed *hIFN-α*

mRNA was expressed in stably transfected bovine fetal fibroblasts, but not in wild-type bovine fetal fibroblasts (Fig. 1D). *LFCIN B* mRNA was not expressed in transfected or wild-type bovine fetal fibroblasts (Fig. 1C). Total protein extracts from bovine fetal fibroblasts were analyzed by western blotting using monoclonal antibodies against hIFN-α-2b and LFCIN B, with hIFN-α protein expression was detected in transfected bovine fetal fibroblasts (Fig. 2A, lane 3, 9), but not in extracts from wild-type bovine fetal fibroblasts (Fig. 2A, lanes 2, 8). LFCIN B expression was not detected in transfected or wild-type bovine fetal fibroblasts (Fig. 2B, lanes 3, 9 and lanes 2, 8).

Antiviral activity of hIFN-α in the transfected bovine fetal fibroblasts

To determine whether the expression of intracellular hIFN-α was sufficient to confer antiviral properties on transfected bovine fetal fibroblasts, fibroblasts were infected with VSV for 2 h. After culture in normal medium for a further 72 h, the concentration of

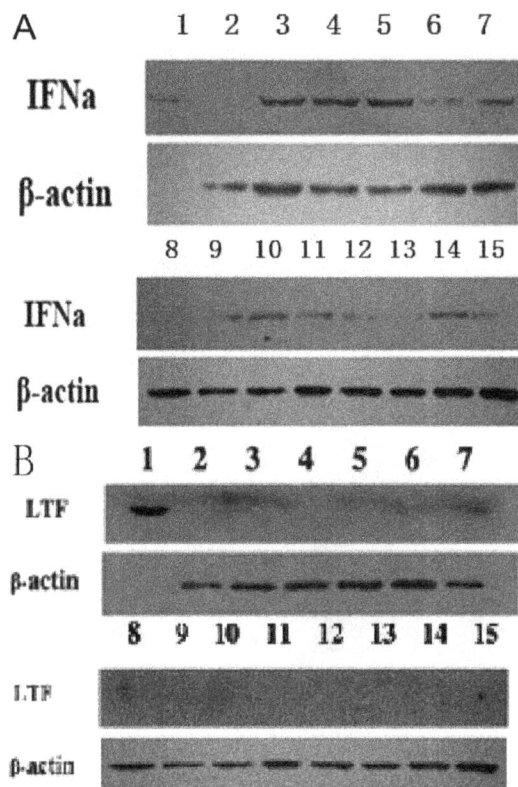

Figure 2. Detection of hIFN-α and LFCIN B (LTF) protein expression in transfected donor cells (transfected bovine fetal fibroblasts) and tissues from the transgenic calf generated by somatic cell nuclear transfer (SCNT). The hIFN-α (A) and LFCIN B (B) protein expression was determined by western blot. Lane 1 in (A), recombinant human IFN-α 2b protein as positive control. Lane 1 in (B), lactoferricin B as positive control; lanes 2 and 8, the same wild-type bovine fetal fibroblasts, as negative controls; lane 3, transfected bovine fetal fibroblast clone 1; lane 9, transfected bovine fetal fibroblast clone 2; lane 4, liver; lane 5, lung; lane 6, spleen; lane 7, heart; lane 10, ear skin; lane 11, muscle; lane 12: stomach; lane 13: intestine; lane 14: colon; lane 15: kidney.

VSV in the supernatants was titrated. Virus titer was markedly reduced in the supernatants of bovine fetal fibroblasts stably transfected with the *hIFN-α* gene compared with those in the supernatants of non-transfected and bovine fetal fibroblasts transfected with the control vector (pEGFP-N1 vector) (both $P < 0.05$) (Fig. 3A). Virus titer was not significantly different between non-transfected and bovine fetal fibroblasts transfected with the control vector ($P > 0.05$), indicating that the expression of hIFN-α conferred protection against viral infection on the transfected bovine fetal fibroblasts.

Expression of intracellular hIFN-α increase transfected bovine fetal fibroblast apoptosis

After infection with VSV for 2 h and culture in normal medium for a further 72 h, cell apoptosis was estimated by staining with annexin V and PI, followed by flow cytometric analysis. Approximately 7% of non-transfected and 11% of transfected bovine fetal fibroblasts carrying the control vector underwent apoptosis, while expression of hIFN-α resulted in 77% apoptosis in infected transfected bovine fetal fibroblast (Fig. 3B).

hIFN-α increases the mRNA expressions of antiviral genes PKR, 2′-5′ OAS, and P53

To test whether the expression of intracellular hIFN-α could trigger antiviral-related signaling events in *hIFN-α* transfected bovine fetal fibroblasts, the expression levels of *IFN*-inducible genes that stimulate antiviral activities, including *PKR*, 2′-5′ *OAS*, and *P53*, in transfected bovine fetal fibroblasts were analyzed using qRT-PCR. The results showed that the expression levels of these genes were significantly increased in *hIFN-α* transfected bovine fetal fibroblasts, compared with those in non-transfected bovine fetal fibroblasts (all $P < 0.05$) (Fig. 3C). There was no significant difference in the expression levels of *PKR* and 2′-5′ *OAS* between bovine fetal fibroblasts transfected with the control vector and non-transfected bovine fetal fibroblasts (both $P > 0.05$). However, the expression of *P53* was significantly increased in bovine fetal fibroblasts transfected with the control vector. To confirm that intracellular *IFN*-α induced biological functions similar to extracellular *IFN*-α, as previously reported [12,13,23], qRT-PCR analysis was used to show that the levels of expression of *PKR*, 2′-5′ *OAS*, and *P53* mRNA were significantly increased in wild-type bovine fetal fibroblasts stimulated with 5×10^{-5} μg extracellular recombinant human IFN-α-2b (rhIFN-α-2b) (Sigma)/mL for 2 h (all $P < 0.05$) (Fig. 3D).

Expression of hIFN-α in *hIFN-α* transgenic SCNT embryos

Transgenic SCNT embryos were produced using an improved SCNT method that combines microextrusion and zona-free fusion, resulting in a higher fusion rate than microinjection (data not shown). Green fluorescent protein (GFP) was observed in transgenic SCNT embryos under fluorescence microscopy (Fig. 4A). The integration of foreign genes within the genome of the SCNT embryos was detected using PCR (Fig. 4B). Analysis by qRT-PCR showed that *hIFN-α* mRNA was expressed in the transgenic SCNT embryos, but that *LFCIN B* mRNA was not (Fig. 4C). IFN-α protein in the transgenic SCNT embryos was also expressed and detected by immunofluorescence (Fig. 4D). To test the potential effect of hIFN-α intracellular expression on the developmental competence of the transgenic SCNT embryos, early development of *hIFN-α* transgenic SCNT embryos and non-transgenic SCNT embryos was compared. No significant differences were found in cleavage rate, blastocyst rate (Table 2), or pregnancy outcomes (Table 3) between them after embryo transfer.

Analysis of hIFN-α expression in the SCNT calf

Forty-six transgenic SCNT embryos were transplanted into surrogate cows, and five cows (10.9%) became pregnant (Table 3). Two male cloned calves were born (Fig. 5A). One calf died of excessive bleeding due to umbilical cord hypogenesis at two days old. GFP was detected by fluorescence microscopy in the ear skin from the transgenic SCNT calf that died at two days of age (Fig. 5B). DNA samples from a wild-type calf and the transgenic SCNT calf that died at two days of age were screened by PCR using primers designed to amplify a 550-bp fragment in the gene cassette. Results showed that blood and the *funiculus umbilicalis* from the transgenic SCNT calf carried the *hIFN-α* gene (Fig. 5C). IFN-α was expressed in the different tissues (liver, lung, spleen, heart, ear skin, muscle, stomach, intestine, colon and kidney) of the transgenic SCNT calf that died at two days, as detected by western blotting (Fig. 2A, lanes 4–7, 10–15). However, LFCIN B was not detected in these tissues (Fig. 2B, lanes 4–7, 10–15).

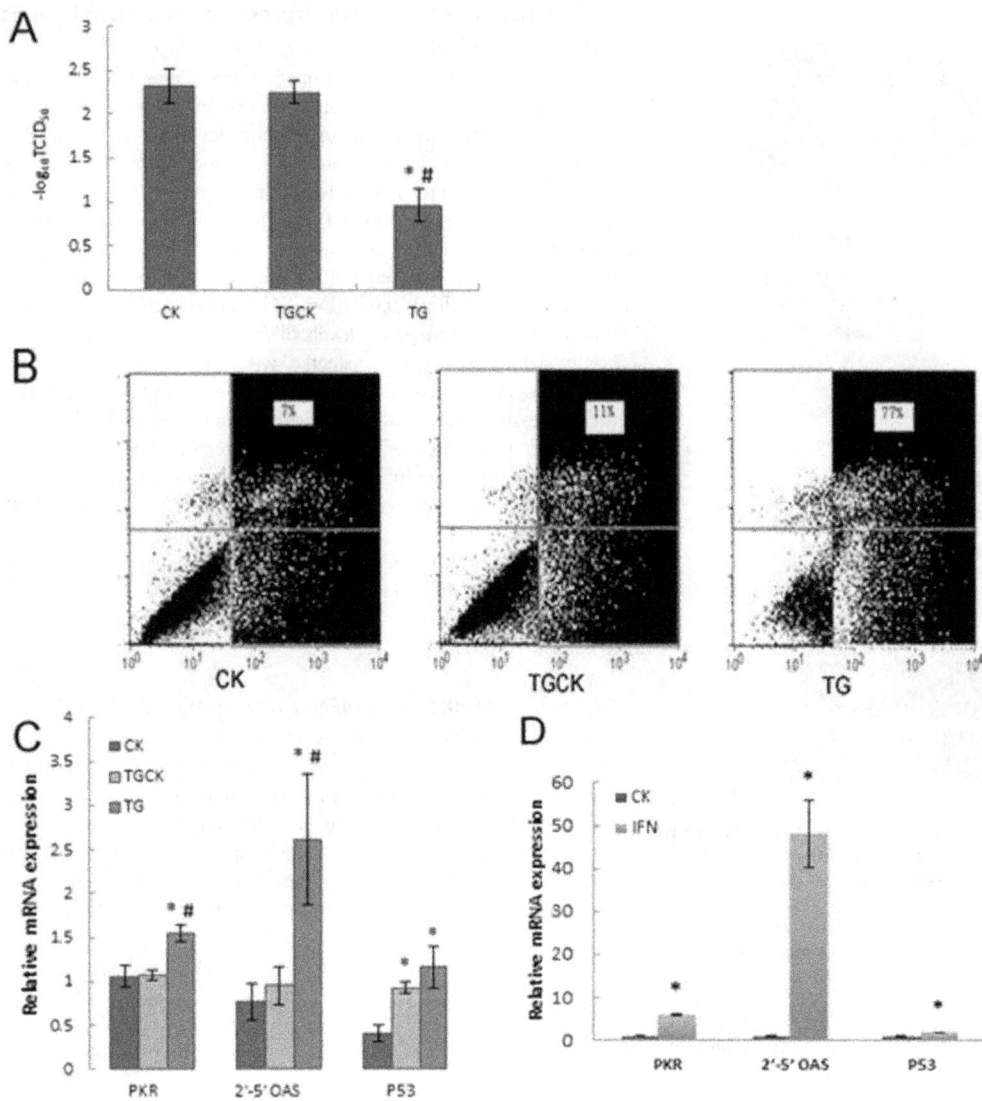

Figure 3. Antiviral activity and apoptosis mediated by hIFN-α in transfected bovine fetal fibroblasts. After VSV infection for 2 h, fibroblasts were cultured in normal medium (without virus) for a further 72 h. (A) Virus titres (TCID 50) in the supernatant of wild-type (CK), control vector (peGFP-N1 vector) transfected (TGCK), and *hIFN-α* transfected (TG) bovine fetal fibroblasts. (B) Apoptosis was determined by FITC-conjugated annexin V and propidium iodide (PI) staining. The y-axis is PI and the x-axis is FITC-annexin V. (C) The expression levels of *PKR*, *2'-5' OAS*, and *P53* mRNA were analyzed using real-time RT-PCR with results normalized to the expression levels of *GAPDH*. (D) The expression levels of *PKR*, *2'-5' OAS*, and *P53* mRNA in wild-type bovine fetal fibroblasts after 5×10^{-5} µg rhIFN-α-2b/mL was added to the culture medium for 2 h, analyzed by qRT-PCR with results normalized to expression levels of *GAPDH*. Data are shown as means ± SD from three independent experiments. *$P<0.05$ *vs.* CK group; #$P<0.05$ TG *vs.* TGCK. *PKR*, double-stranded RNA-activated protein kinase gene; *2'-5' OAS*, 2'-5'-oligoadenylate synthetase gene.

Discussion

In our preliminary study, we focused on biological activities of IFN-α in transfected fetal fibroblasts and transgenic SCNT embryos. We constructed a vector with a bovine *LFCIN B* gene cassette containing a goat β-casein regulatory sequence and a human *IFN-α* (without secretory signal sequence) gene cassette containing the immediate early promoter of HCMV, and hIFN-α was expressed in both transfected bovine fetal fibroblasts and transgenic SCNT embryos, whereas LFCIN B, which was regulated by the goat β-casein promoter was only expected to be expressed during lactation. Two male cloned transgenic calves were born and were not expected to express the *LFCIN B* gene.

To distinguish exogenous *IFN-α* from endogenous *IFN-α* possibly produced by bovine cells, *hIFN-α* was cloned into the

vector. The hIFN-α significantly augmented the expression of IFN-inducible genes, which indicated that exogenous *hIFN-α* triggered the expected signal transduction pathway in bovine cells [24], even though the amino acid sequence of hIFN-α has only 60% identity with that of bovine IFN-α.

Expression of intracellular hIFN-α resulted in antiviral activity, increased apoptosis, and induced the expression of IFN-inducible genes in transfected fetal fibroblasts. Therefore, intracellular hIFN-α had activities similar to those of extracellular IFN-α in bovine cells. This finding is further supported by the observation that rhIFN-α-2b added to the culture medium of wild-type bovine fetal fibroblasts stimulated the expression of IFN-inducible genes. Several studies have indicated that exogenous IFN-α with a deleted secretory signal sequence (i.e., eliminated secretion), which

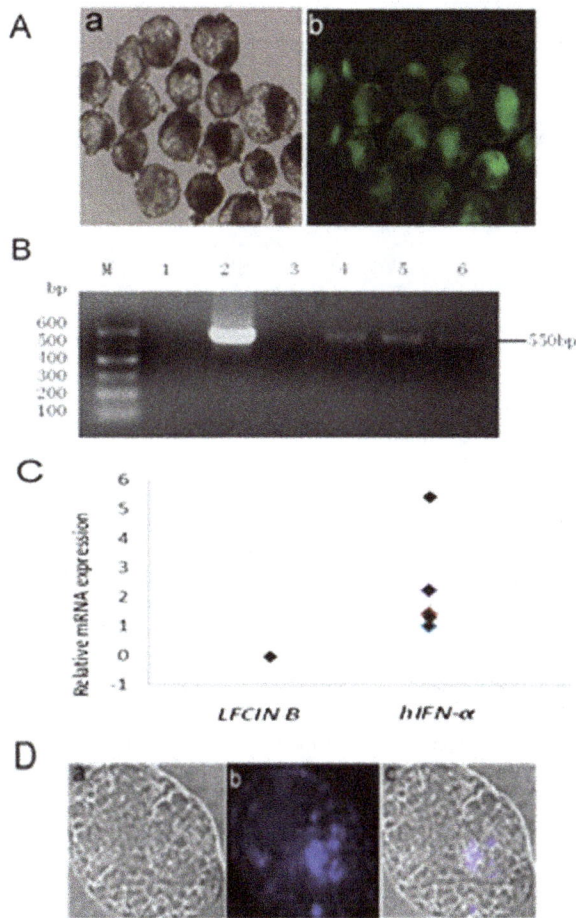

Figure 4. Detection of hIFN-α and LFCIN B expressed in transgenic SCNT embryos by PCR and immunofluorescence. (A) zona-free transgenic SCNT blastocyst (×100). Panel a, under light microscopy; Panel b, under fluorescence microscopy. Green: GFP. (B) PCR analysis of the *hIFN-α* gene in transgenic SCNT embryos. Lane M, 100 bp molecular weight marker; lane 1, water as template; lane 2, pIRSEneo-*IFN*-bCP-*LFCINB-EGFP* as template; lane 3, non-transgenic SCNT embryos DNA as template; lanes 4, 5, and 6, *hIFN-α* transgenic SCNT embryos. (C) Real-time RT-PCR analysis of mRNA expression levels of *hIFN-α* (n = 6) and *LFCIN B* (n = 6) in transgenic SCNT embryos, with results normalized to expression levels of *GAPDH*. (D) Immunofluorescence analysis of hIFN-α expressed in transgenic SCNT embryos (×200, n = 10). Transgenic SCNT blastocysts were stained with V450-conjugated antibody against hIFN-α-2b. Panel a, under light microscopy; Panel b, under fluorescence microscopy with Violet laser Em Max at 450 nm; Panel c, merged images, with blue showing V450 fluorescence.

cannot be secreted, exerts biological functions similar to those of extracellular IFN-α [12,23]. Further studies are needed to elucidate the signaling pathway triggered by the intracellular ligand of IFN-α.

The mechanism of intracellular hIFN-α–mediated suppression of virus replication might have been apoptotic cell death induced by P53, which was significantly induced in *hIFN-α* transfected cells. Prompt induction of apoptosis of virus-infected cells via P53 activation is beneficial to the host in limiting virus replication. However, *P53* mRNA was significantly induced in fetal fibroblasts transfected with the control vector without the *hIFN-α* gene, and there was no significant difference compared to the *hIFN-α* transfected cell group. The induction of *P53* mRNA in transfected fetal fibroblasts containing the control vector might be caused by G418 selection pressure, under which cells that could not express enough neomycin resistance undergo apoptosis. This hypothesis was supported by our cell viability assay; the results showed that more cells went into apoptosis among cells transfected with the control vector than in infected wild-type bovine fetal fibroblasts (11% vs. 7%), but the prevalence apoptosis induced in infected fetal fibroblasts transfected with the control vector was relatively low compared to that of the *hIFN-α* transfected cells (77%). This is consistent with the observation that the virus titer was not significantly different between the non-transfected and transfected control groups, but was significantly reduced in the *hIFN-α* transfected cell group. Therefore, P53 may only partially enhance the apoptotic response in virally infected cells [22].

Other IFN-inducible genes such as *PKR* play important roles in promoting apoptosis of virally infected cells [25,26]. Our qRT-PCR analysis showed that *PKR* expression was significantly induced in the *hIFN-α* transfected cell group, while *PKR* expression was nearly the same between the non-transfected and transfected control groups.

IFN-α has pleiotropic biological effects mediated by hundreds of responsive genes [27]. Therefore, constitutive expression of exogenous IFN-α may have a severely negative impact on a transgenic animal. However, the development of *hIFN-α* transgenic SCNT embryos was not significantly different from that of the control group in our study. Wild-type bovine embryos secrete IFN-τ, a type I IFN, which is an important pregnancy factor known only in ruminants. It has been suggested that extracellular IFN-α elicits the same biologic response as IFN-τ. Both IFN-α and IFN-τ can promote the development of bovine IVF embryos *in vitro* [28], and the growth-promoting effect of hIFN-α was confirmed in our study by adding rhIFN-α-2b to the medium used to culture the bovine IVF embryos (data not shown). However, the intracellular expression of hIFN-α did not significantly promote the development of transgenic SCNT embryos.

This is the first report of the expression of intracelluar IFN-α in embryos. Whether intracelluar IFN-α can elicit the same biological response in embryos as extracellular IFN-α needs to be further investigated. It has been reported that intrauterine application of hIFN-α can change bovine endometrial gene expression in early pregnancy, suggesting that hIFN-α may affect embryo implantation and pregnancy [24]. However, the pregnancy rate of the *hIFN-α* transgenic SCNT embryos was not significantly different from that associated with the control

Table 2. Development of *hIFN-α* transgenic SCNT embryos and non-transgenic SCNT embryos.

Donor cells	No. of embryos reconstructed	Fusion rate (%)	Cleavage rate (%)	Blastocyst rate (%)
WT	248	94.80±1.11	95.09±0.91	34.22±0.91
IFNTG	285	96.04±1.52	93.35±2.01	33.78±6.95

Note: the data are shown as means ± standard deviation (SD). WT: wild-type bovine fetal fibroblasts; IFNTG: *hIFN-α* transfected bovine fetal fibroblasts.

Table 3. Pregnancies of recipients of transgenic SCNT embryos and non-transgenic SCNT embryos.

SCNT embryos	No. transplanted embryos	NO. pregnancy (%)	NO. live-born
NonTG	77	10 (13.0%)	2
IFNTG	46	5 (10.9%)	2

Note: NonTG: non-transgenic SCNT embryos; IFNTG: hIFN-α transgenic SCNT embryos.

embryos in this study. One explanation for this might be that the non-secreted hIFN-α could not elicit a response from the endometrial cells when transgenic SCNT embryos were implanted into surrogate cows. The two live-born hIFN-α transgenic SCNT calves in this study had normal external anatomy and organ development. These results indicate that constitutive expression of intracelluar IFN-α does not have obvious negative effects on the early-stage development of transgenic SCNT calves.

IFN-α can influence the proliferation, differentiation, and function of various types of cells in the immune system and thus influence lympho-hematopoiesis. However, it has been proposed that this effect only takes place under conditions of trauma or inflammation [29]. These factors could explain why expression of hIFN-α did not have obvious adverse effects on transgenic calves in our study. This is supported by a report in which transgenic mice expressing IFN-β (which belongs to the same IFN family as IFN-α) displayed similar behavior, external anatomy, life span, and female fertility as wild type mice, although male transgenic mice displayed reduced fertility [10]. The idiopathic infertilities may be caused by IFNs in the extracellular fluid, which affect the interplay between germ cells and Sertoli cells [22]. It has also been shown that the level of IFN-α in the seminal plasma may be related to sperm production [30]. Intracellular expression of hIFN-α did not result in a high level of IFN-α in the seminal plasma in our study. However, the fertility and the viral resistance of the transgenic calves need to be further determined. The calf is now 17 months old and continued GFP expression has been shown by immunofluorescence in an ear biopsy and in fibroblasts derived from the animal (data not shown).

In conclusion, we constructed a vector with a hIFN-α (without secretory signal sequence) gene cassette. Stably transfected bovine fetal fibroblasts were obtained, and hIFN-α transgenic embryos were produced by SCNT. Two male cloned calves were born. Expression of intracellular hIFN-α conferred viral resistance on transfected bovine fetal fibroblasts and did not significantly affect the full development of SCNT embryos. These data suggest that IFN-α transgenic technology may provide a revolutionary way to achieve disease-control measures through elite breeding of livestock.

Supporting Information

Figure S1 Schematic of the plasmid pIRESneo-IFN-bCP-LFCIN B-EGFP. The backbone of the vector is pIRSEneo. IFN-NEO cassette: CMV is the promoter of IFN and NEO; LFCIN B gene cassette: goat β-casein regulatory sequence is the promoter of LFCIN B; EGFP gene cassette: CMV is the promoter of EGFP. All restriction sites are shown. Amp, ampicillin resistance gene; IVS, synthetic intron; IRES, internal ribosome entry site of encephalomyocarditis virus; Neo, neomycin phosphotransferase gene; polyA, a fragment of bovine growth hormone poly(A) signal; EGFP, enhanced green fluorescent protein; IFN, β-interferon gene.

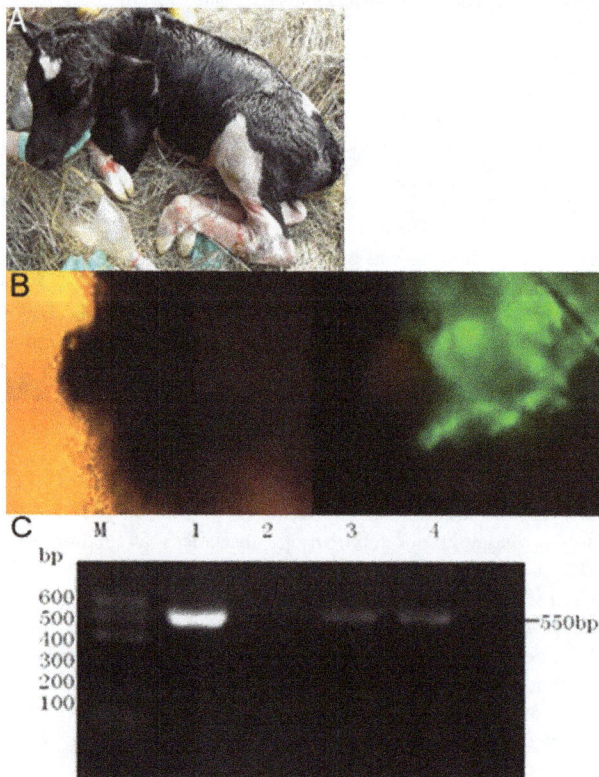

Figure 5. Detection of hIFN-α expression in the transgenic SCNT calf by PCR and GFP expression by fluorescence microscopy. (A) Live-born transgenic calf. (B) GFP expressed in ear skin from the transgenic SCNT calf that died at two days of age under light microscopy (×40) and under fluorescence microscopy (×40). Green: GFP. (C) PCR analysis of the hIFN-α gene expressed in the transgenic SCNT calf that died at two days of age. Lane M, 100 bp molecular weight marker; lane 1, pIRSEneo-IFN-bCP-LFCIN B-EGFP as template; lane 2, DNA from the wild-type calf as a template; lane 3, blood from the transgenic SCNT calf; lane 4, funiculus umbilicalis from the transgenic SCNT calf.

Acknowledgments

The authors thank Yeqing Sun and Dan Wang for technical assistance.

Author Contributions

Conceived and designed the experiments: WHD. Performed the experiments: DWY SFZ JXZ HSH ZXF. Analyzed the data: XMZ YL TQ. Contributed reagents/materials/analysis tools: HBZ. Wrote the paper: DWY WHD.

References

1. Laible G (2009) Enhancing livestock through genetic engineering–recent advances and future prospects. Comp Immunol Microbiol Infect Dis 32: 123–137.

2. Moraes MP, Chinsangaram J, Brum MC, Grubman MJ (2003) Immediate protection of swine from foot-and-mouth disease: a combination of adenoviruses expressing interferon alpha and a foot-and-mouth disease virus subunit vaccine. Vaccine 22: 268–279.

3. Knowles NJ, Samuel AR, Davies PR, Kitching RP, Donaldson AI (2001) Outbreak of foot-and-mouth disease virus serotype O in the UK caused by a pandemic strain. Vet Rec 148: 258–259.

4. Grubman MJ (2003) New approaches to rapidly control foot-and-mouth disease outbreaks. Expert Rev Anti Infect Ther 1: 579–586.

5. Chen W, Liu M, Jiao Y, Yan W, Wei X, et al. (2006) Adenovirus-mediated RNA interference against foot-and-mouth disease virus infection both in vitro and in vivo. J Virol 80: 3559–3566.

6. Li L, Li Q, Bao Y, Li J, Chen Z, et al. (2013) RNAi-based inhibition of porcine reproductive and respiratory syndrome virus replication in transgenic pigs. J Biotechnol 171C: 17–24.

7. Park JH (2013) Requirements for improved vaccines against foot-and-mouth disease epidemics. Clin Exp Vaccine Res 2: 8–18.

8. Platanias LC (2005) Mechanisms of type-I- and type-II-interferon-mediated signalling. Nat Rev Immunol 5: 375–386.

9. Malireddi RK, Kanneganti TD (2013) Role of type I interferons in inflammasome activation, cell death, and disease during microbial infection. Front Cell Infect Microbiol 3: 77.

10. Satie AP, Mazaud-Guittot S, Seif I, Mahe D, He Z, et al. (2011) Excess type I interferon signaling in the mouse seminiferous tubules leads to germ cell loss and sterility. J Biol Chem 286: 23280–23295.

11. Hekman AC, Trapman J, Mulder AH, van Gaalen JL, Zwarthoff EC (1988) Interferon expression in the testes of transgenic mice leads to sterility. J Biol Chem 263: 12151–12155.

12. Shin-Ya M, Hirai H, Satoh E, Kishida T, Asada H, et al. (2005) Intracellular interferon triggers Jak/Stat signaling cascade and induces p53-dependent antiviral protection. Biochem Biophys Res Commun 329: 1139–1146.

13. Ahmed CM, Wills KN, Sugarman BJ, Johnson DE, Ramachandra M, et al. (2001) Selective expression of nonsecreted interferon by an adenoviral vector confers antiproliferative and antiviral properties and causes reduction of tumor growth in nude mice. J Interferon Cytokine Res 21: 399–408.

14. Zhang JX, Zhang SF, Wang TD, Guo XJ, Hu RL (2007) Mammary gland expression of antibacterial peptide genes to inhibit bacterial pathogens causing mastitis. J Dairy Sci 90: 5218–5225.

15. Zhang JX, Zhang SF, Guo XJ, Hu RL (2008) Construction of recombinant retroviral vector of antimicrobial peptide genes and expression in mammary gland. Chinese J Vet Sci 28: 1433–1437.

16. Isakari Y, Harada Y, Ishikawa D, Matsumura-Takeda K, Sogo S, et al. (2007) Efficient gene expression in megakaryocytic cell line using nucleofection. Int J Pharm 338: 157–164.

17. Vajta G, Lewis IM, Trounson AO, Purup S, Maddox-Hyttel P, et al. (2003) Handmade somatic cell cloning in cattle: analysis of factors contributing to high efficiency in vitro. Biol Reprod 68: 571–578.

18. Vajta G, Peura TT, Holm P, Paldi A, Greve T, et al. (2000) New method for culture of zona-included or zona-free embryos: the Well of the Well (WOW) system. Mol Reprod Dev 55: 256–264.

19. Takaoka A, Mitani Y, Suemori H, Sato M, Yokochi T, et al. (2000) Cross talk between interferon-gamma and -alpha/beta signaling components in caveolar membrane domains. Science 288: 2357–2360.

20. Pfaffl MW (2001) A new mathematical model for relative quantification in real-time RT-PCR. Nucleic Acids Res 29: e45.

21. Matoba S, Inoue K, Kohda T, Sugimoto M, Mizutani E, et al. (2011) RNAi-mediated knockdown of Xist can rescue the impaired postimplantation development of cloned mouse embryos. Proc Natl Acad Sci U S A 108: 20621–20626.

22. Takaoka A, Hayakawa S, Yanai H, Stoiber D, Negishi H, et al. (2003) Integration of interferon-alpha/beta signalling to p53 responses in tumour suppression and antiviral defence. Nature 424: 516–523.

23. Will A, Hemmann U, Horn F, Rollinghoff M, Gessner A (1996) Intracellular murine IFN-gamma mediates virus resistance, expression of oligoadenylate synthetase, and activation of STAT transcription factors. J Immunol 157: 4576–4583.

24. Bauersachs S, Ulbrich SE, Reichenbach HD, Reichenbach M, Buttner M, et al. (2012) Comparison of the effects of early pregnancy with human interferon, alpha 2 (IFNA2), on gene expression in bovine endometrium. Biol Reprod 86: 46.

25. Pindel A, Sadler A (2011) The role of protein kinase R in the interferon response. J Interferon Cytokine Res 31: 59–70.

26. Haneji T, Hirashima K, Teramachi J, Morimoto H (2013) Okadaic acid activates the PKR pathway and induces apoptosis through PKR stimulation in MG63 osteoblast-like cells. Int J Oncol 42: 1904–1910.

27. Schoggins JW, Wilson SJ, Panis M, Murphy MY, Jones CT, et al. (2011) A diverse range of gene products are effectors of the type I interferon antiviral response. Nature 472: 481–485.

28. Takahashi M, Takahashi H, Hamano S, Watanabe S, Inumaru S, et al. (2003) Possible role of interferon-tau on in vitro development of bovine embryos. J Reprod Dev 49: 297–305.

29. Oritani K, Kincade PW, Zhang C, Tomiyama Y, Matsuzawa Y (2001) Type I interferons and limitin: a comparison of structures, receptors, and functions. Cytokine Growth Factor Rev 12: 337–348.

30. Fujisawa M, Fujioka H, Tatsumi N, Inaba Y, Okada H, et al. (1998) Levels of interferon alpha and gamma in seminal plasma of normozoospermic, oligozoospermic, and azoospermic men. Arch Androl 40: 211–214.

2

OsRACK1 Is Involved in Abscisic Acid- and H$_2$O$_2$-Mediated Signaling to Regulate Seed Germination in Rice (*Oryza sativa*, L.)

Dongping Zhang[1], Li Chen[1], Dahong Li[2], Bing Lv[1], Yun Chen[1], Jingui Chen[3], XuejiaoYan[1], Jiansheng Liang[1]*

1 Department of Biotechnology, College of Bioscience and Biotechnology, Yangzhou University, Jiangsu, China, **2** Department of Biological Engineering, Huanghuai University, Zhumadian City, Henan Province, China, **3** Biosciences Division, Oak Ridge National Laboratory, Oak Ridge, Tennessee, United States of America

Abstract

The receptor for activated C kinase 1 (RACK1) is one member of the most important WD repeat–containing family of proteins found in all eukaryotes and is involved in multiple signaling pathways. However, compared with the progress in the area of mammalian RACK1, our understanding of the functions and molecular mechanisms of RACK1 in the regulation of plant growth and development is still in its infancy. In the present study, we investigated the roles of rice RACK1A gene (*OsRACK1A*) in controlling seed germination and its molecular mechanisms by generating a series of transgenic rice lines, of which *OsRACK1A* was either over-expressed or under-expressed. Our results showed that *OsRACK1A* positively regulated seed germination and negatively regulated the responses of seed germination to both exogenous ABA and H$_2$O$_2$. Inhibition of ABA biosynthesis had no enhancing effect on germination, whereas inhibition of ABA catabolism significantly suppressed germination. ABA inhibition on seed germination was almost fully recovered by exogenous H$_2$O$_2$ treatment. Quantitative analyses showed that endogenous ABA levels were significantly higher and H$_2$O$_2$ levels significantly lower in *OsRACK1A*-down regulated transgenic lines as compared with those in wildtype or *OsRACK1A*-up regulated lines. Quantitative real-time PCR analyses showed that the transcript levels of *OsRbohs* and amylase genes, *RAmy1A* and *RAmy3D*, were significantly lower in *OsRACK1A*-down regulated transgenic lines. It is concluded that *OsRACK1A* positively regulates seed germination by controlling endogenous levels of ABA and H$_2$O$_2$ and their interaction.

Editor: Tai Wang, Institute of Botany, Chinese Academy of Sciences, China

Funding: This work was supported by the National Science Foundation of China (grant no: 31271622 to JSL) and the Hi-Tech Research and Development Program ("863" project) from the Ministry of Science and Technology of China (grant no: 2008AA10Z120 to JSL), and by the College Postgraduate Research and Innovation Project in Jiangsu Province (to DPZ) and University Innovation Training Program of Jiangsu Province (to SJY) for research. The funders had no role in study design, data collection and analysis, decision to publish, or preparation of the manuscript.

Competing Interests: The authors have declared that no competing interests exist.

* E-mail: jsliang@yzu.edu.cn

Introduction

Seed germination is a complex processes that is under combinatorial control by endogenous and environmental cues. Because seed germination is a critical stage in the plant life cycle and is the first step towards successful plant establishment, it is of importance to unravel the physiological and molecular mechanisms underlying seed germination control in order to genetically manipulate it. A large number of studies have been conducted to explore seed germination mechanisms over the past several decades [1]–[4]. While there is much information with respect to changes in cellular events and molecular processes during germination, neither the key event(s) nor the regulatory network has been identified that results in its completion [5]. Recently, transcriptomic, proteomic and metabolomic analyses have proved that transcriptional, translational and post-translational modification in *Arabidopsis* seeds are more complex and dynamic than previously thought [6]–[9],[3],[10],[11]. Bassel et al.[7], using publicly available gene expression data from *Arabidopsis thaliana*, generated a condition-dependent network model of global transcriptional interactions associated with seed dormancy or germination and identified 1,583 transcripts were associated with germination. He et al.[3] analyzed the protein profiling in the germinating rice seeds through 1-DE via LC MS/MS proteomic shotgun strategy and identified that a total of 673 proteins were involved in seed germination. These proteins could be divided into 14 functional groups and the largest group was metabolism-related. Others included many regulatory proteins which control cellular redox homeostasis and gene expression of the germinating seeds.

In addition to the intensive changes both in gene expression and in metabolic activities during seed germination, the levels of plant hormones and many signaling molecules, such as reactive oxygen species (ROS), NO, etc, also changed drastically during seed development and seed imbibition in response to developmental and environmental cues [2], [12]–[14]. Abscisic acid (ABA) and gibberellins (GAs) were considered as two major hormones controlling seed dormancy and germination [15],[16]. ABA promotes the induction and maintenance of seed dormancy whereas GA is required for the initiation and completion of germination. It is thought that germination is regulated by the antagonistic effects between ABA and GA. However, the regulation of ABA and GA on seed germination was much more complex than it was thought. For example, Okamoto et al. [17] found that, although application of exogenous ABA inhibited germination, the effects of exogenous ABA on ABA-mediated gene transcription differ from those of endogenous ABA. The effects of exogenous ABA were prominent in the expression of several ABA-related genes at a later stage of imbibition, whereas the endogenous ABA affected the expression of critical components, e.g. ABA signaling, photosynthesis, physiological and metabolic genes including a GA biosynthesis enzyme.

Although ROS are long considered hazardous molecules, their functions as cell signaling compounds are now well established and widely studied in plants[4],[18]. Great fluctuations of ROS levels occur continuously during seed development and germination. Growing evidence have shown that ROS may function as messengers or transmitters of environmental cues during seed germination and positively control seed germination [4],[19]–[24]. Little is currently known, however, about ROS biochemistry or their functions or the signaling pathways during these processes.

The receptor for activated C kinase 1 (RACK1) is one member of the most important tryptophan, aspartic acid repeat (WD repeat)–containing family of proteins found in all eukaryotes [25]–[27]. As a scaffolding protein, RACK1 is involved in multiple signaling pathways [27]–[30]. The first plant RACK1 was originally identified as an auxin inducible gene, arcA, in tobacco BY-2 suspension cells in a differential screen for genes involved in auxin-mediated cell division [31],[32], and thereafter, RACK1 gene was isolated from a range of plant species, including Arabidopsis thaliana, Oryza sativa, and found to be expressed ubiquitously in different tissues and organs, including leaf, stem, root, and flower, etc., which implies that RACK1 may play important roles in plant growth and development [28],[33],[34]. The Arabidopsis genome contains three RACK1 orthologues, At1g18080, At1g48630, and At3g18130, designated as RACK1A, RACK1B, and RACK1C, respectively. These three Arabidopsis proteins are approximately 65% identical and 78% similar to mammalian RACK1 [28]. Chen et al. [28] provided direct genetic evidence of the function of RACK1 in plant responses to several plant hormones using the loss-of-function mutants of RACK1 in Arabidopsis. They found that rack1a mutants displayed altered sensitivities to several plant hormones, including hyposensitivity to gibberellic acid and brassinosteroid in seed germination, hyposensitivity to auxin in adventitious and lateral root formation, and hypersensitivity to ABA in seed germination and early seedling development [28]. The ABA-responsive marker genes, RD29B and RAB18, were up-regulated in rack1a mutants and the expression of all three RACK1 genes themselves was down-regulated by ABA [35]. Islas-Flores et al., using an RNAi approach to suppress the RACK1 gene expression in P. vulgaris (PvRACK1), and found that mRNA accumulation of PvRACK1 in roots was induced by auxins, ABA, cytokinin, and gibberellic acid [36],[37]. Our studies on the functions of Arabidopsis and rice RACK1 in the responses to drought stress showed that, when RACK1 gene expression was suppressed, leaf ABA level significantly increased, as a result, the tolerance of seedlings to soil drying increased [38]. Growing evidence has shown that RACK1 is critical regulators of plant development and loss-of-function mutations in RACK1A confer defects in multiple developmental processes including seed germination, leaf production, and flowering [28]. The BLASTP search using Arabidopsis RACK1A protein (NCBI accession number: NP_173248) as a template revealed that, in rice genome,

there are two RACK1 homologous genes, OsRACK1A and OsRACK1B, which are approximately 80% similar to Arabidopsis RACK1 proteins at the amino acid level. Nevertheless, at present, we know little about whether RACK1 is involved in and how it regulates seed germination, especially in rice. Gene expression profile analysis indicated that OsRACK1A gene was highly expressed whereas OsRACK1B was lowly expressed in rice seeds (Figure S1). In this experiment, we firstly generated a series of transgenic rice lines, of which OsRACK1A gene was either over-expressed or under-expressed, and explored the roles of OsRACK1 in regulating seed germination. Analysis of these transgenic lines revealed that OsRACK1A positively regulates seed germination through enhancing ABA catabolism and promoting H_2O_2 production.

Materials and Methods

Plant Materials

Rice (Oryza sativa L. cv. Nipponbare) was used as the wildtype (non-transgenic line, NTL) and in the generation of all transgenic plants. All transgenic rice lines were generated and kept in our laboratory. OsRACK1A over-expression transgenic lines, OeTL3-8 and OeTL4-9, anti-sense transgenic line, AsTL7-6 and RNA-interfered transgenic lines, RiTL3-1, RiTL4-2 and RiTL7-3 were used as experimental materials in this study [38].

Plant Growth Conditions, Seed Germination Assay and Stress Treatments

Seeds collected from the rice plants of wildtype and the 6th generation of transgenic lines were used in this experiment. Fully-filled and uniformed rice seeds were washed with 70% (v/v) ethanol for 30 seconds, and washed three times with sterile water. Sterilized seeds were subsequently sown on sterile filter papers in the petri dishes (size:150 mm diameter) which contained different concentrations of ABA (0~40 μM), NaCl (0~200 mM), Glucose (0~3%) (w/v), and 20 μM fluridone, 100 μM diniconazole, or 20 mM H_2O_2 (Sigma). Stock solution of ABA (mixed isomers, Sigma), fluridone (Sigma) and diniconazole (Sigma) were dissolved in ethanol. Control petri dishes contained equal amount of sterilized water. Seeds were germinated in a growth chamber with 12 h light and 12 h dark at 28°C. Germination (based on radicals >2 mm) was recorded at the indicated time points. For each germination test, fifty seeds per genotype were used, and three experimental replications were performed. The average germination percentage ± SE (standard error) of triplicate experiments was calculated. Seeds imbibed for different time in the presence of water or ABA were collected and stored at −80°C for ABA and H_2O_2 determination, gene expression analysis and enzyme activity analysis.

For gene expression analysis of different organs, well-uniformed rice seedlings were transplanted into ceramic pots (size: 60 cm in height and 30 cm in diameter) and grew under natural conditions. Different organs were sampled at the rapid grain-filing stage (about 10 days after flowering) and stored at −80°C pending assay.

Gene Expression Analysis

Total RNA (from dry or imbibed seeds at indicated times or different plant organs) was extracted by using the RNeasy plant mini kit (Qiagen). DNase I-treated total RNA (2 mg) was denatured and subjected to reverse transcription using Rever-tAid™ first strand cDNA synthesis kit (100 units per reaction; Thermo Scientific). Transcript levels of each gene were measured by qRT-PCR using 7500 Real-Time PCR Systems (ABI) with iTaq™ Universal SYBR® Green Supermix (Bio-Rad). Gene expression was quantified at the logarithmic phase using the expression of the housekeeping *OsActin7* (LOC_Os11g06390) as an internal control. Three biological replicates were performed for each experiment. Primer sequences for qRT-PCR are shown in Table S1.

Protein Blot Analysis

Rice seeds or leaves were homogenized in TEDM buffer (20 mM Tris/HCl, pH 7.5, 1 mM DTT, 5 mM EDTA, and 10 mM MgCl2) containing complete protease inhibitor cocktail (Roche). The homogenate was centrifuged at 6,000 g for 30 min at 4°C to remove cellular debris, and the supernatant was clarified by centrifugation at 5,000 g for 90 min at 4°C. The soluble proteins were separated by SDS–PAGE on a 10% gel and blotted onto Hybond-P membrane (Amersham Pharmacia Biotech). HSP82, OsPIP and eEF1α were used as loading control. The antibodies used in this experiment were bought from Beijing Protein Innovation (BPI).

Fluorescent Localization of OsRACK1A

To determine the intracellular localization of OsRACK1A, *OsRACK1A* was cloned into pXZP008 vector to produce an *OsRACK1A-GFP* fusion construct driven by the maize ubiquitin promoter (*Ubi::OsRACK1A:GFP*). Rice protoplasts prepared from shoots were transformed with *Ubi::OsRACK1A:GFP* by polyethylene glycol treatment. The florescence signal was observed with an Inverted Microscope (Axio Observer A1, Zeiss) at 16 h after transformation.

Quantitative Analyses of the Endogenous Levels of ABA and H_2O_2

Measurement of endogenous ABA levels of imbibed seeds was carried out based on the procedures described by Chen et al. (2006). Briefly, 30 grains of imbibed rice seeds were homogenized in 1 ml of distilled water and then shaken at 4°C overnight. The homogenates were centrifuged at 12,000 g for 10 min at 4°C and the supernatant were used directly for ABA assay. ABA analysis was carried out using the radioimmumoassay (RIA) method as described by Quarrie et al. [39]. The 450 μl reaction mixture contained 200 μl of phosphate buffer (pH 6.0), 100 μl of diluted antibody (Mac 252, Abcam) solution, 100 μl of [³H]ABA (about 8,000 c.p.m.) (Sigma) solution and 50 μl of crude extract. The mixture was then incubated at 4°C for 90 min and the bound radioactivity was measured in 50% saturated $(NH_4)_2SO_4$-precipitated pellets with a liquid scintillation counter (LS6500, Beckman)

For measurement of endogenous H_2O_2 levels of imbibed seeds, 20 grains of seeds were homogenized in 500 μl phosphate buffer (20 mM K_2HPO_4, pH 6.5). After centrifugation, 50 μl of the supernatant was incubated with 0.2 U ml^{-1} horseradish peroxidase and 100 μM Amplex Red reagent (10-acetyl-3,7-dihydrophenoxazine) at room temperature for 30 min in darkness. The fluorescence was quantified using Epoch™ Microplate Reader (BioTek) (excitation at 560 nm and emission at 590 nm).

Enzyme Activity Assay

Frozen seeds (20 grains) were homogenized on ice with 1.5 ml of 100 mM potassium phosphate buffer with protease inhibitor [1 mM EDTA, 10% (v/v) glycerol, 1% (v/v) Triton X 100, 7 mM β-mercaptoethanol, 100 mM NaF, 1 mM NaVO$_3$, 1 mM Na_3VO_4, 10 mM $Na_4P_2O_7$, 10 mM N-ethylmaleimide]. The homogenate was then centrifuged at 12,000 g for 10 min at 4 °C. A 100 μl aliquot of supernatant was taken into the assay buffer

containing 50 mM Na-acetate, pH 5.2, 10 mM CaCl$_2$, 2% boiled soluble potato starch, and incubated at 37 °C for 1 h. Then 100 μl of assay mixture was taken and mixed with 100 μl of reaction termination buffer (0.1 N NaOH) and 100 μl of 3,5-dinitrosalicylic acid (DNS) solution (40 mM DNS, 400 mM NaOH, 1 M K-Na tartrate) for 5 min at 100°C. After dilution with 700 μl of distilled water, the absorbance at 540 nm (A_{540}) was measured, and the reducing power was evaluated with a standard curve obtained with glucose (0–20 μmol). One enzymatic unit is defined as the amount of enzyme releasing 1 mmol of glucose in 1 min.

Statistical Analyses

Data were subjected to analysis of variance and psot-hoc comparisons (Duncan's multiple range test; $P<0.05$ significance level). We used SPPSS versions 13.0 software. Values presented are means ± SD of three replicates and significant difference is labeled as "*" in Figures.

Results

OsRACK1A Expression and Intracellular Localization

In rice, there are two RACK1 homologs, designated as OsRACK1A and OsRACK1B, which are approximately 80% similar to *Arabidopsis* RACK1 proteins at the amino acid level [26]. OsRACK1A and OsRACK1B share over 82.8% identity and 89.3% similarity at the amino acid level when aligned with Blast 2 Sequences [40]. Gene expression profile analysis indicated that *OsRACK1A* gene was highly expressed whereas *OsRACK1B* was lowly expressed and hardly detected in rice seeds during germination (Figure S1). Therefore, to explore the functions of *OsRACK1* in seed germination and seedling growth, we firstly generated a series of transgenic rice lines, of which the expression of *OsRACK1A* gene was either down-regulated using the RNA-interference or anti-sense techniques, or up-regulated using the Ubi-promoter [38]. The homozygous transgenic lines of the 6th generation were used in this experiment. Figure 1 showed the *OsRACK1A* expression in imbibed seeds of the selected transgenic lines. As compared with that of wildtype (non-transgenic line, NTL), the transcript levels of *OsRACK1A* were 30~50% higher in the over-expressed transgenic lines (OeTLs) and 50~70% lower in RNA-interfered transgenic lines (RiTLs) or anti-sense transgenic lines (AsTLs) measured by quantitative RT-PCR (Figure 1A). Consistent with these results, western blot analysis using the OsRACK1A-specific antibody indicated that RACK1A protein was up-regulated in OeTLs and down-regulated in RiTLs (Figure 1B). Meanwhile, the transcript levels of *OsRACK1B* in RiTLs or AsTLs were also obviously lower than those in NTL or OeTLs (Figure S2).

The intracellular localization of OsRACK1A was also investigated using immunoblotting analysis and fluorescence microscopy. Rice protoplasts were transfected with a vector containing *Ubi::OsRACK1A–GFP* or *Ubi::GFP*. Fluorescence microscopic examination revealed that the GFP fluorescence was detected in the cytosol fraction as well as in the plasma membrane and in the nuclei (Figure 2A). To confirm the intracellular localization of OsRACK1A, we further prepared the soluble and insoluble fractions of the cells by differential centrifugation and incubated them with polyclonal antibodies against OsRACK1A. As shown in Figure 2B, heavy protein bands were detected in all fractions by western blot analysis. These results suggested that OsRACK1A was a soluble protein and may anchor in the plasma membrane by means of interacting with other molecules.

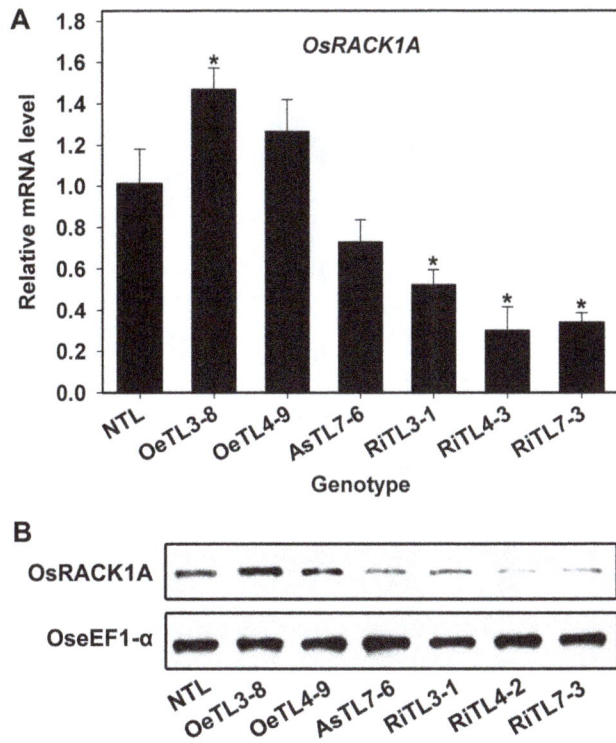

Figure 1. Expressional profile of OsRACK1A genes (A) and proteins (B) in selected transgenic rice lines. A, *OsRACK1A* expression was monitored in seedlings of wildtype (non-transgenic lines, NTL), *OsRACK1A* over-expressed transgenic lines (OeTL), anti-sense transgenic lines (AsTL) and RNA-interfered transgenic lines (RiTLs), respectively. Relative expression levels were calculated and normalized with respect to *OsActin7* (LOC_Os11g06390). Results shown are means of three biological replications±SE. Asterisks (*) indicate significant difference ($P<0.05$) between the levels of expression of transgenic lines compared with the wildtype (NTL). B, *OsRACK1A* expression was analyzed by incubating isolated proteins with polyclonal antibodies against OsRACK1A or OseEF1-α (as reference).

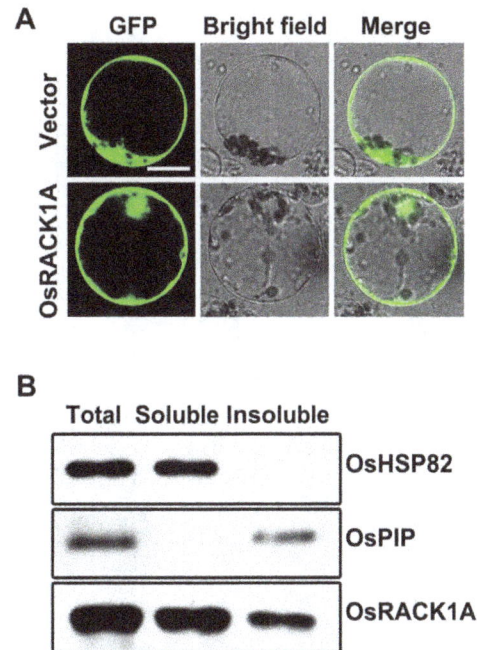

Figure 2. Intracellular localization of OsRACK1A. The intracellular localization of OsRACK1A was monitored with the fluorescence microscopy (A) and using immunoblotting analysis (B). Rice protoplasts were transformed with a vector containing *Ubi::OsRACK1A–GFP* or *Ubi::GFP* by PEG method and the florescence signal was observed with an Inverted Microscope (Axio Observer A1, Zeiss) at 16 h after transformation. For immunoblotting assay, the soluble and insoluble fractions of the cells were isolated by differential centrifugation and incubated with polyclonal antibodies against OsRACK1A or OsHSP82 (as reference) or OsPIP (as reference of insoluble protein).

Seed Germination and Its Responses to ABA and H₂O₂ in Different Transgenic Rice Lines

Quantitative RT-PCR (qRT-PCR) analysis of *OsRACK1A* expression patterns across various tissues and organs indicated that the highest level of *OsRACK1A* transcript was in seeds (Figure S1). Therefore, we wanted to examine whether *OsRACK1A* regulates seed germination. It is well known that seed germination is a multi-stage process and involves in complicated regulatory networks. To assess the roles of *OsRACK1* in controlling germination of rice, seeds of different transgenic rice lines were imbibed under standard germination conditions and germination kinetics was measured. Seed germination started after 36 h imbibition and then the germination rate was increased rapidly for NTLs. A significant delay in germination and a slowly increase in germination rate were observed for seeds of RiTLs (Figure 3A). ABA treatment delayed seed germination of all rice genotypes tested. However, germination of RiTLs or AsTLs was more sensitive to exogenous ABA treatment than those of NTLs and OeTLs (Figure 3B). These results suggested that *OsRACK1A* positively regulates seed germination in rice, as is the case in *Arabidopsis* reported earlier [28], and *OsRACK1A* may involve in ABA-mediated signaling during seed germination. In order to further explore the roles of endogenous ABA in controlling seed germination, fluridone, an inhibitor of ABA biosynthesis, and

diniconazole, a potent competitive inhibitor of ABA catabolism, were applied and seed germination kinetics were measured. As shown in Figures 3C-E, fluridone treatment had no obvious effects on germination of all genotypes, as compared with the control treatment (Figure 3A). A similar result was also observed when comparison of germination was made between the combined treatment of fluridone with exogenous ABA and ABA treatment alone. However, diniconazole treatment significantly delayed seed germination of all genotypes, especially at the earlier stage of seed germination, i.e. within 48 h of imbibition. More severe delay of germination was observed in RiTLs and AsTLs lines (Figure 3E). We also observed that hypersensitivity of AsTLs and RiTLs to ABA inhibition of germination was ABA dosage-dependent (Figure S3). Taken together, these results indicated that the suppression of ABA catabolism rather than an enhancement of ABA biosynthesis may be the major factor for the delay of germination and *OsRACK1A* may exert its effect by controlling ABA metabolism.

It is well known that ABA treatment as well as adverse environmental factors induces production of H₂O₂ that may also involve in the regulation of seed germination [23]. However, it is largely unclear about the molecular mechanisms of H₂O₂ regulation on seed germination. In the present study, we assessed the effects of exogenous H₂O₂ treatment on seed germination of different transgenic rice lines. The result showed that exogenous H₂O₂ treatment significantly stimulated seed germination of all genotypes and after 48 h of imbibition, the germination rate reached approximately 80% for NTL and OeTLs and 45% for RiTLs and AsTLs (Figure 3F), as compared to about 40% of

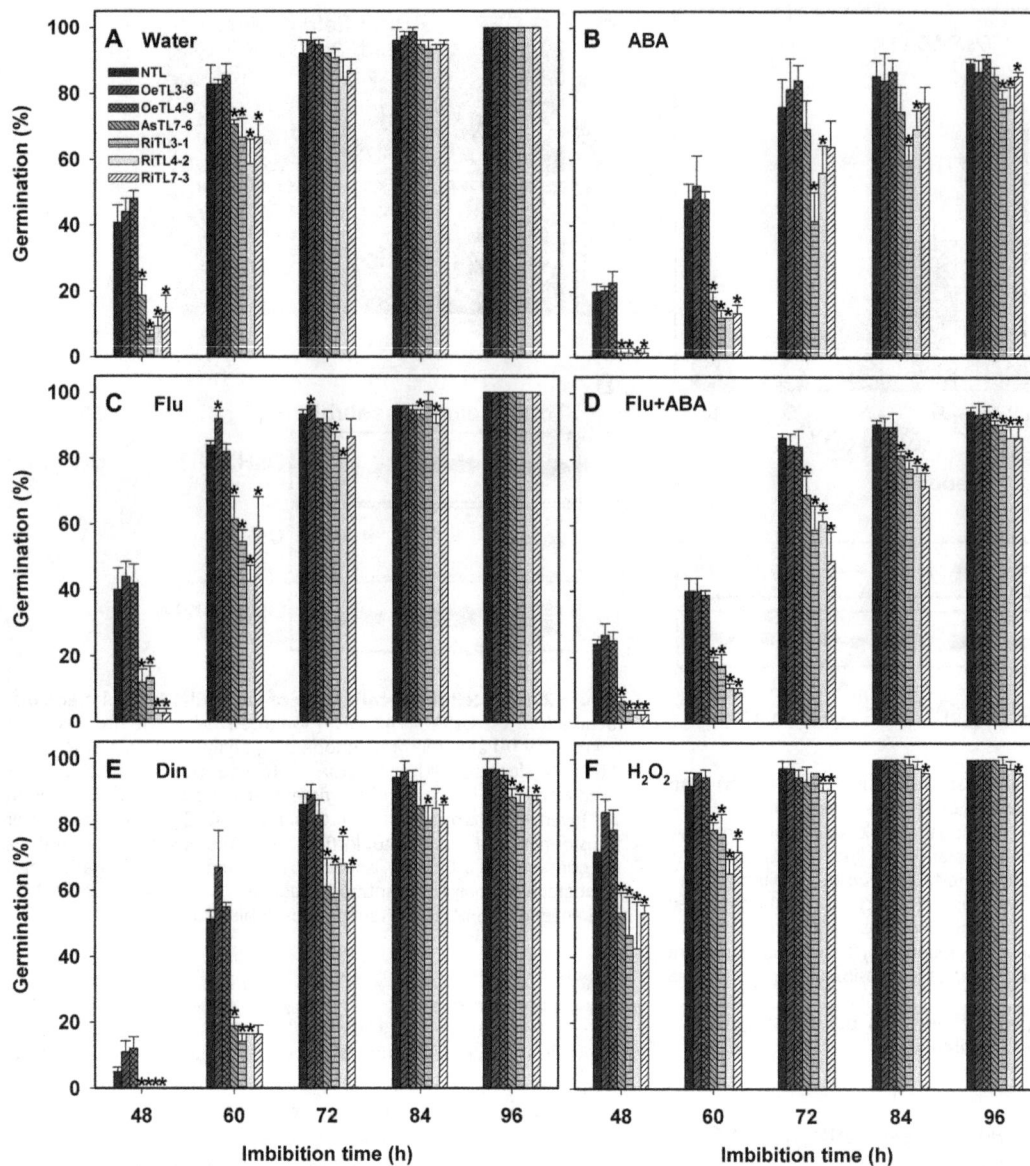

Figure 3. Effects of different treatments on seed germination of different transgenic lines. Sterilized seeds were germinated at 28°C on sterile filter papers in the petri dishes containing different concentrations of ABA (20 μM), fluridone (20 μM), diniconazole (100 μM) or H_2O_2 (20 mM). Germination (based on radicals >2 mm) was recorded at the indicated time points. Fifty seeds per genotype were used. Data shown are means of three biological replications±SE. Asterisks (*) indicate significant difference ($P<0.05$) between the seed germination of transgenic lines compared with the wildtype.

germination for NTL and OeTLs and 15% for RiTLs and AsTLs, and RiTLs and AsTLs were more sensitive to H_2O_2 promotion of seed germination. These results implied that H_2O_2 may also involve in the regulation of *OsRACK1A* on seed germination.

ABA and H_2O_2 Interaction on Seed Germination of Different Transgenic Rice Lines

As described above, ABA had an inhibitory whereas H_2O_2 had a stimulated effect on seed germination. A lot of results have been shown that that ABA interacts with H_2O_2 in many processes, such as stomatal movement [41]. However, it is it is unclear how ABA interacts with H_2O_2 to regulate seed germination and what are their roles in *OsRACK1A*-mediating seed germination. Here we compared germination of different *OsRACK1A* transgenic rice lines

when seeds were treated with ABA or H_2O_2 alone or both. As shown in Figure 4, ABA significantly suppressed and H_2O_2 stimulated seed germination. However, ABA inhibition on seed germination was either almost fully removed for NTL and OeTLs or significantly alleviated for RiTLs and AsTLs by H_2O_2 (Figure 4).

We further analyzed the H_2O_2 levels in different transgenic rice lines and the effects of ABA on H_2O_2 production in imbibed seeds. Great differences in H_2O_2 levels were detected in imbibed seeds of various transgenic rice lines (Figure 5A). As compared to NTL, the H_2O_2 level of OeTL was significantly higher and those of AsTL and RiTL were significantly lower, which may partially explain the difference in germination patterns among different transgenic rice lines. ABA treatment significantly inhibited the H_2O_2 production in germinating seeds, and this inhibition was ABA dosage-dependent (Figure 5B).

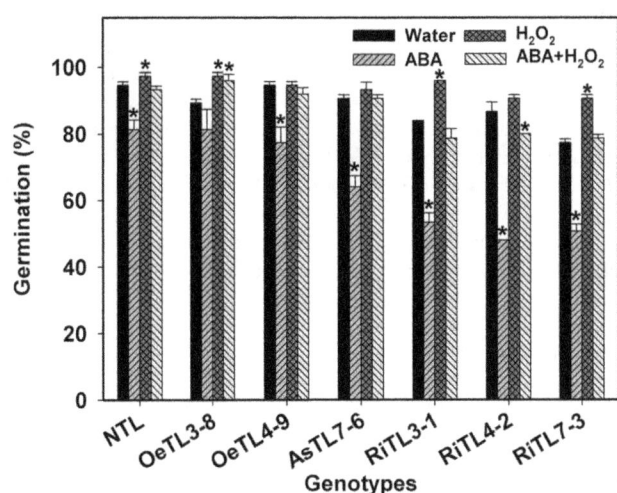

Figure 4. Effects of exogenous ABA and H₂O₂ on seed germination of different transgenic lines. Sterilized seeds were germinated at 28°C on sterile filter papers in the petri dishes containing different concentrations of ABA (20 μM) or H₂O₂ (20 mM) or both. Germination (based on radicals >2 mm) was recorded at 72 h after imbibition. Fifty seeds per genotype were used. Data shown are means of three biological replications±SE. Asterisks (*) indicate significant difference (P<0.05) between the seed germination of transgenic lines compared with the wildtype.

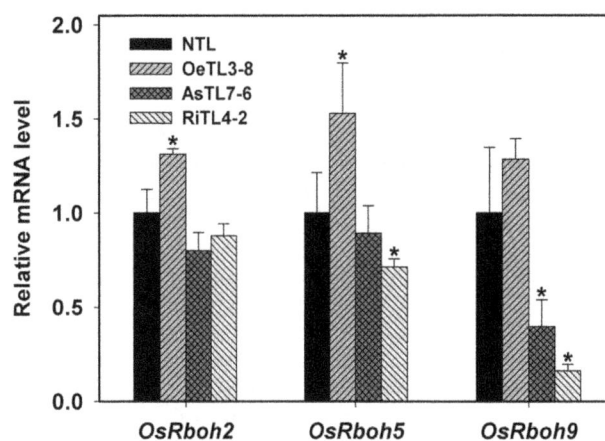

Figure 6. Expressional of *OsRbohs* in imbibed seeds of the selected transgenic rice lines. Expressions of *OsRbohs* were monitored in seeds of NTL, OeTL, AsTL and RiTL, respectively after 48 h of imbibition. Relative expression levels were measured by qRT-PCR and normalized with respect to *OsActin7* (LOC_Os11g06390). Results shown are means of three biological replications±SE. Asterisks (*) indicate significant difference (P<0.05) between the levels of expression of transgenic lines compared with the NTL.

NADPH oxidases have been proposed to be involved in ROS production during seed germination [42]. In rice genome, nine NADPH oxidases genes have been reported [43]. Our preliminary analysis has shown that only three genes, *OsRboh2*, *OsRboh5* and *OsRobh9*, expressed in imbibed seeds of rice. In order to clarify whether the expression of these genes is responsible for ROS (mainly H₂O₂) production and to explain the differences among various genotypes used in this experiment, the transcript levels of these three genes were determined in imbibed rice seeds using

qRT-PCR. As shown in Figure 6, the transcripts of all these three genes could be detected in all genotypes examined. However, the transcript levels of all three genes were substantially higher in the imbibed seeds of OeTLs and significantly lower in those of AsTLs and RiTLs, as compared with those of NTL (Figure 6), which was consistent with the trends of endogenous H₂O₂ levels in these transgenic lines (Figure 5A). These results indicated that *OsRbohs* are responsible for the production of H₂O₂ in imbibed seeds and thus the seed germination.

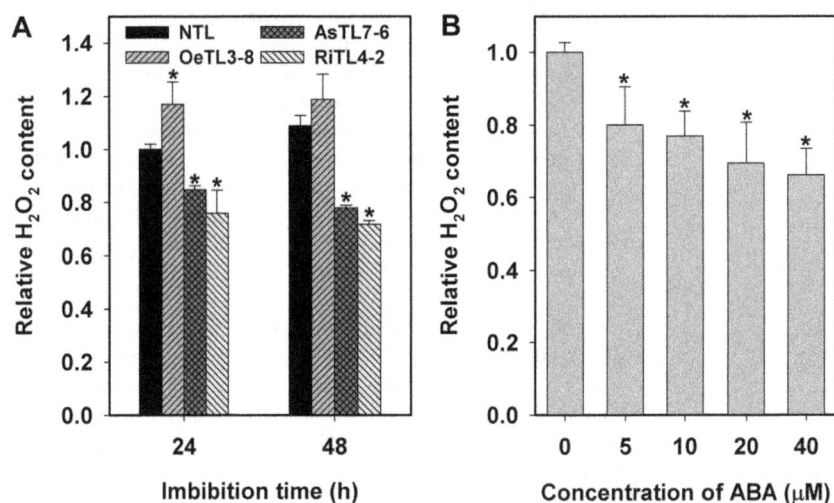

Figure 5. Comparison of relative H₂O₂ contents among different transgenic rice lines and the effect of ABA treatment on it in imbibed seeds. For measurement of endogenous H₂O₂ levels of imbibed seeds, twenty seeds were homogenized in 500 μl phosphate buffer (20 mM K₂HPO₄, pH 6.5). After centrifugation, 50 μl of the supernatant was incubated with 0.2 U ml⁻¹ horseradish peroxidase and 100 μM Amplex Red reagent (10-acetyl-3,7-dihydrophenoxazine) at room temperature for 30 min in darkness. The fluorescence was quantified using EpochTM Microplate Reader (BioTek) (excitation at 560 nm and emission at 590 nm). Data shown are means of three biological replications±SE. Asterisks (*) indicate significant difference (P<0.05) between the relative levels of transgenic lines compared with the wildtype (A) or 0 μM ABA (B).

Changes in Endogenous ABA Levels during Germinating Process

ABA is one of key hormones regulating seed germination. Our results had shown that *OsRACK1A* positively regulate seed germination. In order to investigate whether *OsRACK1* affects endogenous ABA levels, we analyzed the endogenous ABA levels in germinating seeds of different *OsRACK1A* transgenic lines. Seeds of different transgenic rice lines were imbided for different times and ABA levels were detected by radioimmumoassay (RIA) method as described by Quarrie et al. [39]. As shown in Figure 7, significant decreases in endogenous ABA levels were observed with imbibition time proceeded and there were significant differences in ABA levels among different transgenic rice lines. ABA levels in imbided seeds of AsTLs and RiTLs were significantly higher than those of NTL and OeTLs, which may, at least in part, explain the slow initiation and delay of seed germination of AsTLs and RiTLs.

Changes of Amylase Activity and Amylase Gene Expression in Imbibed Seeds of Different Transgenic Rice Lines

Seed germination is a complex physiological and biochemical process and involves the degradation of stored compounds and re-synthesis of new compounds necessary for cell growth and development. Because in mature rice seeds, about 80% of stored carbohydrate is starch, starch degradation is necessary for germination and early growth and development of seedling by providing both energy and carbon cytoskeleton. Amylase is a key enzyme catalyzing starch degradation of germinating seeds. Therefore, understanding the molecular mechanism of amylase in controlling seed germination is of predominant importance in agronomic practice. In rice genome, there are 12 genes putatively encoding alpha–amylases, but only two of them have been identified in rice seeds, *RAmy1A* and *RAmy3D* (unpublished). In the present study, seeds of different *OsRACK1A* transgenic lines imbibed for 24 hours were sampled to analyze the amylase activity and the expression of its encoding genes. A great differences were detected in amylase activity among different transgenic rice lines and the amylase activity of AsTLs and RiTLs

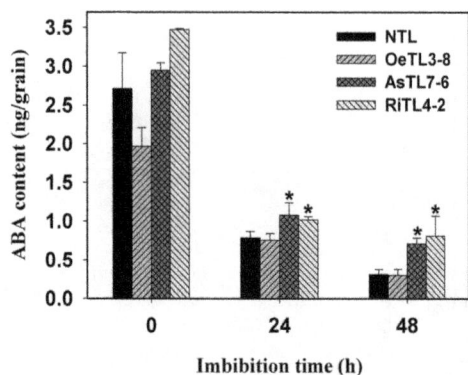

Figure 7. Comparison of endogenous ABA contents in imbibed seeds of different transgenic lines. ABA analysis was carried out using the radioimmumoassay (RIA) method. Thirty imbibed rice seeds were homogenized in 1 ml of distilled water and then shaken at 4°C overnight. The homogenates were centrifuged at 12,000 g for 10 min at 4°C and the supernatant were used directly for ABA assay. Values shown are means of three biological replications±SE. Asterisks (*) indicate significant difference (P<0.05) between the endogenous ABA contents of transgenic lines compared with the wildtype.

was only 15~30% of that NTL and OeTLs (Figure 8A). Among two amylase genes expressed in seeds, the transcript level of *RAmy1A* was about two times higher than that of *RAmy3D*. Furthermore, compared to those in the NTL, the expression of both *RAmy1A* and *RAmy3D* was much lower in AsTLs and RiTLs and much higher in OeTLs (Figure 8B). ABA had significant inhibitory effects on the expression of both *RAmy1A* and *RAmy3D* (Figure 8C and D). Taken together, these results suggested that *OsRACK1A* positively regulates germination through up-regulating amylase activity and the expression of *RAmy1A* and *RAmy3D* in rice seeds.

Discussion

RACK1 is a strongly conserved and scaffolding protein and expresses ubiquitously. It is well recognized that RACK1 is involved in multiple signaling pathways, including growth and development and responses to external environmental cues [28],[34]. However, as compared with the progresses in the field of mammalian RACK1, our understanding of the functions and molecular mechanisms of plant RACK1 is still in its infancy. We and several other research teams have studied the roles of RACK1 in plant growth and development and in the responses to environmental cues and found that RACK1 is involved in regulating cell proliferation and elongation, inoculation and the responses to plant hormones and environmental factors [28],[35], [37], [44], [45]. In rice genome, there are two RACK1 homologous genes, *OsRACK1A* and *OsRACK1B*. Nakashima et al. [34] reported that RACK1A functions in rice innate immunity by interacting with multiple proteins in the Rac1 immune complex. Our study has shown that, although OsRACK1A and Os-RACK1B are high similarity each other, the expression patterns are obviously different and *OsRACK1B* transcript levels are always significantly lower than those of *OsRACK1A*, especially in roots and mature seeds (Figure S1). The highest *OsRACK1A* transcript level and the lowest *OsRACK1B* transcript level are founded in mature seeds. These results imply that *OsRACK1A* is a major gene which may play an important role in controlling seed development and germination. In present study, by generating several transgenic rice lines, of which *OsRACK1A* was either down-regulated or up-regulated, and analyzing the roles of *OsRACK1A* in seed germination and its possible mechanisms, we show that *OsRACK1A* positively regulates seed germination through its enhancement of ABA catabolism and stimulation of H_2O_2 production, and that ABA interacts with H_2O_2 on the regulation of seed germination. As compared with that of wildtype rice (NTL) seeds, when *OsRACK1A* is down-regulated, seed germination is significantly delayed, and over-expression of *OsRACK1A* has, although not obvious, a stimulating effect on germination. This case is very similar to that in *Arabidopsis* seed germination [28]. However, we currently know little about how RACK1 regulates the seed germination. In other words, what are the component(s) in RACK1-mediated signaling during germination? Because ABA is well-known to involve in maintenance of seed dormancy and inhibition of seed germination, it is reasonable to assume that the inhibition of germination as a result of *OsRACK1A* down-regulation may be related to ABA. This assumption is verified in this study. We show that, although exogenous ABA treatment inhibits germination of all genotypes used, this inhibition of germination of *OsRACK1A* down-regulated transgenic rice seeds is much more sensitive to ABA than those of non-transgenic line and of *OsRACK1A* over-expressed transgenic lines (Figures 3, S3). This phenomenon suggests that *OsRACK1A* is a negative regulator of ABA response in seed germination of rice, which is very similar to

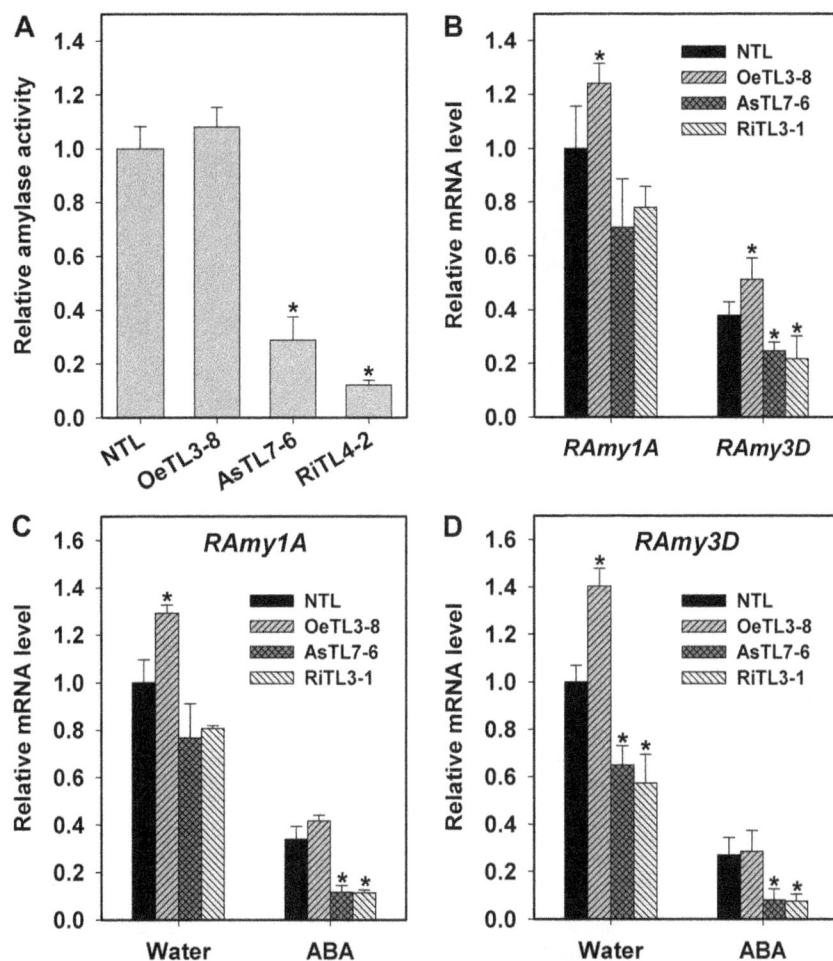

Figure 8. Comparisons of total amylase activity and of expression of two key amylase genes, *RAmy1A* and *RAmy3D*, in imbibed seeds of different transgenic lines and the effects of ABA. A, For amylase activity assay, frozen seeds (20 grains) were homogenized on ice and 100 µl aliquot of supernatant was taken into the assay after centrifugation. The reaction product was monitored using 3,5-dinitrosalicylic acid (DNS) method. B, C and D, Expressions of *RAmy1A* and *RAmy3* were monitored in seeds of NTL, OeTL, AsTL and RiTL, respectively after 48h of imbibition in water or in ABA (20 µM). Relative expression levels were measured by qRT-PCR and normalized with respect to *OsActin7* (LOC_Os11g06390). Results shown are means of three biological replications±SE. Asterisks (*) indicate significant difference (*P*<0.05) between the levels of expression of transgenic lines compared with the NTL.

our previous results that RACK1 negatively regulates the responses of seed germination, cotyledon greening and root growth to ABA treatment in *Arabidopsis* [35]. Furthermore, analysis on the endogenous ABA levels in dry and imbibed seeds shows that the endogenous ABA levels of both dry and imbibed seeds of *OsRACK1A* down-regulated transgenic rice lines are significantly higher than those in seeds of wildtype and *OsRACK1A* over-expressed transgenic lines (Figure 7). These results may explain well why the seed germination of *OsRACK1A* down-regulated transgenic rice lines is delayed under the same conditions.

In *Arabidopsis thaliana*, up-regulation of ABA biosynthesis is suggested as one of the possible mechanisms mediating the glucose-induced delay in seed germination, because the expression of ABA biosynthesis-related genes, such as *ABA2* and *NCED3* increases in responding to glucose signaling [46],[47]. However, our results show that higher endogenous ABA levels and the delay of seed germination of *OsRACK1A* down-regulated transgenic rice lines are not due to an enhancement of ABA biosynthesis because treatment with fluridone, a potent inhibitor of ABA biosynthesis, produces no obvious signs of enhancement of the germination

rate. However, treatment with ABA catabolic inhibitor, diniconazole, dramatically decreased the seed germination rate, as compared with those of control (water) treatment (Figure 3). These results suggest that higher endogenous ABA levels of seeds of *OsRACK1A* down-regulated transgenic rice lines are mainly due to the decrease of ABA catabolism, and thus result in a delay of seed germination. A similar situation was also observed in rice using another experimental system, where, Zhu et al. [2] found that glucose-induced delay of seed germination is a result of the suppression of ABA catabolism rather than any enhancement of ABA biosynthesis during rice seed germination. However, at present, we still know little about the molecular mechanisms of *OsRACK1A* regulation on ABA metabolism during seed germination and it deserves to study further.

ROS change dramatically during seed development and germination, implying that ROS play roles in seed dormancy alleviation, after-ripening, and germination [22],[48]. Although ROS are long considered as hazardous molecules, many recent studies have provided physiological and genetic evidence that ROS may also function as messengers or transmitters of environmental cues during seed germination. There are many reports indicated that seed germination and ROS accumulation appear to be linked, and that seed germination success may be closely associated with internal ROS contents and the activities of ROS-scavenging systems in species such as *Arabidopsis thaliana* [21],[49], sunflower [22], wheat [50], cress [51] and barley [12], soybean [20], rice [24]. Our results present here suggest that *OsRACK1A* may also regulate seed germination by mediating ROS production in imbibed seeds. This finding is supported by several lines of evidence. Externally supplied ROS in the form of H_2O_2 stimulates seed germination of all genotypes used and again the germination of *OsRACK1A* down-regulated transgenic rice lines is more sensitive to H_2O_2 (Figure 3), which is consistent with the cases in several other plant species [20],[21],[52]. Furthermore, comparison of the seed germination rate and the endogenous H_2O_2 levels in imbibed seeds among different rice genotypes, a lower germination rate is paralleled by a lower endogenous level of H_2O_2. For example, after imbibition for 48 h, the germination rate is 40% and around 12% for wildtype and *OsRACK1A* down-regulated transgenic rice lines, respectively, whereas the relative endogenous level of H_2O_2 is 1.14 and 0.78 for wildtype and *OsRACK1A* down-regulated transgenic rice lines, respectively (Figures 3, 5A). These results imply that *OsRACK1A* may also control H_2O_2 production of imbibed seeds, in spite that its molecular mechanism is still unknown. Pharmacological and molecular studies have shown that *Rbohs* genes and NADPH oxidase activity are potentially important to the germination process [21]. Treatment with diphenyleneiodonium (DPI), an inhibitor of *Rbohs*, blocked seed germination, suggesting that internal generation of ROS by *Rbohs* was needed for seed germination [53]. Seeds of the *atrbohB* mutant fail to after-ripen due to its blocked ROS production [53]. Our preliminary studies indicated that only three of nine *Rbohs* genes, i.e. *Rboh2*, *Rboh5* and *Rboh9*, were expressed in imbibed rice seeds. Our results present here show that the transcript levels of these three *Rbohs* genes in imbibied seeds of *OsRACK1A* down-regulated transgenic rice lines are significantly lower than those of wildtype, whereas the highest transcript levels are detected in *OsRACK1A* up-regulated transgenic rice lines. These results are consistent with the results of internal H_2O_2 levels (Figure 5). Taken together, we suggest that the positive control of *OsRACK1A* on seed germination is, at least in part, through up-regulating *Robhs* gene expression, as the results, stimulating internal H_2O_2 production.

There are many reports on the interaction of ROS with plant hormones in several physiological and molecular processes in plants. Involvement of ROS as second messengers in ABA-mediated stomatal movement has been well documented [41]. Generally, ABA induces ROS production in the guard cells as well as in non-seed tissues under adverse conditions. However, in seed physiology, ABA reduces ROS production in germinating seeds (Figures 5B),[24]. On the other hand, exogenous H_2O_2 treatment can mostly recover seed germination inhibited by ABA (Figure 4). Taking it into consideration that a rapid decrease in endogenous ABA level and gradual increase in endogenous H_2O_2 level occur during seed germination, it is understandable that there must be a fine mechanism maintaining a fine balance between ABA and H_2O_2 levels for germination success. This topic is deserved to further investigation.

In summary, it is concluded that *OsRACK1A* positively regulates seed germination by controlling endogenous levels of ABA and ROS and their interaction. Under normal germination conditions, cytosol OsRACK1A moved towards plasma membrane and interacted with OsRAC1 to form a complex [34]. The formed complex then activated downstream OsRboh, and as a result,

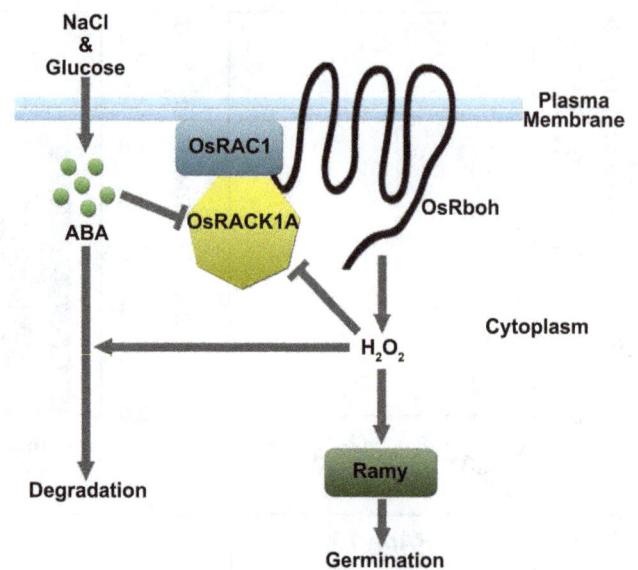

Figure 9. A working model showing proposed molecular steps of OsRACK1A-mediated signal transduction pathway. This model presents a schematic of OsRACK1A and other signal molecules involved in the proposed signal transduction pathway during seed germination. The results presented in this work combined with the results reported by Nakashima et al. (2008) to support this signal transduction. Under normal germination conditions, cytosol OsRACK1A moved towards plasma membrane and interacted with OsRAC1 to form a complex. The formed complex then activated downstream OsRboh, and as a result, stimulated H_2O_2 production and seed germination. When responding to external stimuli, which may induce ABA accumulation (Figure S4; Zhu et al, 2009), excess ABA inhibited *OsRACK1* expression and blocked the formation of OsRACK1A-OsRAC1 complex, and as a consequence, suppressed the expression of *OsRbohs* and the production of H_2O_2 (Figure 5). Excess H_2O_2 in return suppressed *OsRACK1* expression to avoid its hazardous effect (Figure S5). In addition, H_2O_2 may also stimulate ABA catabolism and alleviated the inhibitory effect on seed germination.

stimulated H_2O_2 production and seed germination. When responding to external stimuli, which may induce ABA accumulation (Figure S4),[2], excess ABA inhibited *OsRACK1* expression and blocked the formation of OsRACK1A-OsRAC1 complex, and as a consequence, suppressed the expression of *OsRbohs* and the production of H_2O_2 (Figure 5). Excess H_2O_2 in return suppressed *OsRACK1* expression to avoid its hazardous effect (Figure S5). In addition, H_2O_2 may also stimulate ABA catabolism and alleviated the inhibitory effect on seed germination (Figure 9).

Supporting Information

Figure S1   Quantitative RT-PCR analysis of *OsRACK1A* and *OsRACK1B* expression in rice tissues. Total RNA isolated from root, stem, leaf, panicle, dry seed or seedling (2 weeks) was used as template for qRT-PCR. Relative expression levels were calculated and normalized with respect to OsActin7 (LOC_Os11g06390). Results shown are means of three biological replications±SE.

Figure S2   Expressional profile of *OsRACK1B* in selected transgenic rice lines. *OsRACK1B* expression was monitored in seedlings of wildtype (non-transgenic lines, NTL), *OsRACK1A* over-expressed transgenic lines (OeTL), anti-sense transgenic lines (AsTL) and RNA-interfered transgenic lines (RiTLs), respectively.

Relative expression levels were calculated and normalized with respect to *OsActin7* (LOC_Os11g06390). Results shown are means of three biological replications±SE.

Figure S3 Effects of different concentration of ABA treatments on seed germination of different transgenic lines. A. Morphology of germinating seeds in the different concentration of ABA (0, 5, 10 and 20 μM) treatment for 24 hours. B. Seed germination rates under 0, 5, 10 and 3 μM ABA treatment for 24 hours. Each value is the mean ± standard error of at least 50 seeds. Asterisks (*) indicate significant difference (P<0.05) between the seed germination of transgenic lines compared with the wildtype.

Figure S4 Effects of different concentrations of glucose, NaCl and H2O2 on seed germination of different transgenic lines. Sterilized seeds were germinated at 28°C on sterile filter papers in the petri dishes containing different concentrations of glucose (0, 1, 3 and 6 mM), NaCl (0, 50, 100 and 200 mM) or H2O2 (0, 10, 20 and 40 mM). Germination (based on radicals >2 mm) was recorded at the indicated time points. Fifty seeds per genotype were used. Data shown are means of three biological replications±SE. Asterisks (*) indicate significant difference (P<0.05) between the seed germination of transgenic lines compared with the wildtype.

Figure S5 Effects of ABA and H2O2 on expressional profile of OsRACK1A genes (A) and proteins (B) in selected transgenic rice lines. Two-week-old seedlings were treatment with different concentration of ABA (5, 10, 20 and 40 μM) or H2O2 (10, 20 and 40 mM) for 24 h before RNA or protein extraction. A. Relative gene expression levels were calculated and normalized with respect to *OsActin7* (LOC_Os11g06390). B. OsRACK1A expression was analyzed by incubating isolated proteins with polyclonal antibodies against OsRACK1A or OseEF1-α (as reference).

Table S1 Gene specific primers used in quantitative real-time PCR (qRT-PCR).

Acknowledgments

The authors would like to thank members of the Zhang's lab for helpful comments on the manuscript.

Author Contributions

Conceived and designed the experiments: DPZ JGC JSL YC. Performed the experiments: DPZ DHL YC BL. Analyzed the data: DPZ XJY. Contributed reagents/materials/analysis tools: DPZ DHL YC BL. Contributed to the writing of the manuscript: DPZ JGC JSL.

References

1. Finkelstein R, Reeves W, Ariizumi T, Steber C (2008) Molecular aspects of seed dormancy. Annu Rev Plant Biol 59: 387–415.
2. Zhu G, Ye N, Zhang J (2009) Glucose-induced delay of seed germination in rice is mediated by the suppression of ABA catabolism rather than an enhancement of ABA biosynthesis. Plant Cell Physiol 50(3):644–651.
3. He D, Han C, Yao J, Shen S, Yang P (2011) Constructing the metabolic and regulatory pathways in germinating rice seeds through proteomic approach. Proteomics 11:2693–2713.
4. Diaz-Vivancos P, Barba-Espín G, Hernández JA (2013) Elucidating hormonal/ROS networks during seed germination: insights and perspectives. Plant Cell Rep. doi: 10.1007/s00299-013-1473-7.
5. Nanogaki H, Basel GW, Bewley JD (2010) Germination-still a mystery. Plant Sci 179:574–581.
6. Bassel GW, Fung P, Chow TFF, Foong JA, Provart NJ, et al. (2008) Elucidating the germination transcriptional program using small molecules. Plant Physiol. 147:143–155.
7. Bassel GW, Lan H, Glaab E, Gibbs DJ, Gerjets T, et al. (2011) Genome-wide network model capturing seed germination reveals coordinated regulation of plant cellular phase transitions. Proc Natl Acad Sci USA 108:9709–9714.
8. Belmonte MF, Kirkbride RC, Stone SL, Pelletiera JM, Bui AQ, et al. (2013) Comprehensive developmental profiles of gene activity in regions and subregions of the Arabidopsis seed. Proc Natl Acad Sci USA 14:E435–E444.
9. Gallardo K, Job C, Groot SP, Puype M, Demol H, et al. (2001) Proteomic analysis of Arabidopsis seed germination and priming. Plant Physiol 126:835–848.
10. Howell KA, Narsai R, Carroll A, Ivanova A, Lohse M, et al. (2009) Mapping metabolic and transcript temporal switches during germination in rice highlights specific transcription factors and the role of RNA instability in the germination process. Plant Physiol 149:961–980.
11. Yang P, Li X, Wang X, Chen H, Chen F, et al. (2007) Proteomic analysis of rice (Oryza sativa) seeds during germination. Proteomics 7:3358–3368.
12. Bahin E, Bailly C, Sotta B, Kranner I, Corbineau F, et al. (2011) Crosstalk between reactive oxygen species and hormonal signalling pathway regulates grain dormancy in barley. Plant Cell Environ 34:980–993.
13. Barba-Espín G, Diaz-Vivancos P, Clemente-Moreno MJ, Albacete A, Faize L, et al. (2010) Interaction between hydrogen peroxide and plant hormones during germination and the early growth of pea seedlings. Plant Cell Environ 33:981–994.
14. Kim ST, Kang SY, Wang Y, Kim SG, Hwang DH, et al. (2008) Analysis of embryonic proteome modulation by GA and ABA from germinating rice seeds. Proteomics 8:3577–3587.
15. Finkelstein R, Gampala SS, Rock CD (2002) Abscisic acid signaling in seeds and seedlings. Plant Cell 14 Suppl, S15–45.
16. Holdsworth MJ, Bentsink L, Soppe WJ (2008) Molecular networks regulating Arabidopsis seed maturation, after-ripening, dormancy and germination. New Phytol 179:33–54.
17. Okamoto M, Tatematsu K, Matsui A, Morosawa T, Ishida J, et al. (2010) Genome-wide analysis of endogenous abscisic acid-mediated transcription in dry and imbibed seeds of Arabidopsis using tiling arrays. Plant J 62:39–51.
18. Bailly C (2004) Active oxygen species and antioxidants in seed biology. Seed Sci Res 14:93–107.
19. Bailly C, El-Maarouf-Bouteau H, Corbineau F (2008) From intracellular signaling networks to cell death: the dual role of reactive oxygen species in seed physiology. CR Biol 331:806–814.
20. Ishibashi Y, Koda Y, Zheng S-H, Yuasa T, Iwaya-Inoue M (2013) Regulation of soybean seed germination through ethylene production in response to reactive oxygen species. Ann Bot 111:95–102.
21. Liu Y, Ye N, Liu R, Chen M, Zhang J (2010) H2O2 mediates the regulation of ABA catabolism and GA biosynthesis in Arabidopsis seed dormancy and germination. J Exp Bot 61:2979–2990.
22. Oracz K, El Maarouf-Bouteau H, Farrant JM, Cooper K, Belghazi M, et al. (2007) ROS production and protein oxidation as a novel mechanism for seed dormancy alleviation. Plant J 50:452–465.
23. Oracz K, El Maarouf-Bouteau H, Kranner I, Bogatek R, Corbineau F, et al. (2009) The mechanisms involved in seed dormancy alleviation by hydrogen cyanide unravel the role of reactive oxygen species as key factors of cellular signalling during germination. Plant Physiol 150:494–505.
24. Ye N, Zhu G, Liu Y, Zhang A, Li Y, et al. (2012) Ascorbic acid and reactive oxygen species are involved in the inhibition of seed germination by abscisic acid in rice seeds. J Exp Bot 63:1809–1822.
25. Adams DR, Ron D, Kiely PA (2011) RACK1, A multifaceted scaffolding protein: Structure and function. Cell Comm Signal 9(22):823–830.
26. Guo J, Chen JG, Liang JS (2007) RACK1, a Versatile Scaffold Protein in Plant? Internl J Plant Dev Biol 1:95–105.
27. Smith TF, Gaitatzes C, Saxena K, Neer EJ (1999) The WD repeat: a common architecture for diverse functions. Trends Biochem Sci 24:181–185.
28. Chen JG, Ullah H, Temple B, Liang J, Guo J, et al. (2006) RACK1 mediates multiple hormone responsiveness and developmental processes in Arabidopsis. J Exp Bot 57:2697–2708.
29. Chen S, Dell EJ, Lin F, Sai J, Hamm HE (2004) RACK1 regulates specific functions of Gbetagamma. J Biol Chem 279:17861–17868.
30. Guo J, Wang S, Valerius O, Hall H, Zeng Q, et al. (2011) Involvement of Arabidopsis RACK1 in protein translation and its regulation by abscisic acid. Plant Physiol 155:370–383.
31. Ishida S, Takahashi Y, Nagata T (1993) Isolation of cDNA of an auxin-regulated gene encoding a G protein beta subunit-like protein from tobacco BY-2 cells. Proc Natl Acad Sci USA 90:11152–11156.

32. Ishida S, Takahashi Y, Nagata T (1996) The mode of expression and promoter analysis of the arcA gene, an auxin-regulated gene in tobacco BY-2 cells. Plant Cell Physiol 37:439–448.

33. Iwasaki Y, Komano M, Ishikawa A, Sasaki T, Asahi T (1995) Molecular cloning and characterization of cDNA for a rice protein that contains seven repetitive segments of the Trp-Asp forty-amino-acid repeat (WD-40 repeat). Plant Cell Physiol 36:505–510.

34. Nakashima A, Chen L, Thao NP, Fujiwara M, Wong HL, et al. (2008) RACK1 functions in rice innate immunity by interacting with the Rac1 immune complex. Plant Cell 20:2265–2279.

35. Guo J, Wang J, Xi L, Huang WD, Liang J, et al. (2009) RACK1 is a negative regulator of ABA responses in Arabidopsis. J Exp Bot 60(13):3819–3833.

36. Islas-Flores T, Guillén G, Islas-Flores I, Román-Roque CS, Sánchez F, et al. (2009) Germination behavior, biochemical features and sequence analysis of the RACK1/arcA homolog from Phaseolus vulgaris. Physiol Plant 137(3):264–280.

37. Islas-Flores T, Guillén G, Sánchez F, Villanueva MA (2012) Changes in RACK1 expression induce defects in nodulation and development in Phaseolus vulgaris. Plant Sig Behav 7:1–3.

38. Li DH, Liu H, Yang YL, Zhen PP, Liang JS (2008) Down-Regulated Expression of RACK1 Gene by RNA Interference Enhances Drought Tolerance in Rice. Chin J Rice Sci 22:447–453.

39. Quarrie SA, Whitford PN, Appleford NEJ, Wang TL, Cook SK, et al. (1988) A monoclonal antibody to (S)-abscisic acid: its characterization and use in a radioimmunoassay for measuring abscisic acid in crude extracts of cereal and lupin leaves. Planta 173:330–339.

40. Li DH, Zhang DP, Cao DD, Liang JS (2011) Advances in Plant RACK1 studies. Chin Bull Bot 46 (2): 224–232.

41. Wang P, Song CP (2008) Guard-cell signaling for hydrogen peroxide and abscisic acid. New Phytol 178:703–718.

42. Sagi M, Fluhr R (2006) Production of reactive oxygen species by plant NADPH oxidases. Plant Physiol 141:336–340.

43. Liu J, Zhou J, Xing D (2012) Phosphatidylinositol 3-Kinase Plays a Vital Role in Regulation of Rice Seed Vigor via Altering NADPH Oxidase Activity. PLoS ONE 7(3): e33817. doi:10.1371/journal.pone.0033817.

44. Kundu N, Dozier U, Deslandes L, Somssich IE, Ullah H (2013) Arabidopsis scaffold protein RACK1A interacts with diverse environmental stress and photosynthesis related proteins. Plant Signal Behav 8: e24012.

45. Wamaitha MJ, Yamamoto R, Wong HL, Kawasaki T, Kawano Y, et al. (2012) OsRap2.6 transcription factor contributes to rice innate immunity through its interaction with Receptor for Activated Kinase-C 1 (RACK1). Rice 5:35 doi:10.1186/1939-8433-5-35.

46. Cheng WH, Endo A, Zhou L, Penney J, Chen HC, et al. (2002) A unique short-chain dehydrogenase/reductase in Arabidopsis glucose signaling and abscisic acid biosynthesis and functions. Plant Cell 14:2723–2743.

47. Chen Y, Ji F, Xie H, Liang J, Zhang J (2006) The regulator of G protein signaling proteins involved in sugar and abscisic acid signaling in Arabidopsis seed germination. Plant Physiol 140:302–310.

48. El-Maarouf-Bouteau H, Bailly C (2008) Oxidative signaling in seed germination and dormancy. Plant Signal Behav 3:175–182.

49. Leymarie J, Vitkauskaité G, Hoang HH, Gendreau E, Chazoule V, et al. (2012) Role of reactive oxygen species in the regulation of Arabidopsis seed dormancy. Plant Cell Physiol 53(1):96–106.

50. Ishibashi Y, Yamamoto K, Tawaratsumida T, Yuasa T, Iwaya-Inoue M (2008) Hydrogen peroxide scavenging regulates germination ability during wheat (Triticum aestivum L.) seed maturation. Plant Signal Behav 3(3):183–188.

51. Müller K, Job C, Belghazi M, Job D, Leubner-Metzger G (2010) Proteomics reveal tissue-specific features of the cress (Lepidium sativum L.) endosperm cap proteome and its hormone-induced changes during seed germination. Proteomics 10(3):406–416.

52. Chaudhuri A, Singh KL, Kar RK (2013) Interaction of Hormones with Reactive Oxygen Species in Regulating Seed Germination of Vigna radiata (L.) Wilczek. J Plant Biochem Physiol 1:1–5.

53. Müller K, Carstens AC, Linkies A, Torres MA, Leubner-Metzger G (2009) The NADPH-oxidase AtrbohB plays a role in Arabidopsis seed after-ripening". New Phytol 184(4):885–897.

pSiM24 Is a Novel Versatile Gene Expression Vector for Transient Assays As Well As Stable Expression of Foreign Genes in Plants

Dipak Kumar Sahoo[1]*, **Nrisingha Dey**[2], **Indu Bhushan Maiti**[1]*

1 KTRDC, College of Agriculture, Food and Environment, University of Kentucky, Lexington, Kentucky, United States of America, 2 Department of Gene Function and Regulation, Institute of Life Sciences, Bhubaneswar, Odisha, India

Abstract

We have constructed a small and highly efficient binary Ti vector pSiM24 for plant transformation with maximum efficacy. In the pSiM24 vector, the size of the backbone of the early binary vector pKYLXM24 (GenBank Accession No. HM036220; a derivative of pKYLX71) was reduced from 12.8 kb to 7.1 kb. The binary vector pSiM24 is composed of the following genetic elements: left and right T-DNA borders, a modified full-length transcript promoter (M24) of *Mirabilis mosaic virus* with duplicated enhancer domains, three multiple cloning sites, a 3′rbcSE9 terminator, replication functions for *Escherichia coli* (ColE1) and *Agrobacterium tumefaciens* (pRK2-OriV) and the replicase trfA gene, selectable marker genes for kanamycin resistance (nptII) and ampicillin resistance (bla). The pSiM24 plasmid offers a wide selection of cloning sites, high copy numbers in *E. coli* and a high cloning capacity for easily manipulating different genetic elements. It has been fully tested in transferring transgenes such as green fluorescent protein (GFP) and β-glucuronidase (GUS) both transiently (agro-infiltration, protoplast electroporation and biolistic) and stably in plant systems (*Arabidopsis* and tobacco) using both agrobacterium-mediated transformation and biolistic procedures. Not only reporter genes, several other introduced genes were also effectively expressed using pSiM24 expression vector. Hence, the pSiM24 vector would be useful for various plant biotechnological applications. In addition, the pSiM24 plasmid can act as a platform for other applications, such as gene expression studies and different promoter expressional analyses.

Editor: Jin-Song Zhang, Institute of Genetics and Developmental Biology, Chinese Academy of Sciences, China

Funding: This work was fully supported by the KY state KTRDC grant (no. 1235411320) to IBM. The funders had no role in study design, data collection and analysis, decision to publish, or preparation of the manuscript.

Competing Interests: The authors have declared that no competing interests exist.

* E-mail: dipak_sahoo11@rediffmail.com (DKS); imaiti@uky.edu (IBM)

Introduction

The transfer of foreign genes into higher plants mediated either by *Agrobacterium tumefaciens* or by employing a biolistic process is the core technique used in genetic engineering-based plant modification. Many useful and versatile vectors have been constructed since the birth of the concept and the first generation of binary vectors for plant transformation [1–6]. The general trend in the binary vector development has been to increase the plasmid stability during a long co-cultivation period of *A. tumefaciens* with the target host plant tissues and also to understand the molecular mechanism of broad host-range replication, and to use it to reduce the size of plasmid for ease in cloning and for higher plasmid yield in *Escherichia coli* [7,8]. A number of large (>10 kb), first-generation binary vectors have been constructed for plant transformation, including Ti plasmid [2], pBin19 [1], pKYLX7 [4] and other expression vectors [3]. One of the binary vectors, pBin19 [1], has been modified to pBI121 and pIG121Hm [9,10] to use the β-glucuronidase (*GUS*) reporter gene in plant transformation. Binary vectors include pKYLX expression vectors containing 35S and rbcS promoters that are suitable for constitutive or light-regulated expression of foreign genes [4]. These vectors and their derivatives were soon widely distributed among plant scientists. In addition, another widely used series of vectors includes pPZP vectors [11] and their modified form, pCAMBIA vectors (www.cambia.org). Xiang et al., (1999) constructed a pCB mini-binary vector series [12] from the relatively large, first-generation binary plasmid pBin [1]. Over time, vector technology evolved, and new generations of plant transformation vectors with improved cloning and delivery strategy were introduced, for example, pGreen vectors [13]; pGD or pSITE vectors, which are suitable for the stable integration or transient expression of various autofluorescent protein fusions in plant cells [14,15]; the pCLEAN binary vector system [16]; the pHUGE binary vector system [17]; and binary bacterial artificial chromosome BIBAC vectors [18]. The TMV RNA-based vector pJLTRBO [19]and its derivative pPZPTRBO [20] were reported to produce recombinant proteins in plants without using the RNA-silencing inhibitor P19. Similar expression levels were provided by the pEAQ-HT vector which has an integrated P19 expression cassette [21]. A bean yellow dwarf virus single-stranded DNA-based vector, pBY030-2R was reported to produce high amount of recombinant proteins [22] while the pMAA-Red vector was known for easy production of transgenic *Arabidopsis* overexpression lines with strong expression levels of the gene of interest [23].

The binary vectors widely used for plant transformations vary in size, origin of replication, bacterial selectable markers, T-DNA borders and overall structure. Recent modifications of binary vectors provide a number of user-friendly features, such as a wide selection of cloning sites, high copy numbers in *E. coli*, improved compatibility with strains of choice, a wide pool of selectable markers for plants and a high frequency of plant transformation. Although recent improvements are very useful, the classic vector configuration still appears to be good enough in many occasions. Plasmid manipulations are also easier if the vector replicates in *E. coli* to high copy numbers. Moreover, the efficiency of *in vitro* recombination procedures is inversely proportional to the size of the vector DNA [24]. With an increased requirement for the transfer of large pieces of DNA into plants, the size of binary vectors should be kept to a minimum. The availability of low-molecular-weight, versatile plant expression vectors is currently insufficient in plant molecular biology. For these reasons, we designed a smaller binary vector, pSiM24, which offers a wide selection of cloning sites, high copy numbers in *E. coli* and is fully functional in the transient (using both the gene-gun or Agro-infiltration methods) as well as stable transformation of plants.

Materials and Methods

Chemicals, enzymes, bacterial strains and plasmids

Antibiotics (ampicillin, kanamycin, rifampicin, tetracycline, hygromycin) and chemicals were purchased from Sigma-Aldrich (St. Louis, MO, USA) or Thermo Fisher Scientific (Waltham, MA, USA). All restriction endo-nucleases and DNA-modifying enzymes were obtained from New England Biolab (Beverly, MA, USA) or Invitrogen-Life Technologies (Grand Island, NY USA). The TB1 strain of *E. coli* [25] and the C58C1 [GV3850] strain of *A. tumefaciens* [26] were used. The plasmids pBluescriptIIKS(+) (Genbank Accession no. X52327) from Stratagene (la Jolla, CA, USA) and pKYLX7 [4] or its derivative pKM24KH (GenBank accession no. HM036220.1) were used. Cultures of *E. coli* transformed with pUC-based vectors were grown in the presence of ampicillin (100 µg/ml). Transformed agrobacterium was grown in the presence of kanamycin (25 µg/ml) and rifampicin (100 µg/ml). The Vip3A and KMP-11 antibodies were provided by Dr Raj K. Bhatnagar, ICGEB, New Delhi, India and Dr Shyamal Roy, IICB, Kolkata, India. The IL-10 antibody was obtained from Imgenex, Bhubaneswar, India.

In vitro cloning procedures and DNA sequencing

All in vitro recombination techniques were employed using previously described standard methods [27,28]. For DNA sequencing, a dye terminator labeling procedure was followed using a Genome Lab DTCS-Quick Start kit (Beckman Coulter, USA), and an automated sequencing machine (Beckman Coulter CEQ 8000 Genetic Analysis System, USA) was used in accordance with the manufacturer's instructions.

Construction of plasmid vector pBTdna, pBTdna-rbcT and pBTdna-rbcT-KanR

We designed and generated a 522-bp synthetic DNA fragment containing left and right T-DNA borders and three multiple cloning sites (MCS) of general the structure 5'-BssHI-KpnI-left T-DNA(147-bp)-MCS1(BstXI-StuI-FspAI-PasI-SanDI-BstZ171-SmaI)-MCS2(EcoRI-HindIII-BamHI-XhoI-HpaI-MluI-SalI-SstI-PstI-XbaI)-MCS3 (ClaI-SpeI-BglII-BstEII-EcoNI-FseI-SwaI-NruI-PacI-right T-DNA border (162-bp)-EcoRV-BssHII-3'. This fragment was synthesized by GenArt-Life Technologies (Carlsbad, CA, USA). The 5'-

BssHII-BssHII-3'fragment was cloned into the corresponding sites of pBluescriptIIKS(+), and the resulting plasmid was named pBTdna.

A 657-bp poly(A) signal 3'-rbcS-E9 of the general structure 5'-ClaI-3'-rbcSE9-XbaI-3' was isolated from the binary vector pKM24KH (GenBank Accession No. HM036220) and was inserted into the corresponding site of pBTdna to generate the plasmid pBTdna-rbcT.

A 1343-bp synthetic neomycin phosphotransferase gene/kanamycin resistance gene (nptII/KanR) of the general structure 5'-BglII-Nos-promoter-KanR cDNA-Nos-terminator-SpeI-3' was obtained from GenArt-Life Technologies (Carlsbad, CA, USA). The open reading frame of the KanR gene was optimized for plant codon bias. The KanR gene, with the structure 5'-BglII-SpeI-3', was cloned into the corresponding sites of pBTdna-rbcT to create the plasmid pBTdna-rbcT-KanR (also called pBTRK). The plasmid contains a 2498-bp micro-Tdna fragment of the general structure 5'-BssHII-KpnI-left T-DNA border-MCS1-MCS-2-XbaI-3'rbcS Terminator-ClaI-SpeI-KanR gene (complement)-BglII-MCS3-right T-DNA border- EcoRV-BssHII-3'. The sequence integrity of the fragment was confirmed before further use. The sequence information of these genetic elements is provided in the NCBI database (GenBank accession no. KF032933).

Construction of non-T-DNA plasmids: pBtrfA, pB-oriV-trfA, pBAmpR-ColEI-oriV-trfA

We designed a non-T-DNA plasmid of the general structure 5'-BssHII-KpnI-AmpR gene-ColE1 (origin of replication of pMB1)-ApaI-oriV of pRK2-SalI-trfA gene-EcoRV-BssHII-3'. A synthetic DNA fragment of the physical map 5'-BssHII-KpnI-ApaI-SalI-trfA gene-EcoRV-BssHII was obtained from a commercial supplier (GenArt-Life Technologies, CA, USA). The open reading frame of the trfA gene was optimized with the bias codon of *A. tumefaciens*. This 5'-BssHII-BssHII-3' fragment was cloned into the corresponding sites of pBluescriptIIKS(+) to create the plasmid pBtrfA.

A 642-bp fragment of the replicon OriV of pRK2 was PCR-amplified using appropriately designed forward and reverse primers to insert an ApaI site at the 5'-end and a SalI site at the 3'-end. The gel-purified PCR fragment 5'-ApaI-OriV-SalI-3' was inserted into the corresponding site of pBtrfA to form the plasmid pB-oriV-trfA. A fragment of 1803 bp containing the AmpR gene and the ColE1 replicon in pMA (GeneArt vector, Registry part no. K157000), a pUC derivative, was PCR-amplified using appropriately designed forward and reverse primers to insert the KpnI site at the 5'-end and the ApaI site at the 3'-end. The PCR fragment was digested with KpnI and ApaI, and the gel-purified fragment 5'-KpnI-ApaI-3' was cloned into the corresponding site of pB-oriV-trfA to generate the plasmid pBAmpR-ColEI-oriV-trfA.

Construction of pSi and pSiM24

The T-DNA portion was isolated from pBTdna-rbcT-KanR. First, the pBTdna-rbcT-KanR plasmid was digested with PvuII; the larger band (4235-bp fragment) was isolated and further digested with BssHII to generate a 2492-bp fragment of the general structure (5'-BssHII-KpnI-left T-DNA border-MCS-1-MCS-2-XbaI-3'rbcS Terminator-ClaI-MCS-3-right T-DNA border-EcoRV-BssHII-3'). The non-T-DNA portion of a 3830-bp fragment of the general structure (5'-BssHII-KpnI-AmpRI-ColEI-ApaI-OriV-SalI-trfA-EcoRV-BssHII-3') was isolated from pBAmpR-ColEI-oriV-trfA. Two fragments (T-DNA and nonT-DNA portions) were ligated and circularized to produce the binary vector pSi. The modified full-length transcript promoter (M24) of the *Mirabilis mosaic virus* [27,29] was inserted as 5'-EcoRI-HindIII-

3′ into the corresponding sites of pSi, and the resulting plasmid was named pSiM24. The fully annotated sequence of pSiM24 is available in the NCBI database (GenBank accession no. KF032933).

Construction of plant expression vectors with green fluorescent protein (GFP) and β-glucuronidase (GUS) reporter genes

The M24 promoter fragment 5′-EcoR1-M24-HindIII-3′ and the reporter gene 5′-XhoI-GUS-SstI-3′ or 5′-XhoI-GFP-SstI-3′ were inserted into the corresponding sites of pBTRK and pSi to generate expression constructs pBTRK-M24-GUS/GFP and pSiM24-GUS/GFP, respectively.

Tobacco plant transformation

The plant expression constructs pSiM24-GUS and pSiM24 were introduced into the *A. tumefaciens* strain GV3850 by the freeze-thaw method [30]. Tobacco plants (*Nicotiana tabacum* cv. SamsunNN) were transformed with *Agrobacterium* harboring pSiM24-GUS and pSiM24 constructs as described previously [31] or by the gene-gun method using pSiM24-GUS and pSiM24 constructs [32]. Tobacco shoots and then roots were regenerated from kanamycin-resistant calli derived from independent leaf discs. Ten independent kanamycin-resistant plant lines (R_0 generation, 1st progeny) were generated for the constructs pSiM24-GUS and pSiM24 and were maintained under greenhouse conditions ($30\pm5°C$ with both natural and supplementary lighting of minimum photon flux density, 300 µmole/m^2/s, 17 h day/7 h night cycle). Seeds were collected from self-pollinated primary transformants. Transgenic tobacco seeds (R1 progeny, 2nd generation) were germinated in the presence of Kanamycin (250 mg/L). Positive transformants with a KanR:KanS ratio of 3:1 progeny segregation were selected for further analysis. Transgenic lines (R_1 and R_2 progeny, second and third generation) were screened for gene integration, transcription and translation by polymerase chain reaction (PCR), reverse transcriptase-PCR (RT-PCR), real-time quantitative RT-PCR (qRT-PCR), enzymatic assays and GUS histochemical analysis.

Generation of transgenic *Arabidopsis* plants

The pSiM24 and pSiM24-GUS plasmids introduced into *A. tumefaciens* GV3850 were used to transfer each of these constructs into *Arabidopsis* (*Arabidopsis thaliana* ecotype Columbia-0) by the floral dip method [33]. The transgenic *Arabidopsis* plants were selected and maintained as described previously [34].

Transient Agro-infiltration assay of pSiM24-GUS in tobacco leaves

Suspensions of the *A. tumefaciens* strain GV3850 bearing pSiM24 and pSiM24-GUS constructs were infiltrated into leaves of *Nicotiana benthamiana* as described previously [35]. After two days of agro-infiltration, the transient GUS expression was evaluated by the histochemical GUS staining method [9].

Transient expression analysis in tobacco protoplasts

The isolation of tobacco protoplasts from the suspension cell cultures of *N. tabacum* L. cv Xanthi-Brad and electroporation of tobacco protoplasts with supercoiled plasmid pBTRKM24-GUS/GFP and pBTRKM24 constructs were performed as described previously [36]. After 20 h, protoplasts were harvested for fluorometric GUS enzymatic assay [9]. GUS expression levels were within ±10% for a given construct in this study. All constructs were tested in at least five independent experiments.

Biolistic-onion peel transient assay

Onion tissues were prepared and bombarded with pBTRKM24, pBTRKM24-GUS, pSiM24 and pSiM24-GUS plasmids following a standard protocol [37]. After two days, transient GUS expression was detected by a histochemical method [9] and visualized under an Olympus SZX12 bright-field microscope.

Real-time quantitative reverse transcription polymerase chain reaction (qRT-PCR)

The expression levels of GUS mRNA in transgenic tobacco and *Arabidopsis* plants developed for the plasmids pKCaMV35SGUS and pSiM24GUS were evaluated by real-time quantitative RT-PCR [38] using GUS-specific forward (5′-d-TTACGTCCTGTA-GAAACCCCA-3′) and reverse (5′-d-ACTGCCTGGCACAG-CAAT TGC-3′) primers. The qPCR assays were performed using the iTaq SYBR Green Supermix with ROX (Bio-Rad, USA) according to the manufacturer's instructions. Tobacco tubulin (by using forward 5′-d-ATGAGAGAGTGCATATCGAT-3′ and reverse 5′-d-TTCACTGAAGAAGGTGTTGAA-3′ primers) was used as an internal control to normalize the expression of GUS. The comparative threshold cycle (Ct) method (Applied Biosystems bulletin, part No. 4376784 Rev. C, 04/2007) was used to evaluate the relative expression levels of the transcripts. The threshold cycle was automatically determined for each reaction by the system set with default parameters (Step One Real-Time PCR System, Applied Biosystems). The specificity of the PCR was determined by melting curve analysis of the amplified products using the standard method installed in the system (Step One Real-Time PCR System, Applied Biosystems).

β-Glucuronidase (GUS) assay and histochemical GUS staining

Fluorometric GUS enzymatic assays for measuring GUS activities in tobacco protoplast extracts, *Arabidopsis* and tobacco plant extracts were performed as described previously [9,39]. The total protein content in protoplast and plant extracts was estimated by the Bradford method using BSA as a standard [40]. Histochemical GUS staining was carried out in plants following the published protocol [9,34], and photographs were taken under a bright-field microscope (Olympus SZX12).

GFP detection

GFP fluorescence images of electroporated tobacco protoplasts, onion epidermal cells and transgenic *Arabidopsis* leaves expressing GFP were analyzed using a confocal laser scanning microscope (TCS SP5; Leica Microsystems CMS GmbH, D-68165 Mannheim, Germany) using LAS AF (Leica Application Suite Advanced Fluorescence) 1.8.1 build 1390 software under a PL FLUOTAR objective (10.0X/N.A.0.3 DRY) using a confocal pinhole set of 1 airy unit and a 1× zoom factor for improved 8-bit resolution, as described previously [28,38]. To excite the expressed GFP in transgenic plants, a 488-nm argon laser (30%) with an AOTF (allowing for 40% transmission) was used, and fluorescence emission spectra were collected between 501 and 580 nm with the photomultiplier tube (PMT) detector gain set to 1050 V [28].

Transient expression of GUS using pSiM24 vector through vacuum infiltration method

Suspensions of the *A. tumefaciens* strain GV3850 bearing pSiM24 and pSiM24-GUS constructs were prepared as previously described [35], and the infiltration procedure was conducted following a previously reported protocol [41]. Leaves of *N.*

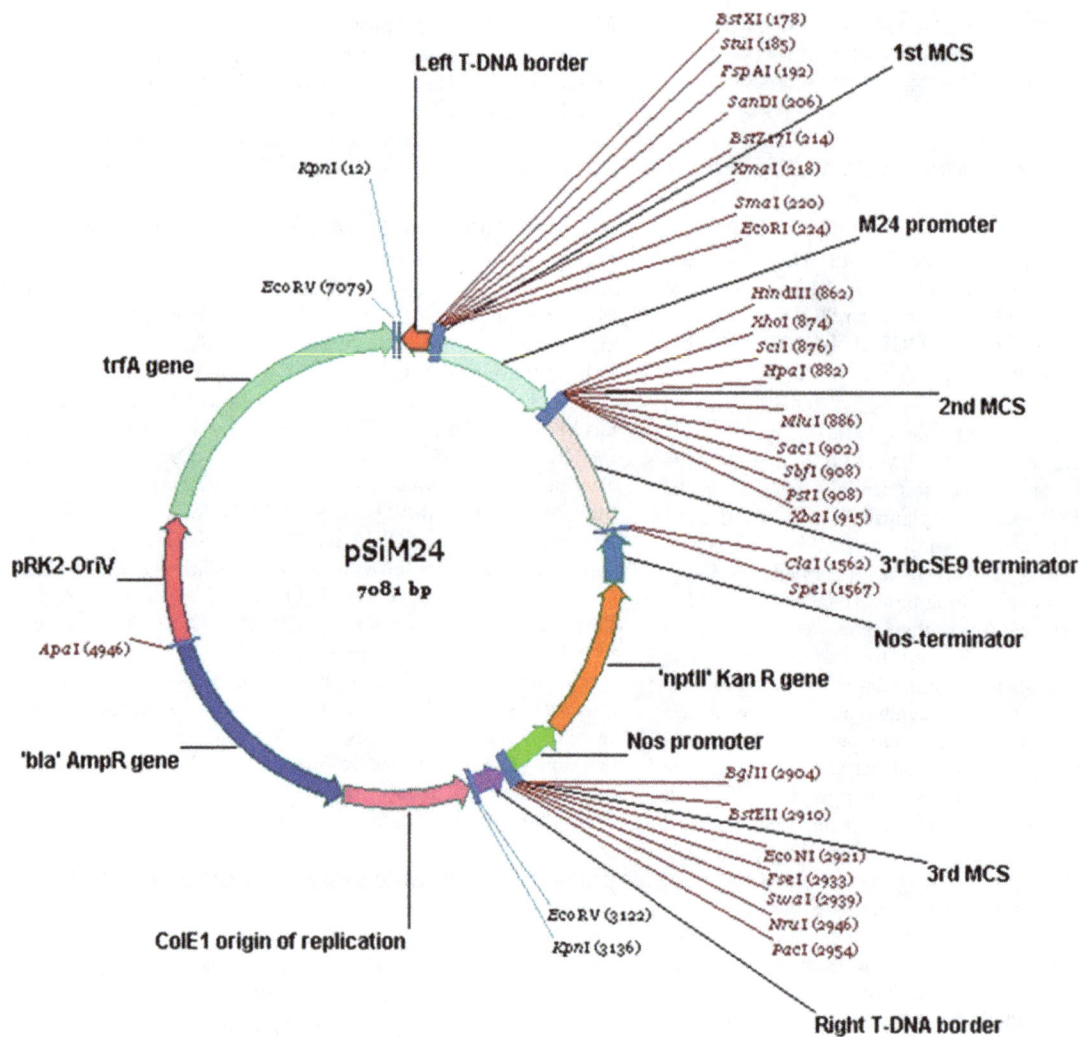

Figure 1. Schematic presentation of binary vector pSiM24. The backbone structure of binary vector pSiM24 (7081-bp) containing the modified full-length transcript promoter (M24) of the *Mirabilis mosaic virus*, which directs the coding sequences of the gene of interest; left T-DNA and right T-DNA borders (Left T-DNA, Right T-DNA); selection marker genes (KanR, neomycin phosphotransferase II, nptII) directed by nopaline synthase promoter (Nos promoter); terminator sequences of ribulose bisphosphate carboxylase small subunits (3'rbcSE9); nopaline synthase terminator (Nos terminator); multiple cloning sites (first MCS, second MCS and third MCS) with various restriction sites; replicon unit pRK2 oriV; trfA gene for agrobacterium; ColE1 origin of replication for *E. coli*; and 'bla' AmpR gene for resistance to ampicillin.

benthamiana were weighed and submerged in a suspension of *A. tumefaciens* strain GV3850 bearing pSiM24 or pSiM24-GUS plasmids. A vacuum level of 760 mm Hg was applied and released several times until the leaves became translucent [41]. Leaves were transferred into MS-media-containing plates and incubated at room temperature for two days. GUS expression in the infiltrated leaves was evaluated by the GUS histochemical staining method and GUS assay [9,39].

Table 1. Transformation frequencies of *Escherichia coli* strain TB1 for pSiM24 binary vector.

Plasmid	Total size (kb)	Amount DNA used (pg)	Average colony (n>4)	Ratio to pCAMBIA control	Ratio to pKM24KH control
pSiM24	7.08	100	634±54[b]	4.73	5.66
pSiM24-GUS	8.89	125	619±46[b]	4.62	5.53
pSiM24-GFP	7.8	110	608±38[b]	4.54	5.43
pCAMBIA2300	8.74	123.4	134±12[a]	1	1.2
pKM24KH	12.94	183	112±13[a]	0.84	1

The bacteria were transformed with equal molar amounts of each of the plasmid DNA. Statistical analysis of the data was performed adopting one way ANOVA analysis (using GraphPad Prism version 5.01) and presented as the means ± S.D. A P value of less than 0.05 was considered significant indicated by different superscript letters.

Table 2. Transformation frequencies of *Agrobacterium tumefaciens* strain GV3850 for pSiM24 binary vector.

Plasmid	Total size (kb)	Amount DNA used (µg)	Average colony (n>4)	Ratio to pCAMBIA control	Ratio to pKM24KH control
pSiM24	7.08	1	546±42[b]	1.4	1.81
pSiM24-GUS	8.89	1.25	534±48[b]	1.37	1.77
pSiM24-GFP	7.8	1.1	542±36[b]	1.39	1.79
pCAMBIA2300	8.74	1.23	391±33[a]	1	1.29
pKM24KH	12.94	1.83	302±20[c]	0.77	1

The bacteria were transformed with equal molar amounts of each of the plasmid DNA. Statistical analysis of the data was performed adopting one way ANOVA analysis (using GraphPad Prism version 5.01) and presented as the means ± S.D. A P value of less than 0.05 was considered significant indicated by different superscript letters.

Analysis GFP-*AtCESA3*$^{ixr1-2}$, Vip3A(a), KMP-11, IL-10 and nat-T-phyllo-GFP after transient expression in tobacco using pSiM24 vector

The GFP fused *Arabidopsis* mutated CESA3 (GFP-*AtCESA3*$^{ixr1-2}$) fragment with Xho I and Sst I sites was obtained from pKM24KH-MD1 (GenBank accession no. JX996118) [42–43] by restriction digestions. Likewise, the native T-phylloplanin fused GFP with the apoplast targeting sequence (nat-T-phyllo-GFP) fragment with Xho I and Sst I sites was obtained from pKM24-ibm8 (GenBank accession no. KF951257) [44] by restriction digestions. Both these fragments were cloned in pSiM24 following standard protocols [28] and the resulted plasmids were named as pSiM24-GFP-*AtCESA3*$^{ixr1-2}$ and pSiM24-nat-T-phyllo-GFP. Suspension of *A. tumefaciens* strain pGV3850 harboring pSiM24, pSiM24-GFP-*AtCESA3*$^{ixr1-2}$ and pSiM24-nat-T-phyllo-GFP constructs were infiltrated into leaves of tobacco plants (*N. tabacum* cv. SamsunNN) as described earlier [35]. After two days of Agro-infiltration the transient *AtCESA3*$^{ixr1-2}$ expression was evaluated by RT-PCR by using gene specific primers and also by Western blotting using *AtCESA3*$^{ixr1-2}$ polyclonal antibody as described earlier [42]. The transient nat-T-phyllo-GFP expression was evaluated by RT-PCR by using gene specific primers and also by confocal microscopy as previously described [44].

Vegetative insecticidal gene, *vip3A(a)* [45,46], kinetoplastid membrane protein-11 (KMP-11) [47] and interleukin-10 (IL-10) [48] were cloned at XhoI and SstI sites in pSiM24 vector to generate pSiM24-vip3A(a), pSiM24-KMP-11 and pSiM24-IL-10 plasmids for transient expression assay in tobacco protoplasts. The isolation of tobacco protoplasts from the suspension cell cultures of *N. tabacum* L. cv Xanthi-Brad and electroporation of tobacco protoplasts with pSiM24-vip3A(a), pSiM24-KMP-11 and pSiM24-IL-10 constructs were performed as described previously [28]. Electroporated protoplasts were incubated for 48 hours and harvested in protein extraction buffer (1X PBS, 0.1% Tween 20 and 1 mM PMSF). Protein samples from pSiM24, pSiM24-vip3A(a), pSiM24-KMP-11 and pSiM24-IL-10 transfected protoplasts were lyophilized and dissolved in protein extraction buffer. The transient *Vip3A(a)* expression was evaluated by Western blotting using Vip3A-specific polyclonal antibody as described earlier [46]. Concentration of transiently expressed KMP-11 and IL-10 was estimated by using anti-KMP-11 and anti-IL-10 antibody following indirect enzyme-linked immunosorbance assay (ELISA) protocol [49].

Results

Features of assembled binary expression vector pSiM24

The binary expression vector pSiM24 was designed to reduce the size of the vector backbone by eliminating non-essential elements of our previous vector pKM24KH (size 12,945-bp, GenBank accession no. HM036220), a derivative of pKYLX7 [4]. The pKM24KH vector is a low-copy-number plasmid. We replaced the *E. coli* replication unit with a high-copy-number replicon ColEI in pSiM24, making the identification and characterization of gene inserts easier. We also modified the agrobacterium replicon unit (Oriv-trfA of pRK2) by optimizing the trfA open reading frame for better expression. The overall DNA yields and transformation frequency of the new vector pSiM24 were several times greater than those of the previous vector pKM24KH in both *E. coli* and *A. tumefaciens*. The binary vector pSiM24 (Figure 1; GenBank Accession no. KF032933) has

Table 3. Binary Ti vectors pSiM24 produced higher plasmid DNA yields in *Escherichia coli* strain TB1 over pCAMBIA.

Plasmid	Average DNA yield (µg) (n>3)	Ratio to pCAMBIA control	Ratio to pKM24KH control
pSiM24	113.5±21.3[b]	3.47	8.11
pSiM24-GUS	96.5±13.2[b]	2.95	6.89
pSiM24-GFP	105.7±16.3[b]	3.21	7.55
pCAMBIA2300	32.74±4.0[a]	1	2.34
pKM24KH	14±1.6[c]	0.43	1

Three single colonies of each plasmid constructs were grown for 16 hrs in 25 ml LB media with 100 mg/L ampicillin (for pSiM24) or 50 mg/L kanamycin (for pCAMBIA) or 15 mg/L tetracycline (for pKM24KH). Plasmid DNA was purified using QIA Midiprep columns. DNA yields represent the average of three independent samples. Statistical analysis of the data was performed adopting one way ANOVA analysis (using GraphPad Prism version 5.01) and presented as the means ± S.D. A P value of less than 0.05 was considered significant indicated by different superscript letters.

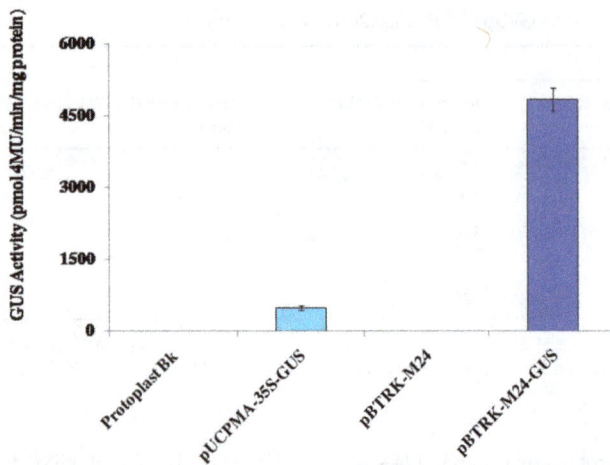

Figure 2. The transient GUS expression analysis of T-DNA assembled fragment of pSiM24 in tobacco protoplast system. Transient GUS expression analysis of pBTRK-M24 (T-DNA assembled fragment of pSiM24), pBTRK-M24-GUS (pBTRK-M24 with GUS reporter gene) constructs in tobacco protoplast. The pUCPMA-35S-GUS construct carries the constitutive CaMV35S promoter. The average GUS activity ± SD is presented in the histogram of three independent experiments replicated three times for each construct. The values significantly differ between tobacco protoplasts with pBTRK-M24-GUS from others at P<0.01 based on Student's t-test.

the following genetic elements: left T-DNA border (coordinates 65 to 90, complement); M24 promoter (223 to 860); three multiple cloning sites; 3′rbcS terminator (915 to 1565); KanR(nptII) gene (1566 to 2911, complement); KanR-terminator (1566 to 1820, complement); Kan R-cDNA (1821 to 2618, complement); KanR-promoter (2619 to 2903, complement); right T-DNA (2957 to 3118), right T-DNA border (3034 to 3059, complement); non-T-DNA portion, ColE1 origin of replication (3137 to 3804, complement); AmpR(bla) gene region (3805 to 4940, complement); terminator (3805 to 3951, complement); AmpR-cDNA (3952 to 4812, complement); AmpR-promoter (4813 to 4940, complement); pRK2-Ori V (coordinates 4947 to 5567); pRK2-trfA gene (5577 to 7081); promoter (5577 to 5880); cDNA (5881 to 7029) and terminator (7030 to 7081).

DNA yield and transformation frequencies of *E. coli* and *A. tumefaciens* with pSiM24 binary vectors

DNA yield and transformation frequencies in *E. coli* and *A. tumefaciens* were evaluated and presented (Tables 1–3). Transformations were performed with equal molar amounts of each of the plasmids to normalize increasing plasmid size as previously described [50]. The transformation frequencies of pSiM24 vectors are four- to six-fold higher than pCAMBIA and pKM24KH vectors, in *E. coli* (Table 1). The transformation frequency of the pSiM24 vector is 1.4- to 1.8-fold higher than conventional pCAMBIA and pKM24KH vectors, in *A. tumefaciens*, although the effect is not as marked as in *E. coli* (Table 2). The DNA yields of pSiM24 vectors were approximately three-fold greater than those of the pCAMBIA vector and seven to eight-fold higher than those of pKM24KH in *E. coli* (Table 3).

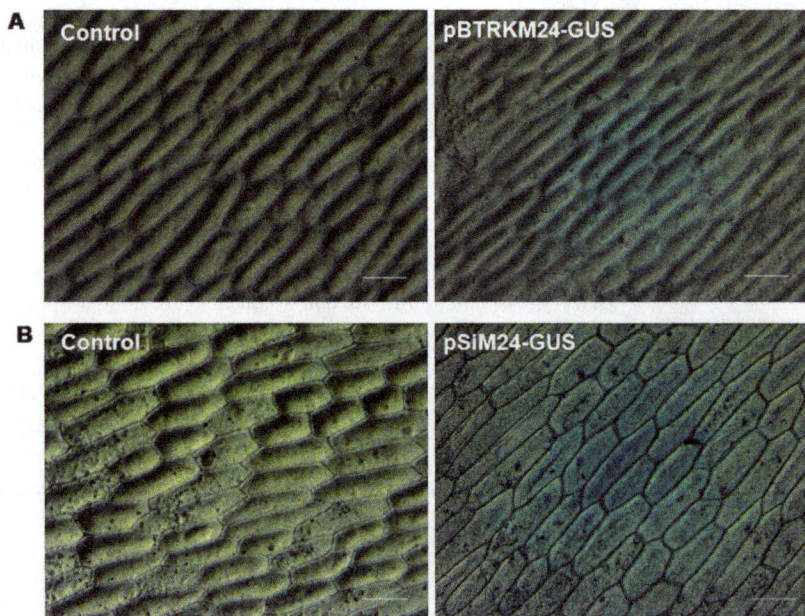

Figure 3. The transient GUS expression analysis of T-DNA assembled fragment of pSiM24 containing GUS and pSiM24-GUS in onion epidermal cells. (A) Light microscopy images of X-gluc-treated onion epidermal cells bombarded with pBTRKM24-GUS (T-DNA assembled fragment of pSiM24 with GUS reporter gene) construct DNA-loaded gold particles are presented. Control represents untransformed onion epidermal cells. These micrographs are representative of data collected after examination of onion epidermal cells from a minimum of four independent experiments. Scale bar, 100 μm. (B) Light microscopy images of X-gluc-treated onion epidermal cells bombarded with pSiM24-GUS (pSiM24 with GUS reporter gene) construct DNA-loaded gold particles are presented. Control represents onion epidermal cells with pSiM24 without GUS reporter gene. These micrographs are representative of data collected after examination of onion epidermal cells from a minimum of four independent experiments. Scale bar represents 100 μm.

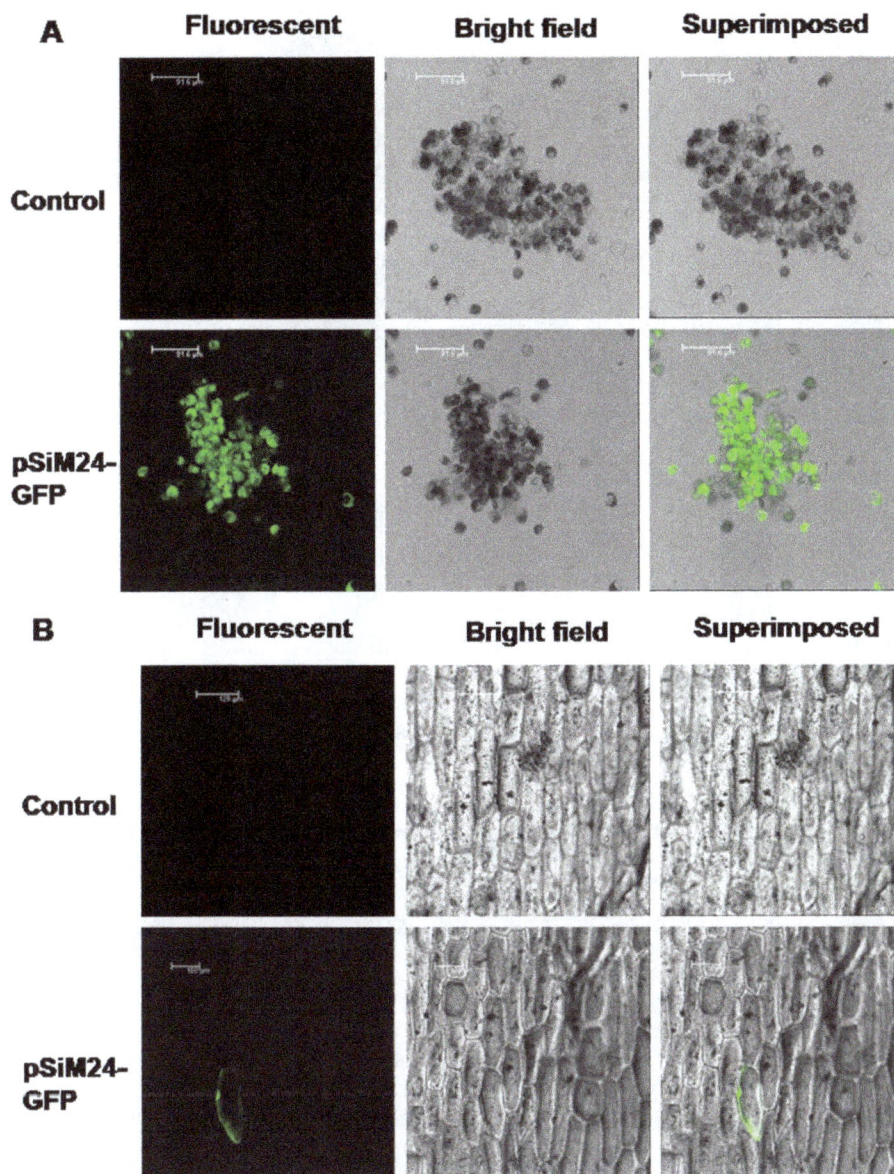

Figure 4. Transient expression of pSiM24-GFP in tobacco protoplasts and onion epidermal cells. (A) Protoplasts were transfected with plasmids pSiM24 (Control) and pSiM24-GFP (having GFP reporter gene). Transformation efficiencies were determined by analyzing the protoplasts with fluorescence after incubation for overnight. Fluorescent, bright-field and superimposed (bright-field and green fluorescent) confocal laser scanning micrographs of tobacco protoplasts are presented. Scale bar, 92 μm. (B) Fluorescent, bright-field and superimposed (bright-field and green fluorescent) confocal laser scanning micrographs of onion epidermal cells bombarded with pSiM24-GFP construct DNA-loaded gold particles are presented. Control represents onion epidermal cells with pSiM24 visualized under CLSM. Scale bar, 100 μm.

Transient expression of the pBTRKM24-GUS and pSiM24-GUS/GFP constructs

The pBluescript-based constructs pBTRKM24-GUS with an M24 promoter and pUCPMA35S-GUS [27] with a 35S promoter were compared by tobacco protoplast transient assay. The M24 promoter showed approximately 10 times higher GUS activity than the CaMV 35S promoter (Figure 2). The pBTRKM24-GUS construct was also evaluated by the biolistic bombardment of epidermal cells of onion peels, showing strong GUS expression, as detected histochemically (Figure 3).

The pSiM24-GFP (with a different reporter gene, i.e., GFP) was studied in a tobacco protoplast system, where GFP fluorescence was visualized by confocal microscopy (Figure 4). The pSiM24-GUS construct was tested in an *Agrobacterium* infiltration assay in *N. benthamiana* leaves. The *A. tumefaciens* (strain C58C1-GV3850) carrying pSiM24 (empty vector), pK-CaMV35S-GUS and pSiM24-GUS constructs was used for agro-infiltration. Transient GUS expression detected histochemically, showed stronger GUS expression in agro-infiltrated patches for pSiM24-GUS construct than for pK-CaMV35S-GUS (Figure 5). The pSiM24-GUS/GFP plasmids were also bombarded in onion cells, and strong GUS or GFP expression was observed in transformed onion epidermal cells (Figure 3-4).

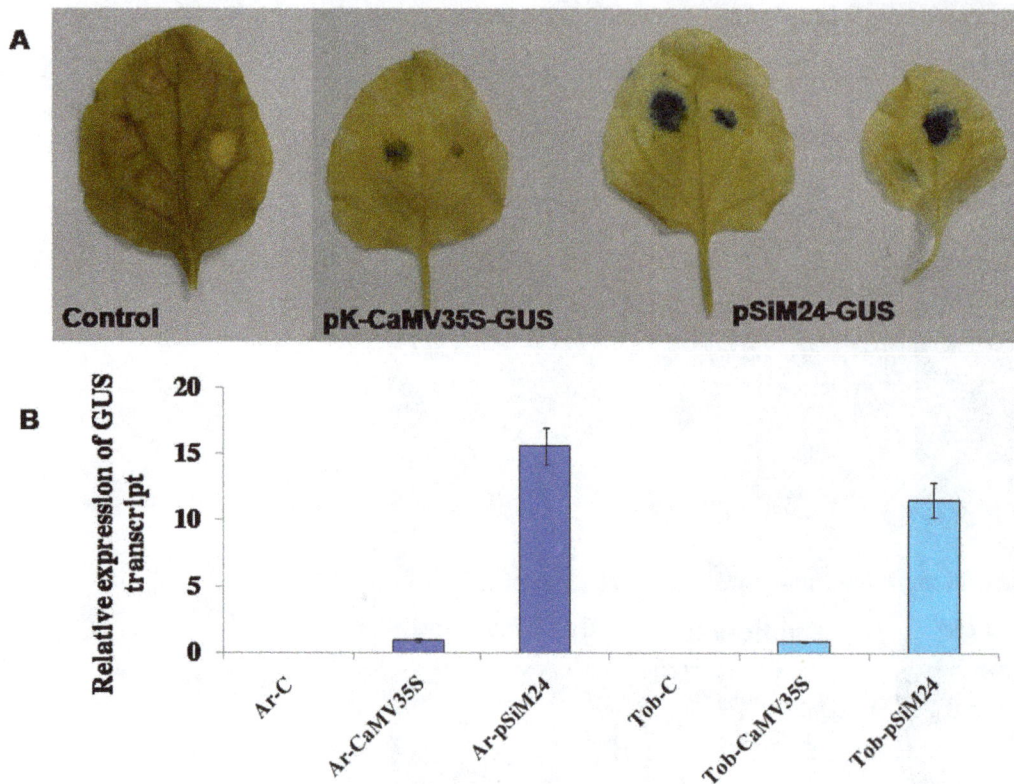

Figure 5. GUS expression analysis of pSiM24-GUS and pKCaMV35S-GUS in transient and transgenic systems. (A) Representative transient GUS expression levels from agrobacterium infiltration assays in *N. benthamiana* leaves are shown for pKCaMV35S-GUS and pSiM24-GUS constructs. GUS was detected histochemically. Control represents pSiM24 without GUS reporter gene. (B) Relative expression of GUS specific transcripts was measured in whole transgenic *Arabidopsis* (second generation, two weeks old; Ar-CaMV35S with pKCaMV35S and Ar-pSiM24 with pSiM24) and tobacco (second generation, three weeks old; Tob-CaMV35S with pKCaMV35S and Tob-pSiM24 with pSiM24) plants. The data represent relative expression of GUS transcript ± S.D. of four independent lines (n = 4) for each construct in which five plants per line were analyzed. The values significantly differ between control and transgenic plants at P<0.01 based on Student's *t*-test.

Transformation ability of pSiM24 vector in tobacco and *Arabidopsis*

Tobacco leaf discs were co-cultivated with *A. tumefaciens* for three days, and transformed leaf discs were selected in the presence of 250 mg/L kanamycin and 500 mg/L of cefotaxime for four weeks. The increase in fresh weight in transformed leaf discs was evaluated as described previously [51]. The increases in fresh weight of four-week-old leaf discs were compared and are presented in Table 4. Leaf discs treated with binary vectors showed a seven- to eight-fold increase in fresh weight over the vector-less control, remaining green with multiple regenerating shoots. In the negative vector-less control, leaf discs did not induce callus and turned yellow within two weeks of antibiotic selection. Thus, the percentage of leaf discs showing an increase in fresh

Table 4. Binary Ti vectors pSiM24 and pSiM24-GUS/GFP conferred a kanamycin-resistant fresh weight (FW) increase in tobacco leaf discs after transformation with *A. tumefaciens*.

Binary Ti Vectors	Mean FW/plate, g	Increase in FW in g per plate over vector-less control	% Leaf discs with increased FW
Vector-less control	0.83 ± 0.09^a	0^c	0
pSiM24	8.12 ± 2.1^b	7.29 ± 2.13^d	88
pSiM24-GUS	6.85 ± 1.8^b	6.02 ± 1.79^d	98
pSiM24-GFP	7.32 ± 2.4^b	6.49 ± 2.42^d	79
pCAMBIA2300	7.85 ± 1.3^b	7.02 ± 1.33^d	83
pKM24KH	7.43 ± 1.7^b	6.6 ± 1.68^d	81

After the co-cultivation with *A. tumefaciens* strain GV3850 at 25°C for 3 days, leaf discs were selected on shooting medium containing 250 mg/L of kanamycin and 500 mg/L of cefotaxime for four weeks. Each treatment involved 5 plates with 10 leaf discs per plate. The same experiment was repeated two more times. Statistical analysis of the data was performed adopting one way ANOVA analysis (using GraphPad Prism version 5.01) and presented as the means ± S.D. A P value of less than 0.05 was considered significant indicated by different superscript letters.

Table 5. Effect of binary Ti vectors pSiM24 and pSiM24-GUS/GFP on transformation frequencies in *A. thaliana*.

Binary Ti Vectors	Transformation frequency (%)
Vector-less control	0^a
pSiM24	2.68 ± 0.29^b
pSiM24-GUS	2.45 ± 0.26^b
pSiM24-GFP	2.38 ± 0.21^b
pCAMBIA2300	1.32 ± 0.12^c
pKM24KH	1.18 ± 0.2^c

A. tumefaciens harboring binary vectors were cultured and floral dipping of *A. thaliana* plants was subsequently performed as described in materials and methods. T_1 seedlings were selected on solid MS medium containing kanamycin. The data was obtained from at least four independent lines developed for each construct. Statistical analysis of the data was performed adopting one way ANOVA analysis (using GraphPad Prism version 5.01) and presented as the means ± S.D. A *P* value of less than 0.05 was considered significant indicated by different superscript letters.

weight over the negative vector-less control indicates the proportion of putatively transformed leaf discs, which ranged from 79 to 98%. There appeared to be no detectable difference between pSiM24 binary vectors and the positive control pCAMBIA and pKM24KH vectors. The effect of pSiM24 binary vector on transformation frequency was also studied in *A. thaliana*. The pSiM24 and pSiM24-GUS/GFP vectors exhibited approximately two-fold more transformation frequency in *A. thaliana* than pCAMBIA2300 and pKM24KH vectors (Table 5).

Expression analysis of pSiM24-GUS/GFP in transgenic plants

Agrobacterium carrying the pSiM24-GUS reporter gene was used to transform *Arabidopsis* and tobacco plants. GUS histochemical analysis confirmed that the pSiM24 vector successfully expressed GUS genes in transgenic *Arabidopsis* and tobacco (both by agrobacterium-mediated transformation as well as by gene-gun methods) plants (Figure 6–7). The *Arabidopsis* pSiM24-GUS and pSiM24-GFP transgenic plants successfully expressed GUS and GFP proteins, as detected by GUS histochemical staining and confocal microscopy of GFP (Figure 7–8). Furthermore, GUS analysis of second-generation plants confirmed the successful

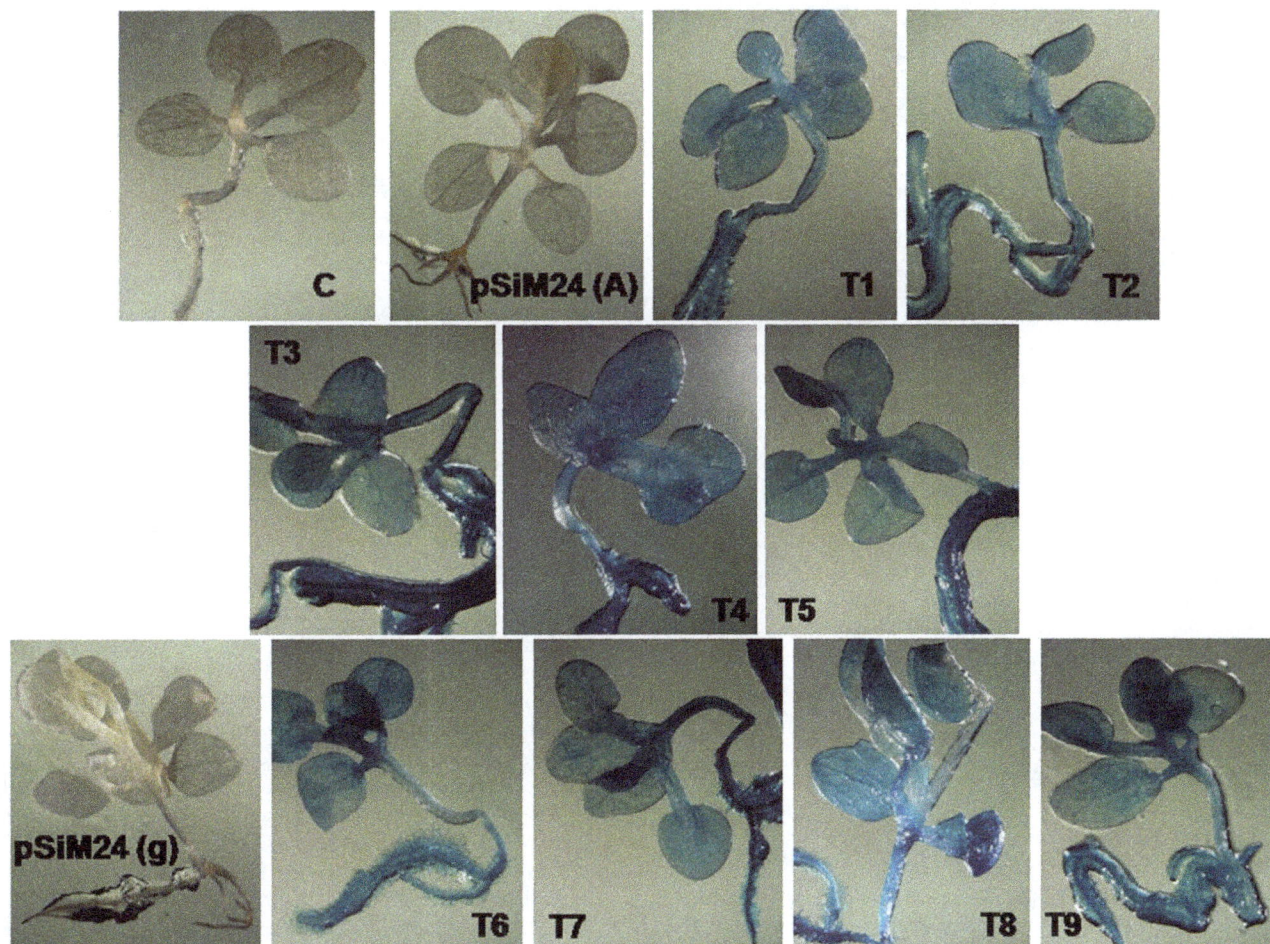

Figure 6. GUS expression in transgenic tobacco plants generated for constructs pSiM24 and pSiM24-GUS. Representative transgenic tobacco plants (second generation, three weeks old) generated by agrobacterium-mediated transformation (T1 to T5 lines) and transformation using gene-gun (T6 to T9 lines) were stained to determine GUS histochemical activity. C: Untransformed tobacco plant; pSiM24 (A): pSiM24 transgenic tobacco plants generated by agrobacterium-mediated transformation; pSiM24 (g): pSiM24 transgenic tobacco plants generated by biolistic bombardment method.

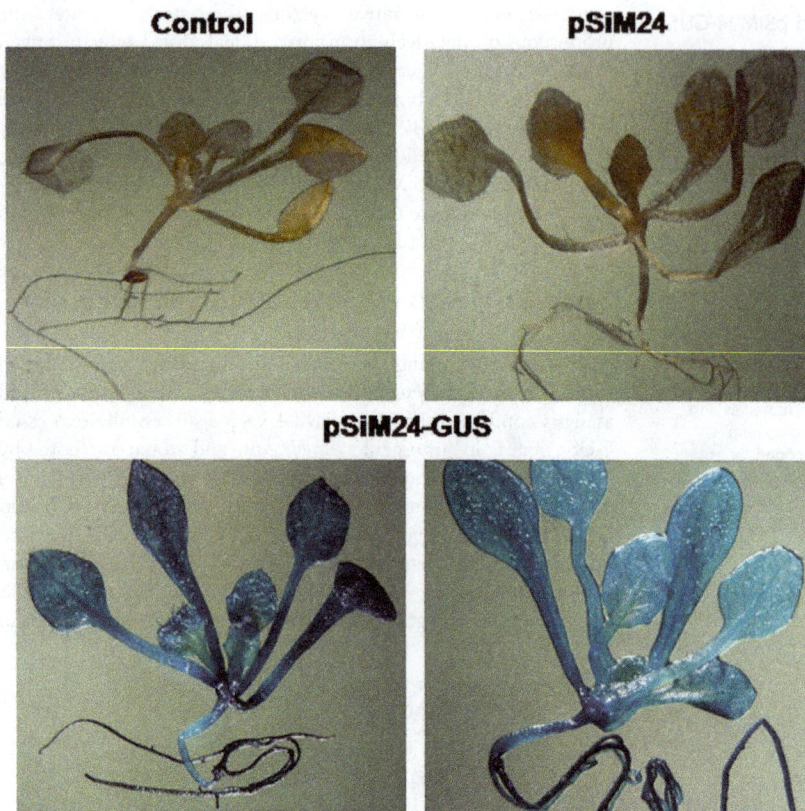

Figure 7. GUS expression in transgenic *Arabidopsis* plants generated for constructs pSiM24 and pSiM24-GUS. Representative transgenic *Arabidopsis* plants (second generation, two weeks old) generated by agrobacterium-mediated transformation were stained to determine GUS histochemical activity. Control: Untransformed *Arabidopsis* plant; pSiM24: pSiM24 transgenic *Arabidopsis* plants; pSiM24-GUS: pSiM24-GUS transgenic *Arabidopsis* plants.

inheritance of the transgene from one generation to another generation in both transgenic tobacco and *Arabidopsis* plants (Figure 9–10). The GUS activity was estimated biochemically in R2-generation transgenic *Arabidopsis* and tobacco plants; it was observed that approximately two times higher GUS activity accumulated in leaf and stem tissues of *Arabidopsis* plants than in tobacco plants (Figure 9). The expression of GUS activities in different tissues of transgenic *Arabidopsis* and tobacco plants containing pSiM24-GUS showed the following pattern: Root > Leaf > Stem (Figure 9). In addition, the histological GUS staining documented that the level of GUS expression was high in the reproductive tissues of transgenic pSiM24-GUS tobacco and *Arabidopsis* plants (Figure 10). Both in tobacco and *Arabidopsis* transgenic plants, GUS transcript levels were higher for the pSiM24 binary vector than for the pKYLX-based expression vector pKCaMV35 (Figure 5).

Transient expression of GUS using pSiM24 vector through vacuum infiltration method

A. tumefaciens carrying pSiM24 and pSiM24-GUS constructs infiltrated *N. benthamiana* leaves were assayed and histochemically stained for GUS enzyme. The completely infiltrated leaves showed approximately 1800 GUS units, whereas the partially infiltrated leaves exhibited approximately 850 GUS units (Figure 11A). One unit of GUS activity was defined as the amount of enzyme that liberated 1 p mol 4-methylumbelliferone mg^{-1} protein min^{-1}[52]. The agro-infiltrated leaves showed strong GUS expression, as

detected by histochemical staining, in leaves of both *N. benthamiana* and *Zea mays* (Figure 11B and Figure 12).

Transient expression of GFP-*AtCESA3*$^{ixr1-2}$, Vip3A(a), KMP-11, IL-10 and nat-T-phyllo-GFP genes using pSiM24 vector

Western blot analysis of pSiM24-GFP-*AtCESA3*$^{ixr1-2}$ agroinfiltrated leaf samples showed the expected bands of size 145 kD for *GFP-AtCESA3*$^{ixr1-2}$ as detected with AtCESA3-specific polyclonal antibodies (Figure 13A). In addition, RT-PCR analysis of agroinfiltrated leaf samples exhibited expected 1318 bp band for a portion of *GFP-AtCESA3*$^{ixr1-2}$ (Figure 13A). The transient expression of Vip3A(a) using pSiM24 expression vector in tobacco protoplast was detected by Vip3A-specific polyclonal antibodies that showed the expected bands of size 88 kD (Figure 13B). Using pSiM24 expression vector KMP-11 and IL-10 were also expressed transiently in tobacco protoplasts and showed expression up to 0.03 mg of KMP-11 and 0.08 mg of IL-10 per mg of protoplast protein samples by indirect ELISA (Figure 13C). The RT-PCR analysis and localization analysis of apoplast targeted nat-T-phyllo-GFP by confocal laser scanning microscopy showed the successful expression of *nat-T-phyllo-GFP* using pSiM24 expression vector (Figure 14).

Discussion

A binary vector, used as a standard tool in the transformation of higher plants mediated by *A. tumefaciens*, consists of T-DNA and

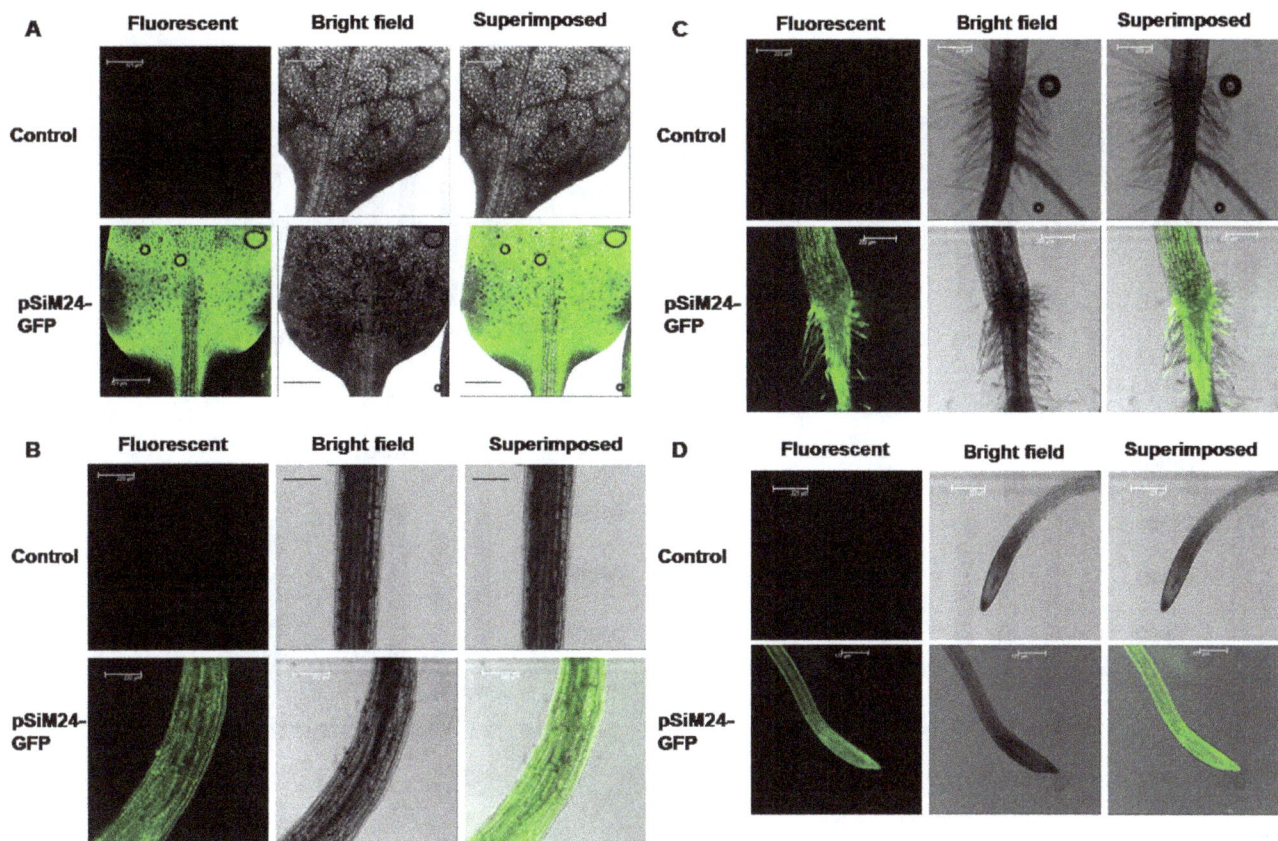

Figure 8. GFP expression in leaf, stem and roots of transgenic *Arabidopsis* plants generated for construct pSiM24-GFP. (A) Representative transgenic *Arabidopsis* plant leaves (second generation, two weeks old) generated by agrobacterium-mediated transformation were imaged to determine GFP activity. Fluorescent, bright-field and superimposed (bright-field and green fluorescent) confocal laser scanning micrographs of transgenic *Arabidopsis* leaves. Scale bar represents 320 μm. (B) Representative transgenic *Arabidopsis* plant stems (second generation, two weeks old) generated by agrobacterium-mediated transformation were imaged to determine GFP activity. Fluorescent, bright-field and superimposed (bright-field and green fluorescent) confocal laser scanning micrographs of transgenic *Arabidopsis* stems. Scale bar represents 220 μm. (C) Representative transgenic *Arabidopsis* plant stem-root junctions (second generation, two weeks old) generated by agrobacterium-mediated transformation were imaged to determine GFP activity. Fluorescent, bright-field and superimposed (bright-field and green fluorescent) confocal laser scanning micrographs of transgenic *Arabidopsis* stem-root junctions. Scale bar represents 220 μm. (D) Representative transgenic *Arabidopsis* plant roots (second generation, two weeks old) generated by agrobacterium-mediated transformation were imaged to determine GFP activity. Fluorescent, bright-field and superimposed (bright-field and green fluorescent) confocal laser scanning micrographs of transgenic *Arabidopsis* roots. Scale bar represents 225 μm.

the vector backbone. T-DNA is the segment delimited by the border sequences, the right border (RB) and the left border (LB), and may contain multiple cloning sites, a selectable marker gene for plants, a reporter gene and other genes of interest [7,53]. The vector backbone carries plasmid replication functions for *E. coli* and *A. tumefaciens*, selectable marker genes for the bacteria, optionally a function for plasmid mobilization between bacteria and other accessory components [7,8].

The binary vector pSiM24 has an overall size of 7.08 kb and carries a plant-gene expression cassette containing a highly active, constitutive promoter (M24) (GenBank Accession no. KF032933). The size of the pSiM24 vector is approximately 2000 bp shorter than the commercially available pCAMBIA vectors (www.cambia.org) and approximately 6000 bp shorter than pKYLX-based vectors [4]. In the pSiM24 binary vector, only the necessary elements were included to attain a minimum size. The right border (RB) and the left border (LB) of pSiM24 are imperfect, direct repeats of 25 bases. The RB and LB are considered to be the only essential cis-elements for T-DNA transfer [54]. The promoter carried by the expression cassettes described here has been studied

in transgenic plants (present study) and is also functional in plants such as tobacco [41–44,55], *Arabidopsis* and corn (Sahoo and Maiti, Unpublished Data). It has been documented that the *Mirabilis mosaic virus* full-length transcript promoter is constitutive in nature, exhibiting 14 to 25 times stronger activity than CaMV35S in the tobacco protoplast transient system and transgenic tobacco plants, respectively [27,34,42]. The modified full-length transcript promoter (M24) of the *Mirabilis mosaic virus* with duplicated enhancer domains [27,29,41–44,55] was used in the pSiM24 vector to evaluate gene expression in plants. In the pSiM24-GUS vector, the coding sequence of GUS was placed between the heterologous M24 promoter and the terminator sequence from the rbcSE9 gene (Figure 1) [43–44]. We showed that microT-DNAs in pSiM24 containing a kanamycin resistance gene and reporter gene (GUS or GFP) were integrated stably in the nuclear chromosomal DNA of transgenic plants for successive generation.

Selectable markers need to be expressed in calli, in cells from those plants that are being regenerated or in germinating embryos to facilitate plant transformation. Therefore, promoters for constitutive expression are preferred. In pSiM24, the Nos

A

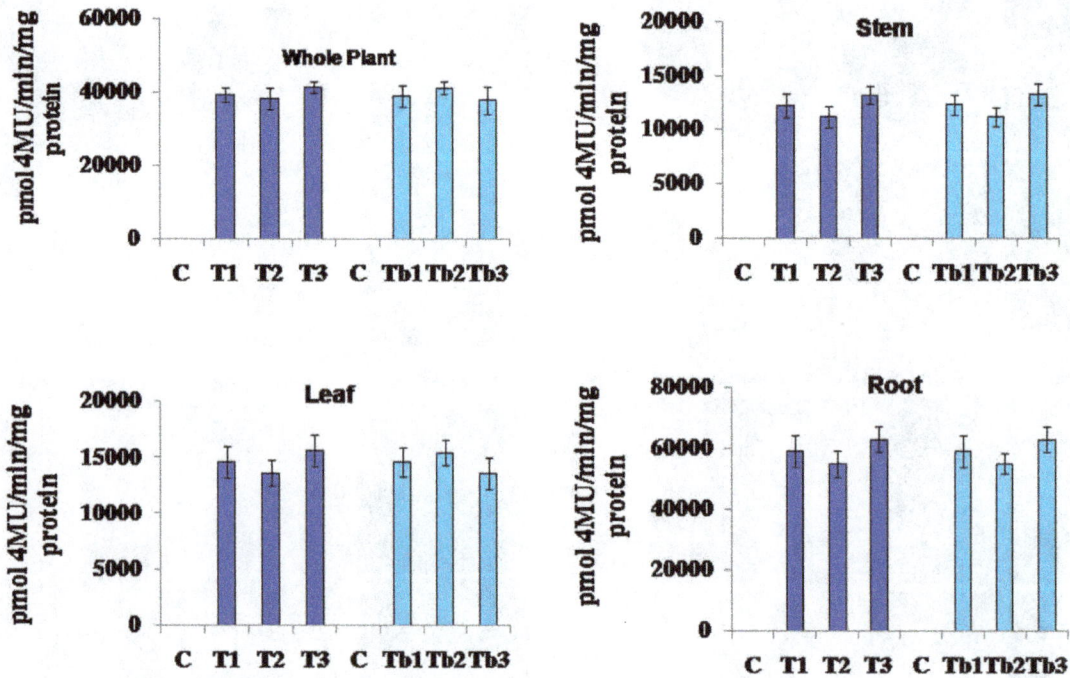

B

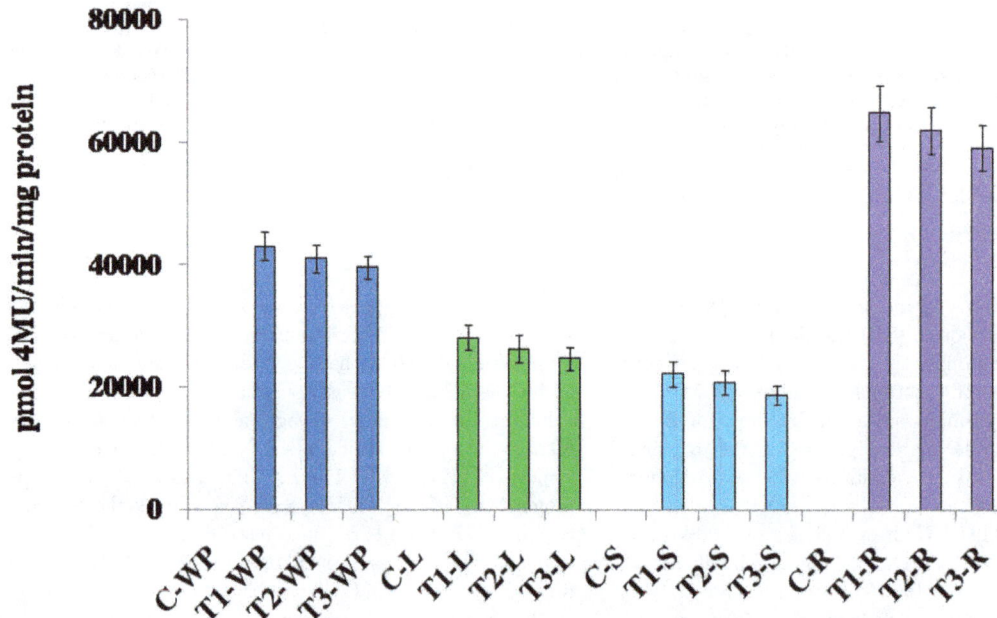

Figure 9. GUS expression in transgenic tobacco and *Arabidopsis* plants generated for constructs pSiM24 and pSiM24-GUS. (A) GUS enzymatic activity of pSiM24-GUS transgenic tobacco (second generation, 3 weeks old) lines was measured in whole plant, leaf, stem and root tissues. Soluble protein extracts isolated from different tissues of plants were used for GUS assay along with the wild-type plants (C). The data represent means ± S.D. of four second generation individuals from one line for each tissue (n = 4). The values significantly differ between control and transgenic plants at P<0.01 based on Student's *t*-test. T1, T2 and T3: Representative transgenic lines generated by agrobacterium-mediated plant transformation procedure; Tb1, Tb2 and Tb3: Representative transgenic lines generated by biolistic plant transformation procedure. (B) GUS enzymatic activity of pSiM24-GUS transgenic *Arabidopsis* (second generation, two weeks old) lines was measured in whole plant (WP), leaf (L), stem (S) and root (R) tissues. Soluble protein extracts isolated from different tissues of plants were used for GUS assay along with the wild-type plants (C).

The data represent means ± S.D. of four second generation individuals from one line for each tissue (n = 4). The values significantly differ between control and transgenic plants at P<0.01 based on Student's t-test. T1, T2 and T3: Representative transgenic lines generated by floral-dip plant transformation procedure.

promoter derived from nopaline synthase (Nos) of *A. tumefaciens* [56] was used to express the selectable marker gene (Figure 1). The choice of promoters responsible for selectable marker gene expression also plays an important role in the efficiency of transformation [57–59]. The use of weak promoters may not always be a bad idea because the levels of expression of marker genes and genes of interest are often linked, and the selection of transformants with weak selectable markers may cause strong expression of the gene of interest to be obtained [8]. It is generally recommended that different promoters be used for the selectable marker and expressing the gene of interest [57–59], as in the pSiM24 vector (Figure 1), which carries the M24 promoter for the expression of the gene of interest (here GUS in pSiM24-GUS) and Nos for the selectable marker (here, KanR). Homology-based gene silencing has been reported to occur extensively in transgenic plants [60]. Gene silencing due to promoter homology can be avoided by either using diverse promoters isolated from different plant and viral genomes or by designing synthetic promoters [27,28,34,38,61–68].

Depending on the plant species to be transformed, the choice of selectable markers greatly affects the efficiency of transformation, and permissive concentrations of selective agents vary considerably among plant species. Genes resistant to antibiotics or herbicides, such as kanamycin, hygromycin, phosphinothricin and glyphosate, are very popular. Kanamycin resistance has been most frequently exploited in the transformation of many dicotyledonous plants such as tobacco, tomato, potato and *Arabidopsis* [34,42,69]. The pSiM24 binary vector contains a synthetic 'nptII' KanR gene (nos

promoter-KanR cDNA-Nos terminator), the open reading frame of which is optimized for plant codon bias; hence, the nptII gene serves both as a selectable marker for the regeneration of plantlets on kanamycin-containing medium (for tobacco 250–300 µg/ml) and as a screenable marker for agrobacterium (25 µg/ml). In the present study, pSiM24-containing nptII gene was used to select the transformed *Arabidopsis* and tobacco plants in 30 µg/ml and 250 µg/ml kanamycin, respectively, (Figure 6–7). Choice of antibiotics is an important factor in plant transformation. For example, kanamycin may not suitable for rice and maize cells, whereas hygromycin resistance (hpt) is very good for rice transformation [10], and the phosphinothricin resistance gene (bar) is efficient for maize and other cereals [70,71]. We also developed a binary vector pKDH, which has a structure similar to that of pSiM24, but the selection marker KanR gene was replaced with a hygromycin resistance (HygR, Hygromycin B transferase, HPH) gene for the selection of transgenic monocot plants, and the sequence information of the binary vector pKDH was provided (Genebank Accession no. KF041008).

The components of the pSiM24 expression system vector are arranged in a modular configuration in which the promoter, terminator and MCS cassettes are flanked by unique restriction endonuclease cleavage sites. The pSiM24 vector provides nine unique cloning sites in the first multiple cloning site (MCS) between the left T-DNA border and the M24 promoter (BstXI, StuI, EspAI, PasI, KflI, Bstz17I, SmaI, XmaI and EcoRI), twelve unique cloning sites in the second MCS between the M24 promoter and the pea rbcSE9 terminator (HindIII, AbsI, PspXI,

Figure 10. GUS expression in flowers of transgenic tobacco and *Arabidopsis* plants generated for constructs pSiM24 and pSiM24-GUS. Histological GUS staining in different floral tissues of untransformed (control), pSiM24 and pSiM24-GUS plants. Histological GUS staining shows strong GUS expression in all floral tissues of transgenic pSiM24-GUS *Arabidopsis* (upper panel) and tobacco (lower panel) plants.

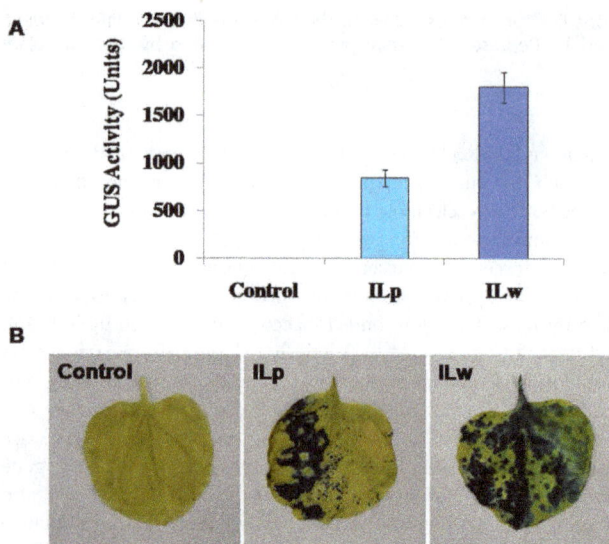

Figure 11. Transient expression of GUS in pSiM24-GUS Agro-infiltrated *N. benthamiana* leaves using vacuum infiltration method. (A) GUS enzymatic activity of pSiM24 and pSiM24-GUS *A. tumefaciens*-infiltrated *N. benthamiana* leaves was measured, and one unit of GUS activity was defined as the amount of enzyme that liberated 1 pmol 4-methylumbelliferone mg^{-1} protein min^{-1}. The data represent means $\pm$ S.D. of four biological replicates for each construct (n = 4). The values significantly differ between control (pSiM24) and pSiM24-GUS agro-infiltrated leaf samples at P<0.01 based on Student's *t*-test. ILw: Whole Infiltrated Leaf; ILp: Partial Infiltrated Leaf. (B) The pSiM24 and pSiM24-GUS *A. tumefaciens*-infiltrated *N. benthamiana* leaves were histochemically stained for GUS enzyme. The pSiM24-GUS agro-infiltrated leaves showed stronger GUS expression, as detected by histochemical staining. Control: *N. benthamiana* leaf infiltrated with *A. tumefaciens* harboring pSiM24 construct; ILw: Whole Infiltrated Leaf; ILp: Partial Infiltrated Leaf. Both ILw and ILp are *N. benthamiana* leaves infiltrated with *A. tumefaciens* carrying pSiM24-GUS construct.

Figure 12. Transient expression of GUS in pSiM24-GUS Agro-infiltrated *Zea mays* leaves using vacuum infiltration method. The representative pSiM24-GUS agro-infiltrated leaf showed strong GUS expression, as detected by histochemical staining. Control: *Z. mays* leaf infiltrated with *A. tumefaciens* carrying pSiM24; pSiM24-GUS: *Z. mays* leaf infiltrated with *A. tumefaciens* harboring pSiM24-GUS construct.

SciI, XhoI, HpaI, MluI, Eco53kI, SacI, SbfI, PstI, and XbaI) and seven unique cloning sites in the third MCS between the Nos promoter and the right T-DNA border (BglII, BstEII, EcoNI, FseI, SwaI, NruI, and PacI). This configuration facilitates the modification or replacement of individual components in the pSiM24 vector. The MCS in pSiM24 contains more additional cleavage sites than that of pUC19. It should be noted that the orientation of the MCS in the pSiM24 plasmid, relative to the rbcS and M24 promoters, is opposite that in pUC19, relative to the lac promoter. The presence of a number of cloning sites unique to the three MCS allow for gene-stacking applications to introduce multiple gene with additional sequences, such as translational initiation, signal and transit peptide sequences and translational termination, into these plasmids. The pSiM24 vector provides a number of options for cloning, transformation and expression strategies. The M24 promoter in the pSiM24 plasmid can be easily replaced with other promoters as EcoRI-HindIII cassettes, thus making different strategies for the regulated expression of foreign genes possible.

Reporter genes, whose expression can be easily monitored, are useful in many ways in plant transformation. Strength and temporal, spatial and other types of regulation of promoters and elements may be conveniently assayed by connecting these elements to the reporter genes. Genes for β-glucuronidase (GUS) [9], luciferase [72] and GFP [73] are popular examples. In the present study, two different reporter genes, GUS and GFP, were introduced into the pSiM24 vector to monitor and analyze their expression under the M24 promoter in both stable and transient

systems. Reporter genes that are connected to constitutive promoters may be used to monitor the process of transformation. The expression of the reporter genes soon after the inoculation of plant cells with *A. tumefaciens*, which is referred to as "transient expression", is a good indication of the transfer of the T-DNA from the bacteria to the nuclei of plant cells. The expression of the reporter genes later in a cluster of cells growing on selection media provides evidence of the integration of the T-DNA in plant chromosomes. A binary vector that carries a constitutive selectable marker and a constitutive reporter is very useful as a control vector both in transformation experiments and in assays of gene expression. Hence, in pSiM24, both "nptII" and GUS/GFP were constitutively expressed by using two different constitutive promoters, i.e., Nos for nptII and M24 for GUS/GFP, for expression in transgenic plants (Figure 6–10).

The rbcSE9 polyadenylation signal used in the pSiM24 vector has previously been used to direct efficient mRNA3′ end formation from chimeric genes in transformed tobacco [74–76]. These 3′ ends are identical to those observed in pea, indicating that this signal is suitable for the predictable expression of foreign genes in plants. The 3′ regions of the cauliflower mosaic virus 35S transcript and the nopaline synthase gene in the wild-type T-DNA of *A. tumefaciens* are frequently used as a 3′ signal to direct selectable marker genes expression.

In pSiM24, the "bla" AmpR gene, which confers resistance to ampicillin, was used as the marker for bacterial selection for *E. coli*. The selectable marker for plants, Nos-nptII, in pSiM24 also provides fair levels of resistance to both *E. coli* and *A. tumefaciens*. Binary vectors need to be replicated both in *E. coli* and *A. tumefaciens*. Hence, the pSiM24 vector carries all of the functions necessary for replication and transfer in *Escherichia coli* and *A. tumefaciens*, which includes a ColE1-replicon and an RK2-replicon derived from pRK2013 [77]. The pSiM24 binary vector carries the origin of vegetative replication (OriV) and the transacting

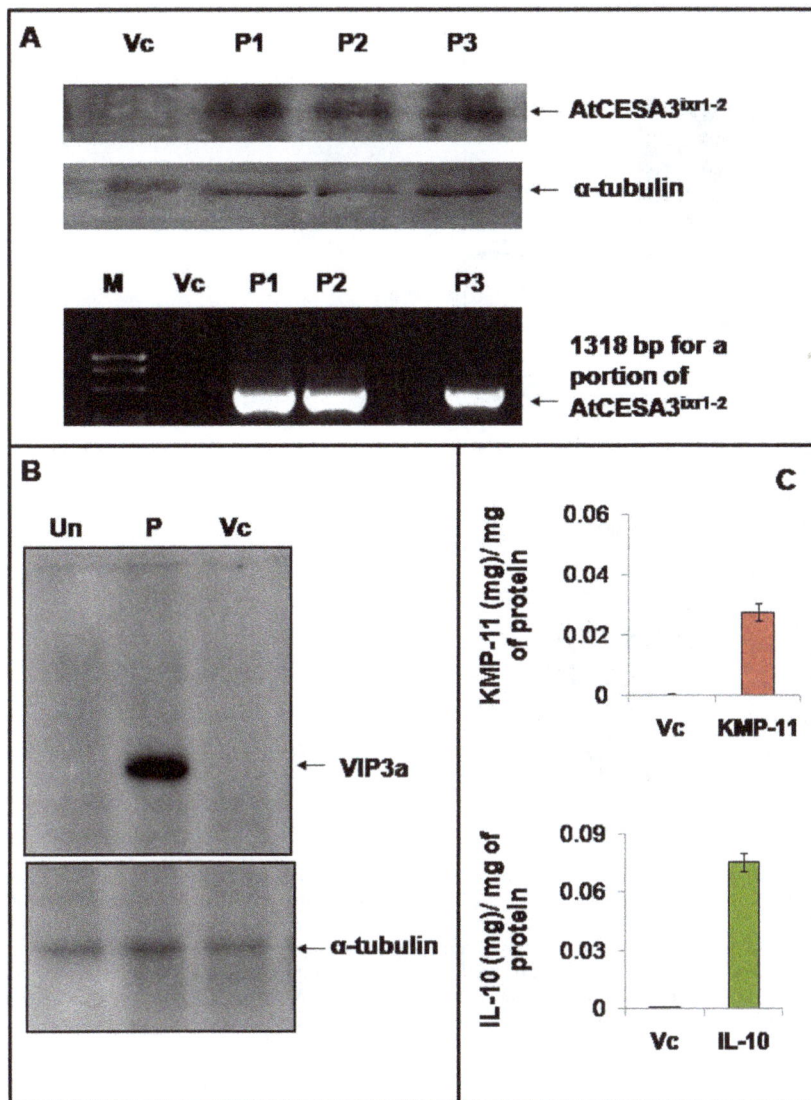

Figure 13. Transient expression of *GFP-AtCESA3^ixr1-2^*, Vip3A(a), KMP-11 and IL-10 genes using pSiM24 expression vector. (A) Western blot analysis of transiently expressed *GFP* fused *Arabidopsis CESA3^ixr1-2^* (*GFP-AtCESA3^ixr1-2^*) detected with AtCESA3-specific polyclonal antibodies showed the expected bands of size 145 kD (Upper panel). Signals were quantitated, normalized to the α-tubulin loading control (Middle panel). RT-PCR products for a portion of *GFP-AtCESA3^ixr1-2^* with the expected 1318 bp band are shown (Lower panel). Transiently *GFP-AtCESA3^ixr1-2^* expressed agro-infiltrated *N. tabacum* L. variety Samsun NN leaf samples using pSiM24-*GFP-AtCESA3^ixr1-2^* (P1, P2 and P3) and pSiM24 (Vc; empty vector control) are presented. (B) Western blot of vegetative insecticidal protein, Vip3A(a) expression in tobacco protoplast. Detection by Vip3A-specific polyclonal antibodies showed the expected bands of size 88 kD (Upper panel). Un: Untransfected control; P: pSiM24-Vip3A(a) transfected protoplast; Vc: pSiM24 vector control transfected protoplast. Signals were quantitated, normalized to the α-tubulin loading control (Lower panel). (C) Estimation of KMP-11 (Kinetoplastid membrane protein-11) concentration (expressed as mg KMP-11 per mg protein) in tobacco protoplasts by ELISA. KMP-11: pSiM24-KMP-11 transfected protoplast; Vc: pSiM24 vector control transfected protoplast. (D) Estimation of Interleukin-10 (IL-10) concentration (expressed as mg IL-10 per mg protein) in tobacco protoplasts by ELISA. IL-10: pSiM24-IL-10 transfected protoplast; Vc: pSiM24 vector control transfected protoplast.

replication functions (Trf) of plasmid incompatibility group P (IncP) plasmids [78]. The types of replication functions exploited determine the copy numbers and stability of the plasmids in bacterial cells. *E. coli* exhibited a transformation frequency up to five- to six-fold higher with pSiM24 than with conventional pCAMBIA and pKM24KH vectors, (Table 1) and the plasmid DNA yields of pSiM24 binary Ti vectors were three-fold and seven- to eight-fold higher in *E. coli* than those of conventional pCAMBIA 2300 and pKM24KH, respectively (Table 3). The pSiM24 binary vector contains the ColE1 replicon without a bom (basis of mobility) sequence, which again reduces its size. The bom

function is necessary for plasmid mobilization from *E. coli* to *A. tumefaciens* [79]. This function is not necessary when vectors are introduced into *A. tumefaciens* by electroporation or freeze-thaw methods.

Not only reporter genes, other introduced genes of size up to 4 kb were also effectively expressed using pSiM24 expression vector. In the present study, nat-T-phyllo-GFP [44] was expressed transiently using pSiM24 expression vector in tobacco leaves. T-phylloplanins have antimicrobial properties and are known to inhibit blue mold disease caused by *Peronospora tabacina* [41,44,80,81]. Both the native and mature tobacco phylloplanin

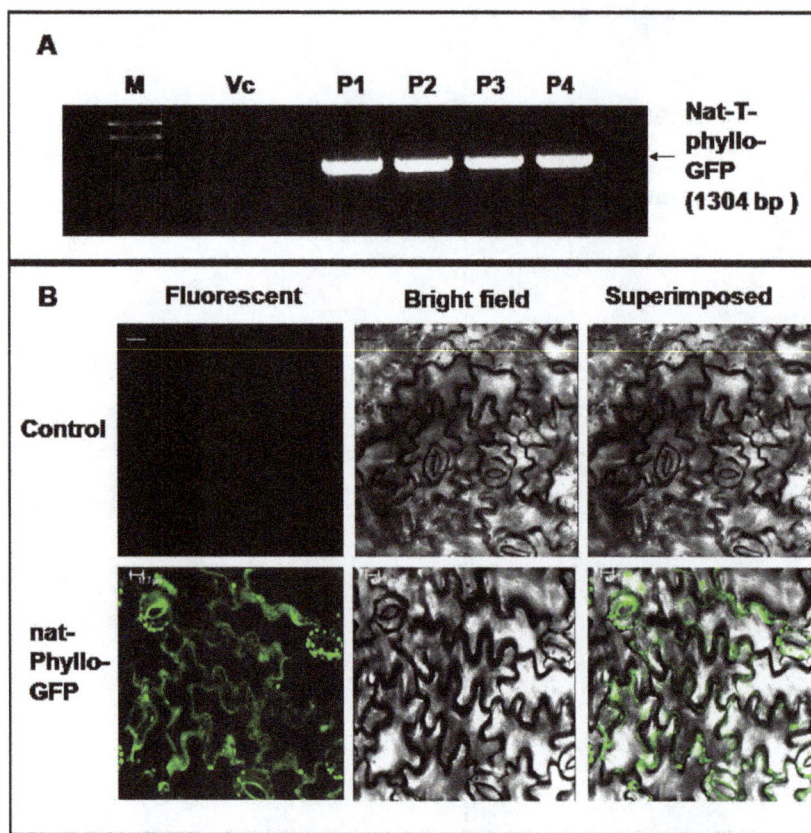

Figure 14. Transient expression of *nat-T-phyllo-GFP* using pSiM24 expression vector. (A) RT-PCR products for native T-phylloplanin fused GFP (*nat-T-phyllo-GFP*) with the expected 1304 bp band are shown. Transiently *nat-T-phyllo-GFP* expressed agro-infiltrated *N. tabacum* L. variety Samsun NN leaf samples using pSiM24-*nat-T-phyllo-GFP* (P1, P2, P3 and P4) and pSiM24 (Vc; empty vector control) are presented. (B) Localization analysis of apoplast targeted nat-T-phyllo-GFP by confocal laser scanning microscopy. Agro-infiltrated *N. tabacum* L. variety Samsun NN leaf cells for pSiM24-nat-T-phyllo-GFP construct expressing GFP (green fluorescence) was visualized by confocal laser scanning microscopy. No GFP fluorescence was detected in agro-infiltrated tobacco leaf cells for pSiM24 construct (Control). Scale bar represents 20 μm on all images. Fluorescent, bright-field and superimposed (bright-field and green fluorescent) confocal laser scanning micrographs of leaf sections are presented.

gene fused with GFP targeted to the apoplasm increases resistance to blue mold disease in tobacco [41,44]. Here, the expression of nat-T-phyllo-GFP using pSiM24 vector was confirmed by the GFP fluorescence in apoplast region of agroinfiltrated plant leaves (Figure 14). Another, chimeric gene (GFP-*AtCESA3*$^{ixr1-2}$) of size 4086-bp fragment was also successfully expressed transiently using pSiM24 following agro-infiltration procedure. The overexpression of GFP fused to the *Arabidopsis CESA3*$^{ixr1-2}$ (GFP-*AtCESA3*$^{ixr1-2}$) gene in transgenic tobacco was known for increasing cellulose digestibility and biomass saccharification [42,43]. Further genes like Vip3A(a), KMP-11 and IL-10 were also successfully expressed transiently in tobacco protoplasts using pSiM24 expression vector (Figures 13). The gene product of novel vegetative insecticidal gene, vip3A(a) shows activity against lepidopteran insects [45,46]. KMP-11, a flagellar protein is known to play an essential role in regulating cytokinesis in both amastigote and promastigote forms of leishmania [47] and is a potential stimulator of T-lymphocyte proliferation [82]. Interleukin-10 (IL-10), an anti-inflammatory cytokine secreted under different conditions of immune activation by a variety of cell types, including T cells, B cells, and monocytes/macrophages [83,84] has been shown to suppress a broad range of inflammatory responses and as an important factor in maintaining homeostasis of overall immune responses [85,86] and thus has been used for developing novel therapies for several human diseases such as allergic responses and autoimmune diseases [87].

The effect of pSiM24 binary vector on transformation frequency studied in *A. thaliana* verified the pSiM24 and pSiM24-GUS/GFP vectors exhibited more transformation frequency in *A. thaliana* than pCAMBIA2300 and pKM24KH vectors (Table 5) however strong GUS transgene expression in pSiM24-GUS transgenic tobacco and *Arabidopsis* than pK-CaMV35S-GUS transgenic plants depends upon the strong M24 promoter (Figure 5) [41,42].

The pSiM24 vector was observed to be active in transferring the transgene both transiently (Figure 3–5, Figure 11–14) and stably (Figure 6–10) in plant systems, making it useful for various plant biotechnological applications. This plasmid has multiple cloning sites and can act as a platform for various applications, such as gene expression studies and different promoter expressional analyses. In addition, pSiM24 offers a wide selection of cloning sites and high copy numbers in *E. coli* for the facile manipulation of different genetic elements. Thus, the pSiM24 binary vector system described in this study has a high degree of flexibility and may serve as a useful tool for the transformation of plants, making it apt for future use in field release experiments.

Acknowledgments

We are very much indebted to the Kentucky Tobacco Research and Development Center (KTRDC) for the provided facilities and support.

The authors would like to thank Ms. Bonnie Kinney for her excellent care of the transgenic tobacco plants and Prof. Arthur G. Hunt, University of Kentucky, Lexington, USA, for critically reading and making necessary corrections to the manuscript. The information reported in this paper (No. 14-17-039) is part of a project of the Kentucky Agricultural Experiment Station and is published with the approval of the Director.

Author Contributions

Conceived and designed the experiments: IBM DKS ND. Performed the experiments: DKS. Analyzed the data: IBM DKS ND. Contributed reagents/materials/analysis tools: IBM DKS ND. Wrote the paper: IBM DKS.

References

1. Bevan M (1984) Binary Agrobacterium vectors for plant transformation. Nucleic Acids Res 12: 8711–8721.
2. Hoekema A, Hirsch PR, Hooykaas PJJ, Schilperoort RA (1983) A Binary Plant Vector Strategy Based on Separation of Vir-Region and T-Region of the Agrobacterium-Tumefaciens Ti-Plasmid. Nature 303: 179–180.
3. Klee HJ, Yanofsky MF, Nester EW (1985) Vectors for Transformation of Higher-Plants. Bio-Technology 3: 637–642.
4. Schardl CL, Byrd AD, Benzion G, Altschuler MA, Hildebrand DF, et al. (1987) Design and Construction of a Versatile System for the Expression of Foreign Genes in Plants. Gene 61: 1–11.
5. Xiang C, Han P, Lutziger I, Wang K, Oliver DJ (1999) A mini binary vector series for plant transformation. Plant Mol Biol 40: 711–717.
6. Zupan JR, Zambryski P (1995) Transfer of T-DNA from Agrobacterium to the plant cell. Plant Physiol 107: 1041–1047.
7. Hellens R, Mullineaux P, Klee H (2000) Technical Focus:a guide to Agrobacterium binary Ti vectors. Trends Plant Sci 5: 446–451.
8. Komari T, Takakura Y, Ueki J, Kato N, Ishida Y, et al. (2006) Binary vectors and super-binary vectors. Methods Mol Biol 343: 15–41.
9. Jefferson RA, Klass M, Wolf N, Hirsh D (1987) Expression of chimeric genes in Caenorhabditis elegans. J Mol Biol 193: 41–46.
10. Hiei Y, Ohta S, Komari T, Kumashiro T (1994) Efficient transformation of rice (Oryza sativa L.) mediated by Agrobacterium and sequence analysis of the boundaries of the T-DNA. Plant J 6: 271–282.
11. Hajdukiewicz P, Svab Z, Maliga P (1994) The small, versatile pPZP family of Agrobacterium binary vectors for plant transformation. Plant Mol Biol 25: 989–994.
12. Xiang CB, Han P, Lutziger I, Wang K, Oliver DJ (1999) A mini binary vector series for plant transformation. Plant Mol Biol 40: 711–717.
13. Hellens RP, Edwards EA, Leyland NR, Bean S, Mullineaux PM (2000) pGreen: a versatile and flexible binary Ti vector for Agrobacterium-mediated plant transformation. Plant Mol Biol 42: 819–832.
14. Chakrabarty R, Banerjee R, Chung SM, Farman M, Citovsky V, et al. (2007) pSITE vectors for stable integration or transient expression of autofluorescent protein fusions in plants: Probing Nicotiana benthamiana-virus interactions. Molecular Plant-Microbe Interactions 20: 740–750.
15. Goodin MM, Dietzgen RG, Schichnes D, Ruzin S, Jackson AO (2002) pGD vectors: versatile tools for the expression of green and red fluorescent protein fusions in agroinfiltrated plant leaves. Plant Journal 31: 375–383.
16. Thole V, Worland B, Snape JW, Vain P (2007) The pCLEAN dual binary vector system for Agrobacterium-mediated plant transformation. Plant Physiol 145: 1211–1219.
17. Untergasser A, Bijl GJ, Liu W, Bisseling T, Schaart JG, et al. (2012) One-step Agrobacterium mediated transformation of eight genes essential for rhizobium symbiotic signaling using the novel binary vector system pHUGE. PLoS One 7: e47885.
18. Takken FLW, van Wijk R, Michielse CB, Houterman PM, Ram AFJ, et al. (2004) A one-step method to convert vectors into binary vectors suited for Agrobacterium-mediated transformation. Current Genetics 45: 242–248.
19. Lindbo JA (2007) TRBO: A high-efficiency tobacco mosaic virus RNA-Based overexpression vector. Plant Physiology 145: 1232–1240.
20. Shah KH, Almaghrabi B, Bohlmann H (2013) Comparison of Expression Vectors for Transient Expression of Recombinant Proteins in Plants. Plant Molecular Biology Reporter 31: 1529–1538.
21. Sainsbury F, Thuenemann EC, Lomonossoff GP (2009) pEAQ: versatile expression vectors for easy and quick transient expression of heterologous proteins in plants. Plant Biotechnology Journal 7: 682–693.
22. Huang Z, Chen Q, Hjelm B, Arntzen C, Mason H (2009) A DNA Replicon System for Rapid High-Level Production of Virus-Like Particles in Plants. Biotechnology and Bioengineering 103: 706–714.
23. Ali MA, Shah KH, Bohlmann H (2012) pMAA-Red: a new pPZP-derived vector for fast visual screening of transgenic Arabidopsis plants at the seed stage. Bmc Biotechnology 12.
24. Wang CT, Yin XL, Kong XX, Li WS, Ma L, et al. (2013) A Series of TA-Based and Zero-Background Vectors for Plant Functional Genomics. Plos One 8.
25. Vieira J, Messing J (1982) The Puc Plasmids, an M13mp7-Derived System for Insertion Mutagenesis and Sequencing with Synthetic Universal Primers. Gene 19: 259–268.
26. Zambryski P, Joos H, Genetello C, Leemans J, Vanmontagu M, et al. (1983) Ti-Plasmid Vector for the Introduction of DNA into Plant-Cells without Alteration of Their Normal Regeneration Capacity. Embo Journal 2: 2143–2150.
27. Dey N, Maiti IB (1999) Structure and promoter/leader deletion analysis of mirabilis mosaic virus (MMV) full-length transcript promoter in transgenic plants. Plant Mol Biol 40: 771–782.
28. Sahoo DK, Ranjan R, Kumar D, Kumar A, Sahoo BS, et al. (2009) An alternative method of promoter assessment by confocal laser scanning microscopy. J Virol Methods 161: 114–121.
29. Dey N, Maiti IB (1999) Further characterization and expression analysis of mirabilis mosaic caulimovirus (MMV) full-length transcript promoter with single and double enhancer domains in transgenic plants. Transgenics 3: 61–+.
30. Hofgen R, Willmitzer L (1988) Storage of Competent Cells for Agrobacterium Transformation. Nucleic Acids Res 16: 9877–9877.
31. Maiti IB, Murphy JF, Shaw JG, Hunt AG (1993) Plants that express a potyvirus proteinase gene are resistant to virus infection. Proc Natl Acad Sci U S A 90: 6110–6114.
32. Svab Z, Hajdukiewicz P, Maliga P (1990) Stable transformation of plastids in higher plants. Proc Natl Acad Sci U S A 87: 8526–8530.
33. Zhang XR, Henriques R, Lin SS, Niu QW, Chua NH (2006) Agrobacterium-mediated transformation of Arabidopsis thaliana using the floral dip method. Nature Protocols 1: 641–646.
34. Kumar D, Patro S, Ranjan R, Sahoo DK, Maiti IB, et al. (2011) Development of useful recombinant promoter and its expression analysis in different plant cells using confocal laser scanning microscopy. PLoS One 6: e24627.
35. Voinnet O, Rivas S, Mestre P, Baulcombe D (2003) An enhanced transient expression system in plants based on suppression of gene silencing by the p19 protein of tomato bushy stunt virus. Plant Journal 33: 949–956.
36. Maiti IB, Gowda S, Kiernan J, Ghosh SK, Shepherd RJ (1997) Promoter/leader deletion analysis and plant expression vectors with the figwort mosaic virus (FMV) full length transcript (FLt) promoter containing single or double enhancer domains. Transgenic Res 6: 143–156.
37. Lu Y, Chen X, Wu Y, Wang Y, He Y (2013) Directly transforming PCR-amplified DNA fragments into plant cells is a versatile system that facilitates the transient expression assay. PLoS One 8: e57171.
38. Banerjee J, Sahoo DK, Dey N, Houtz RL, Maiti IB (2013) An Intergenic Region Shared by At4g35985 and At4g35987 in Arabidopsis thaliana Is a Tissue Specific and Stress Inducible Bidirectional Promoter Analyzed in Transgenic Arabidopsis and Tobacco Plants. PLoS One 8: e79622.
39. Jefferson RA, Kavanagh TA, Bevan MW (1987) Gus Fusions - Beta-Glucuronidase as a Sensitive and Versatile Gene Fusion Marker in Higher-Plants. Embo Journal 6: 3901–3907.
40. Bradford MM (1976) A rapid and sensitive method for the quantitation of microgram quantities of protein utilizing the principle of protein-dye binding. Anal Biochem 72: 248–254.
41. Kroumova AB, Sahoo DK, Raha S, Goodin M, Maiti IB, et al. (2013) Expression of an apoplast-directed, T-phylloplanin-GFP fusion gene confers resistance against Peronospora tabacina disease in a susceptible tobacco. Plant Cell Reports 32: 1771–1782.
42. Sahoo DK, Stork J, DeBolt S, Maiti IB (2013) Manipulating cellulose biosynthesis by expression of mutant Arabidopsis proM24::CESA3(ixr1-2) gene in transgenic tobacco. Plant Biotechnol J 11: 362–372.
43. Sahoo DK, Maiti IB (2014) Biomass derived from transgenic tobacco expressing the Arabidopsis CESA3^{ixr1-2} gene exhibits improved saccharification. Acta Biologica Hungarica 65(2): 189–204 (In Press).
44. Sahoo DK, Raha S, Hall JT, Maiti IB (2014) Over-expression of the synthetic chimeric native-T-phylloplanin-GFP genes optimized for monocot and dicot plants renders enhanced resistance to blue mold disease in tobacco (N. tabacum L.). The Scientific World Journal 2014: Article ID 601314, 12 pages. DOI: http://dx.doi.org/10.1155/2014/601314.
45. Estruch JJ, Warren GW, Mullins MA, Nye GJ, Craig JA, et al. (1996) Vip3A, a novel Bacillus thuringiensis vegetative insecticidal protein with a wide spectrum of activities against lepidopteran insects. Proc Natl Acad Sci U S A 93: 5389–5394.
46. Selvapandiyan A, Arora N, Rajagopal R, Jalali SK, Venkatesan T, et al. (2001) Toxicity analysis of N- and C-terminus-deleted vegetative insecticidal protein from Bacillus thuringiensis. Appl Environ Microbiol 67: 5855–5858.
47. Li Z, Wang CC (2008) KMP-11, a basal body and flagellar protein, is required for cell division in Trypanosoma brucei. Eukaryot Cell 7: 1941–1950.
48. Minter RM, Ferry MA, Rectenwald JE, Bahjat FR, Oberholzer A, et al. (2001) Extended lung expression and increased tissue localization of viral IL-10 with adenoviral gene therapy. Proc Natl Acad Sci U S A 98: 277–282.
49. Song F, Sun X, Wang X, Nai Y, Liu Z (2014) Early diagnosis of tuberculous meningitis by an indirect ELISA protocol based on the detection of the antigen ESAT-6 in cerebrospinal fluid. Ir J Med Sci 183: 85–88.
50. Chan V, Dreolini LF, Flintoff KA, Lloyd SJ, Mattenley AA (2002) The effect of increasing plasmid size on transformation efficiency in Escherichia coli. Journal of Experimental Microbiology and Immunology 2: 207–223.
51. Lee S, Su G, Lasserre E, Aghazadeh MA, Murai N (2012) Small high-yielding binary Ti vectors pLSU with co-directional replicons for Agrobacterium

tumefaciens-mediated transformation of higher plants. Plant Science 187: 49–58.

52. Kusaba M, Takahashi Y, Nagata T (1996) A multiple-stimuli-responsive as-1-related element of parA gene confers responsiveness to cadmium but not to copper. Plant Physiology 111: 1161–1167.

53. Murai N (2013) Review: Plant Binary Vectors of Ti Plasmid in *Agrobacterium tumefaciens* with a Broad Host-Range Replicon of pRK2, pRi, pSa or pVS1. American Journal of Plant Sciences 4: 932–939.

54. Yadav NS, Vanderleyden J, Bennett DR, Barnes WM, Chilton MD (1982) Short Direct Repeats Flank the T-DNA on a Nopaline Ti Plasmid. Proceedings of the National Academy of Sciences of the United States of America-Biological Sciences 79: 6322–6326.

55. Chatterjee A, Das NC, Raha S, Babbit R, Huang QW, et al. (2010) Production of xylanase in transgenic tobacco for industrial use in bioenergy and biofuel applications. In Vitro Cellular & Developmental Biology-Plant 46: 198–209.

56. Breyne P, Gheysen G, Jacobs A, Vanmontagu M, Depicker A (1992) Effect of T-DNA Configuration on Transgene Expression. Molecular & General Genetics 235: 389–396.

57. Hiei Y, Komari T (2006) Improved protocols for transformation of indica rice mediated by Agrobacterium tumefaciens. Plant Cell Tissue and Organ Culture 85: 271–283.

58. Komari T, Hiei Y, Saito Y, Murai N, Kumashiro T (1996) Vectors carrying two separate T-DNAs for co-transformation of higher plants mediated by Agrobacterium tumefaciens and segregation of transformants free from selection markers. Plant Journal 10: 165–174.

59. Komori T, Imayama T, Kato N, Ishida Y, Ueki J, et al. (2007) Current status of binary vectors and superbinary vectors. Plant Physiol 145: 1155–1160.

60. Vaucheret H, Fagard M (2001) Transcriptional gene silencing in plants: targets, inducers and regulators. Trends in Genetics 17: 29–35.

61. Acharya S, Ranjan R, Pattanaik S, Maiti IB, Dey N (2013) Efficient chimeric plant promoters derived from plant infecting viral promoter sequences. Planta.

62. Bhattacharyya S, Dey N, Maiti IB (2002) Analysis of cis-sequence of subgenomic transcript promoter from the Figwort mosaic virus and comparison of promoter activity with the cauliflower mosaic virus promoters in monocot and dicot cells. Virus Research 90: 47–62.

63. Bhullar S, Chakravarthy S, Advani S, Datta S, Pental D, et al. (2003) Strategies for development of functionally equivalent promoters with minimum sequence homology for transgene expression in plants: cis-elements in a novel DNA context versus domain swapping. Plant Physiology 132: 988–998.

64. Kumar D, Patro S, Ghosh J, Das A, Maiti IB, et al. (2012) Development of a salicylic acid inducible minimal sub-genomic transcript promoter from Figwort mosaic virus with enhanced root- and leaf-activity using TGACG motif rearrangement. Gene 503: 36–47.

65. Patro S, Maiti IB, Dey N (2013) Development of an efficient bi-directional promoter with tripartite enhancer employing three viral promoters. Journal of Biotechnology 163: 311–317.

66. Pattanaik S, Dey N, Bhattacharyya S, Maiti IB (2004) Isolation of full-length transcript promoter from the Strawberry vein banding virus (SVBV) and expression analysis by protoplasts transient assays and in transgenic plants. Plant Science 167: 427–438.

67. Ranjan R, Patro S, Kumari S, Kumar D, Dey N, et al. (2011) Efficient chimeric promoters derived from full-length and sub-genomic transcript promoters of Figwort mosaic virus (FMV). Journal of Biotechnology 152: 58–62.

68. Ranjan R, Patro S, Pradhan B, Kumar A, Maiti IB, et al. (2012) Development and Functional Analysis of Novel Genetic Promoters Using DNA Shuffling, Hybridization and a Combination Thereof. PLoS One 7.

69. An G, Watson BD, Chiang CC (1986) Transformation of Tobacco, Tomato, Potato, and Arabidopsis-Thaliana Using a Binary Ti Vector System. Plant Physiol 81: 301–305.

70. Ishida Y, Saito H, Ohta S, Hiei Y, Komari T, et al. (1996) High efficiency transformation of maize (Zea mays L.) mediated by Agrobacterium tumefaciens. Nat Biotechnol 14: 745–750.

71. Vasil IK (1996) Milestones in crop biotechnology—transgenic cassava and Agrobacterium-mediated transformation of maize. Nat Biotechnol 14: 702–703.

72. Ow DW, De Wet JR, Helinski DR, Howell SH, Wood KV, et al. (1986) Transient and stable expression of the firefly luciferase gene in plant cells and transgenic plants. Science 234: 856–859.

73. Pang SZ, DeBoer DL, Wan Y, Ye G, Layton JG, et al. (1996) An improved green fluorescent protein gene as a vital marker in plants. Plant Physiol 112: 893–900.

74. Fluhr R, Chua NH (1986) Developmental Regulation of 2 Genes Encoding Ribulose-Bisphosphate Carboxylase Small Subunit in Pea and Transgenic Petunia Plants - Phytochrome Response and Blue-Light Induction. Proc Natl Acad Sci U S A 83: 2358–2362.

75. Fluhr R, Kuhlemeier C, Nagy F, Chua NH (1986) Organ-Specific and Light-Induced Expression of Plant Genes. Science 232: 1106–1112.

76. Fluhr R, Moses P, Morelli G, Coruzzi G, Chua NH (1986) Expression Dynamics of the Pea Rbcs Multigene Family and Organ Distribution of the Transcripts. Embo Journal 5: 2063–2071.

77. Koncz C, Schell J (1986) The Promoter of Tl-DNA Gene 5 Controls the Tissue-Specific Expression of Chimeric Genes Carried by a Novel Type of Agrobacterium Binary Vector. Molecular & General Genetics 204: 383–396.

78. Pansegrau W, Lanka E, Barth PT, Figurski DH, Guiney DG, et al. (1994) Complete Nucleotide-Sequence of Birmingham Incp-Alpha Plasmids - Compilation and Comparative-Analysis. J Mol Biol 239: 623–663.

79. Lemos ML, Crosa JH (1992) Highly Preferred Site of Insertion of Tn7 into the Chromosome of Vibrio-Anguillarum. Plasmid 27: 161–163.

80. Kroumova AB, Shepherd RW, Wagner GJ (2007) Impacts of T-Phylloplanin gene knockdown and of Helianthus and Datura phylloplanins on Peronospora tabacina spore germination and disease potential. Plant Physiol 144: 1843–1851.

81. Shepherd RW, Bass WT, Houtz RL, Wagner GJ (2005) Phylloplanins of tobacco are defensive proteins deployed on aerial surfaces by short glandular trichomes. Plant Cell 17: 1851–1861.

82. Tolson DL, Jardim A, Schnur LF, Stebeck C, Tuckey C, et al. (1994) The kinetoplastid membrane protein 11 of Leishmania donovani and African trypanosomes is a potent stimulator of T-lymphocyte proliferation. Infect Immun 62: 4893–4899.

83. Filippi CM, von Herrath MG (2008) IL-10 and the resolution of infections. J Pathol 214: 224–230.

84. Sabat R, Grutz G, Warszawska K, Kirsch S, Witte E, et al. (2010) Biology of interleukin-10. Cytokine Growth Factor Rev 21: 331–344.

85. Stober CB, Lange UG, Roberts MT, Alcami A, Blackwell JM (2005) IL-10 from regulatory T cells determines vaccine efficacy in murine Leishmania major infection. J Immunol 175: 2517–2524.

86. Villalta SA, Rinaldi C, Deng B, Liu G, Fedor B, et al. (2011) Interleukin-10 reduces the pathology of mdx muscular dystrophy by deactivating M1 macrophages and modulating macrophage phenotype. Hum Mol Genet 20: 790–805.

87. Gelderblom H, Schmidt J, Londono D, Bai Y, Quandt J, et al. (2007) Role of interleukin 10 during persistent infection with the relapsing fever Spirochete Borrelia turicatae. Am J Pathol 170: 251–262.

4

Expression of a Vacuole-Localized BURP-Domain Protein from Soybean (SALI3-2) Enhances Tolerance to Cadmium and Copper Stresses

Yulin Tang[1]*, **Yan Cao**[2], **Jianbin Qiu**[2], **Zhan Gao**[1], **Zhonghua Ou**[1], **Yajing Wang**[1], **Yizhi Zheng**[1]*

1 Shenzhen Key Laboratory of Microbial and Gene Engineering, College of Life Sciences, Shenzhen University, Shenzhen, Guangdong, People's Republic of China, **2** The Key Laboratory for Marine Bioresource and Eco-environmental Science, College of Life Sciences, Shenzhen University, Shenzhen, Guangdong, People's Republic of China

Abstract

The plant-specific BURP family proteins play diverse roles in plant development and stress responses, but the function mechanism of these proteins is still poorly understood. Proteins in this family are characterized by a highly conserved BURP domain with four conserved Cys-His repeats and two other Cys, indicating that these proteins potentially interacts with metal ions. In this paper, an immobilized metal affinity chromatography (IMAC) assay showed that the soybean BURP protein SALI3-2 could bind soft transition metal ions (Cd^{2+}, Co^{2+}, Ni^{2+}, Zn^{2+} and Cu^{2+}) but not hard metal ions (Ca^{2+} and Mg^{2+}) *in vitro*. A subcellular localization analysis by confocal laser scanning microscopy revealed that the SALI3-2-GFP fusion protein was localized to the vacuoles. Physiological indexes assay showed that *Sali*3-2-transgenic *Arabidopsis thaliana* seedlings were more tolerant to Cu^{2+} or Cd^{2+} stresses than the wild type. An inductively coupled plasma optical emission spectrometry (ICP-OES) analysis illustrated that, compared to the wild type seedlings the *Sali*3-2-transgenic seedlings accumulated more cadmium or copper in the roots but less in the upper ground tissues when the seedlings were exposed to excessive $CuCl_2$ or $CdCl_2$ stress. Therefore, our findings suggest that the SALI3-2 protein may confer cadmium (Cd^{2+}) and copper (Cu^{2+}) tolerance to plants by helping plants to sequester Cd^{2+} or Cu^{2+} in the root and reduce the amount of heavy metals transported to the shoots.

Editor: Abidur Rahman, Iwate University, Japan

Funding: This work was supported by the National Natural Science Foundation of China (30770184). The funders had no role in study design, data collection and analysis, decision to publish, or preparation of the manuscript.

Competing Interests: The authors have declared that no competing interests exist.

* E-mail: yltang@szu.edu.cn (YT); yzzheng@szu.edu.cn (YZ)

Introduction

The BURP domain was defined by Hattori *et al.* following those proteins in which the conserved domain was first identified: the BNM2, USPs, RD22 and PG1beta [1]. The BURP proteins are characterized by the BURP domain in the C terminus and can be classified into four subfamilies: B, U, R and P [2,3]. These proteins are plant kingdom specific-proteins. A number of BURP genes have been identified. In addition to those genes that were sporadically isolated in special tissues, developmental stages or conditions [4,5,6,7,8], 5 members in *Arabidopsis thaliana* [9], 17 in *Oryza sativa* L. [10], 23 in *Glycine max* [11], 15 in *Zea mays*, 11 in *Sorghum vulgare* [12] and 18 in *Populus trichocarpa* [13] were identified genome-widely. The expression of these genes exhibits different temporal and spatial profiles [7,10], and some of them can be regulated by various stress treatments [10,11,14], suggesting that BURP genes might play diverse roles in plant development and stress responses.

A few studies have explored the function of BURP proteins. Some BURP proteins function in particular intracellular compartments. For example, AtUSP in *Arabidopsis thaliana* [9] and BNM2 in *Brassica napus* [15], are most likely localized in cellular compartments such as the Golgi cisternae, dense vesicles, prevacuolar vesicles and protein storage vacuoles. These proteins are related to seed protein synthesis and storage and seed development. Some other BURP proteins are likely localized in the cell wall or apoplasts and interact with cell wall components. For instance, OsBURP16 in rice is localized in the cell wall; its overexpression causes pectin degradation and affects cell wall integrity, resulting in the decreased tolerance to abiotic stresses [16]. The cotton AtRD22-Like 1 protein GhRDL1 interacts with an α- expansin GhEXPA1 in the cell wall and functions in cell wall extension [17]. The soybean GmRD22 interacts with a cell wall peroxidase in apoplasts. This protein affects cell wall integrity and alleviates salinity and osmotic stress in plants [18]. Thus, different BURP proteins present diverse subcellular localizations and play different functional roles. Nevertheless, the molecular functions of most BURP proteins, especially that of their BURP domain, remain unclear.

BURP domain proteins are characterized by their C-terminal BURP domain which possesses conserved features, including four repeated cysteine-histidine motifs that were arranged as: $CH-X_{10}$-$CH-X_{25-27}$-$CH-X_{23-24}$-CH (X can be any amino acid) [1,3] and additional two cysteine residues. In view of these characteristics, we assume that, similar to Cys-rich proteins [19], these proteins might interact with some metal ions with high affinity due to their cysteine residues. SALI3-2 is a member in the U subfamily of the BURP family in soybean. Our previous studies demonstrated that Cu^{2+} could bind to SALI3-2 and alter the conformation of SALI3-

2. In addition, the *Sali3-2*-transgenic Arabidopsis seedlings were more tolerant to excessive Cu^{2+} stress [20], indicating that the SALI3-2 is associated with heavy metal tolerance in plants. However, new questions should be asked and are as following: can SALI3-2 bind other soft transition metal ions besides Cu^{2+}? Can the expression of *Sali3-2* in transgenic plants confer to them tolerance to other heavy metal ions? In this study, we confirmed the *in vitro* interaction of SALI3-2 with either Cu^{2+} and some other soft metal ions (Cd^{2+}, Co^{2+}, Ni^{2+} and Zn^{2+}) using immobilized metal ion affinity chromatography (IMAC). We further analyzed the subcellular localization of SALI3-2 and the tolerance of *Sali3-2*-transgenic Arabidopsis to the stresses of the typical soft metal ions (Cd^{2+} and Cu^{2+}). The mechanisms of SALI3-2 functioning in heavy metal tolerance are discussed.

Materials and Methods

Construction of pET28a/*Sali3-2c* for SALI3-2 Expression in *E. coli*

The region encoding the SALI3-2 fragment without a signal peptid (named SALI3-2C) was amplified by polymerase chain reaction (PCR) with the template pET28a-*Sali3-2* carrying the full-length *Sali3-2* gene and oligonucleotide primers containing the NdeI and a XhoI sites: sense primer 5′- gctcttcatatggagagccatgtc-catgc -3′ and antisense primer 5′- tgcgctcgagttaaacaacaacgttagtct-gatag -3′ (restriction sites underlined). The resulting PCR product as well as the plasmid pET28a were digested with NdeI and XhoI and further ligated. The constructed vector pET28a-*Sali3-2c* was transformed into *E. coli* cells (strain BL21 Star) for protein expression. The encoded protein SALI3-2C was fused with a 6×His tag at its N-terminus (6×His-SALI3-2C) and can be cleaved off by the protease thrombin.

Expression and Purification of the SALI3-2C Protein

E. coli cells transformed with pET28a-*Sali3-2c* were cultured at 37°C and 180 rpm until the OD_{600} reached 0.6~0.8. The 6×His-SALI3-2C protein was then induced by the addition of 0.5 mM isopropyl-β-D thiogalactopyranoside (IPTG) into the cultures. *E. coli* cells were cultured for additional 4 h and then collected by centrifugation at 10,000 rpm for 15 min. The harvested cells were then re-suspended and disrupted in Tris–HCl buffer (50 mM Tris–HCl, 1 mM EDTA and 100 mM NaCl, pH 8.0) by 30 min of sonication, further centrifuged at 10,000 rpm for 15 min. The pellet was re-suspended again in 5 mL of Tris–Urea buffer (0.1 mM Tris–HCl, 6 mM urea, 1 mM PMSF and 1 mM EDTA, pH 8.0) and dissolved at room temperature for at least 1 h. The resulting suspension was centrifuged at 10,000 rpm for 15 min. The supernatant was harvested and passed through a Ni-sephorose column (AKTA, Amersham Biosciences, Tokyo, Japan) and washed with a buffer containing decreasing concentrations of urea (20 mM Tris–HCl, 1 mM EDTA and 6~2 mM urea, pH 8.0) for protein renaturation. Finally, the column was washed with washing buffer (200 mM NaCl, 50 mM Tris–HCl and 100 mM imidazole, pH 8.0) and the 6×His-SALI3-2C protein was eluted with buffer containing 500 mM imidazole. The eluted 6×His-SALI3-2C solution was passed through a desalinization column (AKTA, Amersham Biosciences, Tokyo, Japan) to exclude salt and imidazole.

SALI3-2C protein was derived from the cleavage of the 6×His-SALI3-2C protein by protease thrombin (Sigma). Thrombin was mixed with the 6×His-SALI3-2C protein at a ratio of 3 U/mg and incubated at 4°C for overnight. Thereafter, the mixture was passed through a Ni-sepharose column. Thrombin did not bind to the Ni-sepharose and flowed through the column with the balance

buffer; SALI3-2C was then recovered by passing through a balance buffer with 60 mM imidazole, and the cleaved 6×His tag was bound tightly to the Ni-sepharose which would be eluted by 500 mM imidazole. The protein concentration was determined by the absorbance at 280 nm.

Metal-chelating Affinity Chromatography

The interactions between SALI3-2 and the metal ions were analyzed by IMAC using HiTrap Chelating HP (5 mL, Amersham Pharmacia Biotech, Tokyo, Japan) according to Hara *et al.* [21]. The columns were charged by applying 5 mL 100 mM $MgCl_2$, $CaCl_2$, $CdCl_2$, $CoCl_2$, $NiCl_2$, $ZnCl_2$ or $CuCl_2$. After washing out the excess metal with deionized water, the column was equilibrated with 20 µM Tris–HCl buffer at pH 8.0 containing 500 mM NaCl (equilibrium buffer). A HiTrap Chelating HP column without metal charging was used as a control. A 5 mL sample of SALI3-2C at approximately 2 µM was applied to the column charged with or without metals. The unbound protein was run and washed out with the equilibrium buffer. The bound protein was eluted by the addition of 5 mL 250 mM EDTA. A total of 12 µL of each fraction was subjected to SDS–PAGE analysis.

Subcellular Localization Detection of SALI3-2 in Tobacco Cells

To investigate SALI3-2 localization, the *Sali3-2* gene was inserted into the pCAMBIA1302 vector at the NcoI and SpeI sites to construct the fusion gene *Sali3-2*-m*Gfp* under the control of the CaMV 35S promoter. This construct and pCAMBIA1302 (as control) were introduced into tobacco cells BY2 (kindly provided by Prof. Liwen Jiang of the Department of Biology and Molecular Biotechnology Program, Chinese University of Hong Kong) mediated by *Agrobacterium* LBA 4404 [22]. Using the fluorescent dye FM4-64 as the membrane marker, the expression of SALI3-2-mGFP and mGFP was analyzed by confocal laser scanning microscopy using Olympus FV1000 (Olympus, Tokyo, Japan). The filter sets that were used for mGFP were excitation 488 nm and emission 510 nm and for FM4-64 were excitation 543 nm and emission 572 nm. To visualize the vacuoles, the cells were observed under bright field after a 5 min immersion in 33 µg/mL neutral red solution.

Stable Transformation of *Arabidopsis thaliana* (Ecotype Columbia) Using a 35S-*Sali3-2* Cassette

A cauliflower mosaic virus (CaMV) *35S-Sali3-2* cassette was cloned into the pCAMBIA-1300 vector and transformed into *Arabidopsis thaliana* ecotype Columbia, *via* floral dip [23] mediated by *Agrobacterium tumefaciens* (LBA 4404). The T0 seeds were germinated on 1/2 MS (Murashige–Skoog) medium-containing agar plates under hygromycin selection to obtain T1 resistance plants. The homozygous transgenic lines were further selected according to a segregation ratio of 3:1 of hygromycin resistance of the T2 generation and homozygous in the T3 generation. The expression of *Sali3-2* in homozygous transgenic lines was further confirmed by RT-PCR using total RNA extracted from the young leaves of 3-week-old plants. The primers used in these reactions for *Sali3-2* were 5′-cacacaagcttcaatggaatttcgatgctca-3′ and 5′-tgcgctcgagttaaacaacaacgttagtctgatag-3′ and for tubulin were 5′-ccgatgttgctgtcctcttgg-3′ and 5′-catcaccacggtacatcag-3′.

Seedling Growth Analysis of *Arabidopsis thaliana* after Heavy Metal Exposure

The seeds of the T3 homozygous transgenic lines and wild-type plants were germinated on 1/2 MS medium-containing agar plates (control), or plates that were supplemented with 50 μM or 75 μM $CdCl_2$ or 75 μM or 100 μM $CuCl_2$. The plates were incubated at 21°C under a light intensity of 50 μE m^{-2} s^{-1} at 8 h light/16 h dark. After cultivation for 3 d, the seeds germination ratio was calculated, and after 7 d, the phenotypes of the seedlings, especially the root growth, were recorded.

Measurement of Metal Contents

Two-week-old *Arabidopsis thaliana* seedlings grown on 1/2 MS agar plates were transferred into 1/2 MS solution medium containing 75 μM $CdCl_2$ or 100 μM $CuCl_2$ and grown for an additional 2 days. The shoots and roots of the plants were collected, carefully washed with deionized water and dried. The methods for measuring the contents of copper and cadmium followed those mentioned in Narukawa et al.'s paper [24]. Briefly, the dried samples were digested with 15.3 M HNO_3 in a microwave digestion system ETHOS One (Milestone Inc., Milan, Italy). The digested samples were then diluted with 0.5 N HNO3 and analyzed using an Inductively Coupled Plasma Optical Emission Spectrometer (ICP-OES, OPTIMA 7000DV, Perkin Elmer, NY, USA).

Transcript Measurement

To determine the transcriptional levels, quantitative RT-PCR (qRT-PCR) was performed. The water-soaked seeds of *Glycine max* L. (*var.* Bainong 6#, kindly provided by the Institute of Agriculture Science in Baicheng City, Jilin Province, P. R. China) were germinated on wet filter paper for 4 days, after which the seedlings were transferred into 1/2 Hoagland's nutrient solution for an additional 6 days of growth. The roots, stem and leaves were collected separately for total RNA isolation using TRIzol Reagent (Invitrogen, Carlsbad, CA, USA). cDNA was synthesized from 2 μg RNA using PrimeScript RT reagents Kit (TaKaRa, Dalian, China). qRT-PCR was performed on Real-Time PCR Systems ABI7300 (ABI) with SYBR Premix Ex Taq (TaKaRa, Shiga, Japan) according to the manufacturer's instructions. The gene expression levels were normalized to those of *Tubulin*. The primers of *Tubulin* and *Sali3-2* were the same as those used in RT-PCR described above. Three biological replications were repeated.

Statistical Analysis

The statistical tests were performed using the data analysis program SPSS. The significant differences between the lines were analyzed using a one-way ANOVA Tukey's HSD test.

Results

SALI3-2 is able to Bind Several Types of Soft Metal Ions

A search of the sequence databases with SALI3-2 revealed that SALI3-2 contains 276 amino acids and exhibits a modular structure of a BURP protein consisting of a putative signal peptide of 19 amino acids [25], a variable region of 56 amino acids, and a BURP domain of 188 amino acids followed by a C-terminal tail of 13 amino acids (Figure 1).

For those proteins possessing a signal peptide, the signal peptide is usually cleaved off during protein biogenesis to produce the mature protein. Therefore, to obtain the SALI3-2 protein for function analysis, the cDNA fragment *Sali3-2c* encoding the SALI3-2 fragment without the signal peptide (21–276 aa, named

SALI3-2C, Figure 1) was subcloned into pET28a and transformed into *E. coli* BL21 for protein expression. The isolated fusion protein SALI3-2C with a 6×His tag was re-natured and purified on a Ni-sepharose column with a buffer containing decreasing concentrations of urea. Thrombin was then added to the purified 6×His-SALI3-2C protein to cleave off the N-terminal 6×His peptide (Figure S1). The SALI3-2C protein was then purified and maintained at 20°C for later use.

The highly conserved Cys and Cys-His motifs in the BURP-domain in BURP family proteins indicate that these proteins potentially bind metal ions. To directly address whether SALI3-2 could chelate metal ions, the IMAC was exploited to detect whether SALI3-2C could bind metal ions under high ionic strength (500 mM NaCl). Chromatography was performed at pH 8.0 for each metal ions. The results indicate SALI3-2C was washed off in the equilibrated buffer (W) in the columns without metal ion binding (negative control) or with immobilized Ca^{2+} or Mg^{2+}. In the columns immobilizing Cd^{2+}, Co^{2+}, Ni^{2+}, Zn^{2+} and Cu^{2+}, SALI3-2C was retained in the column and further eluted by the elution solution (E) (Figure 2). These results indicated that SALI3-2 might bind to the soft metal ions Cd^{2+}, Cu^{2+}, Co^{2+}, Ni^{2+} and Zn^{2+} but not to the hard metal ions Ca^{2+} and Mg^{2+}.

SALI3-2-GFP is Localized in Vacuoles

To investigate the subcellular localization of SALI3-2, either the fusion gene *Sali3-2-mGfp* or the *mGfp* in pCAMBIA1302 was introduced into BY2 cells to express the SALI3-2-mGFP fusion protein or mGFP. Using the fluorescent dye FM4-64 to define the cell membrane, the subcellular localization of SALI3-2-mGFP and mGFP in transgenic BY2 cells was analyzed. The green fluorescence signal from mGFP was observed in the cytosol, nuclear matrix and/or cell membrane. And the green fluorescence of the SALI3-2-mGFP fusion protein was present in the cellular compartments that were neutral red-stained, which were mainly vacuoles (Figure 3; additional images are shown in Figure S2). These results demonstrate that SALI3-2 was targeted to cell vacuoles.

Expression of *Sali3-2* Enhances Cd^{2+} and Cu^{2+} Tolerance in *Arabidopsis thaliana*

The binding property of SALI3-2 with soft transition metal ions suggests that SALI3-2 might be functionally regulated by these heavy metal ions or be involved in the plant response to them. To determine whether SALI3-2 contributes to heavy metal resistance in plants, we transformed the *Sali3-2* gene into *Arabidopsis thaliana*. Five homozygous transgenic lines were selected according to the segregation ratio of hygromycin resistance in the T2 and T3 generations and further confirmed by RT-PCR using total RNA extracted from the seedling leaves and two *Sali3-2* -specific primers. The specific products at 830 bp were produced in the *Sali3-2*-transgenic plants but not in the plants transformed with the empty vector pCAMBIA1300 and the wild type Columbia (two lines were shown in Figure S3). No apparent differences between the phenotypes of each line grown in soil were shown (Figure S4).

The T3 seeds of the homozygous transgenic lines were further used to analyze cadmium and copper tolerance. The seeds were germinated on 1/2 MS agar plates in the absence or presence of different concentrations of $CdCl_2$ and $CuCl_2$. The germination ratio of neither the transgenic lines nor the wild type was apparently affected by any of the treatments (Figure S5), but a significant reduction of growth of both the wild-type and transgenic seedlings was caused in the presence of $CdCl_2$ or $CuCl_2$ at the tested concentration. However, compared to the wild type, the transgenic plants displayed less root growth inhibition

A **MEFRCSVISFTILFSLALA**GESHVHASLPEEDYWEAVWPNTPIPTAL
RDVLKPLPAGVEIDQLPKQIDDTQYPKTFFYKEDLHPGKTMKVQFT
KRPYAQPYGVYTWLTDIKDTSKEGYSFEEI*C*IKKEAFEGEEKF*C*AK
SLGTVIGFAISKLGKNIQVLSSSFVNKQEQYTVEGVQNLGDKAVM*C*
*H*GLNFRTAVFY*CH*KVRETTAFVVPLVAGDGTKTQALAV*CH*SDTSG
MNHHILHELMGVDPGTNPV*CH*FLGSKAILWVPNISMDTAYQTNVVV

B

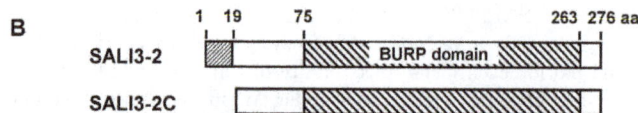

Figure 1. Sequences and structure of SALI3-2 [Glycine max]. A. Sequences of SALI3-2. The accession number is AAB66369. The regions in red, blue and black letters are the putative signal peptide, variable region and BURP domain, respectively. B. Schematic diagram of the structure of SALI3-2 and SALI3-2C.

(Figure 4). Therefore, the *Sali*3-2-expressing seedlings were more tolerant to cadmium and copper exposure.

*Sali*3-2-transgenic *Arabidopsis thaliana* Plants Delivers Less Cadmium and Copper to the Shoot after Corresponding Heavy Metal Ions Exposure

We measured the Cd and Cu contents of the wild type and *Sali*3-2-transgenic plants. Compared to the wild type, the transgenic plants contained less Cd or Cu in the shoots, and more in the roots after treatment with CdCl$_2$ or CuCl$_2$ (Figure 5). The results indicate that the transgenic plants deliver less Cd or Cu to the shoot than the wild type plants do when they experienced Cd^{2+} or Cu^{2+} stresses.

*Sali*3-2 is Expressed at Higher Level in the Roots than in the Upper Ground Tissues in the Native Soybean

Quantitative RT-PCR (qRT-PCR) assays showed that *Sali*3-2 was expressed in the roots at a high level and in the stems and leaves at a low level (Figure 6), indicating that *Sali*3-2 may play a special role in the root in native soybean.

Discussion

In recent years, BURP proteins have attracted attention due to their diverse functions in plant development and stress responses. Here, we cloned the BURP gene *Sali*3-2 from soybean and studied the biochemical functions and mechanism of the SALI3-2 protein in heavy metal resistance. The results that we obtained suggest a role for SALI3-2 in heavy metal conjugation and resistance. First, SALI3-2 can interact with heavy metal ions (Cu^{2+}, Cd^{2+}, *etc.*) *in vitro*. Second, *Sali*3-2-transgenic *Arabidopsis thaliana* seedlings

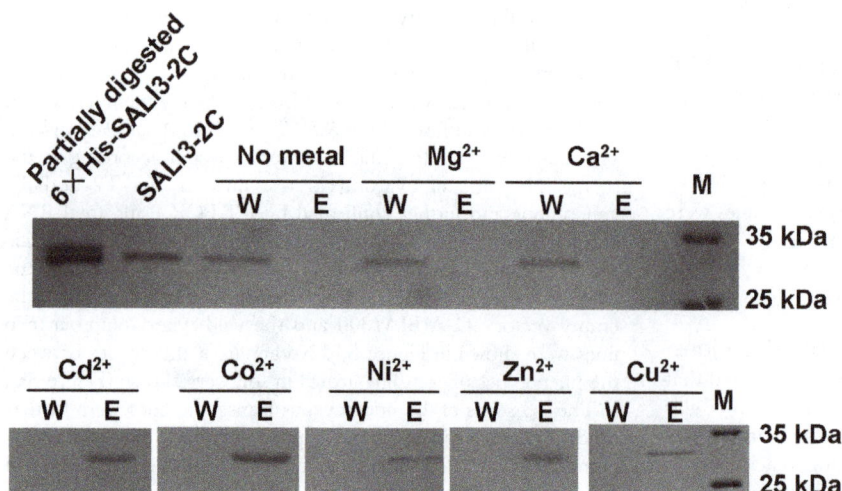

Figure 2. Metal-SALI3-2C binding assay using immobilized metal ion affinity chromatography (IMAC). The columns were charged with no metal, Mg^{2+}, Ca^{2+}, Cd^{2+}, Co^{2+}, Ni^{2+}, Zn^{2+} or Cu^{2+}. The protein was loaded onto a column that was equilibrated with 20 µM Tris–HCl buffer at pH 8.0 containing 500 mM NaCl. The unbound protein on the column was washed out with the equilibrated buffer (W). Then, the bound protein was eluted with 250 mM EDTA (E). The samples were collected, subjected to SDS-PAGE and stained with Coomassie Brilliant Blue. For the gel electrophoresis analysis, the protein marker (M), the partially digested 6×His-SALI3-2C protein and the purified SALI3-2C were also loaded.

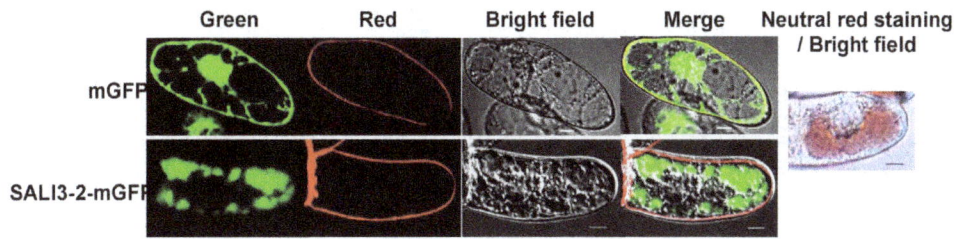

Figure 3. Subcellular localization of SALI3-2. BY2 cells that were expressed with SALI3-2-mGFP or mGFP were observed under confocal microscopy. Confocal fluorescence (green for mGFP, red for the membrane marked by fluorescent dyes FM4-64, and the merged images) and bright-field transmission are shown. As the control, the vacuoles are shown by neutral red staining. Scale bars = 10 μm.

were more tolerant to Cu^{2+} or Cd^{2+} exposure than the wild type were.

Eukaryotic organisms have developed multiple mechanisms for tolerance to heavy metal ion stress during their evolution. One mechanism is to reduce the amount of ions that are transported into cells through the cell wall or by cell surface binding. For example, yeast could prevent copper-induced toxicity by precipitating copper ions on the cell membrane by forming copper

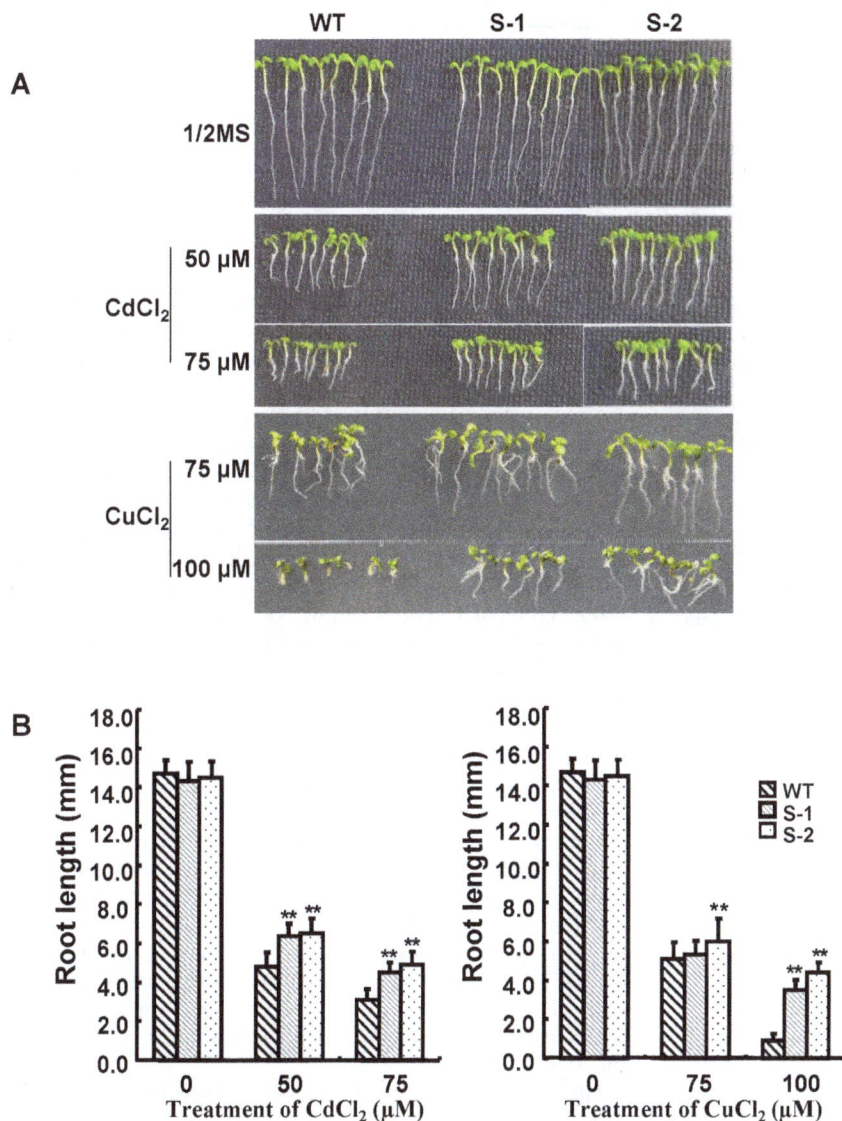

Figure 4. Growth of wild type and *Sali*3-2-transgenic seedlings cultured on Cd^{2+}- or Cu^{2+}-containing medium. The seeds were sown on 1/2 MS agar plates adding different concentrations of CdCl$_2$ or CuCl$_2$. Seven days after sowing, the photographs were taken (A) and the root lengths were measured (B). S-1 and S-2: two different *Sali*3-2-transgenic lines. n = 30~40. The error bars represent SD. **P<0.01 and *P<0.05.

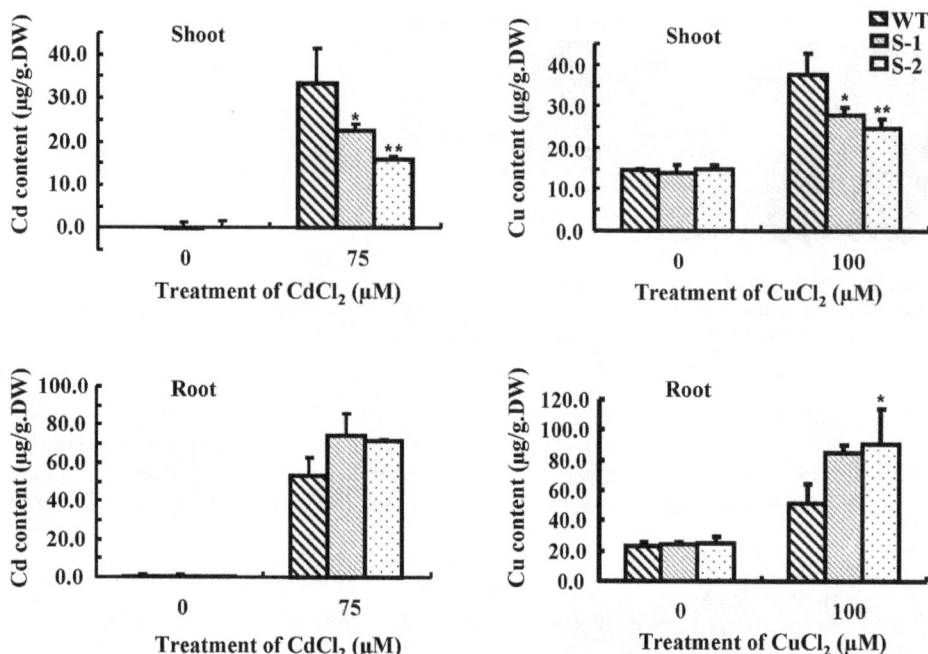

Figure 5. Contents of Cd and Cu in the shoots and the roots of *Sali*3-2-transgenic plants. Wild-type and *Sali*3-2-transgenic seedlings were grown for 2 weeks on 1/2 MS agar plates and were then transferred into 1/2 MS solution medium containing 75 μM $CdCl_2$ or 100 μM $CuCl_2$ for an additional 2 d of growth. The shoots and the roots were harvested to measure the Cd or Cu contents using ICP-OES. A representative from 3 independent experiments is shown. The error bars represent SD. **$P<0.01$ and *$P<0.05$.

sulfide (CuS) mineral lattices [26]. Another mechanism is to maintain ion homeostasis. Ion transporters, such as the copper-transporting P-type ATPases [27,28,29] and the ABC transporter [30], can transport metal ions into certain compartments. Third, intracellular heavy metal ions can be detoxified by organic acids or heavy metal-binding peptides [31]. Metallothioneins (MTs) and phytochelatins (PCs) are two major types of heavy metal-binding peptides [32,33,34,35,36]. MTs have repeated sequence motifs of either Cys-x-Cys or Cys-xx-Cys [37], and PCs have the general structure of (γ-Glu- Cys)n-Gly (n = 2~11) [38,39]. These Cys-rich proteins potentially interact with metals such as Cd^{2+}, Cu^{2+}, or Zn^{2+} of high affinity and with several binding sites *via* their cysteine residues [19]. These proteins provide a transient storage form for the ions and they may play a primary role in heavy metal ion sequestration in vacuoles by binding metal ions [40], while the

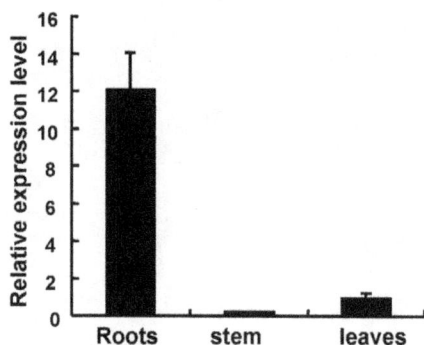

Figure 6. Expression level of *Sali*3-2 in different tissues of soybean seedlings. The expression level in the leaves was set to 1, and the *Tubulin* gene was used as the internal control. The error bars represent SD.

tonoplast transporters ABCCs of the ABC transporter family most likely play an important assisting role in the vacuolar transportation of the peptide-binding heavy metal ions or chelator-ions complex [30].

In general, regarding to the binding of metal ions to proteins, soft metal ions can form strong bonds with soft Lewis bases such as the thiols of cysteine residues and the imidazolium nitrogens of histidine residues. In contrast, hard metal ions bind strongly to the hard Lewis bases, such as the carboxyl oxygens of glutamate or aspartate residues [41]. SALI3-2 is a BURP family member, in its C-termini possessing a conserved BURP domain with four conserved Cys-His repeats and two other Cys residues [1,3]. We previously studied the interaction of Cu^{2+} with SALI3-2 and found that Cu^{2+} binds to SALI3-2 with a relatively high affinity and alters the conformation of SALI3-2 [20]. In this study, we further confirm that SALI3-2 not only interacts with Cu^{2+} but also interacts broadly with other soft metal ions, such as Cd^{2+}, Co^{2+}, Zn^{2+} and Ni^{2+}. Furthermore, SALI3-2 is localized in vacuoles. Therefore, SALI3-2 might function similar to MTs and PCs, playing roles in sequestrating heavy metal ions in vacuoles by binding the ions *via* Cys and His residues and alleviating the toxicity to cells due to these ions, possibly explaining the roots of *Sali*3-2-transgenic seedlings displaying less inhibition than that of the wild type under cadmium and copper stress. Consequently, in the *Sali*3-2-transgenic plants the enhanced retention of metals in root vacuoles may lead plants to accumulate more metal ions in root and reduce the amount of Cd^{2+} or Cu^{2+} that is delivered to shoot [42].

However, the number of binding sites of SALI3-2 with Cu^{2+} is only 1.6 [20]. With such a low bound amount, the mechanism of SALI3-2 in heavy metal ion tolerance would not be similar to that of MTs or PCs in providing a transient storage form for the ions. Another possibility is that SALI3-2 promotes the transport of Cu^{2+}, Cd^{2+}, *etc.* into vacuoles by regulating the action of ion transporters in the vacuole membrane (*i.e.*, the ABCC type proteins [30]) and by maintaining ion homeostasis. To verify this hypothesis, further studies should be performed, for example, identifying the interaction partners of SALI3-2, and analyzing the accumulated amount of heavy metal in vacuoles.

Our qRT-PCR assays showed that *Sali*3-2 was expressed at high levels in roots but at low levels in the upper ground tissues in native soybean. This result is consistent with that reported by Granger *et al.* based on expressed sequence tags (ESTs) analyses [2]. In that paper, the number of ESTs of *Sali*3-2 is also small in pod. These findings also suggest a function of SALI3-2 in sequestering heavy metal ions in the native soybean root. Our results support the exploration of the use of *Sali*3-2 for the bioaccumulation of metal ions in the roots to protect aerial parts from excess metal contamination.

In summary, we found that SALI3-2 is localized in vacuoles and can bind heavy metals (Cd^{2+}, Cu^{2+}, *etc.*) *in vitro*. The heterologous expression of *Sali*3-2 enables plants to sequester Cd^{2+} or Cu^{2+} in the roots and reduce the amount of heavy metals that are transported to the shoot, further enhance the tolerance to excessive copper and cadmium stresses. Therefore, our findings suggest that SALI3-2 may play a role in heavy metal conjugation and resistance in this overexpression system. A possible function of SALI3-2 that is different from those of other BURP proteins is presented.

Supporting Information

Figure S1   SDS-PAGE of protein extracts. M: marker; 1. Purified SALI3-2C; 2 and 3. Purified 6×His-SALI3-2C; 4. 6×His-SALI3-2C partially digested by thrombin for 2 h.

Figure S2   Subcellular localization of SALI3-2. BY2 cells that were expressed with SALI3-2-mGFP or mGFP were observed under confocal microscopy. Confocal green fluorescence and bright-field transmission are shown. Scale bars = 10 μm.

Figure S3   Expression of *Sali*3-2 in transgenic plants as detected by RT-PCR. Total RNA was extracted from the young leaves of 3-week-old plants. RT-PCR was performed using *Tubulin* as the control. WT: wild-type plants; 00: plants that were transformed with the empty vector pCAMBIA1300; S-1 and S-2: two different *Sali*3-2-transgenic lines.

Figure S4   Phenotype of transgenic plants grown in soil. DAS: days after sown.

Figure S5   Seeds germination ratio of different lines on Cd^{2+}- or Cu^{2+}-containing medium. More than 60 seeds of each line were tested in each independent treatment. A representative from 3 independent experiments is shown. The error bars represent SD.

Acknowledgments

We thank Prof. Liwen Jiang at the Chinese University of Hong Kong for kindly providing tobacco BY2 cells. We also acknowledge the useful discussions from our colleague Prof. Beixin Mo, Prof. Qiong Liu and Dr. Guobao Liu and the technical help from Prof. Xu Deng in the measurement of metal contents using ICP-OES.

Author Contributions

Conceived and designed the experiments: YT YZ. Performed the experiments: YC ZG ZO YW JQ. Analyzed the data: YT YC ZG ZO. Contributed reagents/materials/analysis tools: YC ZG ZO YW. Wrote the paper: YT YZ.

References

1. Hattori J, Boutilier KA, van Lookeren CM, Miki BL (1998) A conserved BURP domain defines a novel group of plant proteins with unusual primary structures. Mol Gen Genet 259: 424–428.
2. Granger C, Coryell V, Khanna A, Keim P, Vodkin L, et al. (2002) Identification, structure, and differential expression of members of a BURP domain containing protein family in soybean. Genome 45: 693–701.
3. Tang Y, Wang Y, Cai X, Zheng Y (2009) A BURP-domain family of plant proteins. Prog Nat Sci 19: 241–247.
4. Malik MR, Wang F, Dirpaul JM, Zhou N, Hammerlindl J, et al. (2008) Isolation of an embryogenic line from non-embryogenic Brassica napus cv. Westar through microspore embryogenesis. J Exp Bot 59: 2857–2873.
5. Fernandez L, Torregrosa L, Terrier N, Sreekantan L, Grimplet J, et al. (2007) Identification of genes associated with flesh morphogenesis during grapevine fruit development. Plant Mol Biol 63: 307–323.
6. Wang A, Xia Q, Xie W, Datla R, Selvaraj G (2003) The classical Ubisch bodies carry a sporophytically produced structural protein (RAFTIN) that is essential for pollen development. Proc Natl Acad Sci U S A 100: 14487–14492.
7. Batchelor AK, Boutilier K, Miller SS, Hattori J, Bowman LA, et al. (2002) SCB1, a BURP-domain protein gene, from developing soybean seed coats. Planta 215: 523–532.
8. Zheng L, Watson CF, DellaPenna D (1994) Differential Expression of the Two Subunits of Tomato Polygalacturonase Isoenzyme 1 in Wild-Type and rin Tomato Fruit. Plant Physiol 105: 1189–1195.
9. Van Son L, Tiedemann J, Rutten T, Hillmer S, Hinz G, et al. (2009) The BURP domain protein AtUSPL1 of Arabidopsis thaliana is destined to the protein storage vacuoles and overexpression of the cognate gene distorts seed development. Plant Mol Biol 71: 319–329.
10. Ding X, Hou X, Xie K, Xiong L (2009) Genome-wide identification of BURP domain-containing genes in rice reveals a gene family with diverse structures and responses to abiotic stresses. Planta 230: 149–163.
11. Xu H, Li Y, Yan Y, Wang K, Gao Y, et al. (2010) Genome-scale identification of soybean BURP domain-containing genes and their expression under stress treatments. BMC Plant Biol 10: 197.
12. Gan D, Jiang H, Zhang J, Zhao Y, Zhu S, et al. (2011) Genome-wide analysis of BURP domain-containing genes in maize and sorghum. Mol Biol Rep 38: 4553–4563.
13. Shao Y, Wei G, Wang L, Dong Q, Zhao Y, et al. (2011) Genome-wide analysis of BURP domain-containing genes in Populus trichocarpa. J Integr Plant Biol 53: 743–755.
14. Yue Y, Zhang M, Zhang J, Tian X, Duan L, et al. (2012) Overexpression of the AtLOS5 gene increased abscisic acid level and drought tolerance in transgenic cotton. J Exp Bot 63: 3741–3748.
15. Teerawanichpan P, Xia Q, Caldwell SJ, Datla R, Selvaraj G (2009) Protein storage vacuoles of Brassica napus zygotic embryos accumulate a BURP domain protein and perturbation of its production distorts the PSV. Plant Mol Biol 71: 331–343.
16. Liu H, Ma Y, Chen N, Guo S, Liu H, et al. (2013) Overexpression of stress-inducible OsBURP16, the beta-subunit of polygalacturonase 1, decreases pectin contents and cell adhesion, and increases abiotic stress sensitivity in rice. Plant Cell Environ.
17. Xu B, Gou JY, Li FG, Shangguan XX, Zhao B, et al. (2013) A cotton BURP domain protein interacts with alpha-expansin and their co-expression promotes plant growth and fruit production. Mol Plant 6: 945–958.
18. Wang H, Zhou L, Fu Y, Cheung MY, Wong FL, et al. (2012) Expression of an apoplast-localized BURP-domain protein from soybean (GmRD22) enhances tolerance towards abiotic stress. Plant Cell Environ 35: 1932–1947.
19. Griffin BA, Adams SR, Tsien RY (1998) Specific covalent labeling of recombinant protein molecules inside live cells. Science 281: 269–272.
20. Tang Y, Gao Z, Xu H, He J, Dong Y, Zheng Y (2013) Interaction of Copper(II) Ion with SALI3-2. Chemical Journal of Chinese Universities 1: 128–134.
21. Hara M, Fujinaga M, Kuboi T (2005) Metal binding by citrus dehydrin with histidine-rich domains. J Exp Bot 56: 2695–2703.
22. Wang J, Ding Y, Wang J, Hillmer S, Miao Y, et al. (2010) EXPO, an exocyst-positive organelle distinct from multivesicular endosomes and autophagosomes, mediates cytosol to cell wall exocytosis in Arabidopsis and tobacco cells. Plant Cell 22: 4009–4030.
23. Clough SJ, Bent AF (1998) Floral dip: a simplified method for Agrobacterium-mediated transformation of Arabidopsis thaliana. Plant J 16: 735–743.
24. Narukawa T, Matsumoto E, Nishimura T, Hioki A (2014) Determination of sixteen elements and arsenic species in brown, polished and milled rice. Anal Sci 30: 245–250.

25. Petersen TN, Brunak S, von Heijne G, Nielsen H (2011) SignalP 4.0: discriminating signal peptides from transmembrane regions. Nat Methods 8: 785–786.

26. Yu W, Farrell RA, Stillman DJ, Winge DR (1996) Identification of SLF1 as a new copper homeostasis gene involved in copper sulfide mineralization in Saccharomyces cerevisiae. Mol Cell Biol 16: 2464–2472.

27. Lutsenko S, Petris MJ (2003) Function and regulation of the mammalian copper-transporting ATPases: insights from biochemical and cell biological approaches. J Membr Biol 191: 1–12.

28. Shikanai T, Muller-Moule P, Munekage Y, Niyogi KK, Pilon M (2003) PAA1, a P-type ATPase of Arabidopsis, functions in copper transport in chloroplasts. Plant Cell 15: 1333–1346.

29. Abdel-Ghany SE, Muller-Moule P, Niyogi KK, Pilon M, Shikanai T (2005) Two P-type ATPases are required for copper delivery in Arabidopsis thaliana chloroplasts. Plant Cell 17: 1233–1251.

30. Song WY, Park J, Mendoza-Cozatl DG, Suter-Grotemeyer M, Shim D, et al. (2010) Arsenic tolerance in Arabidopsis is mediated by two ABCC-type phytochelatin transporters. Proc Natl Acad Sci U S A 107: 21187–21192.

31. Memon AR, Schroder P (2009) Implications of metal accumulation mechanisms to phytoremediation. Environ Sci Pollut Res Int 16: 162–175.

32. Zhou J, Goldsbrough PB (1994) Functional homologs of fungal metallothionein genes from Arabidopsis. Plant Cell 6: 875–884.

33. Murphy A, Zhou J, Goldsbrough PB, Taiz L (1997) Purification and immunological identification of metallothioneins 1 and 2 from Arabidopsis thaliana. Plant Physiol 113: 1293–1301.

34. Cobbett CS (2000) Phytochelatins and their roles in heavy metal detoxification. Plant Physiol 123: 825–832.

35. Cobbett C, Goldsbrough P (2002) Phytochelatins and metallothioneins: roles in heavy metal detoxification and homeostasis. Annu Rev Plant Biol 53: 159–182.

36. Verbruggen N, Hermans C, Schat H (2009) Mechanisms to cope with arsenic or cadmium excess in plants. Curr Opin Plant Biol 12: 364–372.

37. Romero-Isart N, Vasak M (2002) Advances in the structure and chemistry of metallothioneins. J Inorg Biochem 88: 388–396.

38. Grill E, Winnacker EL, Zenk MH (1985) Phytochelatins: the principal heavy-metal complexing peptides of higher plants. Science 230: 674–676.

39. Zenk MH (1996) Heavy metal detoxification in higher plants–a review. Gene 179: 21–30.

40. Huang J, Zhang Y, Peng JS, Zhong C, Yi HY, et al. (2012) Fission yeast HMT1 lowers seed cadmium through phytochelatin-dependent vacuolar sequestration in Arabidopsis. Plant Physiol 158: 1779–1788.

41. Rensing C, Ghosh M, Rosen BP (1999) Families of soft-metal-ion-transporting ATPases. J Bacteriol 181: 5891–5897.

42. Mendoza-Cozatl DG, Jobe TO, Hauser F, Schroeder JI (2011) Long-distance transport, vacuolar sequestration, tolerance, and transcriptional responses induced by cadmium and arsenic. Curr Opin Plant Biol 14: 554–562.

5

Platelet Specific Promoters Are Insufficient to Express Protease Activated Receptor 1 (PAR1) Transgene in Mouse Platelets

Amal Arachiche, María de la Fuente, Marvin T. Nieman*

Department of Pharmacology, Case Western Reserve University, Cleveland, Ohio, United States of America

Abstract

The *in vivo* study of protease activated receptors (PARs) in platelets is complicated due to species specific expression profiles. Human platelets express PAR1 and PAR4 whereas mouse platelets express PAR3 and PAR4. Further, PAR subtypes interact with one another to influence activation and signaling. The goal of the current study was to generate mice expressing PAR1 on their platelets using transgenic approaches to mimic PAR expression found in human platelets. This system would allow us to examine specific signaling from PAR1 and the PAR1-PAR4 heterodimer *in vivo*. Our first approach used the mouse GPIbα promoter to drive expression of mouse PAR1 in platelets (GPIbα-Tg-mPAR1). We obtained the expected frequency of founders carrying the transgene and had the expected Mendelian distribution of the transgene in multiple founders. However, we did not observe expression or a functional response of PAR1. As a second approach, we targeted human PAR1 with the same promoter (GPIbα-Tg-hPAR1). Once again we observed the expected frequency and distributing of the transgene. Human PAR1 expression was detected in platelets from the GPIbα-Tg-hPAR1 mice by flow cytometry, however, at a lower level than for human platelets. Despite a low level of PAR1 expression, platelets from the GPIbα-Tg-hPAR1 mice did not respond to the PAR1 agonist peptide (SFLLRN). In addition, they did not respond to thrombin when crossed to the PAR4$^{-/-}$ mice. Finally, we used an alternative platelet specific promoter, human α$_{IIb}$, to express human PAR1 (α$_{IIb}$-Tg-hPAR1). Similar to our previous attempts, we obtained the expected number of founders but did not detect PAR1 expression or response in platelets from α$_{IIb}$-Tg-hPAR1 mice. Although unsuccessful, the experiments described in this report provide a resource for future efforts in generating mice expressing PAR1 on their platelets. We provide an experimental framework and offer considerations that will save time and research funds.

Editor: Wilbur Lam, Emory University/Georgia Insititute of Technology, United States of America

Funding: The work was supported by grants from the American Heart Association (AHA Scientist Development Grant, 10SDG2600021) and the NIH (HL098217) to MTN. This research was also supported by the Cytometry and Light Microscopy Core Facility and the Transgenic & Targeting Core Facility of the Comprehensive Cancer Center of Case Western Reserve University and University Hospitals of Cleveland (P30 CA43703). The funders had no role in study design, data collection and analysis, decision to publish, or preparation of the manuscript.

Competing Interests: The authors have declared that no competing interests exist.

* E-mail: nieman@case.edu

Introduction

Protease activated receptors (PARs) are G-protein-coupled receptors (GPCR) that are activated by proteolytic cleavage of the N-terminus. Human platelets express PAR1 and PAR4 and both contribute to thrombin signaling [1]. In contrast, mouse platelets express PAR3 and PAR4, and thrombin signaling is mediated entirely by PAR4. Although PAR3 does not signal, it facilitates the cleavage of PAR4 at low thrombin concentrations by serving as a cofactor [2,3]. In addition, PAR3 can modulate the signaling from PAR4 at high agonist concentrations [4].

PAR4 is expressed on the platelets of most species, whereas, PAR1 expression in platelets is species specific. PAR1 is expressed on human, monkey, and guinea pig platelets, but not on canine, rat, murine, and rabbit platelets; these species do express PAR1 in other tissues [5,6]. The platelets from guinea pig express PAR1, PAR3, and PAR4. The expression of PAR3 complicates the translation to human disease since human platelets do not express PAR3 and, in mice, PAR3 influences PAR4 signaling [4]. Further,

some antagonists have species-specific interactions. For example, the PAR1 antagonist vorapaxar does not interact with mouse PAR1 [7]. Mice expressing human proteins would circumvent this potential issue. Finally, there are well-described mouse models of thrombosis in which one could test the specific role of the interaction between PAR1 and PAR4 or PAR1 specific signaling. The PAR profile on platelets from cynomolgus monkeys is comparable to that of human platelets, with the expression of PAR1 and PAR4, and no PAR3 [5]. However, the specific requirements for primate studies make this model impractical for initial preclinical studies. The selective expression of PAR1 has limited the opportunities for assessing the role of a PAR1 antagonist as antithrombotic agents with *in vivo* models.

In the present study we aimed to generate a mouse model that expressed PAR1 in mouse platelets to give us a unique opportunity to examine the individual roles of PAR1 and the PAR1-PAR4 heterodimer in platelet signaling *in vivo* by endogenous agonists. Mice expressing PAR1 on their platelets would also allow investigations into specific signaling pathways by crossing these mice with strains that have genetically altered signaling pathways. These studies would also generate a tool to characterize novel PAR antagonists. We chose a transgenic approach using three separate constructs and two different promoters (GPIbα and αIIb) that were ultimately unsuccessful at achieving sufficient PAR1 expression in mouse platelets. In each case we obtained the expected number of genetically positive founders. In this report we detail the transgenic approach that was unsuccessful in generating mice expressing PAR1 on their platelets and offer alternative strategies to generate an extremely valuable tool for the cardiovascular field.

Materials and Methods

Ethics statement

All animal studies were approved by the Institutional Animal Care and Use Committee at Case Western Reserve University School of Medicine. Human platelets were used as controls in some experiments. These studies were approved by the Case Western Reserve University Institutional Review Board and written informed consent was obtained from all donors.

Reagents and antibodies

Human α-thrombin (specific activity of 5380 NIH units/mg) was purchased from Haematological Technologies (Essex Junction, VT). PAR1 activating peptide (SFLLRN-NH2) was synthesized at PolyPeptide Laboratories (San Diego, CA). Fura-2AM was purchased from Invitrogen. Prostaglandin I2 was purchased from Calbiochem. The anti-PAR1-PE (WEDE15-PE) antibody was purchased from Beckman Coulter. The anti-P-selectin-FITC (CD62P-FITC) antibodies and JON/A-PE antibodies were purchased from (Emfret Analytics, Germany).

Plasmid construction

The vector containing mouse GPIbα promoter driving Factor VII was obtained from Dr. Mortimer Poncz (Children's Hospital of Philidelphia) [8]. The cDNA for Factor VII was replaced with that of mouse or human PAR1 (see Figure 1 and Figure 2). The human αIIb promoter was kindly provided by Dr. David Wilcox (Medical College of Wisconsin) [9]. The vector for PAR1 expression driven by the αIIb promoter in platelets was generated by replacing the GPIbα promoter. In addition, the Kozak sequence was placed in front of the PAR1 cDNA to enhance expression.

Animals

Transgenic mice were generated at the Case Western Reserve University Transgenic and Targeting Core Facility. The 4.2 kb transgene (promoter, cDNA, and poly(A)) were released from the vector backbone with a SalI digest, purified, and injected into fertilized eggs. The transgenic animals with mouse PAR1 targeted to platelets under control of the GPIbα promoter were generated on a B6SJL background. The transgenic animals with human PAR1 targeted to platelets under the control of GPIbα or αIIb promoter were generated on a C57BL6/J background. In all cases, founders were bred with C57BL6/J. C57BL6/J mice were purchased from The Jackson Laboratory (Bar Harbor, Maine). PAR3$^{-/-}$ and PAR4$^{-/-}$ mice were obtained from the Mutant Mouse Regional Resource Center (MMRRC) (Chapel Hill, NC). The PAR3$^{-/-}$ PAR4$^{-/-}$ double knockout mice were generated through breeding. All animals were genotyped with PCR analysis using primers specific for each gene. For PAR1 transgenic mice, multiple sets of primers were used to confirm genotyping results. A detailed list of primers used is available upon request.

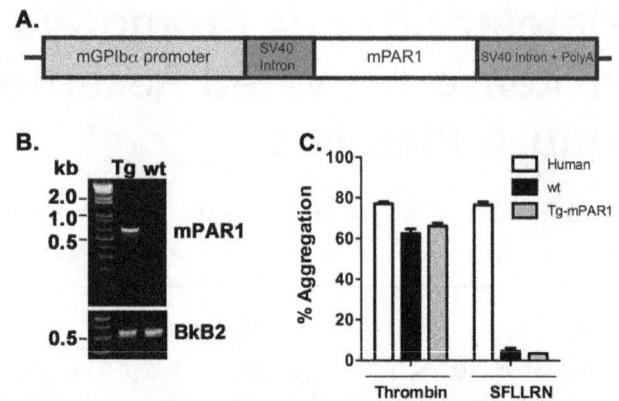

Figure 1. Generation and characterization of transgenic mice expressing mouse PAR1 transgene under control of mouse GPIbα promoter (GPIbα-Tg-mPAR1). (**A**) Schematic representation of the transgene construct. The cDNA for mouse PAR1 was inserted into a vector containing the mouse GPIbα, promoter small-t intron of simian virus 40 (SV40) in the 5′-untranslated region (UTR) and SV40 polyadenylation (polyA) sequence in the 3′-UTR. (**B**) Representative genotyping from GPIbα-Tg-mPAR1. The control PCR reactions used primers specific for the bradykinin B2 receptor (BkB2). (**C**) Platelet aggregation in response to thrombin (100 nM) or SFLLRN (50 μM) expressed as a percentage of the maximal light transmission. The results are the mean of six independent experiments.

Preparation of mouse and human platelets

Mouse platelet isolation was carried out as previously described [4]. Human platelets were obtained from healthy donors. Whole blood was collected into the anticoagulant acid citrate dextrose (ACD) (2.5% sodium citrate, 71.4 mM citric acid, 2% D-glucose) and centrifuged at 250×g for 10 min to isolate PRP. One-third volume of ACD and 1 μM prostaglandin I2 (PGI$_2$) were added to PRP and the preparation was centrifuged at 750×g for 10 min at room temperature. The platelet pellet was washed once in HEPES-Tyrode's buffer (pH 7.4) containing (1/5) volume of ACD and 1 μM PGI$_2$. Washed human platelets were counted on a Hemavet 950FS (Drew Scientific Inc, Waterbury, CT, USA) and the final platelet count adjusted with HEPES-Tyrode's buffer.

Platelet aggregation

Washed platelets were adjusted to a final concentration of 2×10^8 platelets/mL. Platelets aggregations were analyzed in an optical aggregometer (Bio/Data Corporation) at 37°C under constant stirring at 1200 rpm.

Measurement of PAR1 expression in mouse and human platelets

Mouse and human platelets were adjusted to final concentration of 40×10^6/mL in HEPES-Tyrode's buffer (pH 7.4). Platelets (10×10^5) were incubated with 20 μL anti-PAR1-PE (WEDE15-PE) antibody at room temperature for 20 min. Platelets samples were diluted to (1:8), acquired on a Beckman Coulter LSRII (Case Comprehensive Cancer Center Flow Core) and the total of 10,000 events were collected for each sample and analyzed with Flowjo software.

A.

B.

C.

D.

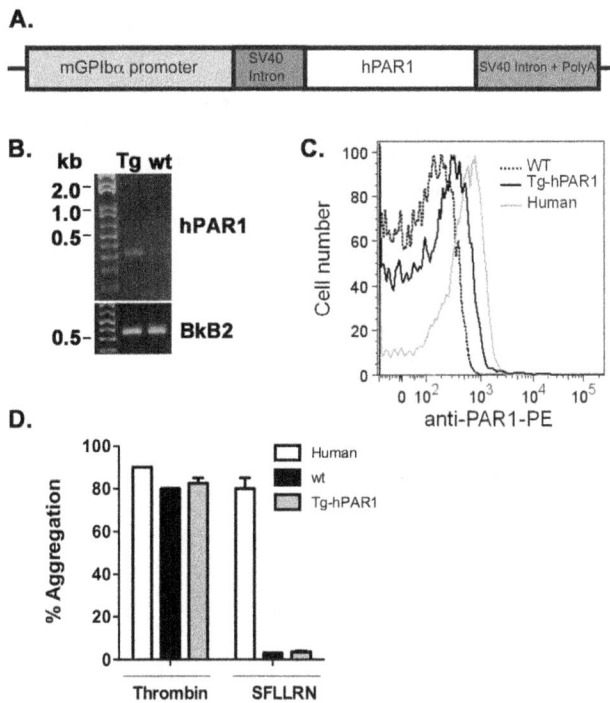

Figure 2. Generation and characterization of transgenic mice expressing human PAR1 transgene under control of mouse GPIbα promoter (GPIbα-Tg-hPAR1). (**A**) Schematic representation of the transgene construct. The coding sequence of the mouse PAR1 was replaced with human PAR1 in the GPIbα vector. (**B**) Representative genotyping from GPIbα-Tg-hPAR1. The control PCR reactions used primers specific for the bradykinin B2 receptor (BkB2). (**C**) Expression of hPAR1 was measured on the surface of platelets from human, wild type mice (wt) and transgenic mice (GPIb-Tg-hPAR1) using anti-PAR1-PE (WEDE15) and analyzed by flow cytometry. (**D**) Platelet aggregation in response to thrombin (100 nM) or SFLLRN (50 µM) expressed as a percentage of the maximal light transmission. The results are the mean of three independent experiments.

Measurement of the concentration of free intracellular calcium ([Ca^{2+}]$_i$)

Intracellular calcium mobilization in response to thrombin or PAR1 agonist peptide were measured fluorometrically in Fura-2-labelled washed mouse platelets as described previously [4].

Measurement of P-selectin expression and integrin αIIbβ3 activation

Washed platelets were adjusted to final concentration of 40×10^6 platelets/mL in HEPES-Tyrode's buffer (pH 7.4). Twenty microliter aliquots were activated with agonists for 5 min at 37°C and incubated with 0.5 µg/mL anti-P-selectin-FITC (CD62P-FITC) and JON/A-PE antibodies at room temperature for 15 min and analyzed by flow cytometery as described above.

Results

Characterization of transgenic mice expressing mouse PAR1 transgene under control of mouse GPIbα promoter (GPIbα-Tg-mPAR1)

Our initial efforts to generate mice expressing PAR1 on platelets used the mouse glycoprotein Ibα (GPIbα) promoter (kindly provided by Dr.Mortimer Poncz, University of Pennsylvania) [8]. The construct used to generate this transgenic mice was

represented in the Figure 1A. At the time of our experiments, the gene targeting and transgenic facility at Case Western Reserve University had higher success rates generating transgenic mice on the B6SJL background. Therefore, our experimental strategy was to generate the transgenic mice on the B6SJLF1/J background and backcross the mice to the C57BL6/J, which would also allow us to breed the mice with the PAR3 knockout (PAR3$^{-/-}$) and PAR4 knockout (PAR4$^{-/-}$) mice. We obtained 65 potential transgenic mice, from which we had 15 founder mice (7 males and 8 females) that were identified as positive for the transgene by PCR analysis (Figure 1B). Primers directed to the bradykinin B2 receptors were used as positive control for the PCR reaction. The founders were bred to C57BL6/J. The litters had the expected number of pups with an even distribution of males and females, and the predicted 50% of pups positive for the *mPAR1* gene.

To characterize transgenic mice (GPIbα-Tg-mPAR1) that were positive for the *mPAR1* transgene we examined mPAR1 protein expression and platelet function. We were unable to detect expression of mPAR1 protein in platelets by flow cytometry or Western blot with multiple PAR1 antibodies (data not shown). Since there are examples of transgenic mice in which extremely low protein levels were able to elicit a functional response [10,11], we tested for PAR1 function by measuring platelet activation in response to thrombin and PAR1 agonist peptide (SFLLRN). Platelets from the GPIbα-Tg-mPAR1 mice did not respond to 50 µM SFLLRN (Figure 1C), whereas the same concentration of SFLLRN induced human platelet aggregation. Platelets from GPIbα-Tg-mPAR1 mice responded to 100 nM thrombin similar to human and wild type mouse platelets (Figure 1C) indicating that they were functional. In addition, SFLLRN did not stimulate P-selectin surface expression or integrin α$_{IIb}$β$_3$ activation in platelets from GPIbα-Tg-mPAR1 mice as measured by flow cytometry (data not shown). These data indicate that the mice that were positive for the *mPAR1* transgene did not express mPAR1 protein.

Characterization of transgenic mice expressing human PAR1 transgene under control of mouse GPIbαpromoter (GPIbα-Tg-hPAR1)

Since we were unsuccessful in generating transgenic mice with mPAR1 expressed on platelets, we altered the transgene to express human PAR1 (hPAR1). The cDNA for mouse PAR1 was replaced with human PAR1 in the mouse GPIbα vector as shown in Figure 2A. Due to improvements in the efficiency of generating transgenic mice on a C57BL6/J background, we chose this approach to eliminate the need for backcrossing. After screening of 16 potential transgenic mice by PCR analysis, we identified 4 founder mice that were positive for the transgene (Figure 2B). The founders were bred to C57BL6/J. As with the previous transgenic animals, the litters had the expected number of pups with an even distribution of males and females, and the predicted 50% of pups positive for the *hPAR1* gene. We determined the surface expression of hPAR1 in platelets from GPIbα-Tg-hPAR1 mice by flow cytometry (Figure 2C). GPIbα-Tg-hPAR1 mice expressed hPAR1 on their platelets; however the expression level of hPAR1 was very low compared to the human platelets. As expected, PAR1 was not detected on platelets from wild type mice. To investigate whether hPAR1 was functional in mouse platelets, we measured platelet aggregation in response to thrombin and SFLLRN (Figure 2D). Platelets from GPIbα-Tg-hPAR1 or wild type mice did not aggregate in response to 50 µM SFLLRN however, they did respond to thrombin indicating the platelets were functional.

To be certain that the hPAR1 was not functional in the platelets from the GPIbα-Tg-hPAR1 mice, we wanted to test the response to the endogenous agonist thrombin. To do this without the influence of other PARs, we crossed the GPIbα-Tg-hPAR1 onto the PAR3$^{-/-}$-PAR4$^{-/-}$ background (GPIbα-Tg-hPAR1-PAR3$^{-/-}$-PAR4$^{-/-}$). The PAR3$^{-/-}$-PAR4$^{-/-}$ mice that were identified as positive for the hPAR1 transgene by PCR analysis (data not shown). Calcium mobilization was measured in platelets from GPIbα-Tg-hPAR1-PAR3$^{-/-}$-PAR4$^{-/-}$ mice (Figure 3). In contrast to wild type, Tg-hPAR1-PAR3$^{-/-}$-PAR4$^{-/-}$ platelets did not increase their intracellular calcium in response to 30 nM thrombin. These data confirm that hPAR1 is not functional in the transgenic mice, likely due to low expression.

Characterization of transgenic mice expressing human PAR1 transgene under control of human α$_{IIb}$ promoter (α$_{IIb}$-Tg-hPAR1)

The GPIbα promoter has been shown to efficiently drive the expression of transgenes in mouse platelets [8]. However, we were unable to detect sufficient expression of either murine or human PAR1 in mouse platelets using GPIbα promoter. Next, we used an alternative platelet-specific promoter human α$_{IIb}$ (kindly provided by Dr. David A Wilcox, Medical College of Wisconsin) [9]. The cDNA for human PAR1 was inserted to the α$_{IIb}$ vector containing a small-t intron of simian virus 40 (SV40), Kozak sequence (GCCGCCACC) at the 5'-untraslated region (UTR), and SV40 polyadenylation (polyA) sequence at the 3'-UTR (Figure 4A). The Kozak sequence was added upstream of the start codon to increase and improve expression level of hPAR1 in mouse platelets. α$_{IIb}$-Tg-hPAR1 mice were generated on a C57BL6/J background to eliminate the need for backcrossing. We received 19 potential transgenic mice, from which we identified 4 founder mice (2 males and 2 females) that were positive for the transgene by PCR analysis (Figure 4B). As with the previous transgenic animals, the litters had the expected number of pups with an even distribution of males and females and the predicted 50% of pups positive for the *hPAR1 trans*gene. Human PAR1 expression was examined by flow cytometry (Figure 4C). There was no detectable hPAR1 in the genetically positive mice. To fully characterize these mice, functional studies were performed on the platelets from α$_{IIb}$-Tg-hPAR1 mice (Figure 4D and E). In response to 1 nM thrombin the intracellular calcium was increased in platelets from α$_{IIb}$-Tg-hPAR1 mice to the same level as the wild type (Figure 4D). However, there was no increase in the intracellular calcium in

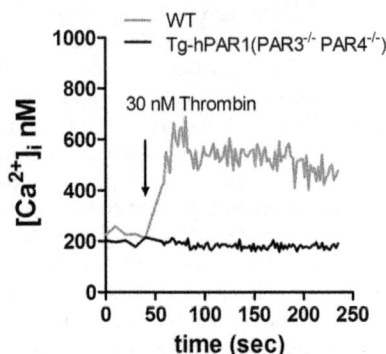

Figure 3. GPIbα-Tg-hPAR1 transgenic mice on a PAR3-PAR4 double knockout (PAR3$^{-/-}$, PAR4$^{-/-}$) background do not respond to thrombin. Intracellular calcium mobilization was measured in platelets from wild type (wt) (gray line) and GPIbα-Tg-hPAR1- PAR3$^{-/-}$ -PAR4$^{-/-}$ (black line) mice in response to thrombin (30 nM). The tracings are representative of three independent experiments.

response to 100 µM SFLLRN in platelets from α$_{IIb}$-Tg-hPAR1 mice (Figure 4E). As expected wild type platelets did not increase the intracellular calcium in response to the same concentration of SFLLRN (Figure 4E).

We were unable to detect the expression of hPAR1 in the transgenic mice platelets using the promoter α$_{IIb}$. We decided to examine the functional response of hPAR1 to thrombin by generating transgenic mice with PAR4 knockout (PAR4$^{-/-}$) background (α$_{IIb}$-Tg-hPAR1-PAR4$^{-/-}$). The α$_{IIb}$-Tg-hPAR1-PAR4$^{-/-}$ mice that were identified as positive for the transgene by PCR analysis (data not shown). Activation of platelets from α$_{IIb}$-Tg-hPAR1-PAR4$^{-/-}$ mice with 10 nM thrombin did not induced intracellular calcium mobilization (Figure 5A). In addition, 10 nM thrombin or 30 µM SFLLRN did not stimulate P-selectin surface expression (Figure 5B) or integrin α$_{IIb}$β$_3$ activation (Figure 5C) as measured by flow cytometry. As expected, platelets from wild type responded to thrombin, but not SFLLRN (Figure 5B and C).

Discussion

It has been well documented that human platelets express PAR1 and PAR4, whereas mouse platelets express PAR3 and PAR4. This species-specific expression profile has limited the ability to use mouse models to examine the individual roles of PAR1 and PAR4 on platelets and preclinical testing of pharmacologic agents. The aim of the current study was to generate a mouse model expressing PAR1 in platelets. These mice would allow us to determine the role of PAR1 in primary hemostasis and platelet thrombus formation *in vivo*. These mice would also allow us to examine the interplay between PAR1 and other platelet receptors in vivo. Although we were unsuccessful at generating mice with sufficient expression of PAR1 on their platelets, our studies provide useful insights and potential alternative methods for future studies generating mice with humanized PAR1 expression on their platelets.

In the current study, we have used mouse glycoprotein Ibα (mGPIbα) promoter to direct the expression of mouse or human PAR1 into mouse platelets (Figure 1A and Figure 2A) or the human α$_{IIb}$ promoter to express human PAR1 (Figure 4A). Although we obtained the expected number of genetically positive animals in each of the transgenic lines, we did not have any mice positive for PAR1 expression or function, except for GPIbα-Tg-hPAR1, where the expression level of PAR1 was detected in mouse platelets but at lower level compared to PAR1 expression in human platelets. Our transgene construct using the α$_{IIb}$ promoter added the Kozak sequence upstream of the start codon in order to optimize hPAR1 translation [12]. This modification did not increase PAR1 expression. There are multiple reasons for having a varied level of expression in transgenic animals. First, the transgene may insert into a locus that is subject to gene silencing. However, we have examined multiple transgenic mice using three separate transgenes making this unlikely. Second, PAR1 may be inherently difficult to express in exogenous systems. In agreement with this, our previous studies have used multiple promoters and constructs in several cell lines and we consistently have lower expression of PAR1 than other GPCRs [4,13,14]. In addition, we have difficulties generating stable cell lines expressing PAR1 [14]. A third possibility is that PAR1 expression in platelets may alter cell fate during development [15,16]. We have bred our transgenic mice to PAR3$^{-/-}$, PAR4$^{-/-}$, or the double knockout (PAR3$^{-/-}$-PAR4$^{-/-}$) and observed the expected litter size and transgenic ratios for each line of animals. Therefore, it is also unlikely that PAR1 is altering cell fate during development. Based on our data,

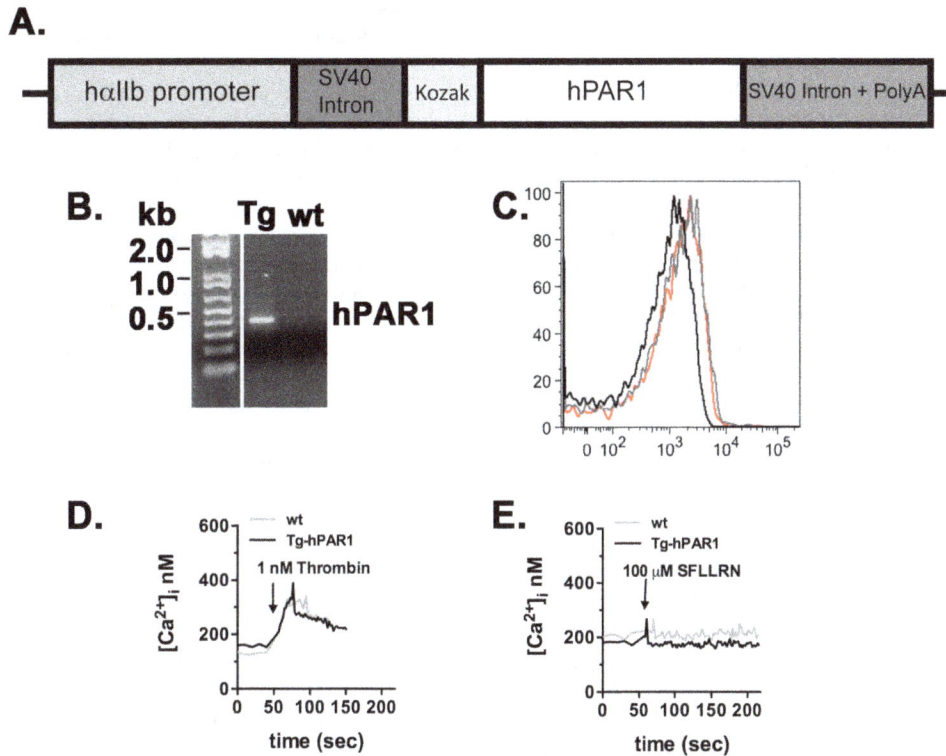

Figure 4. Generation and characterization of transgenic mice (Tg-hPAR1) expressing human PAR1 transgene under control of human α_{IIb} promoter. (A) Schematic representation of the transgene construct. The cDNA for human PAR1 was inserted into a vector containing the human α_{IIb} promoter, a small-t intron of simian virus 40 (SV40), and a Kozak sequence at the 5'-untranslated region (UTR) and SV40 polyadenylation (polyA) sequence at the 3'-UTR. (B) Representative genotyping from α_{IIb}-Tg-hPAR1. (C) Expression of hPAR1 was measured on the surface of platelets from human, wild type mice (wt, black line) and transgenic mice (α_{IIb}-Tg-hPAR1) using anti-PAR1-PE (WEDE15, red line) or IgG-PE (gray line) and analyzed by flow cytometry. (D and E) Intracellular calcium mobilization was measured in platelets from wild type (wt) (gray line) and α_{IIb}-Tg-hPAR1 (black line) mice in response to thrombin (1 nM) (D) or SFLLRN (100 µM) (E). The calcium tracings are representative of three independent experiments.

we expect that PAR1 is not stable when exogenously expressed in platelets similar to our observation in cell lines. The primary goal of generating mice expressing PAR1 on their platelets was to develop a research tool to determine specific PAR1 signaling events on platelets *in vivo* in thrombosis models, therefore we did not further explore the precise mechanism for the absence of PAR1 expression in mouse platelets.

The GPIbα and α_{IIb} promoters have been successfully used to generate platelet specific expression of transgenes in previous studies [8,9]. An alternative strategy would be to use another

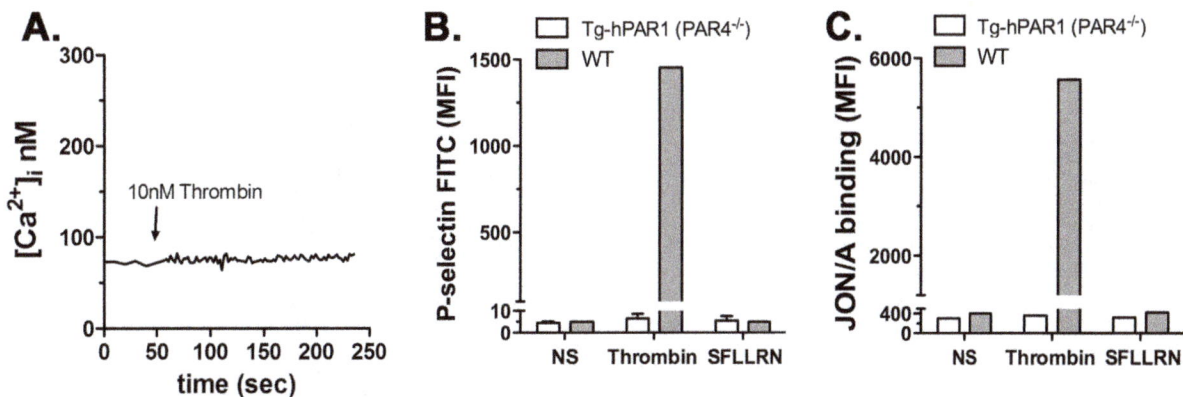

Figure 5. α_{IIb}-Tg-hPAR1 transgenic mice on a PAR4$^{-/-}$ background do not respond to thrombin. (A) Intracellular calcium mobilization was measured in platelets from α_{IIb}-Tg-hPAR1-PAR4$^{-/-}$ mice in response to thrombin (10 nM). The tracing is representative of three independent experiments. (B) P-selectin expression was measured in the surface of platelets from wild type (wt) (gray bars) or α_{IIb}-Tg-hPAR1-PAR4$^{-/-}$ (white bars) mice by flow cytometry using FITC conjugated P-selectin antibody in response to thrombin (10 nM) or SFLLRN (30 µM). (C) Platelets were treated as in (B) and integrin $\alpha_{IIb}\beta_3$ activation was measured using PE conjugated JON/A antibody. The results are the mean of three independent experiments.

platelet specific promoter such as the platelet factor 4 (PF4) promoter. However, given our difficulties with expressing PAR1 with multiple platelet specific promoters the PF4 promoter is also unlikely to give sufficient expression. Bacterial artificial chromosome (BAC) transgenic can also offer increased expression of the transgene over traditional transgenic approaches. Attempts to express the factor V (FV) in platelets using the PF4 BAC transgene offered only marginal improvements in expression [10,11]. There have been successful studies generating PAR1 transgenic animals in other tissues [17,18]. Therefore, the difficulties in expression may be due to the platelet specific promoters used in this study. One could envision using a global gene such as actin to deliver platelet specific expression. The resulting animals would have PAR1 overexpressed in other tissues, which may have consequences for *in vivo* experiments. More complex strategies have been described to generate inducible expression of platelet specific transgenes and could be applied to PAR1 [19]. Since mice express PAR3, but not PAR1 and human platelet express PAR1 but not PAR3, an elegant approach would be to knock human PAR1 into the PAR3 locus. This approach would have the benefit of simultaneously inserting human PAR1 and deleting mouse PAR3. Recent advances in gene targeting, such as TALEN or CRISPR approaches make this strategy more attracting than in the past. Finally, human PAR1 could be expressed in mouse platelets using a fetal liver cell transplant. With this approach, fetal liver cells are isolated from donor mice and transduced with lentivirus or retrovirus to express hPAR1. The transduced cells are transplanted to irradiated recipient mice. This approach has also been unsuccessful in our hands. The difficulty appears to be in generating cells that stably express PAR1. These data are similar to our cell line experiments described above.

We have discussed alternative methods for generating mice expressing PAR1 on their platelets. There are other considerations that must be taken into account in order for these mice to reflect human disease. For example, we may need to replace mouse PAR4 with human PAR4 in mouse platelets. We have shown in previous work that the calcium signaling is increased in PAR3 knockout mouse platelets expressing mouse PAR4 alone compared to wild type platelets [4]. These data demonstrate that the receptors interact to influence signaling. However, whether PAR3 interact with human PAR4 and influence signaling is not known. Further, during the course of the current study, antagonists to platelet GPCRs have been developed that target the C-terminus of the receptors [20]. The sensitivity of this antagonist was limited to GPCRs that contain a palmitoylation site at the C-terminus of helix 8. In this study, Dowal et al. demonstrated that the antagonist interacts and inhibits mouse PAR4 but not human PAR4. Mouse PAR4 contains a Cys residue at the C-terminal end of helix 8 that is expected to be palmitoylated. In contrast, human PAR4 does not have the Cys residue and contains Gly residues, which can disrupt alpha helices. Given these differences, the best approach may be to completely humanize mice with respect to PARs by also knocking in human PAR4. In this light, the extensive efforts that will be required to generate such animals (double knock-in of hPAR1 and hPAR4) will have to be carefully considered in regards to the potential benefits.

In summary, our extensive efforts to generate mice expressing PAR1 on their platelets using a transgenic approach were unsuccessful. In this report we describe our attempts to express PAR1 in mouse platelets with the hope it will guide others in the field that are interested in generating these mice. Furthermore, we offer alternative strategies and considerations that will be required for interpreting data using a mixture of human and mouse PARs on platelets.

Author Contributions

Conceived and designed the experiments: AA MF MTN. Performed the experiments: AA MF MTN. Analyzed the data: AA MF MTN. Wrote the paper: AA MF MTN.

References

1. Coughlin SR (2000) Thrombin signalling and protease-activated receptors. Nature 407: 258–264.
2. Nakanishi-Matsui M, Zheng YW, Sulciner DJ, Weiss EJ, Ludeman MJ, et al. (2000) PAR3 is a cofactor for PAR4 activation by thrombin. Nature 404: 609–613.
3. Sambrano GR, Weiss EJ, Zheng YW, Huang W, Coughlin SR (2001) Role of thrombin signalling in platelets in haemostasis and thrombosis. Nature 413: 74–78.
4. Arachiche A, de la Fuente M, Nieman MT (2013) Calcium Mobilization And Protein Kinase C Activation Downstream Of Protease Activated Receptor 4 (PAR4) Is Negatively Regulated By PAR3 In Mouse Platelets. PLoS One 8: e55740.
5. Derian CK, Santulli RJ, Tomko KA, Haertlein BJ, Andrade-Gordon P (1995) Species differences in platelet responses to thrombin and SFLLRN. receptor-mediated calcium mobilization and aggregation, and regulation by protein kinases. Thromb Res 78: 505–519.
6. Connolly TM, Condra C, Feng DM, Cook JJ, Stranieri MT, et al. (1994) Species variability in platelet and other cellular responsiveness to thrombin receptor-derived peptides. Thromb Haemost 72: 627–633.
7. Zhang C, Srinivasan Y, Arlow DH, Fung JJ, Palmer D, et al. (2012) High-resolution crystal structure of human protease-activated receptor 1. Nature 492: 387–392.
8. Yarovoi HV, Kufrin D, Eslin DE, Thornton MA, Haberichter SL, et al. (2003) Factor VIII ectopically expressed in platelets: efficacy in hemophilia A treatment. Blood 102: 4006–4013.
9. Wilcox DA, Olsen JC, Ishizawa L, Griffith M, White GC 2nd (1999) Integrin alphaIIb promoter-targeted expression of gene products in megakaryocytes derived from retrovirus-transduced human hematopoietic cells. Proc Natl Acad Sci U S A 96: 9654–9659.
10. Sun H, Yang TL, Yang A, Wang X, Ginsburg D (2003) The murine platelet and plasma factor V pools are biosynthetically distinct and sufficient for minimal hemostasis. Blood 102: 2856–2861.
11. Yang TL, Cui J, Taylor JM, Yang A, Gruber SB, et al. (2000) Rescue of fatal neonatal hemorrhage in factor V deficient mice by low level transgene expression. Thromb Haemost 83: 70–77.
12. Kozak M, Shatkin AJ (1978) Identification of features in 5′ terminal fragments from reovirus mRNA which are important for ribosome binding. Cell 13: 201–212.
13. de la Fuente M, Noble DN, Verma S, Nieman MT (2012) Mapping human protease-activated receptor 4 (PAR4) homodimer interface to transmembrane helix 4. J Biol Chem 287: 10414–10423.
14. Arachiche A, Mumaw MM, de la Fuente M, Nieman MT (2013) Protease-activated Receptor 1 (PAR1) and PAR4 Heterodimers Are Required for PAR1-enhanced Cleavage of PAR4 by α-Thrombin. J Biol Chem 288: 32553–32562.
15. Yue R, Li H, Liu H, Li Y, Wei B, et al. (2012) Thrombin receptor regulates hematopoiesis and endothelial-to-hematopoietic transition. Dev Cell 22: 1092–1100.
16. Aronovich A, Nur Y, Shezen E, Rosen C, Zlotnikov Klionsky Y, et al. (2013) A novel role for factor VIII and thrombin/PAR1 in regulating hematopoiesis and its interplay with the bone structure. Blood 122: 2562–2571.
17. Yin YJ, Katz V, Salah Z, Maoz M, Cohen I, et al. (2006) Mammary gland tissue targeted overexpression of human protease-activated receptor 1 reveals a novel link to beta-catenin stabilization. Cancer Res 66: 5224–5233.
18. Pawlinski R, Tencati M, Hampton CR, Shishido T, Bullard TA, et al. (2007) Protease-activated receptor-1 contributes to cardiac remodeling and hypertrophy. Circulation 116: 2298–2306.
19. Zhang Y, Ye J, Hu L, Zhang S, Zhang SH, et al. (2012) Increased platelet activation and thrombosis in transgenic mice expressing constitutively active P2Y12. J Thromb Haemost 10: 2149–2157.
20. Dowal L, Sim DS, Dilks JR, Blair P, Beaudry S, et al. (2011) Identification of an antithrombotic allosteric modulator that acts through helix 8 of PAR1. Proc Natl Acad Sci U S A 108: 2951–2956.

6

GmFT2a and GmFT5a Redundantly and Differentially Regulate Flowering through Interaction with and Upregulation of the bZIP Transcription Factor GmFDL19 in Soybean

Haiyang Nan[1,2,9], Dong Cao[1,9], Dayong Zhang[3,9], Ying Li[4,9], Sijia Lu[1,2], Lili Tang[1], Xiaohui Yuan[1], Baohui Liu[1]*, Fanjiang Kong[1]*

1 The Key of Soybean Molecular Design Breeding, Northeast Institute of Geography and Agroecology, Chinese Academy of Sciences, Nangang District, Harbin, China, 2 University of Chinese Academy of Sciences, Beijing, China, 3 Institute of Biotechnology, Jiangsu Academy of Agricultural Sciences, Nanjing, China, 4 State Key Laboratory of Tree Genetics and Breeding, Northeast Forestry University, Harbin, China

Abstract

FLOWERING LOCUS T (FT) is the key flowering integrator in Arabidopsis (*Arabidopsis thaliana*), and its homologs encode florigens in many plant species regardless of their photoperiodic response. Two FT homologs, GmFT2a and GmFT5a, are involved in photoperiod-regulated flowering and coordinately control flowering in soybean. However, the molecular and genetic understanding of the roles played by GmFT2a and GmFT5a in photoperiod-regulated flowering in soybean is very limited. In this study, we demonstrated that GmFT2a and GmFT5a were able to promote early flowering in soybean by overexpressing these two genes in the soybean cultivar Williams 82 under noninductive long-day (LD) conditions. The soybean homologs of several floral identity genes, such as *GmAP1*, *GmSOC1* and *GmLFY*, were significantly upregulated by GmFT2a and GmFT5a in a redundant and differential pattern. A bZIP transcription factor, GmFDL19, was identified as interacting with both GmFT2a and GmFT5a, and this interaction was confirmed by yeast two-hybridization and bimolecular fluorescence complementation (BiFC). The overexpression of *GmFDL19* in soybean caused early flowering, and the transcription levels of the flowering identity genes were also upregulated by GmFDL19, as was consistent with the upregulation of GmFT2a and GmFT5a. The transcription of *GmFDL19* was also induced by GmFT2a. The results of the electrophoretic mobility shift assay (EMSA) indicated that GmFDL19 was able to bind with the cis-elements in the promoter of *GmAP1a*. Taken together, our results suggest that GmFT2a and GmFT5a redundantly and differentially control photoperiod-regulated flowering in soybean through both physical interaction with and transcriptional upregulation of the bZIP transcription factor GmFDL19, thereby inducing the expression of floral identity genes.

Editor: Erik Souer, Vrije Universiteit Amsterdam, Netherlands

Funding: National Natural Science Foundation of China (31071445, 31171579, 31201222 and 31371643); the Open Foundation of the Key Laboratory of Soybean Molecular Design Breeding, Chinese Academy of Sciences; the "Hundred Talents" Program of the Chinese Academy of Sciences; the Strategic Action Plan for Science and Technology Innovation of the Chinese Academy of Sciences (XDA08030108); and the Heilongjiang Natural Science Foundation of China (ZD201001, JC201313). The funders had no role in study design, data collection and analysis, decision to publish, or preparation of the manuscript.

Competing Interests: The authors have declared that no competing interests exist.

* E-mail: liubh@iga.ac.cn (BL); kongfj@iga.ac.cn (FK)

⑨ These authors contributed equally to this work.

Introduction

Plants integrate various environmental signals, such as photoperiod and temperature, to ensure flowering under those conditions that optimize seed production. In *Arabidopsis thaliana* (Arabidopsis), multiple pathways converge on a small number of floral integrator genes, which include the floral promoters *FLOWERING LOCUS T (FT)* and *TWIN SISTER OF FT (TSF)*, to integrate photoperiod, temperature, vernalization, and light quality signaling [1]. FT and TSF are members of a family of proteins similar to the mammalian phosphatidylethanolamine-binding domain protein (PEBP) [2], [3]. In addition to the FT-like proteins, the plant PEBP family consists of two other phylogenet-

ically distinct groups of proteins, the TERMINAL FLOWER 1 (TFL1)-like proteins and the MOTHER OF FT AND TFL (MFT)-like proteins [4]–[8]. FT and TSF act redundantly to promote flowering under long-day (LD) photoperiods[7], [9], [10]. Arabidopsis FT and TSF proteins produced in the phloem [7], [11] and are transported to the shoot apex, where they dimerize with the bZIP transcription factor FD to activate the expression of *SUPPRESSOR OF OVEREXPRESSION OF CONSTANS 1 (SOC1)* [9], [12] and the floral meristem identity genes *APETALA1 (AP1)* and *LEAFY (LFY)* [13], [14]. FT-like proteins from various species function in a manner similar to that of FT regarding the induction of flowering, transport in the phloem, and interaction with FD-like proteins [15]–[18], suggesting that this general mechanism is likely widely conserved across flowering plants. However, the rice FT ortholog Hd3a interacts with OsFD1 indirectly through a 14-3-3 protein to form a ternary trimer known as the florigen activation complex in the nuclei of the shoot apex, where it activates the expression of *OsMADS15*, an *AP1* homolog that regulates flowering [19], [20].

Soybean, *Glycine max* (L.) Merr., is basically a short-day (SD) plant: its flowering is induced when the day length becomes shorter than a critical length. Soybean is grown at a wide range of latitudes, from at least North 50° to South 35°, although the cultivation area of each soybean cultivar is restricted to a very narrow range of latitudes. This wide adaptability has most likely been generated by genetic diversity at a large number of the major genes and quantitative trait loci that control flowering behavior. Nine major genes, *E1* to *E8* and *J*, that control flowering time and maturity have been previously identified in soybean [21]–[28]. Among these genes, *E1* has been cloned using a map-based approach and is assumed to be a legume-specific transcription factor containing a putative nuclear localization signal and a B3 distantly related domain [29]; *E2* has been identified as an ortholog of the Arabidopsis *GIGANTEA* gene [30]; and *E3* and *E4* have been confirmed as *PHYA* homologs using map-based cloning [31] and a candidate gene approach [32], respectively. Many allelic variations occur at the *E1*, *E3* and *E4* loci, and their allelic combinations condition soybean flowering time, regulate preflowering and postflowering photoperiod responses, and contribute greatly to the wide adaptability of soybean [33], [34]. Two *FT* homologs, *GmFT2a* and *GmFT5a*, are involved in the transition to flowering and these two genes coordinately control flowering in soybean [35]. The maturity genes *E1*, *E2*, *E3* and *E4* downregulate *GmFT2a* and *GmFT5a* expression to delay flowering and maturation under LD conditions, suggesting that GmFT2a and GmFT5a are the soybean flowering integrators and major targets in the control of flowering [29], [30], [35], [36]. In addition, two *SOC1* homologs, *GmSOC1* and *GmSOC1-like*, have been molecular characterized in soybean: the overexpression of *GmSOC1* partially rescued the late-flowering phenotype of the *soc1-1* Arabidopsis mutant under LD conditions [37], while the overexpression of *GmSOC1-like* promoted flowering in *Lotus corniculatus* [38]. These results suggest that the two soybean *SOC1* homologs may function as floral activators in soybean. A soybean *AP1* homolog, *GmAP1*, has also been isolated in soybean and is specifically expressed in the flower, especially in the sepals and petals. This gene caused early flowering and the alteration of floral organ patterns when ectopically expressed in tobacco [39]. Despite the economic importance of soybean, knowledge regarding its molecular mechanisms of flowering remains limited. Here, we report that the overexpression of *GmFT2a* and *GmFT5a* in soybean can promote early flowering by activating the expression of floral identity gene homologs such as *GmAP1*, *GmLFY* and *GmSOC1*. The GmFT2a and GmFT5a proteins interact with the bZIP transcription factor GmFDL19. The overexpression of *GmFDL19* in soybean can also induce the expression of floral identity genes. Additionally, we show that the bZIP transcription factor GmFDL19 is able to bind with the ACGT cis-element of the *GmAP1* promoter. Our results suggest that the putative flowering model FT/FD-AP1 is well conserved in the legume soybean and that GmFDL19 may act as the key component in the photoperiod-regulated flowering pathway controlled by GmFT2a and GmFT5a.

Materials and Methods

Plant materials and growth conditions

The soybean cultivars Harosoy, Williams 82 and Dongnong 50 were used in this study. All plants were grown in a growth chamber (Conviron ADAPTIS-A1000, Canada) at a consistent temperature of 25°C and an average photon flux of 300 µmol m^{-2}s^{-1}, supplied by T5 fluorescent lamps. Day length regimes were 12L/12D for SD and 16L/8D or 18L/6D for LD. Tissue-specific expression was analyzed using the cultivar Harosoy grown under SD. Total RNA was isolated from trifoliate leaves, shoot apices, roots, flowers, flower buds, and roots. For the temporal expression analysis, pieces of young, fully developed trifoliate leaves and shoot apices were bulk sampled at 4 hours after dawn from 4 individual plants grown under SD every five days from 10 DAE until 25 DAE. The trifoliate leaves and shoot apices from 4 plants of both transgenic and untransformed lines were bulk sampled at 4 hours after dawn at 20 DAE under the LD condition and stored until total RNA extraction.

Soybean genetic transformation. The cDNA sequences of Harosoy *GmFT2a/5a* and *GmFDL19* were first cloned into the pEASY-T1 vector (Transgene, Beijing, China). XbaI/SacI-digested fragments were then inserted at multiple cloning sites in the pTF101.1 vector, and the transgenes were driven by the cauliflower mosaic virus 35S promoter [40]. The *GmFT2a/5a*-pTF101 and *GmFDL19*-pTF101 constructs were used to transform the cultivars Williams 82 and Dongnong 50, respectively, following the cotyledonary node method [41]. T0, T1 and T2 transformants were screened by daubing 160 mg/L glufosinate into the preliminary leaves of the seedlings. Herbicide-resistant T2 plants were subjected to molecular and phenotypic analysis.

RT-PCR and quantitative RT-PCR analyses

Total RNA was isolated and cDNA was synthesized as described in Kong et al. [35]. RT-PCR of *GmFDL19*, *GmAP1* (*a, b, c, d*), *GmSOC1a*, *GmSOC1b*, *GmLFY1*, *GmLFY2* and *Tubulin* (as an internal control) was conducted using cDNAs synthesized from total RNA. PCR conditions were as follow: one cycle of 5 min at 94°C; 30 cycles of 30 sec at 94°C, 30 sec at 55°C to 60°C (depending on the gene), and 30 sec at 72°C; and a final extension of 10 min at 72°C. RT-PCR was performed using homolog-specific primers to easily separate the RT-PCR products (approximately 500 bp) from the fragments amplified from genomic DNAs (>1 kb). The RT-PCR products were separated by electrophoresis in a 1% agarose gel and visualized with EtBr under UV light. Quantitative RT-PCR was performed as described in [35]. The quantitative RT-PCR mixture was prepared by mixing a 1 µl aliquot of the reaction mixture from the cDNA synthesis, 5 µl of 1.2 µM primer premix, 10 µl SYBR Premix ExTaq Perfect Real Time (TaKaRa Bio), and water to a final volume of 20 µl. The analysis was conducted using the DNA Engine Opticon 2 System (Bio-Rad). The PCR cycling conditions were as follow: 95°C for 10 sec, 55°C to 60°C (depending on the gene) for 20 sec, 72°C for 20 sec, and 78°C for 2 sec. This cycle was repeated 40 times. Fluorescence quantification was conducted before and after the incubation at 78°C to monitor the formation of primer dimers. The mRNA level of the *Tubulin* gene was used as a control for the analysis. A reaction mixture without reverse transcriptase was also used as a control to confirm that no amplification occurred from genomic DNA contaminants in the RNA sample. In all of the PCR experiments, the amplification of a single DNA species was confirmed using both melting curve analysis of the quantitative PCR and gel electrophoresis of the

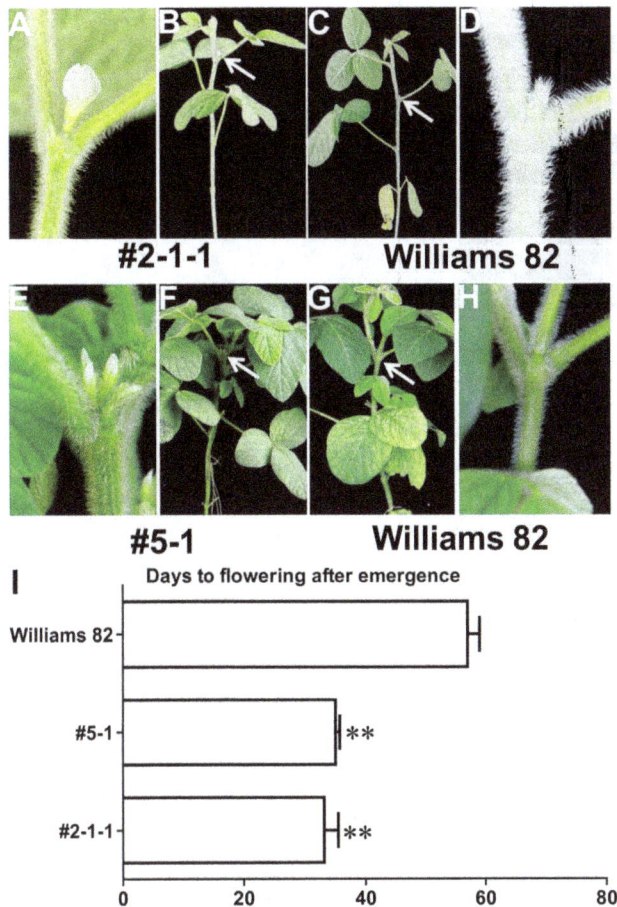

Figure 1. Overexpression of *GmFT2a* and *GmFT5a* causes precocious flowering in the soybean cultivar Williams 82. (A) The close shot of the transgenic plant in (B) shows the precocious flowers at the axils of the trifoliate leaves. (B) A transgenic *GmFT2a* plant showing precocious flowering at the axils of the trifoliate leaves. (C) A wild-type Williams 82 plant. (D) The close shot of the wild type Williams 82 plant in (C) does not show flowers at the axils of the trifoliate leaves. (E) The close shot of the transgenic plant in (F) shows the precocious flowers at the axils of the trifoliate leaves. (F) A transgenic *GmFT5a* plant showing precocious flowering at the axils of the trifoliate leaves. (G) A wild-type Williams 82 plant. (H) The close shot of the wild-type Williams 82 plant in (G) does not show flowers at the axils of the trifoliate leaves. (I) Days to flowering from the emergence of the transgenic plants and wild-type plants. Averages and standard errors are calculated from four T2 plants for each construct and 5 Williams 82 plants. Double asterisks indicate significant differences from the corresponding wild-type Williams 82 at $P<0.01$.

PCR products. The primers used for qRT-PCR and RT-PCR are listed in Table S1 and Table S2, respectively.

Identification of soybean *FD*, *AP1*, *LFY* and *SOC1* homologs. The database used for these searches is available at Phytozome (http://www.phytozome.net/soybean). Starting with the Arabidopsis FD, AP1, LFY and SOC1 protein sequences, TBLASTN searches were conducted against the soybean (*Glycine max*) gene index (release 1.0). The top 18 *FD*-like gene sequences producing high-scoring segment pairs were chosen and investigated further. Primers were designed to amplify cDNAs for each of the top 18 *FD*-like genes (Table S2). Seven pairs of full-length CDS PCR (Table S3) primers were designed to amplify the seven expressed *FD*-Like genes, and restriction sites (underlined) were included in the oligos to facilitate the cloning of the PCR products

into the yeast vector pGADT7. A multiple sequence alignment and a neighbor-joining phylogenetic tree were constructed using DDBJ (http://clustalw.ddbj.nig.ac.jp/) online ClustalW software and Treeview 2.0. The tree was based on the full-length amino acid region including the bZIP domain and the SAP motif (Figure S1). The bootstrap percentage supports are indicated at the branches of the tree.

Yeast two-hybridization assays

The yeast cloning vectors pGBKT7 and pGADT7, the control vectors pGADT7-T and pGBKT7-53, and the yeast strain Y2H used in the yeast-two hybridization assays were obtained from the Clontech company (http://www.clontech.com/). The yeast two-hybridization assays were performed according to the manufacturer's instructions. Soybean full-length CDS of *GmFT2a* and *GmFT5a* were inserted into pGBKT7 vectors to generate fused GAL4 DNA binding domains as the soybean baits. Full-length CDS of the three soybean *FD*-like genes containing the SAP motif (*GmFDL19*, *GmFDL08* and *GmFDL15*) were cloned into pGADT7 to generate fused GAL4 DNA activation domains as the soybean preys, and the full-length CDS of Arabidopsis *FD* was also cloned into pGADT7 to generate a positive control. Table S4 lists the primers and restriction sites used to generate the yeast bait and prey constructs. The bait and prey plasmids were cotransformed into the yeast strain Y2H using the lithium acetate method and selected on SD medium lacking leucine (Leu) and tryptophan (Trp). After 4 days of incubation at 30°C, the yeast cells were replated on selection plates with SD medium lacking Leu, Trp, histidine (His) and adenine (Ade) but including the X-α-gal substrate for the interaction test.

Bimolecular fluorescence complementation

The full-length CDS of *GmFT2a* and *GmFT5a* were amplified using the primer pairs *GmFT2a*-NE-F/R and *GmFT5a*-NE-F/R, respectively (Table S5), and were then inserted into the pUC_-SPYNE [42] vector, which contains the DNA encoding the N-terminus of YFP. The full-length cDNAs of *GmFDL08*, *GmFDL15*, *GmFDL19* and *FD* were amplified using the primer pairs listed in Table S5 and then inserted into the pUC_SPYCE [42] vector, which contains the DNA encoding the C-terminus of YFP. The recombined pUC_SPYNE/CE plasmids were cotransformed into Arabidopsis protoplasts using polyethylene glycol–mediated transfection, as described previously [43]. YFP-dependent fluorescence was detected 24 h after transfection using a confocal laser-scanning microscope (Zeiss LSM 510 Meta).

Electrophoretic mobility shift assay

The full-length coding region of *GmFDL19* was amplified by PCR using the primer pair *GmFDL19*-29b-F/R (Table S6). The PCR product and the pET29b plasmid (Novagene, WI, USA) were digested with *NdeI* and *SalI*. After ligation, the construct was transformed into the *E. coli* competent cell line BL21 (DE3) (Transgene, Beijing, China) according to the manufacturer's instructions. The recombinant GmFDL19 protein was purified using the His tag purification nickel ion system (Kangweishiji, Beijing, China). EMSA was conducted using the recombinant GmFDL19 protein and the DNA products of the *GmAP1a* promoter obtained by hybridizing the forward and reverse complementary oligos containing the ACGT core sequence (Table S6). The EMSA assay was conducted using the EMSA kit (Invitrogen, www.Invitrogen.com, Cat #E33075). The DNA-protein complex samples were loaded into a TBE gradient 6% polyacrylamide native gel (Bio-Rad Laboratories, www.bio-rad.com) at 200 V for 45 minutes. The DNA in the gel was stained

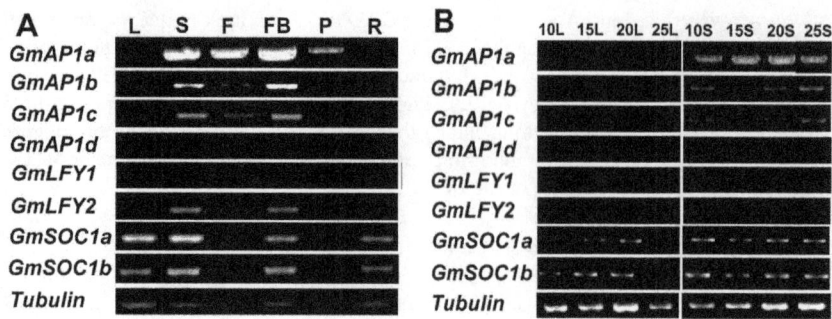

Figure 2. Temporal and spatial expression of soybean flowering-related genes. (A) Transcript levels of eight soybean flowering-related genes (*GmAP1a*, *GmAP1b*, *GmAP1c*, *GmAP1d*, *GmSOC1a*, *GmSOC1b*, *GmLFY1*, *GmLFY2*) in leaves and shoot apices of the soybean cultivar Harosoy under SD (12L/12D) conditions; *Tubulin* is included as an endogenous control. Samples were collected from 10 DAE to 25 DAE. DAE: days after emergence. L: leaves; S: shoot apex. (B) Tissue-specific expression analyses of eight flowering-related genes by RT-PCR under SD (12L/12D) conditions. L: leaves, S: shoot apices, F: flowers, FB: flower buds, P: pods, R: roots.

using SYBR Green, provided in the same kit, and visualized using the GE Typhoon LFA 9500 Imaging System (GE Healthcare Life Science, USA).

Results

Overexpression of GmFT2a and GmFT5a causes precocious flowering in soybean

Two *FT* homologs, *GmFT2a* and *GmFT5a*, are involved in photoperiod-regulated flowering, and these two genes coordinately control flowering in soybean [35]. To determine how these two *FT* homologs regulate soybean flowering, *GmFT2a* and *GmFT5a* were genetically transformed into the soybean cultivar Williams 82 under the control of the cauliflower mosaic virus (CaMV) 35S promoter. The overexpression of *GmFT2a* and *GmFT5a* caused the early flowering of Williams 82 even under noninductive LD (16L/8D) conditions (Figure 1A, B, C, D and Figure 1E, F, G, H). The transgenic *GmFT2a* T2 overexpression line #2-1-1 flowered at approximately 33 days after emergence (DAE) and the transgenic *GmFT5a* T2 overexpression line #5-1 flowered at approximately 35 DAE; however, the untransformed Williams 82 flowered at approximately 57 DAE (Figure 1I). These data suggested that both *GmFT2a* and *GmFT5a* are able to induce early flowering in soybean under noninductive LD conditions.

GmFT2a and GmFT5a upregulate floral meristem identity genes

All flowering pathways converge onto floral integrators, including *FT* and *SOC1*, and induce the expression of floral meristem identity genes, including *AP1* and *LFY* [44], [45]. Several genes involved in the determination of flowering time have recently been isolated and characterized in soybean, including *GmAP1*, *GmSOC1*, *GmSOC1-like* and *GmLFY* [38], [39], [46]. By searching the soybean reference genome using Phytozome (http://www.phytozome.net/soybean), four *AP1* homologs were identified and designated as *GmAP1b* (Glyma01g08150), *GmAP1c* (Glyma08g36380), *GmAP1d* (Glyma02g13420) and *GmAP1a* (Glyma16g13070, *GmAP1*), the last of which has been characterized previously [39]. Spatial RT-PCR analyses suggested that *GmAP1a*, *GmAP1b* and *GmAP1c* were mainly transcribed in reproductive organs such as shoot apices, flower buds and flowers under SD (12L/12D) conditions in the cultivar Harosoy, with the transcription of *GmAP1a* being the most prominent; however, the transcription of *GmAP1d* was not detected in any tissues (Figure 2A). Two *LFY* homologs, designated as *GmLFY1*

(Glyma04g37900) [46] and *GmLFY2* (Glyma06g17170), were also identified in the soybean genome. *GmLFY1* was transcribed mainly in developing pods and seeds and was not detected in leaves or the SAM (shoot apex meristem) (Figure 2A), suggesting that the gene might contribute to seed development in soybean instead of flowering, as was previously reported [46]. Similarly to the *AP1* homologs, *GmLFY2* was also transcribed in shoot apices, flower buds and flowers (Figure 2A). Two soybean *SOC1* homologs, *GmSOC1a* (Glyma18g45780) [39] and *GmSOC1b* (Glyma09g40230) [38], could be identified from the soybean genome. Both of these genes were highly expressed in shoot apices, leaves, flower buds and roots but were weakly expressed in flowers and pods (Figure 2A), in agreement with the expression patterns observed for *SOC1* in multiple organs of Arabidopsis [47]. In total, six floral meristem identity genes, *GmAP1a*, *GmAP1b*, *GmAP1c*, *GmSOC1a*, *GmSOC1b* and *GmLFY2*, were constantly expressed in the shoot apices of the cultivar Harosoy from 10 DAE to 25 DAE under SD (12L/12D) conditions before the floral bud formation stage (floral buds formed at 25 DAE), and *GmSOC1a* and *GmSOC1b* were also constantly expressed in the leaves of the cultivar (Figure 2B). The constantly high expression levels of these genes in shoot apices before the floral bud formation stage most likely indicate their involvement in the flowering transition of soybean.

In Arabidopsis, FT controls photoperiod-regulated flowering by activating the downstream flower meristem identity genes *AP1*, *LFY* and *SOC1* [12]–[14], [48]–[50]. To determine whether GmFT2a and GmFT5a induce early flowering in soybean by regulating the orthologs of *AP1*, *LFY* and *SOC1*, the expression levels of *GmAP1* (*a*, *b*, *c*), *GmSOC1a*, *GmSOC1b* and *GmLFY2* were determined using quantitative PCR in the leaves and shoot apices of the transgenic *GmFT2a* and *GmFT5a* overexpression soybean lines. In the transgenic *GmFT2a* soybean line #2-1-1, the expression levels of *GmFT2a*, *GmAP1* (*a*, *b*, *c*), *GmSOC1a*, *GmSOC1b* and *GmLFY2* were significantly higher in the shoot apices than were levels in untransformed Williams 82 (Figure 3A), while the expression of *GmFT5a* remained unchanged. However, in the transgenic *GmFT5a* soybean line #5-1, the expression levels of *GmFT5a*, *GmAP1* (*a*, *b*, *c*), and *GmSOC1b* were significantly higher in the shoot apices than were levels in untransformed Williams 82 (Figure 3B), while the expression levels of *GmFT2a*, *GmSOC1a* and *GmLFY2* were unchanged (Figure 3B). These results suggest that the flowering genes *GmAP1*s, *GmSOC1a*, *GmSOC1b* and *GmLFY2* are downstream of *GmFT2a* and *GmFT5a* and that these genes are most likely differentially involved in the GmFT2a and GmFT5a–induced early flowering of soybean. In addition, the expression of

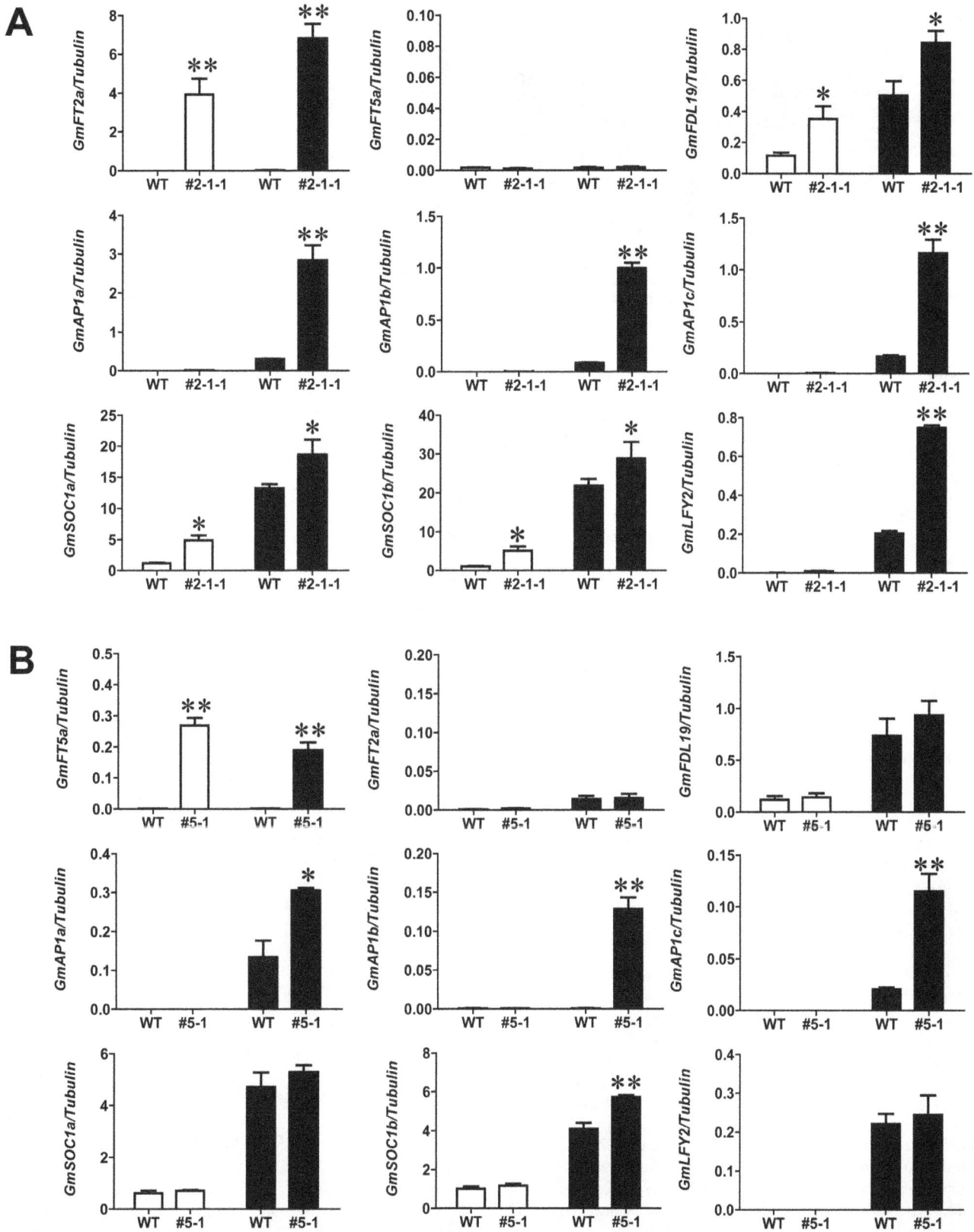

Figure 3. *GmFT2a* and *GmFT5a* promote the expression of soybean flowering-related genes. (A) Expression analyses of *GmFT2a*, *GmFT5a* and flowering-related genes in transgenic *GmFT2a* plants (#2-1-1) and wild-type Williams 82 plants (WT). (B) Expression analyses of *GmFT5a*, *GmFT2a* and flowering-related genes in transgenic *GmFT5a* plants (#5-1) and wild-type Williams 82 plants (WT). The white and black columns represent relative expression in leaves and shoot apices, respectively. Asterisks and double asterisks indicate significant differences between transgenic and WT plants at $0.01 < P < 0.05$ and $P < 0.01$, respectively.

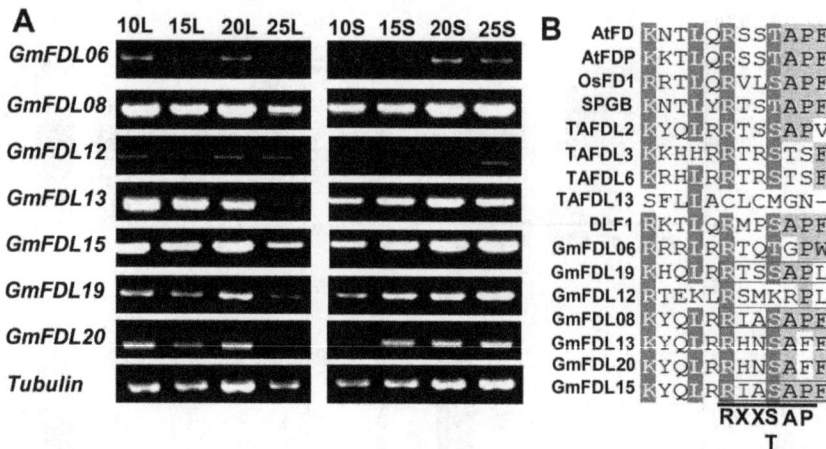

Figure 4. Seven *GmFDLs* (soybean *FD*-like genes) transcribed in leaves and shoot apices. (A) Transcript levels of seven *GmFDLs* in leaves and shoot apices of the soybean cultivar Harosoy under SD (12L/12D) conditions; *Tubulin* is included as an endogenous control. Samples were collected from 10 DAE to 25 DAE. L: leaves; S: shoot apices. (B) Multiple alignment of the amino acid sequences in the SAP motif region of the FDs from soybean and other species. The SAP motif is a putative sequence for FT binding.

GmFT2a in the *GmFT5a* transgenic line and the expression of *GmFT5a* in the *GmFT2a* transgenic line were both unchanged in the leaves and shot apices, suggesting that *GmFT2a* and *GmFT5a* promote soybean flowering in a redundant manner (Figure 3).

GmFT2a and GmFT5a interact with GmFDL19

FT and its homologs are widely understood to move from leaves to shoot apices, where they interact with FDs to form FT/FD complexes that bind to the promoter of *AP1* [13], [18], [19]. In this study, we found that GmFT2a and GmFT5a promote flowering by inducing the expression of *GmAP1s*, *GmSCO1s* and *GmLFY2*, and it is easily assumed that GmFTs also require a partner such as GmFD to regulate the downstream flowering genes in soybean. Taking the amino acid sequence of FD from Arabidopsis as the query, we searched for orthologs of FD in the soybean genome using Phytozome and identified 18 high-scoring candidate GmFDLs (GmFD-Like) genes (Figure S1). Expression analyses were conducted using RT-PCR for all eighteen selected *GmFDLs*, of which seven *GmFDLs* were transcribed both in leaves and shoot apices (Figure 4A). A multiple sequence alignment of the seven soybean FD-like proteins with the FDs from other species shows a conserved bZIP domain of 42 amino acids (N-X7-R-X9-L-X6-L-X6-L) (Figure S2) and an SAP (RXXS/TAP) motif (Figure 4B), SAP motif has been reported as a putative binding sequence for FT [13]. Phylogenetic analysis was conducted based on the amino acid sequences of the 18 candidate GmFDLs, and the proteins were grouped into three clades (Figure S1, Table S7). GmFDL02, GmFDL04 and GmFDL0602 were divided into the FD clade, but these three GmFDLs did not transcribe or share an SAP motif. GmFDL08, GmFDL13, GmFDL15, GmFDL19 and GmFDL20 were divided into the wheat TaFDL2 clade, with GmFDL08, GmFDL15 and GmFDL19 sharing the classic SAP motif; these three GmFDLs were therefore tested for interactions with soybean GmFT2a and GmFT5a. GmFDL06 and GmFDL12 were divided into the AREB and ABI5 cluster; these two proteins may represent the stress-related bZIP transcription factors in soybean.

GmFT2a and GmFT5a promoted early flowering in Arabidopsis [35], [36], so we first examined the interactions of FD with GmFT2a and GmFT5a using the yeast two-hybridization assay. The results indicated that both GmFT2a and GmFT5a were able

to interact with FD (Figure 5A). Autoactivation tests of GmFT2a and GmFT5a in yeast confirmed that both proteins were unable to activate the reporter genes when used alone as bait (data not presented). The yeast two-hybridization assays of the three SAP motif proteins among the GmFDLs, GmFDL08, GmFDL15 and GmFDL19, with GmFT2a and GmFT5a revealed that only GmFDL19 was able to interact with GmFT2a and GmFT5a (Figure 5A). To validate the results of the yeast two-hybridization tests, *in vivo* BiFC analyses were conducted. These results confirmed that both FD and GmFDL19 could interact with GmFT2a and GmFT5a in the nuclei of Arabidopsis protoplasts (Figure 5B). Our results suggest that GmFT2a and GmFT5a promote early flowering in soybean, most likely through interacting with GmFDL19 and upregulating downstream floral identity genes in a manner similar to that of FT in Arabidopsis. *GmFDL19* was constantly transcribed both in leaves and shoot apices, with increasing expression levels in the shoot apices following the growth stage from 10 DAE to 25 DAE before floral bud formation (25 DAE) (Figure 4A). In addition, the expression of *GmFDL19* was upregulated by GmFT2a in the transgenic overexpression line #2-1-1, while the transcription of *GmFDL19* was unchanged in the transgenic GmFT5a overexpression line #5-1 (Figure 3A, B). The expression patterns and protein interactions of GmFDL19 strongly support this protein as the candidate soybean FD ortholog, which may participate differentially in the early flowering of soybean promoted by GmFT2a and GmFT5a. That is, GmFT2a promotes soybean early flowering through both transcriptional upregulation of and physical interaction with GmFDL19, but GmFT5a only promotes flowering through physical interaction with the protein.

Overexpression of *GmFDL19* causes early flowering in soybean

The expression patterns and protein interactions of GmFDL19 suggest that the protein might be involved in the GmFT2a- and GmFT5a-regulated flowering pathway in soybean. To determine the functions of GmFDL19 in soybean flowering, overexpression of *GmFDL19* driven by the 35S promoter was genetically transformed into the cultivar Dongnong 50 (Figure 6A, B). Under LD (16L/8D) conditions, this cultivar flowered early, at approximately 30 DAE. To observe clearer flowering differences between the transgenic T2 line #12-1 and untransformed Dongnong 50,

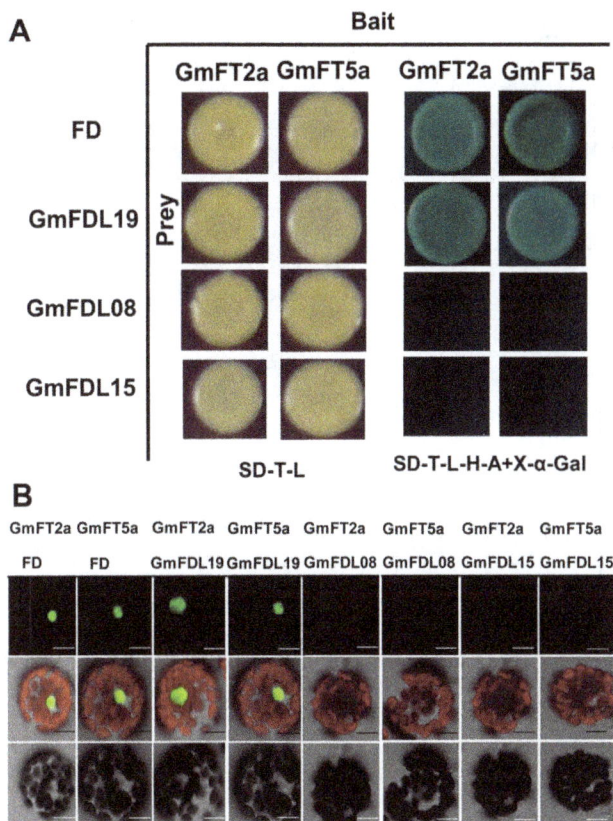

Figure 5. Interactions of GmFDLs with GmFT2a and GmFT5a.
(A) Yeast two-hybridization assays; FD was also included because both GmFT2a and GmFT5a promoted early flowering in Arabidopsis. After cotransformation of the baits and preys, an equal amount of yeast clones were plated on SD-Leu-Trp and SD-Leu-Trp-His-Ade+X-α-gal selective plates, and the plates were incubated at 30°C until the emergence of the yeast clones. (B) BiFC (bimolecular fluorescence complementation) assays to confirm the results of the yeast two-hybridization assays. Arabidopsis protoplasts cotransformed with constructs of FTs or FDs fused to the N-terminal (YN) and C-terminal (YC) halves of YFP, respectively (as indicated), were imaged using a confocal microscope after incubation at room temperature (20°C to 25°C) over 18 hours. Images are shown as YFP, merged YFP and bright field. Scale bars indicate 20 μm.

we evaluated the flowering times of both lines under longer photoperiod LD (18L/6D) conditions. Under these conditions, the transgenic line #12-1 flowered, on average, at approximately 43 DAE, while Dongnong 50 flowered at 55 DAE, indicating that GmFDL19 is able to promote flowering in soybean (Figure 6A, B, C, D, E).

We have demonstrated that GmFT2a and GmFT5a promote flowering by inducing the expression of flowering-related genes and that *GmFDL19* transgenic plants flower earlier than untransformed Dongnong 50 plants under an LD photoperiod. We next wished to determine whether these flowering-related genes were also upregulated in the shoot apices of *GmFDL19* transgenic plants. As expected, the transcription levels of *GmFDL19*, as well as those of the flowering related genes *GmAP1s*, *GmSOC1s* and *GmLFY2*, were significantly higher in the shoot apices of the transgenic *GmFDL19* line #12-1 than in those of untransformed Dongnong 50 (Figure 6F). However, the expression levels of *GmFT2a* and *GmFT5a* in the transgenic *GmFDL19* line #12-1 were unchanged and were only faintly detected in both #12-1 and Dongnong 50

under the LD (18L/6D) conditions (Figure 6F). Considering the upregulation of *GmFDL19* by GmFT2a, as well as the fact that GmFDL19 has the same effect on the upregulation of *GmAP1s*, *GmSOC1s* and *GmLFY2* as do GmFT2a, the results suggest that *GmFDL19* may act downstream of GmFT2a in the regulation of the flowering transition in soybean. In addition to the transcriptional upregulation of *GmFDL19* by GmFT2a, GmFDL19 interacts with both GmFT2a and GmFT5a, suggesting that GmFDL19 may be required for GmFT2a- and GmFT5a-regulated flowering in soybean.

GmFDL19 binds to the *GmAP1a* promoter *in vitro*

FT-like proteins interact with FD-like proteins to form FT/FD complexes, which bind to the core ACGT cis-elements located at the promoters of *AP1*-like genes in Arabidopsis, rice and wheat [14], [18], [19]. To determine whether this mechanism is conserved in soybean, we conducted an electrophoretic mobility shift assay (EMSA) to test the binding of GmFDL19 with the promoter of *GmAP1a*. *GmAP1a* was selected for the binding assay because it showed a higher expression level than did the other two homologs, *GmAP1b* and *GmAP1c* (Figure 2, Figure 3 and Figure 6F). This gene contains seven ACGT core elements and one CArG box, representing the putative binding site for MADS domain transcription factors in its promoter (Figure 7). The EMSA results demonstrated that GmFD19 binds to the ACGT core sequences *in vitro*. As the negative control, the CArG box could not be bound by GmFDL19 (Figure 7). These results suggest that the FT/FD-AP1 pathway is well conserved in soybean and that GmFDL19 serves as an important component of GmFT2a- and GmFT5a-regulated flowering in the legume.

Discussion

Photoperiod-regulated flowering is integrated through GmFT2a and GmFT5a in soybean

Soybean is adapted to wide range of latitudes, from at least North 50° to South 35°, and its wide adaptability can most likely be attributed to the genetic diversity at many of the major genes (*E1* to *E8* and *J*) and unclassified quantitative trait loci controlling its flowering and maturity. Of the major maturity genes, *E1* to *E4* delay flowering and maturity under LD but have no effects on flowering and maturity under SD [51]. The molecular identities of *E1* to *E4* have recently been characterized [52]. The *E1* gene has the largest effect of the maturity genes on delaying flowering and maturity under LD conditions [29], [33], [51], [53] and has been cloned using a map-based approach, revealing it to be a legume-specific transcription repressor with a putative nuclear localization signal (NLS) and a B3 distantly related domain [29]. The repression of flowering by E1 is most likely due to its suppression of the transcription of *GmFT2a* and *GmFT5a* under LD conditions [29], [36]. *E1* is transcribed mainly in vegetative organs such as cotyledons and leaves [29], which is consistent with the transcription sites of *GmFT2a* and *GmFT5a* [35]. The diurnal circadian rhythm of *E1* transcription contains two peaks in the leaves within 24 hours [29], of which the first transcription peak overlaps with the transcription peaks of *GmFT2a* and *GmFT5a* at 4 hours after dawn [35]. These results suggest that *GmFT2a* and *GmFT5a* are most likely the direct targets of E1 regulation, but this hypothesis requires further evidence for verification. *E3* and *E4* encode the light receptor phytochrome A (PHYA) proteins GmPHYA3 and GmPHYA2, respectively [31], [32]. The expression levels of *GmFT2a* and *GmFT5a* were additively suppressed by E3 and E4 [35], perhaps indirectly via E1 function, as shown in genetic studies revealing that *E3* and *E4* have epistatic

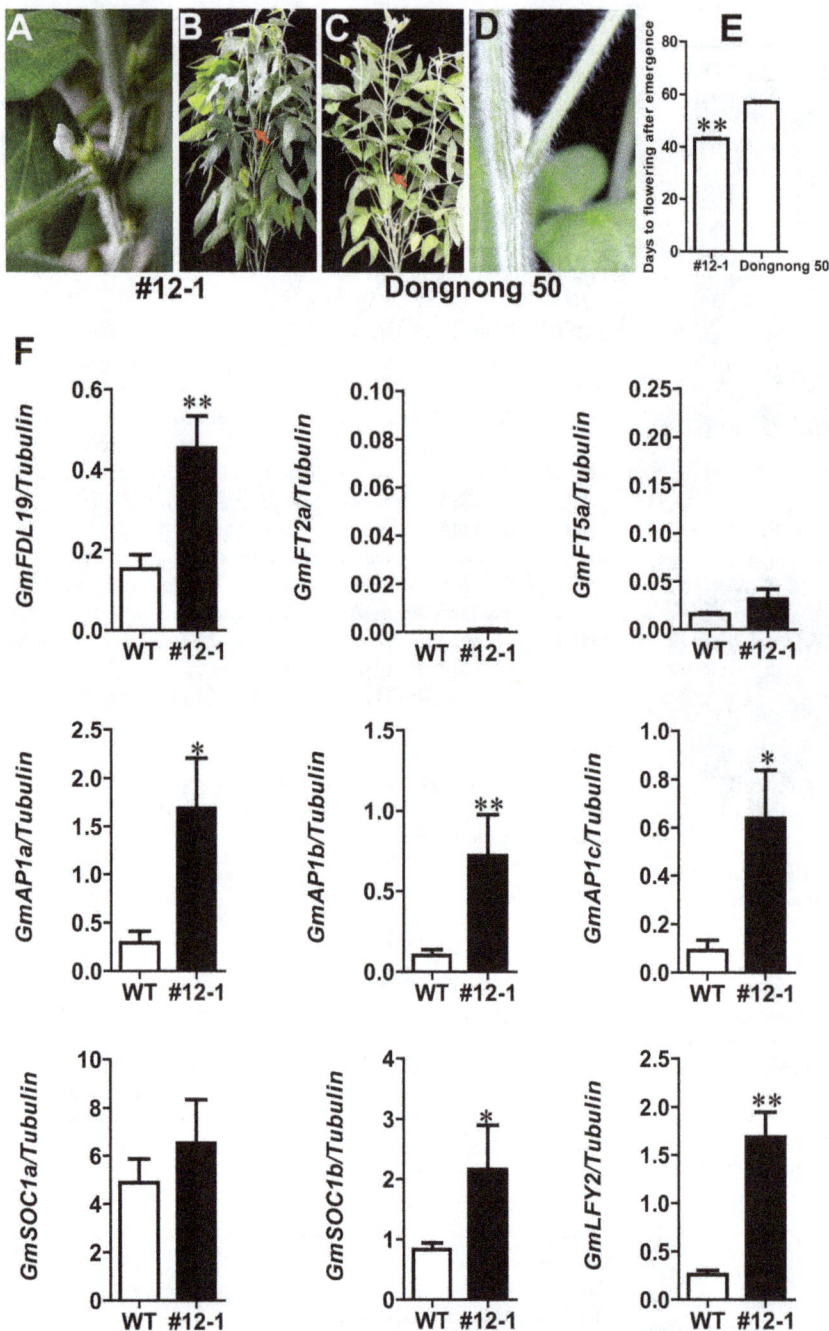

Figure 6. Overexpression of *GmFDL19* in the soybean cultivar Dongnong 50 causes early flowering. (A) The close shot of the transgenic plant in (B) shows the precocious flowers at the axils of the trifoliate leaves (B) A transgenic *GmFDL19* plant showing precocious flowering at the axils of the trifoliate leaves. (C) A wild-type Dongnong 50 plant. (D) The close shot of the wild-type Dongnong 50 plant in (C) does not show flowers at the axils of the trifoliate leaves. (E) Number of days to flowering in transgenic and wild-type plants. Averages and standard errors are calculated from five independent T2 plants and five Dongnong 50 plants. (F) Expression analyses of *GmFDL19* and flowering-related genes in transgenic *GmFDL19* plants (#12-1) and wild-type Dongnong 50 plants (WT); because these flowering related genes are transcribed mostly in shoot apices, shoot apex samples were collected from transgenic and wild-type Dongnong 50 plants at 40 DAE under LD (18L/6D) conditions. Asterisks and double asterisks indicate significant differences from the corresponding wild-type Dongnong 50 at $0.01 < P < 0.05$ and $P < 0.01$, respectively.

effects on *E1* under LD conditions [29]. *E2* was identified molecularly as an ortholog of the Arabidopsis *GIGANTEA* gene [30]. The *E2* gene mainly controls flowering time through the regulation of *GmFT2a*, not *GmFT5a* [30]. Taken together, these results indicate that the photoperiod-regulated flowering pathway in soybean converges at GmFT2a and GmFT5a.

GmFT2a and GmFT5a redundantly and differentially regulate flowering in soybean

An SD-to-LD transfer experiment demonstrated the differences in photoperiod response between *GmFT2a* and *GmFT5a*. The expression of *GmFT2a* was strictly regulated by photoperiodic changes from SD to LD, whereas the response of *GmFT5a* to

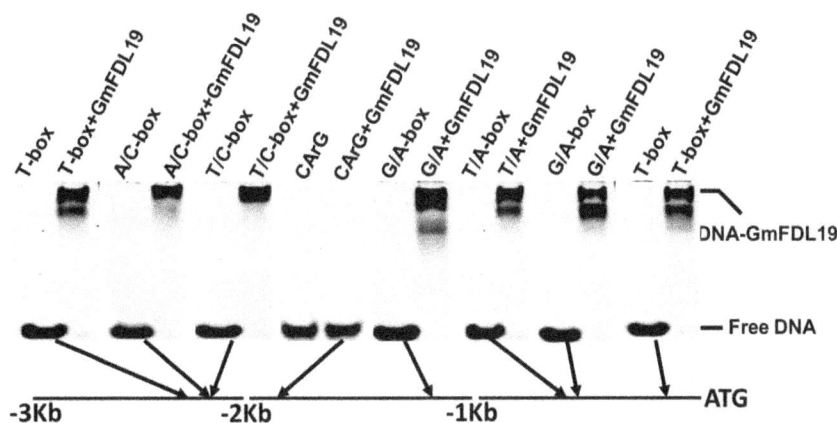

Figure 7. The GmFDL19 protein specifically binds with the ACGT core sequence *in vitro*. Potential bZIP binding sites presented in the *GmAP1a* promoter were used as probes in binding reactions with the purified recombinant GmFDL19 protein. The eight probes included seven potential bZIP binding sites and a CArG-box as negative control. (1) T-box (AACGTT), (2) A/C hybrid of A and C-box (TACGTC), (3) T/C hybrid of T and C-box (AACGTC), (4) CArG-box (CCNNNNNNNNNGG), (5) G/A hybrid of G and A-box (CACGTA), (6) T/A hybrid of T and A-box (AACGTA), (7) G/A hybrid of G and A-box (CACGTA), (8) T-box (AACGTT). The scheme below indicates the positions of the various bZIP binding sites.

photoperiodic changes was gradual, and its expression was maintained at low levels even after the plants were transferred to LD [35]. These findings suggest that, in addition to the phyA-mediated photoperiod response, a second regulatory mechanism may be involved in the differences of expression pattern between *GmFT2a* and *GmFT5a*. In addition to *E3* and *E4*, *E2* influences the mRNA abundance of *FT* homologs. Watanabe et al. (2011) found a clear association between flowering time and *GmFT2a* expression in two sets of near isogenic lines (NILs) for the *E2* locus [30]; dysfunctional *e2* alleles promoted *GmFT2a* expression and conditioned earlier flowering. However, these authors did not observe significant differences in *GmFT5a* expression between the NILs. These results suggest that the *E2* gene (*GmGIa*) mainly controls flowering time through the regulation of *GmFT2a* [30]. The different responses to photoperiodic changes observed between *GmFT2a* and *GmFT5a* [35] may thus be caused by the involvement of the GI (E2)-regulated pathway in *GmFT2a* expression, but not in *GmFT5a* expression. Under the phyA (E3 and E4)-mediated photoperiodic regulation system, GmFT2a and GmFT5a may redundantly and strongly induce flowering under shorter day lengths, but GmFT5a alone may promote flowering in a photoperiod-independent manner under longer day lengths.

The expression analyses of *GmFT2a* and *GmFT5a* in their respective transgenic overexpression lines further demonstrate that GmFT2a and GmFT5a are not regulated by each other, suggesting that GmFT2a and GmFT5a induce soybean flowering redundantly (Figure 3A, B). GmFT2a significantly upregulates downstream floral identity genes such as *GmAP1* (*a*, *b*, *c*), *GmSOC1a*, *GmSOC1b* and *GmLFY2*. However, GmFT5a only upregulates *GmAP1* (*a*, *b*, *c*) and *GmSOC1b* and has no effect on *GmSOC1a* and *GmLFY2* (Figure 3A, B). In addition, the expression of *GmFDL19* was upregulated by GmFT2a in the transgenic overexpression line #2-1-1 while the transcription of the gene was unchanged in the transgenic GmFT5a overexpression line #5-1 (Figure 3A, B). A hypothesis for the system was developed: GmFT2a, in combination with GmFDL19, triggers the upregulation of *GmLFY2*, and *GmLFY2* then feeds back directly to further upregulate *GmFDL19*. However, this hypothesis requires further confirmation. These results suggest that GmFT2a and GmFT5a induce soybean flowering differentially and redundantly. The GmFT2a-regulated flowering pathway and the GmFT5a-regulated flowering pathway

may be integrated in the SAM and are redundantly balanced in a very complex manner to ensure precise flowering in paleopoly-ploid soybean. These two FT homologs may therefore coordinately and redundantly control flowering in soybean.

GmFDL19 may be involved in GmFT2a- and GmFT5a-regulated flowering in soybean

FT and its homologs are widely known to move from leaves to shoot apices, where they interact with FDs to form FT/FD complexes that bind to the promoter of *AP1* and induce flowering in many plant species [13], [18], [19]. In this study, we report that the bZIP transcription factor GmFDL19 is able to physically interact with two soybean FT homologs, GmFT2a and GmFT5a, as confirmed by both yeast two-hybridization *in vitro* and BiFC *in vivo*. The binding of GmFDL19 with the cis-elements in the promoter of the *AP1* soybean ortholog *GmAP1a* was further confirmed by EMSA *in vitro*. Our results further extend the regulatory module of FT/FD-AP1 in the legume species soybean. In addition to the interaction of GmFDL19 with GmFT2a and GmFT5a, GmFT2a is able to upregulate the transcription of *GmFDL19* in shoot apices. The transcription levels of GmFT2a and GmFT5a are not regulated by GmFDL19, suggesting that GmFDL19 functions downstream of GmFT2a and GmFT5a. The floral identity genes *GmAP1* (*a*, *b*, *c*), *GmSOC1a*, *GmSOC1b* and *GmLFY2* are upregulated by GmFDL19, GmFT2a and GmFT5a in their respective transgenic overexpression lines. Taken together, these results suggest that GmFDL19 may be involved in GmFT2a- and GmFT5a-regulated flowering in soybean.

In summary, we propose a molecular network of photoperiod-regulated flowering in soybean (Figure 8). Under SD conditions, the *E3*, *E4* and *E1* genes do not express in leaves where *GmFT2a* and *GmFT5a* are able to transcribe at high levels, and the GmFT2a and GmFT5a proteins are then transported from the leaves to the shoot apices, where they bind with GmFDL19 to induce the expression of flowering-related genes (*GmAP1*, *GmSOC1*, *GmLFY*), thus leading to early flowering. Under LD conditions, *E3* and *E4* genes are highly transcribed in leaves, where they epistatically induce the high expression of the *E1* gene, thereby suppressing the expression levels of *GmFT2a* and *GmFT5a*. The expression levels of flowering related genes are not

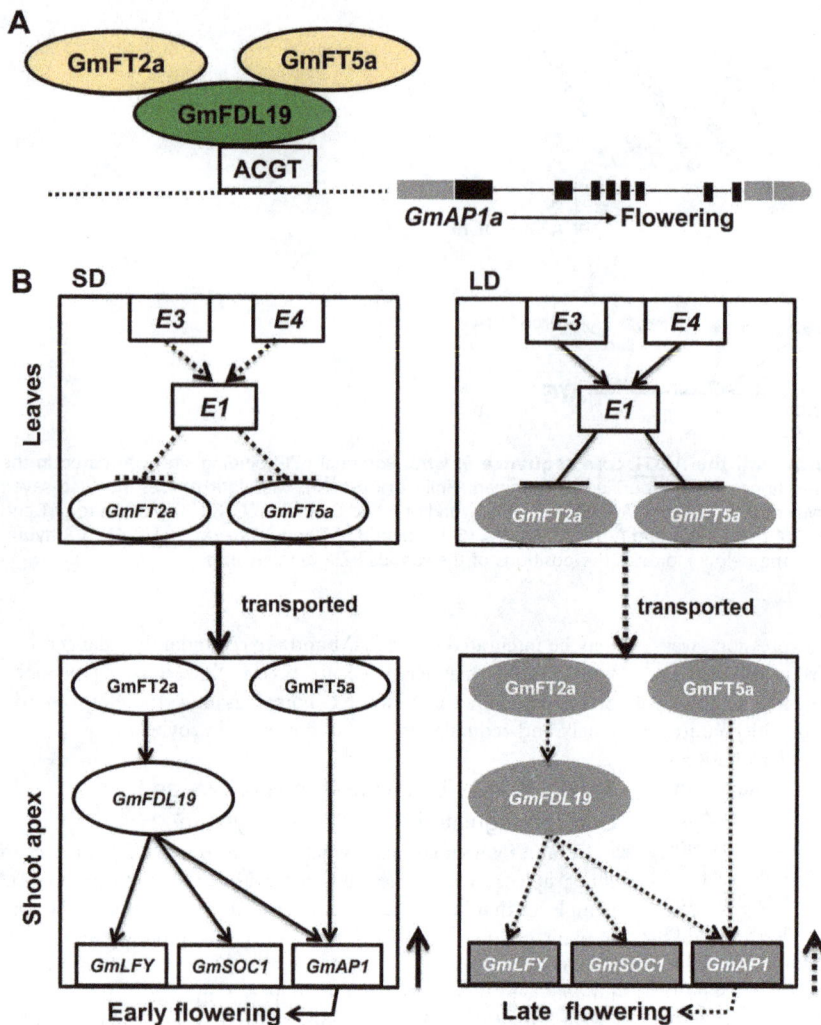

Figure 8. Proposed mechanism of photoperiod-regulated flowering controlled by GmFT2a and GmFT5a in soybean. (A) Model of GmFT2a and GmFT5a regulating the expression of *GmAP1a*. The horizontal dotted line represents the *GmAP1a* promoter, and the black vertical bars indicate the eight exons of *GmAP1a*. The green oval represents the GmFDL19 protein, and this protein can bind to the T-box, G-box or hybrid box (white rectangle) in the *GmAP1a* promoter. The orange oval represents the interactions of GmFT2a and GmFT5a with GmFDL19. (B) A proposed molecular network for photoperiod-regulated flowering in soybean.

upregulated due to lack of FT proteins, leading to a late-flowering phenotype.

Supporting Information

Figure S1 Phylogenetic relationship of soybean FD-like proteins and FDs from other species constructed using the neighbor-joining method with the program CLUSTAL W. Bootstrap percentage supports are indicated at the branches of the tree. The seven red filled rectangles indicate the bZIP domain of seven expressed *FD*-like genes in soybean, and the red rectangles indicated the SAP motif contained in soybean FD-like proteins and FDs from other species. The locus IDs or accession numbers of these FDs are presented in Table S7.

Acknowledgments

We thank Dr. Kan Wang, Iowa State University of Science and Technology for providing soybean transformation vector pTF101.1 and Agrobacterium strain EHA101; Miss Dandan Zhang, Miss Jiajia Deng, Mr. Hang Ren and Miss Lingli Kong for generating transgenic soybean lines.

Author Contributions

Conceived and designed the experiments: HYN BHL FJK. Performed the experiments: HYN DC DYZ YL SJL LLT. Analyzed the data: HYN XHY BHL FJK. Contributed reagents/materials/analysis tools: XHY BHL FJK. Contributed to the writing of the manuscript: HYN FJK.

References

1. Amasino R (2010) Seasonal and developmental timing of flowering. Plant J 61: 1001–1013.
2. Kardailsky I, Shukla VK, Ahn JH, Dagenais N, Christensen SK, et al. (1999) Activation tagging of the floral inducer *FT*. Science 286: 1962–1965.
3. Kobayashi Y, Kaya H, Goto K, Iwabuchi M, Araki T (1999) A pair of related genes with antagonistic roles in mediating flowering signals. Science 286: 1960–1962.
4. Bradley D, Ratcliffe O, Vincent C, Carpenter R, Coen E (1997) Inflorescence commitment and architecture in *Arabidopsis*. Science 275: 80–83.
5. Mimida N, Goto K, Kobayashi Y, Araki T, Ahn JH, et al. (2001) Functional divergence of the *TFL1*-like gene family in *Arabidopsis* revealed by characterization of a novel homologue. Genes Cells 6: 327–336.
6. Yoo SY, Kardailsky I, Lee JS, Weigel D, Ahn JH (2004) Acceleration of flowering by overexpression of *MFT* (*MOTHER OF FT AND TFL1*). Mol Cells 17.
7. Yamaguchi A, Kobayashi Y, Goto K, Abe M, Araki T (2005) *TWIN SISTER OF FT* (*TSF*) acts as a floral pathway integrator redundantly with *FT*. Plant Cell Physiol 46: 1175–1189.
8. Yoo SJ, Chung KS, Jung SH, Yoo SY, Lee JS, et al. (2010) *BROTHER OF FT AND TFL1* (*BFT*) has *TFL1*-like activity and functions redundantly with *TFL1* in inflorescence meristem development in *Arabidopsis*. Plant J 63: 241–253.
9. Michaels SD, Himelblau E, Kim SY, Schomburg FM, Amasino RM (2005) Integration of flowering signals in winter-annual *Arabidopsis*. Plant Physiol 137: 149–156.
10. Jang S, Torti S, Coupland G (2009) Genetic and spatial interactions between *FT*, *TSF* and *SVP* during the early stages of floral induction in Arabidopsis. Plant J 60: 614–625.
11. Takada S, Goto K (2003) TERMINAL FLOWER2, an *Arabidopsis* homolog of HETEROCHROMATIN PROTEIN1, counteracts the activation of *FLOWERING LOCUS T* by CONSTANS in the vascular tissues of leaves to regulate flowering time. Plant Cell 15: 2856–2865.
12. Yoo SK, Chung KS, Kim J, Lee JH, Hong SM, et al. (2005) Constans activates *suppressor of overexpression of constans 1* through *Flowering Locus T* to promote flowering in *Arabidopsis*. Plant Physiol 139: 770–778.
13. Abe M, Kobayashi Y, Yamamoto S, Daimon Y, Yamaguchi A, et al. (2005) FD, a bZIP protein mediating signals from the floral pathway integrator FT at the shoot apex. Science 309: 1052–1056.
14. Wigge PA, Kim MC, Jaeger KE, Busch W, Schmid M, et al. (2005) Integration of spatial and temporal information during floral induction in *Arabidopsis*. Science 309: 1056–1059.
15. Lifschitz E, Eviatar T, Rozman A, Shalit A, Goldshmidt A, et al. (2006) The tomato *FT* ortholog triggers systemic signals that regulate growth and flowering and substitute for diverse environmental stimuli. Proc Natl Acad Sci USA 103: 6398–6403.
16. Lin MK, Belanger H, Lee YJ, Varkonyi-Gasic E, Taoka KI, et al. (2007) FLOWERING LOCUS T protein may act as the long-distance florigenic signal in the cucurbits. Plant Cell 19: 1488–1506.
17. Tamaki S, Matsuo S, Wong HL, Yokoi S, Shimamoto K (2007) Hd3a protein is a mobile flowering signal in rice. Science 316: 1033–1036.
18. Li C, Dubcovsky J (2008) Wheat FT protein regulates *VRN1* transcription through interactions with FDL2. Plant J 55: 543–554.
19. Taoka K, Ohki I, Tsuji H, Furuita K, Hayashi K, et al. (2011) 14-3-3 proteins act as intracellular receptors for rice Hd3a florigen. Nature 476: 332–335.
20. Kobayashi K, Yasuno N, Sato Y, Yoda M, Yamazaki R, et al. (2012) Inflorescence meristem identity in rice is specified by overlapping functions of three *AP1/FUL*-like MADS box genes and *PAP2*, a *SEPALLATA* MADS box gene. Plant Cell 24: 1848–1859.
21. Bernard R (1971) Two major genes for time of flowering and maturity in soybeans. Crop sci 11: 242–244.
22. Buzzell R (1971) Inheritance of a soybean flowering response to fluorescent-daylength conditions. Can J Genet Cytol 13: 703–707.
23. Buzzell R, Voldeng H (1980) Inheritance of insensitivity to long daylength. Soybean Genet Newsl 7: 26–29.
24. McBlain B, Bernard R (1987) A new gene affecting the time of flowering and maturity in soybeans. J Hered 78: 160–162.
25. Bonato ER, Vello NA (1999) *E6*, a dominant gene conditioning early flowering and maturity in soybeans. Genet Mol Biol 22: 229–232.
26. Cober ER, Voldeng HD (2001) A new soybean maturity and photoperiod-sensitivity locus linked to *E1* and *T*. Crop Sci 41: 698–701.
27. Cober ER, Molnar SJ, Charette M, Voldeng HD (2010) A new locus for early maturity in soybean. Crop sci 50: 524–527.
28. Ray JD, Hinson K, Mankono J, Malo MF (1995) Genetic control of a long-juvenile trait in soybean. Crop sci 35: 1001–1006.
29. Xia Z, Watanabe S, Yamada T, Tsubokura Y, Nakashima H, et al. (2012) Positional cloning and characterization reveal the molecular basis for soybean maturity locus *E1* that regulates photoperiodic flowering. Proc Natl Acad Sci USA 109: E2155–E2164.
30. Watanabe S, Xia Z, Hideshima R, Tsubokura Y, Sato S, et al. (2011) A map-based cloning strategy employing a residual heterozygous line reveals that the *GIGANTEA* gene is involved in soybean maturity and flowering. Genetics 188: 395–407.
31. Watanabe S, Hideshima R, Xia Z, Tsubokura Y, Sato S, et al. (2009) Map-based cloning of the gene associated with the soybean maturity locus *E3*. Genetics 182: 1251–1262.
32. Liu B, Kanazawa A, Matsumura H, Takahashi R, Harada K, et al. (2008) Genetic redundancy in soybean photoresponses associated with duplication of the *phytochrome A* gene. Genetics 180: 995–1007.
33. Tsubokura Y, Matsumura H, Xu M, Liu B, Nakashima H, et al. (2013) Genetic variation in soybean at the maturity locus *E4* is involved in adaptation to long days at high latitudes. Agronomy 3: 117–134.
34. Xu M, Xu Z, Liu B, Kong F, Tsubokura Y, et al. (2013) Genetic variation in four maturity genes affects photoperiod insensitivity and PHYA-regulated post-flowering responses of soybean. BMC Plant Biol 13: 91.
35. Kong F, Liu B, Xia Z, Sato S, Kim BM, et al. (2010) Two coordinately regulated homologs of *FLOWERING LOCUS T* are involved in the control of photoperiodic flowering in soybean. Plant physiol 154: 1220–1231.
36. Thakare D, Kumudini S, Dinkins RD (2011) The alleles at the *E1* locus impact the expression pattern of two soybean *FT*-like genes shown to induce flowering in *Arabidopsis*. Planta 234: 933–943.
37. Zhong X, Dai X, Xv J, Wu H, Liu B, et al. (2012) Cloning and expression analysis of *GmGAL1*, *SOC1* homolog gene in soybean. Mol Biol Rep 39: 6967–6974.
38. Na X, Jian B, Yao W, Wu C, Hou W, et al. (2013) Cloning and functional analysis of the flowering gene *GmSOC1*-like, a putative *SUPPRESSOR OF OVEREXPRESSION CO1/AGAMOUS*-LIKE 20 (*SOC1/AGL20*) ortholog in soybean. Plant Cell Rep 32: 1219–1229.
39. Chi Y, Huang F, Liu H, Yang S, Yu D (2011) An *APETALA1*-like gene of soybean regulates flowering time and specifies floral organs. J Plant Physiol 168: 2251–2259.
40. Paz MM, Shou H, Guo Z, Zhang Z, Banerjee AK, et al. (2004) Assessment of conditions affecting agrobacterium-mediated soybean transformation using the cotyledonary node explant. Euphytica 136: 167–179.
41. Flores T, Karpova O, Su X, Zeng P, Bilyeu K, et al. (2008) Silencing of *GmFAD3* gene by siRNA leads to low α-linolenic acids (18: 3) of *fad3*-mutant phenotype in soybean [*Glycine max* (Merr.)]. Transgenic Res 17: 839–850.
42. Walter M, Chaban C, Schütze K, Batistic O, Weckermann K, et al. (2004) Visualization of protein interactions in living plant cells using bimolecular fluorescence complementation. Plant J 40: 428–438.
43. Sheen J (2002) Phosphorelay and transcription control in cytokinin signal transduction. Science 296: 1650–1652.
44. Blázquez MA (2000) Flower development pathways. J Cell Sci 113: 3547–3548.
45. Lee JH, Yoo SJ, Park SH, Hwang I, Lee JS, et al. (2007) Role of *SVP* in the control of flowering time by ambient temperature in *Arabidopsis*. Genes Dev 21: 397–402.
46. Meng Q, Zhang C, Huang F, Gai J, Yu D (2007) Molecular cloning and characterization of a *LEAFY*-like gene highly expressed in developing soybean seeds. Seed Sci Res 17: 297–302.
47. Lee J, Lee I (2010) Regulation and function of *SOC1*, a flowering pathway integrator. J Exp Bot 61: 2247–2254.
48. Ruiz-García L, Madueño F, Wilkinson M, Haughn G, Salinas J, et al. (1997) Different roles of flowering-time genes in the activation of floral initiation genes in *Arabidopsis*. Plant Cell 9: 1921–1934.
49. Moon J, Lee H, Kim M, Lee I (2005) Analysis of flowering pathway integrators in *Arabidopsis*. Plant Cell Physiol 46: 292–299.
50. Searle I, He Y, Turck F, Vincent C, Fornara F, et al. (2006) The transcription factor *FLC* confers a flowering response to vernalization by repressing meristem competence and systemic signaling in *Arabidopsis*. Genes Dev 20: 898–912.
51. Cober E, Tanner J, Voldeng H (1996) Genetic control of photoperiod response in early-maturing, near-isogenic soybean lines. Crop sci 36: 601–605.
52. Xia Z, Zhai H, Liu B, Kong F, Yuan X, et al. (2012) Molecular identification of genes controlling flowering time, maturity, and photoperiod response in soybean. Plant Syst Evol 298: 1217–1227.
53. Liu B, Fujita T, Yan Z-H, Sakamoto S, Xu D, et al. (2007) QTL mapping of domestication-related traits in soybean (*Glycine max*). Ann Bot 100: 1027–1038.

7

Pea p68, a DEAD-Box Helicase, Provides Salinity Stress Tolerance in Transgenic Tobacco by Reducing Oxidative Stress and Improving Photosynthesis Machinery

Narendra Tuteja[1]*, Mst. Sufara Akhter Banu[1], Kazi Md. Kamrul Huda[1], Sarvajeet Singh Gill[2], Parul Jain[1], Xuan Hoi Pham[1¤], Renu Tuteja[1]

1 International Centre for Genetic Engineering and Biotechnology, Aruna Asaf Ali Marg, New Delhi, India, 2 Stress Physiology and Molecular Biology Lab, Centre for Biotechnology, MD University, Rohtak, India

Abstract

Background: The DEAD-box helicases are required mostly in all aspects of RNA and DNA metabolism and they play a significant role in various abiotic stresses, including salinity. The *p68* is an important member of the DEAD-box proteins family and, in animal system, it is involved in RNA metabolism including pre-RNA processing and splicing. In plant system, it has not been well characterized. Here we report the cloning and characterization of *p68* from pea (*Pisum sativum*) and its novel function in salinity stress tolerance in plant.

Results: The pea p68 protein self-interacts and is localized in the cytosol as well as the surrounding of cell nucleus. The transcript of pea *p68* is upregulated in response to high salinity stress in pea. Overexpression of *p68* driven by constitutive cauliflower mosaic virus-35S promoter in tobacco transgenic plants confers enhanced tolerances to salinity stress by improving the growth, photosynthesis and antioxidant machinery. Under stress treatment, pea *p68* overexpressing tobacco accumulated higher K^+ and lower Na^+ level than the wild-type plants. Reactive oxygen species (ROS) accumulation was remarkably regulated by the overexpression of pea *p68* under salinity stress conditions, as shown from TBARS content, electrolyte leakage, hydrogen peroxide accumulation and 8-OHdG content and antioxidant enzyme activities.

Conclusions: To the best of our knowledge this is the first direct report, which provides the novel function of pea *p68* helicase in salinity stress tolerance. The results suggest that p68 can also be exploited for engineering abiotic stress tolerance in crop plants of economic importance.

Editor: Jin-Song Zhang, Institute of Genetics and Developmental Biology, Chinese Academy of Sciences, China

Funding: Work on RNA/DNA metabolism and plant abiotic stress tolerance in N.T.'s laboratory is partially supported by Department of Biotechnology (DBT), Government of India and Department of Science and Technology (DST), Government of India. S.S.G. acknowledges the receipt of research grants from CSIR, New Delhi for helicase work. The funders had no role in study design, data collection and analysis, decision to publish, or preparation of the manuscript.

Competing Interests: The authors have declared that no competing interests exist.

* E-mail: narendra@icgeb.res.in

¤ Current address: Department of Plant Molecular Pathology and Abiotic Stress, The Institute of Agricultural Genetics, Hanoi, Vietnam

Introduction

The DEAD-box families of proteins are helicases conserved from bacteria to humans and are involved in a variety of nucleic acid metabolic processes such as replication, repair, recombination, transcription, pre-mRNA processing, RNA degradation, RNA export, ribosome assembly and translation [1–3]. At the sequence level, helicases have been classified into five superfamilies (SF1–SF5). The largest of these groups are SF1 and SF2. Most of these contain nine conserved helicase motifs, Q, I, Ia, Ib, II, III, IV, V and VI. All the helicases exhibit nucleic acid dependent ATPase activity which provides energy for the helicase action [3,4].

The p68 is a prototype member of DEAD-box family and is one of the best characterized helicases and it plays a very important role in cell/organ development and participates in a variety of

biological processes including pre-mRNA and pre-rRNA processing [5–6], rearrangement of RNA secondary structures [6], RNA splicing [7–8] and gene transcription. In human, *p68* is a RNA-binding protein endowed with an ATP-dependent RNA helicase, RNA-dependent ATPase and RNA-protein complex remodeling activities. In human malaria parasite *Plasmodium falciparum*, p68 has also been reported as a dual helicase and its helicase and ATPase activities are stimulated after phosphorylation with protein kinase C [9–10].

The DEAD-box RNA helicases are becoming a subject of attention as they play a significant role during development and stress responses in plants [11–14]. Each DEAD box RNA helicase is thought to be differentially regulated during development and in response to environmental stresses in plants [2,13]. *Arabidopsis* LOS4 and RCF1 (a DEAD-box RNA helicase) were reported to regulate gene expression in response to chilling stress [8,15]. The overexpression of pea DNA helicase (PDH45) was shown to confer salt tolerance in tobacco and rice [12,14]. A rice DEAD-box RNA helicase (OsBIRH1) has been reported to function in defense responses against pathogen and oxidative stresses [16]. OsSUV3, a member of DEAD-box RNA helicase, has been shown to be involved in salinity stress tolerance recently [17]. Splicing factors are also known to affect the alternative splicing of many genes which lead to alter the gene expression through splicing factor networks and thereby involving in plant stress tolerance [18]. The plant homologue of animal splicing factor p68 has not been characterized in detail till date. The transcript of *Arabiopsis thaliana p68* DEAD-box RNA helicase (AtDRH1) was reported to be accumulated at a high level and almost equally in every part of the *Arabidopsis* plant [19]. Studies on MA16 (maize RNA-binding protein) and ZmDRH1 (*Z. mays* DEAD-box RNA helicase 1) revealed that these proteins might be part of a ribonucleoproteins complex involved in ribosomal RNA (rRNA) metabolism [20]. However, the precise role of DEAD-box RNA helicase, especially the role of *p68* in abiotic stress tolerance, has not been reported so far. Here, we report detailed characterization of *p68* from pea (*Pisum sativum*). Our results show that pea p68 self-interacts and the overexpression of pea p68 enhances the salinity stress tolerance in tobacco by controlling the generation of stress-induced reactive oxygen species (ROS) through modulating antioxidative defence machinery thereby protecting the photosynthesis and yield.

Materials and Methods

Construction of Plasmid for Tobacco Transformation

The *pea p68* gene was isolated by screening of pea cDNA library using a heterologous probe from *Arabidopsis thaliana*. The complete ORF of *p68* was cloned into pGEMT easy vector and sequenced (Accession number: AF271892.1). The insert (1.8 kb) was release from pGEMT-*p68* by digesting with EcoRI and BamHI restriction enzymes and ligated into the MCS of pRT101. The CaMV35S-*p68*-polyA cassette generated in pRT101 was then cut with Hind III and ligated to the MCS of pCAMBIA-1301 binary vector containing GUS as the reporter and hygromycin phosphotrans-ferase as the selection marker gene. The pCAMBIA1301-*p68* clone was then transformed into *Agrobacterium* LBA4404 strain for generation of transgenic tobacco using standard protocol [21]. Positive transgenic lines were confirmed by PCR using gene specific primers (FP: 5′- **GAATTC**ATGTCGTATGTTCCTC-CACAC-3/EcoRI site bold) and (RP: 5′- **GGATCC**CCAT-TACCTACAAACATGACTGAT-3/BamHI site bold). The transgenic tobacco plants were analyzed in two independent experiments with three technical replicates.

Transcript Analysis and Yeast-two-hybrid Assay

To analyze the expression of pea *p68* transcripts, northern blots analysis was performed as described earlier [11]. The pGBKT7-*p68* and pGADT7-*p68* were generated by inserting a PCR fragment encoding the complete ORF of *p68*. The empty yeast strain AH109, pGADT7, pGBKT7, pGADT7/pGBKT7, pGADT7/pGBKT7-*p68*, pGADT7-*p68*/pGBKT7 and pGBKT7-*p68*/pGADT7-*p68* were transformed separately on yeast cells. Yeast cells carrying all the plasmids were selected on the synthetic medium lacking Leu and Trp (SD-Leu-Trp-). The yeast cells were then streaked on SD medium [(Leu), (Trp), (His)] containing 30 mM 3-AT (3-Amino-1, 2, 4-triazole) to determine the expression of HIS3 nutritional reporter. The β-galactosidase expression of the fusion proteins encoded by pGBKT7-*p68*, pGADT7-*p68* constructs was assayed by colony filter lift assays as per manufacturer instruction (Clontech).

In vivo Localization by Immunofluorescence Staining and Confocal Microscopy

Polyclonal antibody against pea p68 was raised in rabbit as per the standard procedure. For *in vivo* localization, exponentially growing tobacco BY2 [11] suspension cells were fixed in 4% formaldehyde, permeabilized by cellulase and layered onto poly-L-lysine-coated cover slips. The cells were immunostained with p68-specific primary rabbit antibody in 1:2000 dilutions and Alexafluor 488-labeled goat antirabbit secondary antibody (Molecular Probes, Eugene, OR, USA) in 1:1000 dilutions. Counter staining of the cells with DAPI (4/, 6-diamidino-2-phenylindole), confocal laser scanning microscopy and image processing was carried out in a method described earlier [11].

Southern and Western Blots Analysis

Genomic DNA was isolated by CTAB method from PCR positive *p68* tobacco transgenic lines and WT. For southern blots analysis, ~20 μg of genomic DNA was digested with HindIII and resolved on agarose gel. The DNA was transferred to nylon membrane (Hybond N, Amersham Pharmacia, http://www.gelifesciences.com/), and hybridized with radiolabelled *p68* cDNA as described previously [12]. For western blot analysis, the total soluble proteins were isolated from the unstressed tissue samples of transgenic lines as well as WT plants and separated on 12% SDS-PAGE. Western blot analysis was performed by using anti-*p68* (1:5,000 dilutions) as primary and anti-rabbit (alkaline phosphatase conjugated antirabbit antibody-Sigma) as secondary antibody (1:12,500 dilutions). The blot was developed as per manufacturer's protocol (Sigma, USA).

Histochemical GUS Staining and Morphological Characterization of Transgenic Plants

The seedlings of pea *p68* overexpressing tobacco transgenic lines and WT were vacuum infiltrated for 10 m and histochemical *GUS* staining was performed by a method described earlier [22]. For *in vitro* pollen germination, mature pollen was isolated aseptically and cultured on pollen germination media (SMM: 0.3 M sucrose, 1.6 mM H_3BO_3, 3 mM $Ca(NO_3)_2$ $4H_2O$, 0.8 mM $MgSO_4.7H_2O$, 1 mM KNO_3) supplemented with 100 and 200 mM NaCl and incubated at 26°C in the dark. The germination status of the pollens was monitored and scored during a 7 d experimentation period. The sensitivity of seed germination to NaCl was assayed on MS agar plates saturated with 200 mM NaCl and incubated at 26°C under cool-white light for germination. Leaf disks assays, measurement of growth characteristics like shoot length, root length, leaf area and plant dry weight, measurement of tolerance index, determination of the total chlorophyll content and yield characteristics of transgenic and wild-type (WT) plants were performed as described earlier [14].

Measurement of Photosynthetic Characteristics and Photosystem II Activity (F_v/F_m)

The photosynthetic characteristics like net photosynthetic rate (P_N), stomatal conductance (gs), intercellular CO_2 concentration (Ci) and chlorophyll fluorescence (F_v/F_m) of transgenic lines and WT plants were recorded on the fully expanded leaves using infra-red gas analyzer (Li–6400, Li–COR, Lincoln, NE, USA) between 11:00 and 12:00 h. The conditions during the measurement were photosynthetically active radiation (PAR) 945±8 μmol m^{-2} s^{-1}, relative humidity 75±6%, temperature 28±2°C and an ambient CO_2 concentration of 350 μmol mol^{-1}. The chlorophyll fluores-cence i.e. maximal efficiency of PSII photochemistry (F_v/F_m) was also determined on the same leaves used for photosynthetic measurements after dark adaptation for 30 min.

Measurement of Ion Content

The Na$^+$ and K$^+$ ion content was measured as described earlier [23]. Salinity treated (0, 100 or 200 mM NaCl) leaves of the transgenic lines and WT plants were collected and rinsed with deionised water thoroughly. The fresh weight was determined for each sample. After drying (70°C for 48 h), dry weight was also measured. The samples are then subjected to an overnight digestion with HNO$_3$/H$_2$O$_2$. The materials were picked in 2 M HCl, and Na$^+$ and K$^+$ ion content was analyzed by using simultaneous inductively coupled plasma emission spectrometry (ICP trace analyzer, Labtam, Braeside, Australia).

Measurement of Oxidative Stress, Enzymatic Antioxidants (SOD, CAT, APX, GR) and Non-enzymatic Antioxidants (AsA and GSH) in p68 Transgenic Lines and WT

Oxidative stress was detected by measuring thiobarbituric acid reactive substances (TBARS), hydrogen peroxide (H$_2$O$_2$) content, electrolyte leakage and oxidative DNA damage (8-OHdG) in the leaves of *p68* overexpressing transgenic lines and WT plants by a previously described method [14]. The salinity stressed (0, 100 or 200 mM NaCl) and fully expanded leaves from transgenic and WT plants were used for the measurement of enzymatic antioxidants (superoxide dismutase, catalase, ascorbate peroxidase, glutathione reductase) and non-enzymatic antioxidants (ascorbate and glutathione). Leaf samples (transgenic lines and WT) were homogenized with an extraction buffer containing 100 mM potassium phosphate buffer (pH 7.0), 0.5% Triton X-100 and 1% polyvinylpyrrolidone (PVP) using pre-chilled mortar and pestle. The homogenate was centrifuged at 15,000×g for 20 min at 4°C. The supernatant obtained after centrifugation was used for enzyme assays. Measurement of enzymatic antioxidants (SOD, CAT, APX, and GR) and non-enzymatic antioxidants (AsA and GSH) was performed as described previously [14]. All the measurements were carried out 3 weeks after initiating the NaCl treatment.

Transgenic Plants and Salinity Stress Tolerance

For the assay of sensitivity to salinity stress, 14 d-old seedlings of WT and the transgenic plants (grown on vermiculite pots) were transferred to nutrient solutions containing 200 mM NaCl. After 2 d, growth status of the transgenic lines was observed. In another experiment, 40 plants of each line and WT were grown in vermiculite pots and watered for 10 d and then 200 mM NaCl solution was irrigated for every 3 d interval up to 12 d. After treatments, morphological changes were observed. In order to investigate the effect of salinity stress, 35-day-old plants were also subjected to salt stress for 28 d and growth was observed till maturity and photographs were taken.

Statistical Analysis

All the treatments were performed in three independent trials with consistent results. The results from only one representative experiment are shown, expressed as means± standard errors. Analysis of one-way variance (ANOVA) was performed on the data using SPSS (12.0 Inc., USA) to determine the least significant difference (LSD) for the significant data to identify the differences in the mean among the treatments. The means were separated by Duncan's multiple range tests (DMRT). The graphs were prepared using Sigmaplot Ver. 11. Different letters indicate significant difference at $P<0.05$.

Results

Isolation and Sequence Analysis of Pea p68 cDNA

The pea cDNA library was screened using a 1.9 kb cDNA fragment of *p68* from *Arabidopsis thaliana* (kindly provided by Tetsuo Meshi of Kyoto University, Japan) as a probe. This resulted in the isolation of a positive clone of pea *p68* (Figure S1A). The sequence analysis shows that the open reading frame (ORF) of pea *p68* cDNA (1.8 kb) encodes a protein of 622 amino acid residues, with a calculated molecular mass of 67.65 kDa and a pI of 6.46. It also exhibits all the known canonical helicase motifs (Q, I, Ia, Ib, II-VI) (Figure S1B). Phylogenetic analysis identified the closest orthologs of pea *p68* as *Arabidopsis thaliana* and *Saccharomyces cerevisiae p68* followed by *Oryza sativa p68* (Figure S1C).

Tissue Specific Distribution and Regulation of Pea p68 Transcript in Response to Abiotic Stresses

To investigate the expression pattern of the transcript level of pea *p68* in different parts of pea plant, the total RNA was isolated from root, leaf, tendril and flower and was subjected to northern blot analysis using 1.8 kb pea *p68* cDNA as the probe. For the internal control, 18S ribosome probe was used to show the equal loading. A single transcript of the expected size for pea *p68* mRNAs was present in all the tissues (Figure 1A). Further, to analyze the expression of pea *p68* under salinity stress, 7-day old pea seedlings were treated with 200 mM NaCl for 24 h and used for northern blot analysis. The results show that the pea *p68* transcript was dramatically up-regulated (~4.8 fold) in response to salinity stress (Figure 1B).

Pea p68 Interacts with Itself in the Yeast Two-hybrid System

To verify the self-interaction, we co-expressed BD-pea p68 with a construct encoding the full ORF of pea *p68* fused to the GAL4 activation domain (AD-Pea p68). Yeast cells carrying both the plasmids of BD-pea p68+ AD-Pea p68 were able to grow on 3DO media [Figure 1C (ii)]. All the clones including negative and positive control transformed in yeast cells were grown in YPD medium [Figure 1C (iii)]. Yeast cell carrying the plasmids of BD-pea p68+ AD-Pea p68 and Gβ-pGADT7+Gβ-pGBKT7 exhibited β-galactosidase activity, indicating the expression of reporter genes [Figure 1C (iv)]. These results confirm that the full-length pea p68 is able to self-interact in the yeast two-hybrid system.

In vivo Localization of p68

The *in vivo* localization of p68 was also analyzed by immuno-fluorescence labeling and observed under confocal microscopy. Immunofluorescent labeling of tobacco BY2 cells with anti-pea p68 antibodies showed p68 protein exclusively localized in the cytoplasm as well as in the surrounding of cell nucleus (Figure 2).

Overexpression of p68 and Molecular Analysis of Transgenic Tobacco Lines

To establish the functional significance of the *p68* gene, the complete ORF of the gene was cloned in Hind III site of pCAMBIA1301 (Figure 3A). Pea *p68*-pCAMBIA-1301 construct was overexpressed in tobacco plants by using *Agrobacterium*-mediated transformation. The integration of the transgene was confirmed by PCR (Figure 3B). The stable integration was also confirmed by Southern blot analysis (Figure 3C). Western blot detected ~68 kDa band in transgenic lines (Figure 3D). The GUS expression was positive for all the three transgenic lines while no GUS expression was observed in WT plant (Figure 3E). The

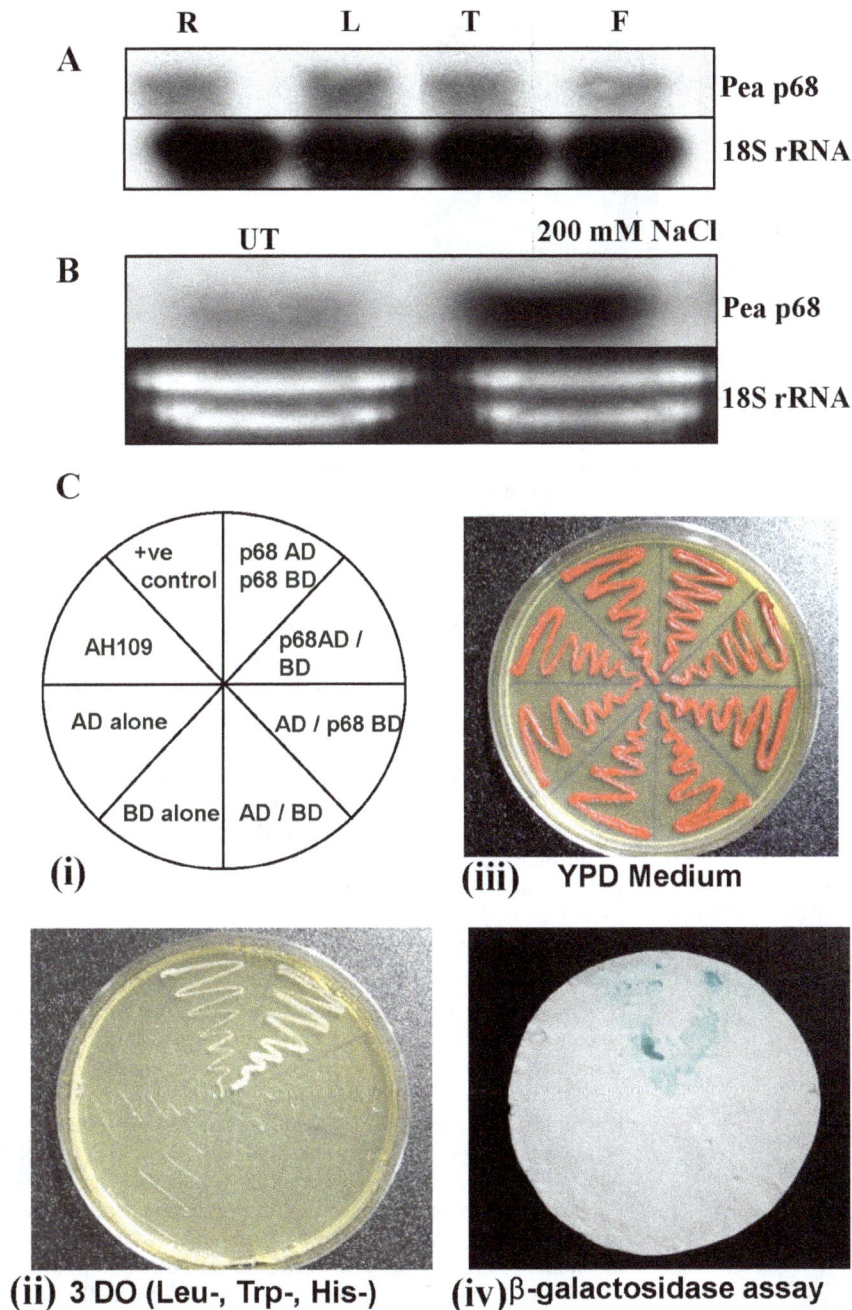

Figure 1. Expression analysis and self-interaction assay for pea *p68*. (A). Transcript analysis of pea p68 by Northern blot analysis (R: root, S: shoot, T: tendril and F: flower tissue respectively). (B) Transcript level in response 200 mM NaCl stress respectively (UT: untreated samples). About 30 μg of total was separated by electrophoresis, blotted and hybridized with the [32P]-labeled ORF of pea p68 cDNA (1.8 kb). For equal loading of RNA in each lane, the same blot was hybridized with the 18S rRNA, as shown in bottom of the each panel. (C) Yeast two-hybrid system-based self interaction showing pea p68 interacts with pea p68 *in vivo* (i) Template showing the organization of Y2H experiment (ii) Showing phenotypes on a YPD plate (iii) Yeast growing on a synthetic dextrose plate lacking leucine, tryptophan and histidine (3 DO) and (iv) β-galactosidase filter lift assay showing positive pea p68 self interaction. Yeast strain (AH109) carrying and G*β*-AD+G*γ*-BD used as a positive control for yeast two-hybrid assay.

germination efficiency of the seeds was also tested. The seeds of pea *p68* transgenic lines and WT were grown on MS media containing 200 mM NaCl (Figure 3F). In NaCl containing medium, the seeds of pea *p68* transgenic plants started to germinate at 3 d with germination efficiency of more than 70% (data not shown). In contrast, the germination of WT seeds was observed at 9 d on the NaCl-containing medium. The highest germination rates of WT seeds were 35.6% on the NaCl-

containing MS medium (data not shown), which was lower than those of pea *p68* transgenic seeds, suggesting the susceptibility of WT plants to 200 mM NaCl. The seedlings of *p68* transgenic S11 and WT were transferred to pots supplemented with 200 mM NaCl, WT plants died after sometime whereas S11 survived and resumed the growth (Figure 3G). Salinity stress tolerance was also tested by leaf disk senescence assay. Leaf disks of *p68* transgenic line S11 and WT were floated separately on 0 (H₂O only), 100 or

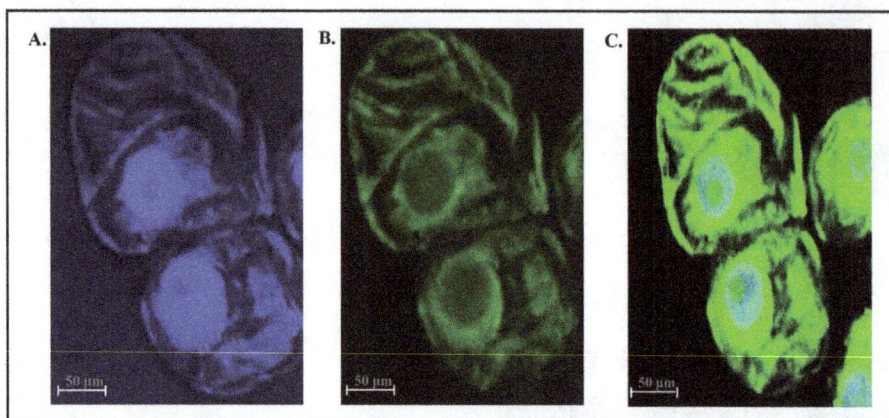

Figure 2. *In vivo* localization of the p68 in tobacco BY2 cells. The tobacco BY2 cells were fixed, permeabilized and immunostained with primary antibodies against p68 followed by Alexafluor 488-labeled secondary antibody and then counterstained with DAPI. A single confocal image is shown. (A) Image of cell stained with DAPI (blue). (B) Immunofluorescently stained cell (green). Anti-p68 labeling is restricted to the nucleus and cytosol. (C) Superimposed image of cell.

200 mM NaCl for 72 h. The salinity induced damage was reflected in the degree of bleaching observed in the leaf tissue after 72 h. The leaves of WT plants were bleached, whereas, the leaf disks of S11 retained chlorophyll (Figure 3H). The measurement of salinity stress tolerance index of the 200 mM NaCl treated *p68* transgenic line S11 and WT plants was made using the data of plant dry weight and it was noted that the tolerance potential of *p68* transgenic line S11 was 86.11%, whereas, it was only 30.47% in WT plants (Figure 3I). The performance of *p68* transgenic lines in the presence or absence of salt was similar to the WT plants, which revealed the potential of *p68* in salinity stress tolerance (Figure 3J).

The *p68* transgenic S11 line and WT plants were phenotypically similar when grown in absence of NaCl. To assess the effect of high salt (200 mM NaCl) on growth, morphology and development of *p68* overexpressing and WT plants, the three-week old seedlings were grown in the presence of continuous salt (200 mM NaCl). In the presence of NaCl some stress induced symptoms appeared on the transgenic plants but it still grew normally and set viable seeds, whereas the WT plants could not survive under continuous salinity stress (Figure 3K). In response to stress, the *p68* transgenic set healthy pods similar to H_2O grown WT plants (Figure 3L). The results show that the *p68* overexpressing transgenic lines have better ability to tolerate salinity stress. p68 overexpressing transgenic lines were further analysed for physiological and biochemical parameters to understand the mechanism of salinity stress tolerance in *p68* overexpressing transgenic lines.

Segregation Ratio and Seedling Survival of Transgenic and WT Plants Under Salinity Stress

To determine whether the salinity tolerance imparted by *p68* is functionally and genetically stable, the homozygous T_2 progeny was analyzed. Seeds from the T_0 plants, when plated onto hygromycin containing medium, segregated in 3:1 ratio (Table S1). The percent seedlings survival of *p68* overexpressing transgenic lines and WT were also observed and it was found that 200 mM NaCl did not affect the seedlings survival and there was no difference when compared with H_2O grown WT plants (Table S1).

Effect of Salinity on Germination of Pollens and Seeds of Pea *p68* Transgenic Tobacco Plants

To compare the germination efficiency, first pollens of WT and *p68* transgenic lines were sown on pollen germination medium (PGM) alone or on medium supplemented with either 100 or 200 mM NaCl. Without addition of NaCl, the germination of the pollen of pea *p68* transgenic was similar to that of WT pollen (Figure S2). The germination of the WT pollen was repressed in the medium containing 100 mM NaCl while no germination was observed in 200 mM NaCl (Figure S2 B, C).

Effect of Salinity on Growth Performance and Yield Transgenic and WT Plants

Significant growth reduction was noted for the WT plants in response to 200 mM NaCl treatment while *p68* transgenic lines resist the adverse effects of stress by maintaining vigorous growth. Growth performance measured in terms of shoot length, root length, leaf area and plant dry weight remained almost similar in *p68* overexpressing transgenic lines and WT plants under 0 mM NaCl (Figure S3). High concentration of salt (200 mM NaCl) significantly reduced the shoot length, root length, leaf area and plant dry weight of WT plants by 51.70, 57.31, 64.10 and 61.96%, in comparison to 0 mM NaCl. However, the decrease was 16.67, 19.55, 18.82 and 17.35% in the case of S9; 14.50, 18.04, 18.22 and 13.31% in the case of S11 and 18.85, 18.01, 16.92 and 15.62% in the case of S26 in comparison to their respective controls (Figure S3). Under salinity stress, transgenic lines maintained yield contributing parameters including time required for flowering, number of pods per plant, seed number per pod and seed weight per pod and therefore set normal seeds (Table S1).

Ion (Na$^+$ and K$^+$) Accumulation in Transgenic and WT Plants under Salinity Stress

To observe Na$^+$ and K$^+$ accumulation, WT and *p68* transgenic plants were exposed to salt stress. No significant difference was observed in the accumulation of Na$^+$ in between WT and *p68* overexpressing transgenic lines without NaCl treatment. However, with the increasing salt concentration (100 or 200 mM NaCl), the accumulation of Na$^+$ was significantly increased in WT plants while *p68* transformed plants accumulated less Na$^+$ (Figure 4A). The pattern of K$^+$ accumulation was similar in the leaves of *p68*

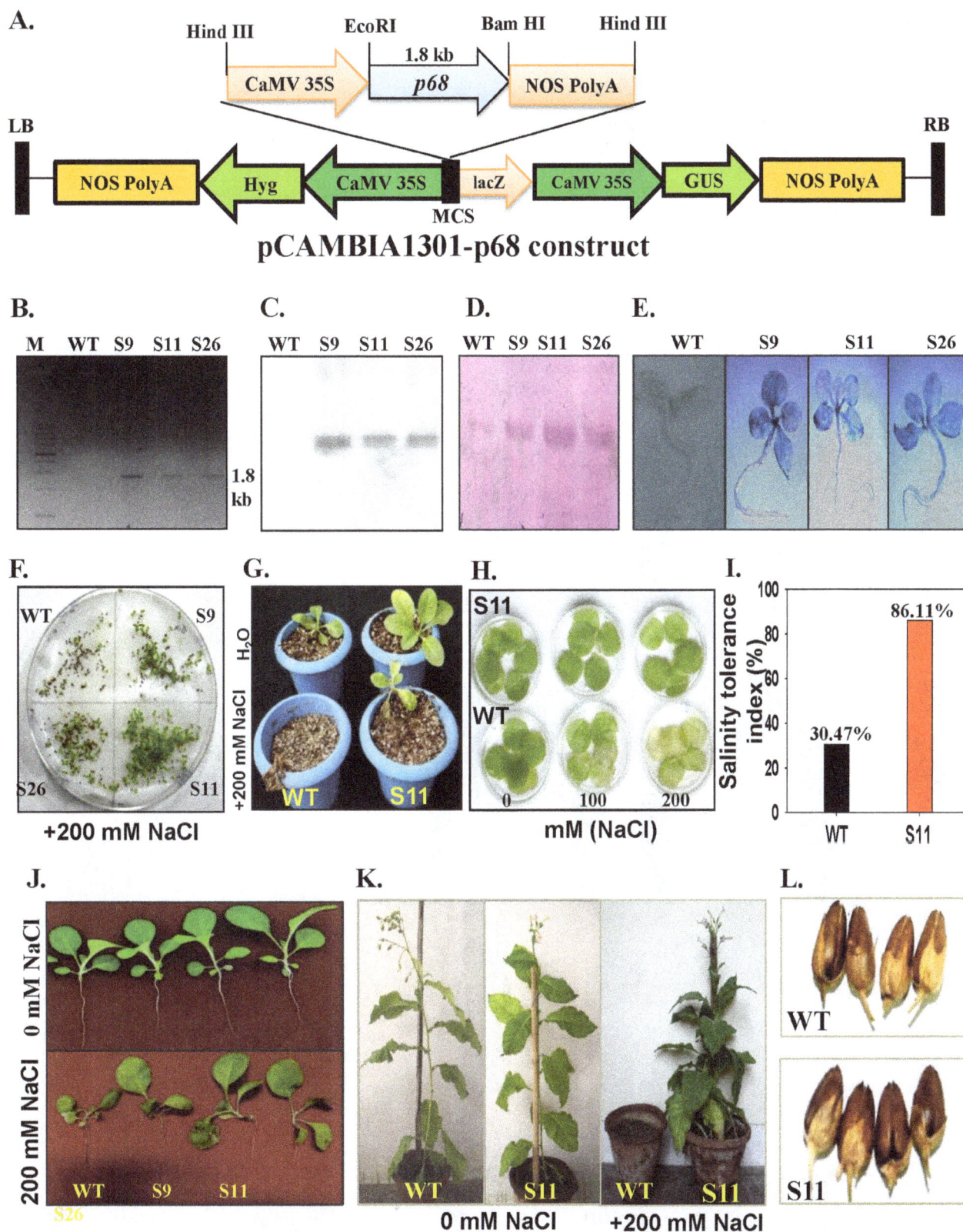

Figure 3. Molecular and morphophysiological analysis of pea *p68* overexpressing transgenic tobacco plants. (A) The map of the construct of pCAMBIA1301 containing the *p68* gene (pCAMBIA1301-*p68*). (B) PCR analysis using gene-specific primers. (C) Southern blot analysis for the integration of pea *p68* gene in tobacco genome. (D) Western blot analysis of each transgenic lines using anti-pea *p68* polyclonal antibody. (E) Histochemical GUS staining of each transgenic line. (F) Comparison of seeds germination of the WT and transgenic lines in response to 200 mM NaCl. (G) Phenotypic comparison of WT and transgenic line (S11) in response to 200 mM NaCl stress. 21 d-old-seedling growing in vermiculite pots and supplied with 200 mM NaCl solution for 5 d. (H) Leaf-disk senescence assay for salinity stress tolerance in transgenic line (S11). (I) Salinity tolerance index potential of WT and transgenic line (S11). (J) Stress responses of WT and *p68* overexpressing transgenic lines. 14 d-old vermiculite grown

seedlings were transferred to without or with NaCl in nutrient solution (0 and 200 mM NaCl). (K) WT and *p68* overexpressing plant (S11) in soil pots supplied without or with 200 mM NaCl solution. Note that the WT plant could not sustain growth under salinity stress. (L) Phenotypic comparison of pods grown in water and stress condition of WT and transgenic line (S11).

overexpressing transgenic lines and WT in response to 0 mM NaCl treatment (Figure 4B) but in response to salinity stress, *p68* overexpressing transgenic lines retained more K^+ compared to the WT plants (Figure 4B). Furthermore, *p68* overexpressing transgenic lines showed lower Na^+/K^+ ratio in comparison to WT plants (Figure 4C), which reflect the potential of transgenic lines to tolerate salinity stress.

Effect of Salinity on Chlorophyll Content, Photosynthesis and Chlorophyll Fluorescence of Transgenic and WT Plants

To access the effect of salinity on Chlorophyll (Chl), Chl a, Chl b, total Chl and Chl a: b was measured in *p68* overexpressing transgenic lines and WT plants (Figure 5A–D). Salinity stress (100 or 200 mM NaCl) significantly reduced the Chla, Chlb and total

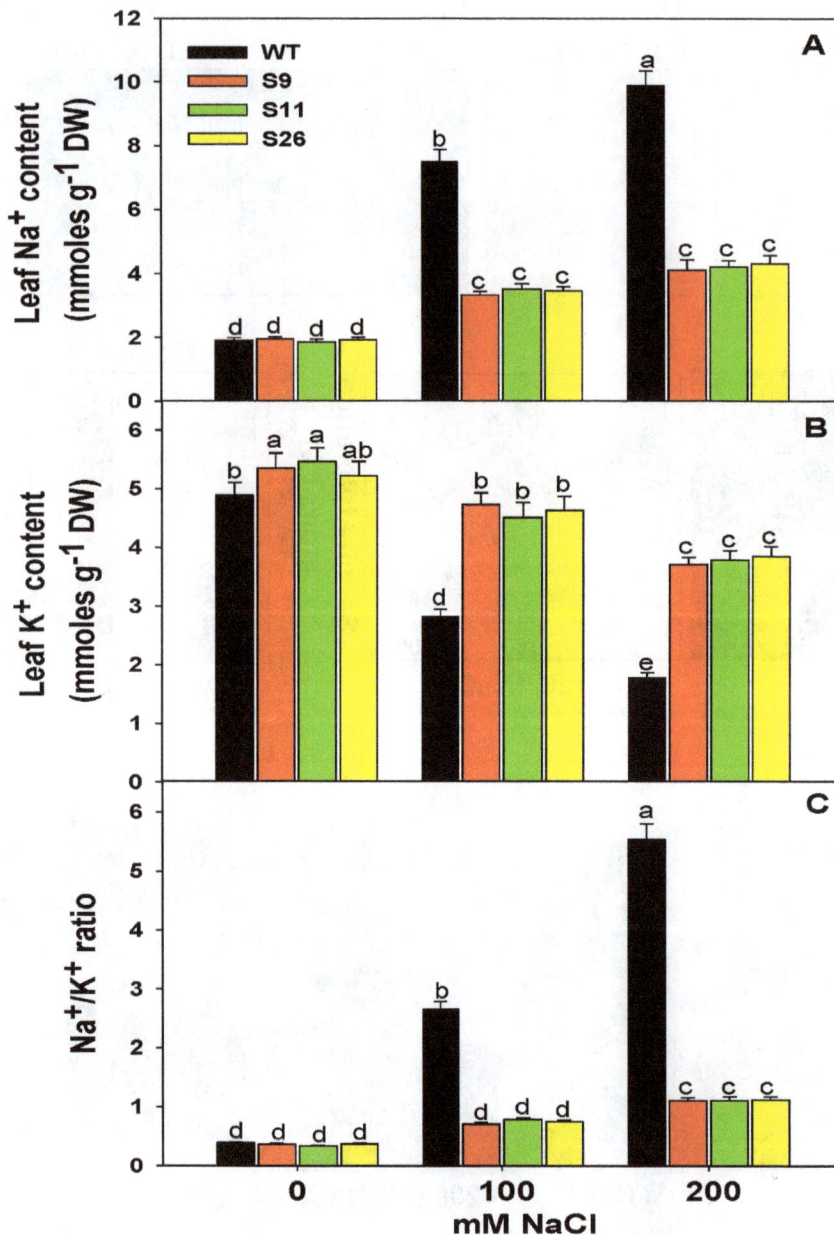

Figure 4. Analysis of ion content in transgenic and WT tobacco plants. (A) Na^+ content in the leaves of transgenic and WT plants. (B) K^+ content in the leaves of transgenic and WT plants. (C) Na^+/K^+ ratio in the leaves of transgenic and WT plants. The leaves were exposed to 0, 100 or 200 mM NaCl for 3 weeks of salinity treatment. Values are mean $\pm$ SE (n = 3). Different letters on the top of bars indicate significant differences at P< 0.05 level as determined by Duncan's multiple range test (DMRT). The results are representative of similar results obtained from two independent experiments.

Chl in transgenic lines and WT plants but the extent of reduction was higher in WT than *p68* overexpressing transgenic lines. The reduction in Chl a, Chl b and total Chl under 200 mM NaCl in WT was 46.56, 70.12 and 54.18% in comparison to 0 mM NaCl, whereas, the reduction was 16.65, 26.24 and 19.73% in the case of S9; 15.71, 21.61 and 17.59% in the case of S11 and 14.83, 23.73 and 17.69% in the case of S26 in comparison to their respective controls (Figure 5A–C). Under salinity stress the Chl content remained significantly higher in transgenic than WT plants. In WT plants the Chl a:b ratio followed the reverse pattern as of Chl content and significantly increased with the increasing salt concentration, whereas, no significant change was noted in the case of transgenic lines (Figure 5D). It reflects that Chl b was severely affected by salinity stress than Chl a under increasing salt concentration in WT plants than transgenic lines which led to significant increase in Chla: b ratio.

Salinity stress affects the photosynthetic functions at various levels, therefore, the photosynthetic parameters like net photosynthetic rate (P_N), stomatal conductance (gs) and internal CO_2 (*Ci*) were measured in *p68* overexpressing transgenic lines (S9, S11 & S26) and WT plants under different salinity levels (0, 100 or 200 mM NaCl) (Figure 5E–G). High level of salinity (200 mM NaCl) significantly reduced the photosynthetic parameters but the extent of reduction was several folds higher in WT plants than transgenic lines. It is interesting to note that *p68* overexpressing transgenic lines maintained higher photosynthesis than WT even under 0 mM NaCl. The reduction in P_N, gs and *Ci* of WT plants was 53.75, 49.32 and 55.44% under 200 mM NaCl in comparison to their controls, whereas, the decrease in P_N, gs and *Ci* of S9, S11 and S26 was 17.67, 16.77, 14.92%; 16.94, 15.87, 16.88% and 18.85, 18.54, 16.99%, respectively in comparison to their controls.

Chlorophyll fluorescence measurement is one of the most commonly used parameters to study the ecophysiology of plants under salinity stress. To understand the response of salt stress, we measured maximal efficiency of PSII photochemistry (F_v/F_m) in *p68* overexpressing transgenic lines and WT plants (Figure 5H). High salinity stress (100 or 200 mM NaCl) significantly reduced the F_v/F_m, whereas, it remained unaltered in transgenic lines. Statistically insignificant change in F_v/F_m in *p68* overexpressing transgenic lines reflects that PSII complex did not suffer damage under NaCl stress.

Less Oxidative Stress in *p68* Transgenic Lines than WT under Salinity Stress

Abiotic stresses including salinity cause overproduction of ROS, which leads to oxidative stress in plants. Therefore, the indicators of oxidative stress such as lipid peroxidation (TBARS content), H_2O_2 content, electrolyte leakage and oxidative DNA damage were studied in *p68* overexpressing transgenic lines and WT plants (Figure 6A–D). High concentration of salt (100 or 200 mM NaCl) significantly increased the extent of oxidative damage and it was significantly higher in WT as compared to *p68* transgenic lines. The response of oxidative stress parameters reflects non-significant difference in the values of TBARS content, H_2O_2 content, electrolyte leakage and oxidative DNA damage (8-OHdG) between *p68* overexpressing transgenic lines and WT plants under no salt i.e. 0 mM NaCl. The increase in TBARS, H_2O_2, electrolyte leakage and 8-OHdG under 200 mM NaCl in WT was 600.00, 337.44, 181.03 and 197.59% in comparison to 0 mM NaCl, whereas, the increase was 86.61, 72.76, 79.53 and 37.67% in the case of S9; 74.48, 102.39, 107.08 and 32.48% in the case of S11 and 70.20, 93.78, 84.48 and 53.62% in the case of S26 in comparison to their respective controls (Figure 6A–D). Overall, it is noted that WT plants suffered maximum oxidative damage

reflected in terms of peroxidation of lipids, nucleic acid and electrolyte leakage, whereas, *p68* overexpressing transgenic lines experienced less oxidative stress, therefore less oxidative damage.

Overexpression of *p68* Enhances ROS Scavenging Capacity in *p68* Transgenic Lines

Salinity stress is known to cause ROS induced oxidative damage in plant cells. Therefore, we analyzed the response of enzymatic (SOD, CAT, APX and GR) and non-enzymatic antioxidants (AsA and GSH) in *p68* overexpressing transgenic lines and WT plants under salinity stress. Antioxidant defense machinery protects the plant cells from ROS induced oxidative damage. SOD constitutes the primary step of cellular defense, where SOD dismutates $O_2^{\bullet}$ to H_2O_2 and O_2. The increase in SOD activity was noted in both *p68* transgenic lines and WT plants but activity of SOD was several folds higher as compared to WT plants. Significant increases in SOD activity in *p68* overexpressing transgenic lines under 200 mM NaCl was 59.92, 73.13, 61.12%, whereas, it was just 33.92% in WT in comparison to their respective controls (Figure 6E). The H_2O_2 generated by the action of SOD is restricted through the action of CAT and APX, which reduces H_2O_2 to water. In the present study, reverse to SOD activity, the CAT and APX activity decreased significantly in WT plants but CAT and APX activity showed significant increase in *p68* overexpressing transgenic lines under 200 mM NaCl (Figure 6F–G). The decrease in CAT and APX activity was 34.80 and 34.12% in WT plants under 200 mM NaCl. The activity of CAT and APX increased in *p68* overexpressing transgenic lines under 200 mM NaCl by 25.29, 153.06%; 35.68, 111.07% and 39.88, 154.19%, respectively, in comparison to their respective controls (Figure 6F–G). GR catalyzes the NADPH-dependent reduction of oxidized GSSG to the reduced GSH. Here, GR activity showed an increasing trend under salinity stress in both *p68* overexpressing transgenic lines and WT plants but maximum significant increase was seen in *p68* transgenic lines (Figure 6H). Increase in GR activity in *p68* overexpressing transgenic lines under 200 mM NaCl was 100.00, 101.54 and 95.31%, respectively, in comparison to their respective controls, whereas, it was just 44.26% in WT plants. Overall, the upregulation of ROS scavenging antioxidant enzymes in *p68* overexpressing transgenic lines protected the plant cells from the damaging effect of ROS generated by salinity stress.

AsA is the only antioxidant buffer in the apoplast and a key antioxidant that reacts with superoxide and hydroxyl radicals, whereas, GSH is a major non-enzymatic scavenger of ROS. The efficient coordination of AsA and GSH in AsA-GSH cycle can protect the plants from salinity induced oxidative damage. Under control condition (0 mM NaCl), ASC and GSH contents were same in control and transformed plants. However, *p68* overexpression enhances the pool of AsA and GSH in the transgenic lines. The decrease in AsA and GSH content was 48.74 and 50.00% in WT plants under 200 mM NaCl. The content of AsA and GSH increased in *p68* overexpressing transgenic lines under 200 mM NaCl by 65.32, 193.53%; 61.84, 233.33% and 72.35, 226.86%, respectively, in comparison to their respective controls (Figure 6F–G).

Discussion

Abiotic stress is known to affect the cellular gene-expression that limits crop productivity worldwide. So the molecules that are involved in nucleic acid processing, such as helicases, are expected to be affected in response to stress as well. It is evident that stress triggers the expression of many genes including DEAD-box helicases, which play a crucial role in various abiotic stresses

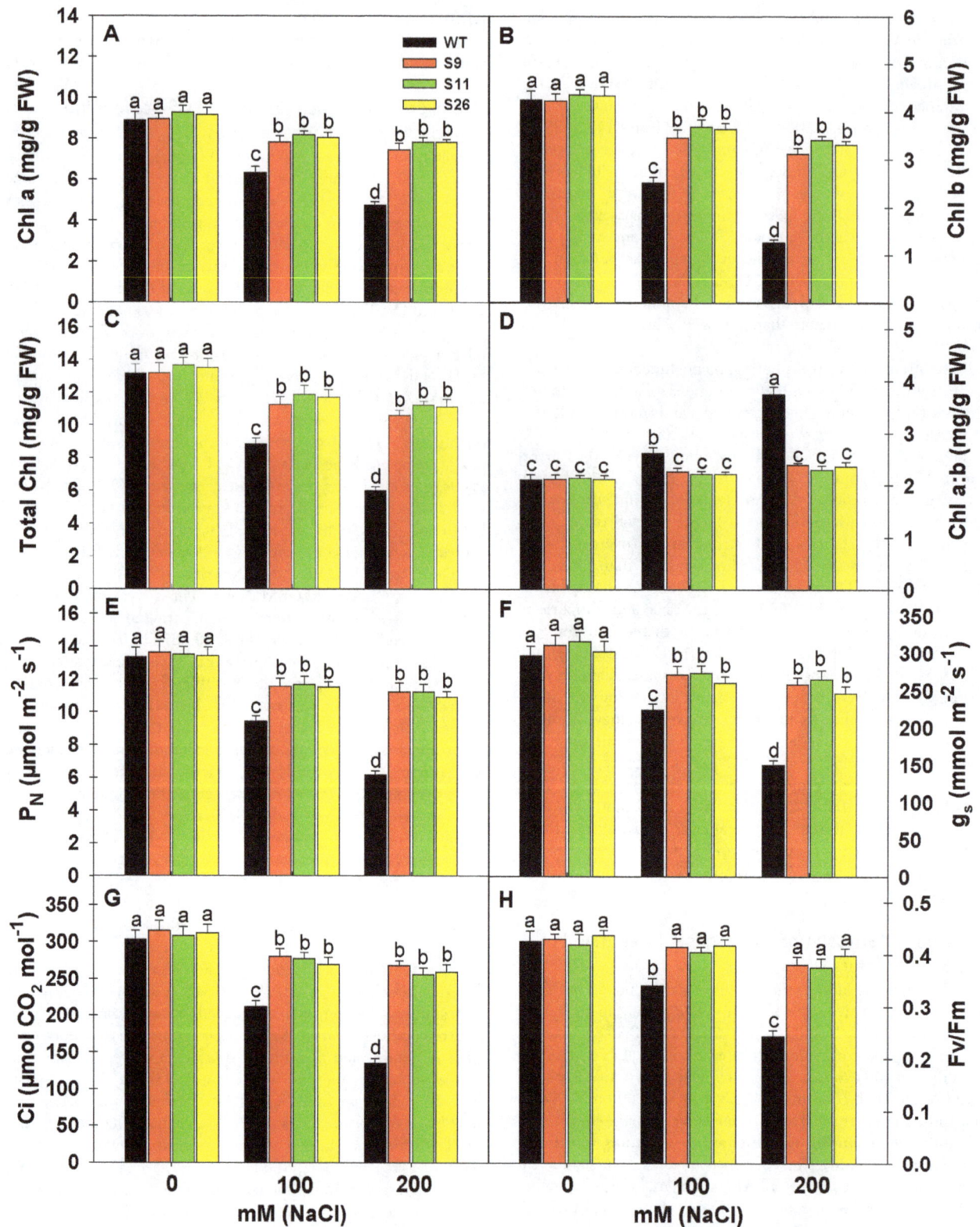

Figure 5. Effects of salinity stress on chlorophyll content, photosynthesis and chlorophyll fluorescence of transgenic and WT tobacco plants. (A) Chlorophyll a content. (B) Chlorophyll b content. (C) Total Chlorophyll content. (D) Ratio of Chlorophyll a and b. (E) Measurement of Net photosynthetic rate (P_N). (F) Measurement of stomatal conductance (gs). (G) Measurement of internal CO_2 concentration [(C_i)]. (H) Measurement of Chlorophyll fluorescence (F_v/F_m). Data's were recorded from WT and *p68* overexpressing transgenic tobacco lines after 3 weeks exposure to 0, 100 or 200 mM NaCl treatment. Each value represents mean of three replicates ± SE. Means were compared using ANOVA. Data followed by the same letters are not significantly different at P<0.05 as determined by least significant difference (LSD) test. [a, b, c] indicate significant differences at P<0.05 level as determined by Duncan's multiple range test (DMRT).

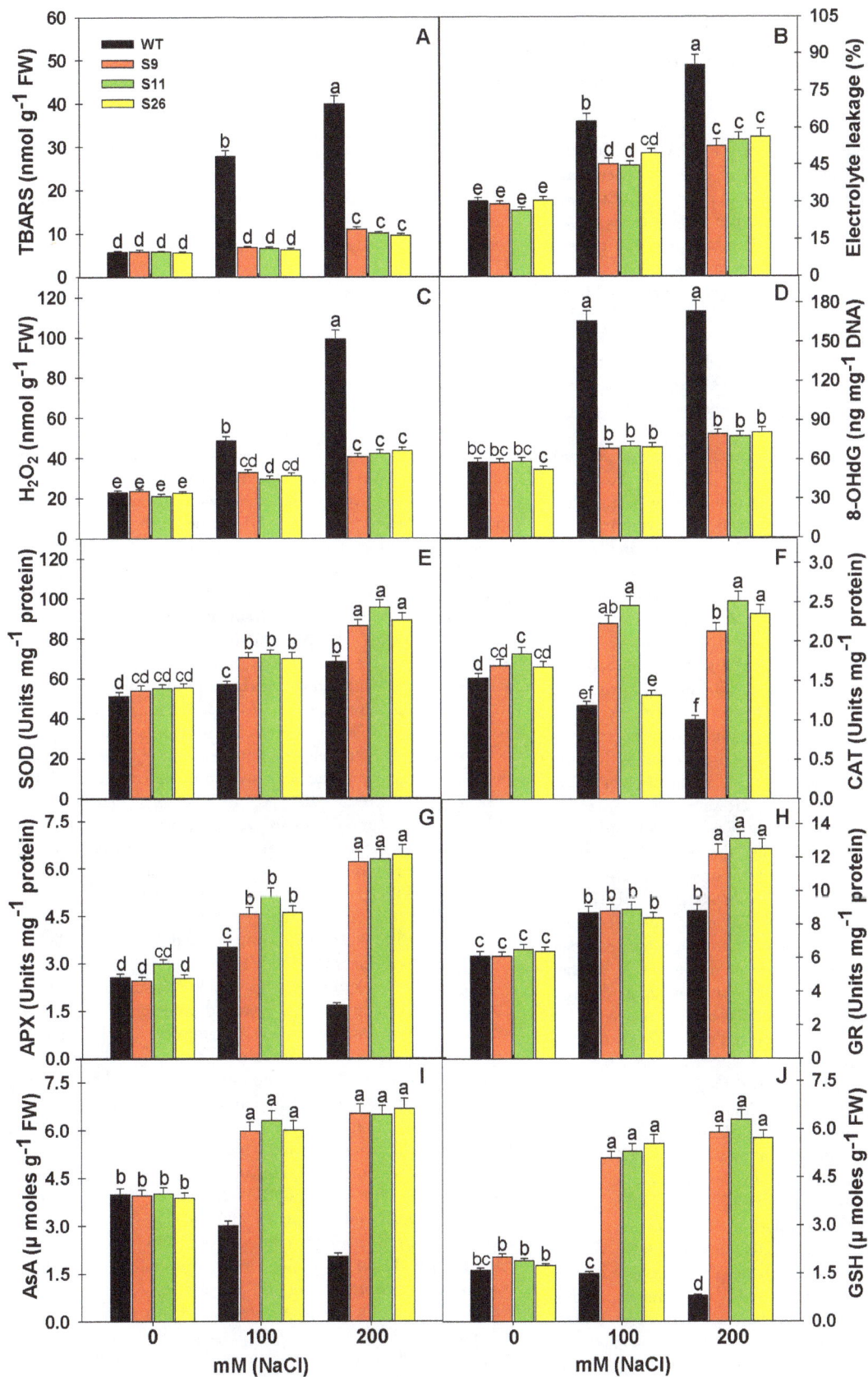

Figure 6. Overexpression of pea *p68* showed less oxidative damage by modulating the ROS machinery under salinity stress.
Measurement thiobarbituric acid reactive substance (A), Electrolyte leakage (B), hydrogen peroxide content (C) and oxidative DNA damage in terms of 8-OhdG (D) in the leaves of WT and *p68* overexpressing transgenic tobacco exposed to 0, 100 or 200 mM NaCl after 3 weeks of salinity treatment.

Measurement of the activities of superoxide dismutase (E), Catalase (F), ascorbate peroxidase (G) guaiacol peroxidase (H) ascorbate content (I) and glutathione content (J) in the leaves of WT and p68 overexpressing transgenic tobacco exposed to 0, 100 or 200 mM NaCl after 3 weeks of salinity treatment. Values are mean ± SE (n = 3). Different letters on the top of bars indicate significant differences at P<0.05 level as determined by Duncan's multiple range test (DMRT). The results are representative of similar results obtained from two independent experiments.

[8,11–12,14,17,24]. In this study, a novel DEAD-box helicase gene (*p68*) was isolated from pea plant which is specifically upregulated in response to salinity. The transcript of pea *p68* is also accumulated at a high level and almost equally in every part (roots, leaves, tendrils and flowers) of the pea plant. This result is consistent with the earlier report of transcript analysis of *AtDRH1* gene expression in *A. thaliana* [19]. Therefore, this gene could be a potential candidate for developing stress-tolerant transgenic plants.

The pea p68 protein contains all conserved domains that are characteristic of the DEAD-box proteins including 'Q' and 'GG' motifs [25]. The ability of p68 to interact with itself indicated that the oligomerization of p68 may be essential for its proper functioning. The p68 is exclusively localized to the cytoplasm and also seems to surround the nucleus of cell. Similar results were reported for another pea helicase (PDH45) [26]. Previously nuclear localization of some DEAD-box helicases (eg. mammalian *eIF4AIII* and *A. thaliana* UAP56) showed the involvement in nonsense-mediated mRNA decay, mRNA splicing and export [27–28]. RNA splicing process involves the association and dissociation of the pre-mRNA with snRNAs, which may be facilitated by RNA helicases. Previous report in animal system has shown that p68 and p72 RNA helicases are the crucial factors required for efficient RNA splicing [29–30]. Both p68 and p72 interact with the U1 small nuclear ribonucleoprotein that recognizes the 5′ splice site [31]. The p68 protein, devoid of RNA helicase or ATPase activity also inhibited the dissociation of U1 from 5′ splice site, and downregulation of DDX5 resulted in the accumulation of unspliced RNA [29]. The role of plant p68 in splicing has not been yet reported. However, there may be possibility that the p68 is sustaining the splicing activity during the stress condition and therefore allowing the p68 overexpressing tobacco plant to survive under the salinity stress condition.

Previously a number of studies demonstrated that DEAD-box RNA helicases are involved in regulating many stages of plant development processes including plant morphogenesis, embryogenesis, pollen tube guidance, floral meristems, flowering, plastids and seed development [32–34]. In plant the first report of stress induced helicase gene came by cDNA microarray analysis of 1300 *Arabidopsis* genes where the authors reported a DEAD-box helicase gene (accession number AB050574) as a cold stress-inducible gene suggesting a new role of helicases in stress signaling [35]. Later, many plant DEAD-box helicases were identified and found to be activated in response to changing environmental conditions [2,13–14,36]. Evidence is accumulating that transcript of PDH45 and PDH47 was found to be induced in response to high salt, dehydration and cold stresses [11,13–14,17,26]. In barley, a salt-responsive transcript HVD1 is induced under salt stress, cold stress, and ABA treatment [37]. AvDH1 is another DEAD-box helicase gene from the halophyte dogbane plant that also strongly upregulated in response to salinity and low temperature [24]. Under normal growth conditions relatively high level of basal expression of the pea *p68* gene in different plant parts implies its function in growth and/or development processes. Under salt treatment, a single species of pea *p68* mRNA was detected abundantly and constitutively in the tissues examined. This indicated that basic activity of cells might be regulated by pea *p68* under salt stress.

Genome-wide expression analysis of many DEAD-box helicase genes have been identified and suggested that these genes might be stress regulated [38]. Overexpression analysis in different DEAD-box helicases has been shown to provide multiple abiotic stress tolerance in crop plants by regulating different signalling pathways [11–12,17]. For example, overexpression of *PDH45* and *OsSUV3* gene provided salinity stress tolerance in tobacco and rice respectively [12,17]. LOS4 and RCF1 mutant analysis in *Arabidopsis* was found to play an important role in response to cold and heat stress [8,15]. Our study showed that overexpression of pea *p68* provides salinity stress tolerance in tobacco.

The reduction in leaf chlorophyll content under abiotic stress has been attributed to the destruction of chlorophyll pigments in various crop plants [3,39–40]. We observed that stress-induced chlorophyll loss was enhanced in WT plant while transgenic lines retained more chlorophyll. This finding has strong correlation with the previous studies in other DEAD-box helicases [12,41–42]. Hence it indicated that overexpression of pea *p68* could have positive effects on the growth and photosynthetic metabolism process. Under stress condition, maintenance of the Na^+, K^+ levels and Na^+/K^+ ratio are important indicators for plant stress tolerance [43–45]. The pea *p68* overexpressing tobacco plants accumulated less Na^+ and more K^+ as compared to the WT plants. Higher K^+ content implies delayed leaf senescence in the transgenic lines. Previously it was reported that lower cytosolic K^+ content controlled endonuclease and caspase-like proteases activity in the cells causing leaf senescence under stress conditions [40,46]. Transgenic tobacco plants also extruded more Na^+ from cells and, as a result, low Na^+ content was detected which indicated that overexpression of pea *p68* enhances stress tolerance in transgenic plants. In wheat increased K^+ uptake and Na^+ extrusion was reported in response to salinity stress [47]. The pea p68 overexpressing transgenic plants are also capable of absorbing more water and diluting the Na^+ content. Earlier transgenic approaches with transporter proteins and DESD-box helicase (PDH45) showed that lower Na^+/K^+ ratio helps plants to respond to salinity stress tolerance [14,45,48]. We found lower Na^+/K^+ ratio in transgenic lines which suggested overexpression of pea *p68* might restrict the entry of Na^+ ions into the cells thereby protecting photosynthetic machinery from abiotic stresses.

Stress also leads to the rapid production of ROS including H_2O_2 in plant tissues that ultimately cause damages to the cell membrane and other cellular components such as plasma membrane, mitochondria and chloroplasts [14,40]. Hence, to avoid any stress-induced injuries plant needs to develop efficient mechanism to remove excess ROS from cells. Enzymatic ROS-scavenging and non-enzymatic antioxidants system are such mechanisms in the plant cells that prevent ROS induced oxidative damage [49–51]. In the present study, stress-induced high H_2O_2 accumulation was noted more in WT plants in comparison to the transgenic lines. Therefore, we speculated that transgenic plants are capable of removing excess H_2O_2 from the cells, hence preventing cellular damages. Catalase, ascorbate and peroxidase are the major enzymes that are known to be involved in scavenging of cellular production of H_2O_2 [52–53]. Interestingly in this study the activity of these enzymes increased in transgenic lines in response to stress treatment. This indicated that overexpressing lines could readily scavenge H_2O_2 either decomposing it through increased activity of catalase or by ascorbate through the ascorbate/glutathione cycle. Previously, a number of overexpression studies have shown an increased activity of catalase, ascorbate and peroxidase in response to abiotic stress treatment [14,54–56].

The involvement of DEAD-box helicases in various metabolic processes in plant cells might have general implications. The present study provides new insights into the novel function of p68 DEAD-box protein in conferring salinity stress tolerance in transgenic tobacco plants without affecting yield. The overexpression of stress-induced DEAD-box *p68* can also provide a good example of the exploitation of factors of RNA metabolism pathways including splicing factor for enhanced agricultural production of economically important crops under stress conditions.

Acknowledgments

We thank Mr Kazi Mostaque Ahmed for helpful corrections.

Author Contributions

Conceived and designed the experiments: NT RT. Performed the experiments: MSAB KMKH PJ SSG XHP. Analyzed the data: NT MSAB KMKH SSG. Contributed reagents/materials/analysis tools: NT. Wrote the paper: NT MSAB KMKH RT.

References

1. Tanner NK, Linder P (2001) DExD/H box RNA helicases: from generic motors to specific dissociation functions. Mol Cell 8: 251–262.
2. Owttrim GW (2006) RNA helicases and abiotic stress. Nucleic Acids Res 34: 3220–3230.
3. Tuteja N, Gill SS, Tuteja R (2012) Helicases in Improving Abiotic Stress Tolerance in Crop Plants. In: Tuteja N, Gill SS, Tiburcio AF, Tuteja R (eds) Improving crop resistance to abiotic stress. Wiley-VCH Verlag GmbH and Co, KGaA, Germany, 433–445.
4. Tuteja N (2000) Plant cell and viral helicases: essential enzymes for nucleic acid transactions. Cri Rev Plant Sci 19: 449–478.
5. Bates GJ, Nicol SM, Wilson BJ, Jacobs AM, Bourdon JC, et al. (2005) The DEAD box protein p68: a novel transcriptional coactivator of the p53 tumour suppressor. EMBO J 24: 543–553.
6. Fuller-Pace FV (2006) DExD/H box RNA helicases: multifunctional proteins with important roles in transcriptional regulation. Nucleic Acids Res 34: 4206–4215.
7. Ishizuka A, Siomi MC, Siomi H (2002) A Drosophila fragile X protein interacts with components of RNAi and ribosomal proteins. Genes Dev 16: 2497–2508.
8. Guan Q, Wu J, Zhang Y, Jiang C, Liu R, et al. (2013) A DEAD box RNA helicase is critical for pre-mRNA splicing, cold-responsive gene regulation, and cold tolerance in *Arabidopsis*. Plant Cell 25: 342–356.
9. Pradhan A, Chauhan VS, Tuteja R (2005a) A novel 'DEAD box' DNA helicase from Plasmodium falciparum is homologous to p68. Mol Biochem Parasitol 140: 55–60.
10. Pradhan A, Chauhan VS, Tuteja R (2005b) Plasmodium falciparum DNA helicase 60 is a schizont stage specific, bipolar and dual helicase stimulated by PKC phosphorylation. Mol Biochem Parasitol 144: 133–141.
11. Vashisht A, Pradhan A, Tuteja R, Tuteja N (2005) Cold and salinity stress-induced pea DNA helicase 47 is involved in protein synthesis and stimulated by phosphorylation with protein kinase C. Plant J 44: 76–87.
12. Sanan-Mishra N, Pham XH, Sopory SK, Tuteja N (2005) Pea DNA helicase 45 overexpression in tobacco confers high salinity tolerance without affecting yield. Proc. Natl. Acad. Sci. USA 102: 509–514.
13. Vashisht A, Tuteja N (2006) Stress responsive DEAD box helicases: a new pathway to engineer plant stress tolerance. J Photochem Photobiol 84: 150–160.
14. Gill SS, Tajrishi M, Madan M, Tuteja N (2013) A DESD-box helicase functions in salinity stress tolerance by improving photosynthesis and antioxidant machinery in rice (Oryza sativa L. cv. PB1). Plant Mol Biol 82: 1–22.
15. Gong Z, Dong CH, Lee H, Zhu J, Xiong L, et al. (2005) A DEAD box RNA helicase is essential for mRNA export and important for development and stress responses in *Arabidopsis*. Plant Cell 17: 256–267.
16. Li D, Zhang H, Wang X, Song F (2008) OsBIRH1, a DEAD box RNA helicase with functions in modulating defence responses against pathogen infection and oxidative stress. J Exp Bot 59: 2133–2146.
17. Tuteja N, Sahoo RK, Garg B, Tuteja R (2013) OsSUV3 dual helicase functions in salinity stress tolerance by maintaining photosynthesis and antioxidant machinery in rice (Oryza sativa L. cv. IR64). Plant J 76: 115–127.
18. Staiger D, Brown JW (2013) Alternative splicing at the intersection of biological timing, development, and stress responses. Plant Cell 25: 3640–3656.
19. Okanami M, Meshi T, Iwabuchi M (1998) Characterization of a DEAD box ATPase/RNA helicase protein of *Arabidopsis thaliana*. Nucleic Acids Res 26: 2638–2643.
20. Gendra E, Moreno A, Alba MM, Pages M (2004) Interaction of the plant glycine-rich RNA-binding protein MA16 with a novel nucleolar DEAD box RNA helicase protein from Zea mays. Plant J 38: 875–886.
21. Horsch RB, Fry JE, Hoffmann NL, Eichholtz D, Rogers SG, et al. (1985) A Simple and General Method for Transferring Genes into Plants. Science 227: 1229–1231.
22. Jefferson RA, Kavanagh TA, Bevan MW (1987) GUS fusions β-glucuronidase as a sensitive and versatile gene fusion marker in higher plants. EMBO J 6: 3901–3907.
23. Munns R, Wallace PA, Teakle NL, Colmer TD (2010) Measuring soluble ion concentrations (Na⁺, K⁺, Cl⁻) in salt-treated plants. In: Sunkar R (ed) Methods in molecular biology, vol 639. Plant stress tolerance: methods and protocols. Humana Press. Springer, New York, 371–382.
24. Liu HH, Liu J, Fan SL, Song MZ, Han XL, et al. (2008) Molecular cloning and characterization of a salinity stress-induced gene encoding DEAD-box helicase from the halophyte Apocynum venetum. J Exp Bot 59: 633–644.
25. Tanner NK, Cordin O, Banroques J, Doe're M, Linder P (2003) The Q motif: a newly identified motif in DEAD box helicases may regulate ATP binding and hydrolysis. Mol Cell 11: 127–138.
26. Pham XH, Reddy MK, Ehtesham NZ, Matta B, Tuteja N (2000) A DNA helicase from *Pisum sativum* is homologous to translation initiation factor and stimulates topoisomerase I activity. Plant J 24: 219–229.
27. Ferraiuolo MA, Lee C, Ler LW, Hsu JL, Costa-Mattioli M, et al. (2004) A nuclear translation-like factor eIF4AIII is recruited to the mRNA during splicing and functions in nonsense-mediated decay. Proc. Natl Acad. Sci. USA 101: 4118–23.
28. Kammel C, Thomaier M, Sørensen BB, Schubert T, Längst G, et al. (2013) *Arabidopsis* DEAD-Box RNA Helicase UAP56 Interacts with Both RNA and DNA as well as with mRNA Export Factors. PLoS ONE 8: e60644.
29. Lin C, Yang L, Yang JJ, Huang Y, Liu ZR (2005) ATPase/helicase activities of p68 RNA helicase are required for pre-mRNA splicing but not for assembly of the spliceosome. Mol. Cell Biol 25: 7484–7493.
30. Janknecht R (2010) Multi-talented DEAD-box proteins and potential tumor promoters: p68 RNA helicase (DDX5) and its paralog, p72 RNA helicase (DDX17). American J Transl Res 2: 223–234.
31. Liu ZR (2002) p68 RNA helicase is an essential human splicing factor that acts at the U1 snRNA- 5′ splice site duplex. Mol Cell Biol 22: 5443–5450.
32. Shimizu KK, Ito T, Ishiguro S, Okada K (2008) MAA3 (MAGATAMA3) helicase gene is required for female gametophyte development and pollen tube guidance in *Arabidopsis thaliana*. Plant Cell Physiol 49: 1478–1483.
33. Ohtani M, Demura T, Sugiyama M (2013) *Arabidopsis* root initiation defective1, a DEAH-box RNA helicase involved in pre-mRNA splicing, is essential for plant development. Plant Cell 25: 2056–2069.
34. Kanai M, Hayashi M, Kondo M, Nishimura M (2013) The plastidic DEAD-box RNA helicase 22, HS3, is essential for plastid functions both in seed development and in seedling growth. Plant Cell Physiol 54: 1431–1440.
35. Seki M, Narusaka M, Abe H, Kasuga M, Yamaguchi-Shinozaki K, et al. (2001) Monitoring the expression pattern of 1300 *Arabidopsis* genes under drought and cold stresses by using a full-length cDNA microarray. Plant Cell 13: 61–72.
36. Mahajan S, Tuteja N (2005) Cold, salinity and drought stresses: an overview. Arch Biochem Biophys 444: 139–158.
37. Nakamura T, Muramoto Y, Takabe T (2004) Structural and transcriptional characterization of a salt-responsive gene encoding putative ATP-dependent RNA helicase in barley. Plant Sci 167: 63–70.

38. Kant P, Kant S, Gordon M, Shaked R, Barak S (2007) STRS1 and STRS2, two DEAD-box RNA helicases that attenuate *Arabidopsis* responses to multiple abiotic stresses. Plant Physiol 145: 814–830.

39. Zhang ZH, Liu Q, Song HX, Rong XM, Ismail AM (2012) Responses of different rice (*Oryza sativa* L.) genotypes to salt stress and relation to carbohydrate metabolism and chlorophyll content. Afr J Agric Res 7: 19–27.

40. Huda KM, Banu MS, Garg B, Tula S, Tuteja R, et al. (2013) OsACA6, a P-type IIB Ca^{2+}ATPase promotes salinity and drought stress tolerance in tobacco by ROS scavenging and enhancing the expression of stress-responsive genes. Plant J 76: 997–1015.

41. Dang HQ, Tran NQ, Gill SS, Tuteja R, Tuteja N (2011) A single subunit MCM6 from pea promotes salinity stress tolerance without affecting yield. Plant Mol Biol 76: 19–34.

42. Sahoo RK, Gill SS, Tuteja N (2012) Pea DNA helicase 45 promotes salinity stress tolerance in IR64 rice with improved yield. Plant Signal Behav 7: 1037–1041.

43. Cuin TA, Betts SA, Chalmandrier R, Shabala S (20080 A root's ability to retain K$^+$ correlates with salt tolerance in wheat. J Exp Bot 59: 2697–2706.

44. Gao Z, He X, Zhao B, Zhou C, Liang Y, et al. (2010) Overexpressing a putative aquaporin gene from wheat, TaNIP, enhances salt tolerance in transgenic *Arabidopsis*. Plant Cell Physiol 51: 767–775.

45. Hill CB, Jha D, Bacic A, Tester M, Roessner U (2012) Characterization of ion contents and metabolic responses to salt stress of different *Arabidopsis* AtHKT1;1 genotypes and their parental strains. Mol Plant 6: 350–368.

46. Shabala S (2009) Salinity and programmed cell death: unraveling mechanisms for ion specific signalling. J Exp Bot 60: 709–712.

47. Munns R, James RA, Lauchli A (2006) Approaches to increasing the salt tolerance of wheat and other cereals. J Ex Bot 57: 1025–1043.

48. Rajagopal D, Agarwal P, Tyagi W, Singla-Pareek SL, Reddy MK, et al. (2007) Pennisetum glaucum Na$^+$/H$^+$ antiporter confers high level of salinity tolerance in transgenic *Brassica juncea*. Mol. Breed 19: 137–151.

49. Gill SS, Tuteja N (2010) Reactive oxygen species and antioxidant machinery in abiotic stress tolerance in crop plants. Plant Physiol Biochem 48: 909–939.

50. Gill SS, Singh LP, Gill R, Tuteja N (2012) Generation and scavenging of reactive oxygen species in plants under stress. In: Tuteja N, Gill SS, Tiburcio AF, Tuteja R (eds) Improving crop resistance to abiotic stress. Wiley-VCH Verlag GmbH and Co. KGaA, Germany, 49–70.

51. Bhattacharjee S (2012) The language of reactive oxygen species signaling in plants. J Bot. doi:10.1155/2012/985298.

52. Willekens H, Langebartels C, Tire C, Van Montagu M, Inze D, et al. (1994) Differential expression of catalase genes in *Nicotiana plumbaginifolia* (L.). Proc. Natl Acad. Sci. USA 91: 10450–10454.

53. Noctor G, Foyer CH (1998) Ascorbate glutathione: Keeping active oxygen under control. Annu Rev Plant Physiol Plant Mol Biol 49: 249–279.

54. Jiang M, Zhang J (2002) Water stress-induced abscisic acid accumulation triggers the increased generation of reactive oxygen species and upregulates the activities of antioxidant enzymes in maize leaves. J Exp Bot 53: 2401–2410.

55. Luna CM, Pastori GM, Driscoll S, Groten K, Bernard S, et al. (2004) Drought controls on H$_2$O$_2$ accumulation, catalase (CAT) activity and CAT gene expression in wheat. J Exp Bot 56: 417–423.

56. Mhamdi A, Queval G, Chaouch S, Vanderauwera S, Van Breusegem F, et al. (2010) Catalase function in plants: a focus on *Arabidopsis* mutants as stress-mimic models. J Exp Bot 61: 4107–4320.

8

Transgenic Mice Convert Carbohydrates to Essential Fatty Acids

Victor J. Pai[1][9]**, Bin Wang**[1][9]**, Xiangyong Li**[1]**, Lin Wu**[2]**, Jing X. Kang**[1]*****

1 Laboratory for Lipid Medicine and Technology (LLMT), Massachusetts General Hospital and Harvard Medical School, Boston, Massachusetts, United States of America,
2 Cutaneous Biology Research Center, Massachusetts General Hospital and Harvard Medical School, Boston, Massachusetts, United States of America

Abstract

Transgenic mice (named "Omega mice") were engineered to carry both optimized *fat-1* and *fat-2* genes from the roundworm *Caenorhabditis elegans* and are capable of producing essential omega-6 and omega-3 fatty acids from saturated fats or carbohydrates. When maintained on a high-saturated fat diet lacking essential fatty acids or a high-carbohydrate, no-fat diet, the Omega mice exhibit high tissue levels of both omega-6 and omega-3 fatty acids, with a ratio of ~1:1. This study thus presents an innovative technology for the production of both omega-6 and omega-3 essential fatty acids, as well as a new animal model for understanding the true impact of fat on human health.

Editor: Thierry Alquier, CRCHUM-Montreal Diabetes Research Center, Canada

Funding: This study was supported by funding from the Fortune Education Foundation and Sansun Life Sciences. The funders had no role in study design, data collection and analysis, decision to publish, or preparation of the manuscript.

Competing Interests: JXK is the inventor and co-applicant (with Massachusetts General Hospital, a non-profit organization) of a patent relating to compositions and methods for modifying the content of polyunsaturated fatty acids in mammalian cells (PCT/US2002/007649). The authors have no other competing interests to declare.

* E-mail: kang.jing@mgh.harvard.edu

[9] These authors contributed equally to this work.

Introduction

There are several classes of fats that differentially affect human health, such as saturated fatty acids (SFA), monounsaturated fatty acids (MUFA), and polyunsaturated fatty acids (PUFA), including both essential omega-6 (n-6) and omega-3 (n-3) PUFA. Recent research has shown that our health is impacted not just by the quantity of fat consumed, but more significantly by the types of fat consumed [1],[2]. For example, the worldwide trend of increased SFA and n-6 PUFA intake and decreased n-3 PUFA intake has coincided with the growing prevalence of chronic diseases, such as heart disease, cancer, Alzheimer's, and diabetes [2].

Normally, mammals readily obtain SFA from either the diet or endogenous synthesis from glucose or amino acids [3], and MUFA can also be obtained from the diet or converted from SFA by the stearoyl-CoA desaturase-1 (SCD-1) gene [4]. On the other hand, n-6 and n-3 PUFA cannot be inter-converted or synthesized *de novo* in mammals and are mainly acquired through the diet [5]. The primary essential fatty acids in the diet are n-6 linoleic acid (18:2n-6, LA) and n-3 α-linolenic acid (18:3n-3, ALA), which can be converted into longer-chain n-6 PUFA arachidonic acid (20:4n-6, AA) and n-3 PUFA eicosapentaenoic acid (20:5n-3, EPA) and docosahexaenoic acid (22:6n-3, DHA), respectively, through a series of desaturation and chain-elongation enzyme systems [5]. The metabolites derived from n-6 and n-3 PUFA, namely eicosanoids, are functionally distinct and have important opposing physiological effects [6]. For example, n-6-derived eicosanoids generally promote inflammation, while n-3-derived eicosanoids have anti-inflammatory properties. Since the synthesis of n-6 and n-3 long-chain PUFA and their metabolites compete for the same enzymes, their ratio in body tissues determines the eicosanoid profile. Recent studies have indicated that the tissue n-6/n-3 PUFA ratio plays an important role in the pathogenesis of many chronic diseases [2],[7]. Many global health organizations now recommend an increased daily intake of n-3 PUFA, particularly EPA and DHA, but these n-3 PUFA are largely limited to marine sources (such as fish and algae) that can be less affordable or accessible [8]. Thus, it is often challenging to balance the different types of dietary fat in a way that optimizes health.

Unlike mammals, the roundworm *C. elegans* is capable of producing all classes of fatty acids as they possess the required genes. For example, the *fat-2* gene encodes a desaturase that catalyzes the conversion of MUFA to n-6 PUFA, and the *fat-1* gene enables the conversion of n-6 to n-3 PUFA [9]. Our laboratory previously created the *fat-1* transgenic technology to express the *fat-1* gene in mammals (both mice and livestock), which successfully demonstrated the conversion of dietary n-6 to n-3 PUFA and exhibited a balanced n-6/n-3 PUFA ratio in body tissues, regardless of diet [10],[11]. Our earlier work generated a unique *fat-1* mouse model for n-3 PUFA research and also provided a sustainable, land-based strategy for the production of n-3 PUFA from n-6 PUFA [12],[13],[14].

The objective of this study was to generate a novel transgenic mouse model that can endogenously synthesize all essential fatty acids. Our strategy was to first create a *fat-2* transgenic mouse, possessing the *C. elegans fat-2* gene encoding an enzyme that converts MUFA into n-6 LA [9], and then cross the *fat-2* transgenic mice with *fat-1* transgenic mice, which we generated previously to possess the *C. elegans fat-1* gene, encoding an enzyme that converts n-6 to n-3 PUFA [10]. Through this procedure, we can generate a compound *fat-1/fat-2* transgenic mouse – hereafter

A

B

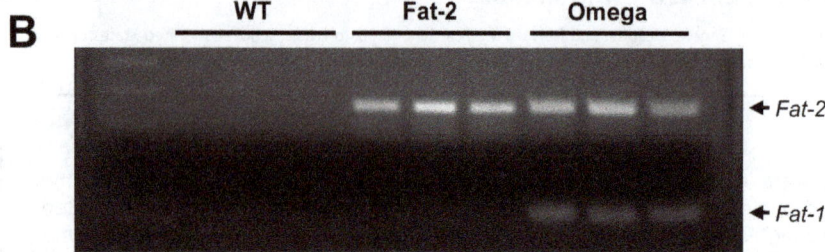

C

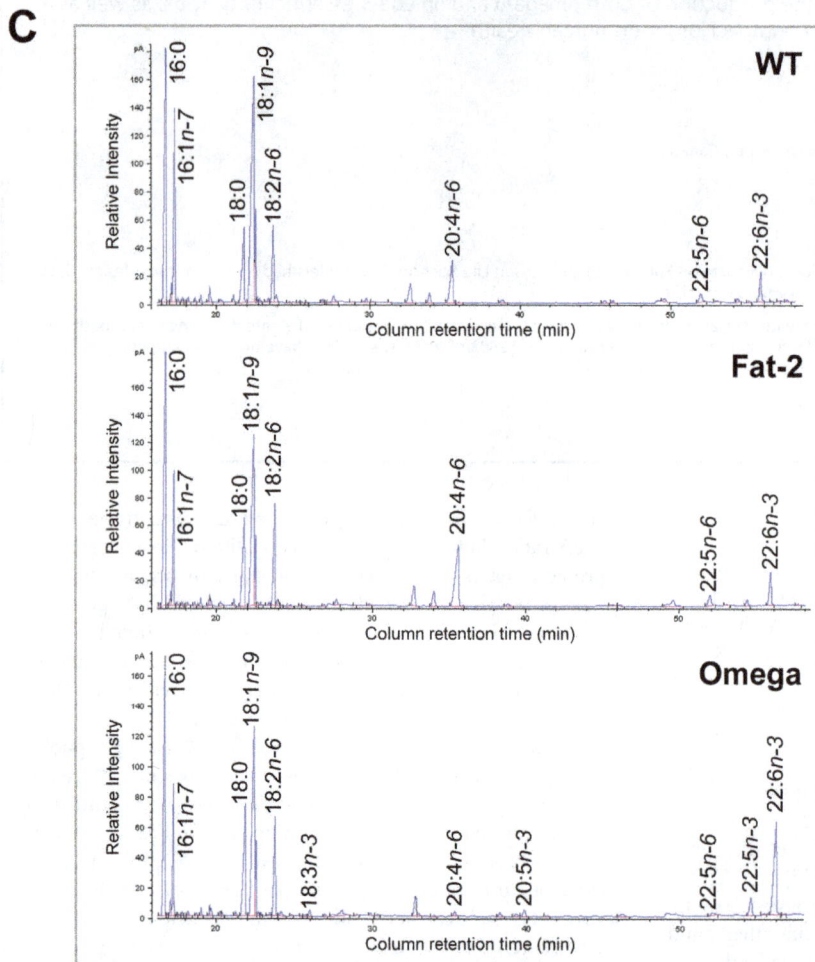

D

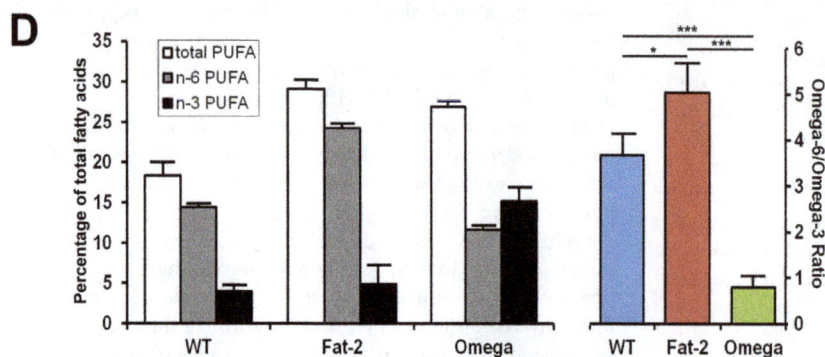

Figure 1. Generation, genotyping, and phenotyping of the *fat-2* and Omega transgenic mice. (**A**) Roadmap for the conversion of essential fatty acids from non-essential nutrients in Omega mice. In mammals, carbohydrates can be converted to SFA, and SFA can be converted into MUFA by SCD-1. Introduction of the *fat-2* and *fat-1* transgenes allows mammals to further convert MUFA into n-6 PUFA, and n-6 PUFA into n-3 PUFA, respectively. (**B**) Validation of *fat-1* and *fat-2* transgene expression in wild-type (WT), *fat-2*, and Omega mouse littermates by PCR. (**C**) Partial gas chromatograph traces showing the fatty acid profiles of total lipids extracted from skeletal muscles of a wild-type mouse (WT, upper panel), a *fat-2* transgenic mouse (Fat-2, middle panel), and an Omega transgenic mouse (Omega, lower panel). All mice were 10-week-old males and fed with the same diet high in SFA and carbohydrates and low in n-6 PUFA. (**D**) Quantification of PUFA from muscle tissue of WT, *fat-2*, and Omega mice (left). For significance values, refer to **Table 1** and **Table S2**. Comparison of the n-6/n-3 PUFA ratio among the phenotypes (right). Values expressed as mean ± s.d. (n = 3 per group; *P<0.05, **P<0.01).

referred to as the "Omega" mouse – that is capable of producing both n-6 and n-3 PUFA from a diet containing only saturated fat or carbohydrates (**Figure 1A**).

Results and Discussion

To enable the higher expression of the *C. elegans fat-2* gene in mammals, we first optimized the codon of the *fat-2* sequence based on the mammalian desaturase sequence (**Figure S1**) and utilized a chicken beta-actin promoter to build a transgene construct (**Figure S2**), and then used microinjection to create the *fat-2* transgenic mouse. After the *fat-2* phenotype and genotype were confirmed, the heterozygous *fat-2* transgenic mice were backcrossed with C57BL6 wild-type (WT) mice for 5 generations, and then crossbred with heterozygous *fat-1* transgenic mice. The resulting offspring consisted of WT, *fat-1*, *fat-2*, and Omega mice. The *fat-1* mouse phenotype has been reported previously [7],[10],[12]. Expression of the *fat-1* and *fat-2* transgenes was

validated by PCR, confirming that WT mice do not express any of the transgenes, the *fat-2* mice carry only the *fat-2* transgene, and the Omega mice express both the *fat-1* and *fat-2* transgenes (**Figure 1B**).

When WT, *fat-1*, *fat-2*, and Omega mouse littermates were fed the same diet high in saturated fat and carbohydrates and low in n-6 PUFA (**Table S1**), phenotype validation by gas chromatography (GC) revealed four distinct tissue fatty acid profiles (**Tables 1 and 2**). The difference in fatty acid profiles for the new *fat-2* and Omega mice are highlighted in **Figure 1C and D**. As expected, the WT mice exhibited high levels of saturated fat and very low levels of essential fatty acids, primarily n-6 PUFA, with an n-6/n-3 PUFA ratio of roughly 3.5. The *fat-2* transgenic mice displayed a significant increase in total tissue PUFA content, with the n-6 PUFA content in the muscle tissue doubling from about 700 μg/g to 1350 μg/g due to the conversion of oleic acid into n-6 LA and AA, without much change in n-3 PUFA levels, resulting in an n-6/n-3 PUFA ratio of about 5. The Omega transgenic mice also

Table 1. Comparison of the fatty acid profiles of various tissues among the four genotypes.

Muscle

	SFA	MUFA	Total PUFA	n-6 PUFA	n-3 PUFA	n-6/n-3
WT	38.96±0.98	42.87±1.99##△△	18.19±1.29##△△	14.23±1.59##＊※	3.96±0.48△△＊※	3.60±0.95△△＊※
Fat-1	40.51±0.35	42.28±1.81▲▲＊＊	16.89±1.55▲▲＊＊	6.23±1.16※※▲▲＊＊	10.66±0.76※※▲▲＊＊	0.59±0.10※※▲▲
Fat-2	41.15±2.02	29.71±2.13##▲▲	29.01±0.99##★▲▲	24.29±0.93##★★▲▲	4.72±0.40★★▲▲	5.18±0.52★★▲▲
Omega	41.08±0.47	33.03±2.12△△＊＊	25.90±1.95△△★＊＊	11.23±2.43★★＊＊	14.67±2.09△△★★＊＊	0.78±0.24△△★★

Liver

	SFA	MUFA	Total PUFA	n-6 PUFA	n-3 PUFA	n-6/n-3
WT	36.70±0.83	48.78±1.00##△△	14.44±1.32##△△	11.86±1.33##△△※※	2.57±0.07#△△※※	4.62±0.55##△△※※
Fat-1	37.49±1.27	49.38±0.72▲▲＊＊	13.16±0.77▲▲＊＊	7.28±0.58※※▲▲＊＊	5.88±0.40※※▲▲＊＊	1.24±0.11※※▲▲
Fat-2	37.33±2.25	33.87±3.05##▲▲	28.82±1.55##▲▲	25.03±1.69##★★▲▲	3.78±0.29★★▲▲	6.65±0.82##★★▲▲
Omega	37.90±2.09	34.21±3.75△△＊＊	27.90±2.50△△＊＊	17.38±1.76△△★★＊＊	10.52±0.74△△★★＊＊	1.65±0.05△△★★

Tail

	SFA	MUFA	Total PUFA	n-6 PUFA	n-3 PUFA	n-6/n-3
WT	29.41±1.55	62.39±2.19##△△	8.20±0.63##△△	7.27±0.41##△△※	0.93±0.23△△※※	8.07±1.38##△△△※
Fat-1	26.99±1.0▲＊	64.96±1.38▲▲＊＊	8.05±0.38▲▲＊＊	2.49±0.32※※▲▲＊＊	5.55±0.70※※▲▲＊＊	0.46±0.11※※▲▲
Fat-2	33.16±3.42▲	47.65±3.55##▲▲	19.20±1.57##▲▲	17.85±1.50##★★▲▲	1.35±0.33★★▲▲	13.81±3.87##★★▲▲
Omega	32.79±2.73*	49.02±2.15△△＊＊	18.20±1.18△△＊＊	10.16±1.62△★★＊＊	8.04±0.44△△★★＊＊	1.27±0.27△△★★

The four genotypes of mice were fed the same low-PUFA diet for about two months and tissue samples were subject to lipid analysis by gas chromatography. WT: Wild-type; SFA: saturated fatty acids; MUFA: monounsaturated fatty acids; PUFA: polyunsaturated fatty acids; n-6: omega-6; n-3: omega-3; n = 3 for each group;
※(WT vs Fat-1),
#(WT vs Fat-2),
△(WT vs Omega),
▲(Fat-1 vs Fat-2),
*(Fat-1 vs Omega),
★(Fat-2 vs Omega), One symbol = P<0.05, Two symbols = P<0.01.

exhibited significantly increased total PUFA content, but the tissue content of n-3 PUFA was increased by roughly five-fold compared to their WT and *fat-2* littermates, due to the conversion of almost half the n-6 PUFA content into n-3 PUFA, with a markedly decreased n-6/n-3 PUFA ratio of 0.75 (**Figure 1D**). MUFA levels were accordingly reduced in the *fat-2* and Omega mice, showing that n-6 PUFA had been converted from MUFA. No significant differences in SFA levels were observed among the phenotypes. Detailed fatty acid profiles for the four genotypes (WT, *fat-1*, *fat-2*, and Omega) are shown in **Tables S2–5**. Furthermore, when fed a high-carbohydrate, no-fat diet (**Table S1**), the *fat-2* and Omega mice still exhibited significant tissue levels of n-6 and n-3 PUFA, respectively, confirming their ability to produce essential fatty acids when lacking dietary fat and given only carbohydrates (**Table 2**). The tissue abundance of n-6 PUFA in the *fat-2* mouse and of both n-3 and n-6 PUFA in the Omega mouse, despite the diet containing little of these fatty acids, thus demonstrates the capability of our transgenic mice to produce essential fatty acids from non-essential nutrients – MUFA, SFA, and even carbohydrates.

The Western human diet today is fundamentally different than it was throughout the majority of human evolution. Among the many shifts in dietary nutrients that have occurred over the last few decades, key changes include increases in saturated fat, carbohydrates, and n-6 PUFA intake with a decrease in n-3 PUFA intake [15],[16],[17]. As a result, modern humans have an n-6 to n-3 PUFA ratio that favors n-6 PUFA by as much as 20:1; evolutionarily, this ratio would have been closer to 1:1, and the discrepancy is thought to have profound physiological consequences [15],[16]. This study presents a new transgenic technology that could produce essential fatty acids, especially the beneficial n-3 PUFA, in animal products (such as meat, milk, eggs, etc.) by feeding animals with just carbohydrates and/or saturated fat. Given that n-3 PUFA are largely limited to marine sources (such as fish and algae), this technology could therefore generate sustainable and accessible n-3 PUFA resources, especially where only carbohydrates are available.

The transgenic technology presented in this study will also be of great utility in elucidating the impact of essential fatty acids on health. Conventionally, dietary modification is used to investigate the effects of different fatty acid profiles, which requires feeding animals with different diets. However, this method is problematic since the diets may not only contain different fatty acid compositions, but also variations in impurities, flavor, calories, or other components used between study groups, ultimately leading to confounding factors that complicate interpretation of results. Our transgenic technology allows us to create four different tissue fatty acid profiles by using a single diet, which makes it possible to evaluate the true health effects of different fats without the confounding factors of diet. This is very important for identification of metabolic biomarkers related to lipid metabolism.

This study clearly demonstrates the feasibility of producing mammals with the capability to convert SFA and MUFA to the essential n-6 and n-3 PUFA. The transgenic mice created by this project will serve as unique animal models, free of dietary confounding factors, for the reliable study of the biological effects of different fatty acid profiles. Ultimately, this transgenic technology serves as a new biotechnology for the production of essential fatty acids, especially n-3 EPA and DHA, to meet increasing demand.

Materials and Methods

Ethics statement

All animal procedures in this study were reviewed and approved by the Massachusetts General Hospital (MGH) Subcommittee on Research Animal Care (SRAC).

Codon optimization

In order to efficiently express a gene from the lower-life C. elegans in mammals *in vivo*, the codon usage by *C. elegans* must be adjusted to match those used by mammals. The *fat-2* gene sequence was obtained from the gene bank (GenBank accession number NM_070159). We used mammalian desaturases as well as our previously optimized *fat-1* gene as references to determine the differences in codon usages of desaturases between C. elegans and mammals. We then manually adjusted the codon to achieve over 80% optimization (**Figure S1**).

Gene synthesis

After modifying the gene sequence to optimize the codon usage, the *fat-2* gene was synthesized by GenScript (Piscataway, NJ). The gene was synthesized to be flanked with EcoR I digestion sites and delivered on pUC57 plasmid. After amplification, the *fat-2* sequence was inserted into a pCAGGS expression plasmid containing the chicken beta-actin promoter and cytomegalovirus enhancer (kindly provided by Dr. J Miyazaki, Osaka University Medical School). After ligation, the orientation was confirmed and the plasmid was amplified. Finally, the fragment containing the

Table 2. Fatty acid profile of tail tissue from mice fed with a high-carbohydrate, no-fat diet.

	SFA	MUFA	Total PUFA	n-6 PUFA	n-3 PUFA	n-6/n-3
WT	28.85±7.64	64.09±8.13	7.06±0.89##△△	6.4±0.75※※##△△	0.66±0.14△△※※	9.93±1.2##△△※※
Fat-1	28.24±0.33	65.46±1.05▲*	6.32±0.71▲▲**	2.32±0.38※※▲▲**	4±0.66※※▲▲**	0.59±0.16※※▲▲
Fat-2	27.78±1.56	57.14±2.01▲	15.1±0.65##▲▲	14.44±0.74##★★▲▲	0.65±0.09★★▲▲	22.44±3.82##★★▲▲
Omega	29.06±3.43	56.98±2.65*	13.97±0.78△△**	8.82±0.75△△★★**	5.16±0.28△△★★**	1.71±0.18△△★★

The four genotypes of mice were fed the same no-fat diet for about two months and tail tissue was subject to lipid analysis by gas chromatography. WT: Wild-type; SFA: saturated fatty acids; MUFA: monounsaturated fatty acids; PUFA: polyunsaturated fatty acids; n-6: omega-6; n-3: omega-3; n = 3 for each group;
※(WT vs Fat-1),
#(WT vs Fat-2),
△(WT vs Omega),
▲(Fat-1 vs Fat-2),
*(Fat-1 vs Omega),
★(Fat-2 vs Omega), One symbol = P<0.05, Two symbols = P<0.01.

promoter, the *fat-2* sequence, and the polyA sequence was excised with SspI and Sfi I for microinjection (**Figure S2**).

Microinjection

Transgenic mouse lines were produced at the MGH Transgenic Core Facility by injecting the purified *Ssp* I and *Sfi* I fragment into fertilized eggs from C57BL/6 X C3H mice. The fertilized eggs containing the transgene were transferred to pseudo-pregnant mice (B6CF1) to produce transgenic mice. The founder transgenic mice were then subjected to genotyping and phenotyping.

Genotyping and phenotyping

Genotyping was carried out by removing the tip of the tail to acquire a DNA sample for RT-PCR, which was performed with the following primers: *fat-2* forward, GCGGCCA GACCCA-GACCATC; and *fat-2* reverse, GGGCGAC GTGACCGTTGGTA. PCR products were run through gel electrophoresis on 2% agarose gel. Phenotyping by fatty acid composition analysis using gas chromatography (GC) was performed as previously described [18]. The *fat-2* mice were maintained after weaning on a low-PUFA diet (**Table S1**). GC was carried out after 20 days on the diet to allow for clearer phenotypes. Tissue samples were ground to powder under liquid nitrogen and total lipids were extracted using chloroform/methanol (2:1, v/v). Fatty acids were then methylated by heating them at 100°C for 1 hour under 14% boron trifluoride (BF3)-methanol reagent (Sigma-Aldrich, St. Louis, MO) and hexane (Sigma-Aldrich). Fatty acid methyl esters were analyzed by GC using a fully automated 6890N Network GC System (Agilent Technologies, Santa Clara, CA) equipped with a flame-ionization detector and an Omegawax 250 capillary column (30 m×0.25 mm ID). Fatty acid standards (Nu-chek-Prep, Elysian, MN) were used to identify peaks of resolved fatty acids, and area percentages for all resolved peaks were analyzed using GC ChemStation Software (Agilent). The fatty acid C23:0 (20 µg/sample) was used as an internal standard to calculate the amount of each fatty acid measured.

After identifying the genotype and phenotype, the *fat-2* mice were mated with WT C57BL6 mice to create the F1 generation. The F1 generation was then backcrossed with WT C57BL6 mice at least 5 times in order to verify that the gene is transmittable as well as to establish a pure background, so that *fat-2* lines could be maintained with a significant phenotype. Each generation was subjected to genotyping by RT-PCR and phenotyping by GC.

Generation of Omega mice

The compound *fat-1/fat-2* transgenic mice were created by crossbreeding heterozygous *fat-2* transgenic mice with heterozygous *fat-1* transgenic mice, which were previously generated by our group [10]. After weaning, the offspring were maintained either on low-PUFA diet or a no-fat diet (**Table S1**), and then genotyped by RT-PCR and phenotyped by GC. Genotyping by RT-PCR of the Omega mice was carried out with the following primers: *fat-1* forward, TGTTCATGCCTTCT TCTTTTTCC; *fat-1* reverse, GCGACCATACC TCAAACTTGGA; *fat-2* forward, GCGGCCA GACCCAGACCATC; *fat-2* reverse, GGGCGAC GTGACCGTTGGTA. Phenotyping by fatty acid composition analysis using GC was performed as previously described [18].

Statistical analysis

GraphPad Prism 5 (GraphPad Software, San Diego, CA) was used for all statistical analyses. Data sets were analyzed by F-test to verify normal distribution. One-way ANOVA followed by the Tukey test was used to determine statistical significance, set at $*P< 0.05$ and $**P<0.01$.

Supporting Information

Figure S1 Optimized fat-2 sequence.

Figure S2 The pCAGGS plasmid and fat-2 construct for microinjection.

Table S1 Composition of the low-PUFA diet and the no-fat diet.

Table S2 Comparison of the muscle fatty acid profile of mice fed with a low-PUFA diet among the four genotypes.

Table S3 Comparison of the liver fatty acid profile of mice fed with a low-PUFA diet among the four genotypes.

Table S4 Comparison of the tail fatty acid profile of mice fed with a low-PUFA diet among the four genotypes.

Table S5 Comparison of the tail fatty acid profile of mice fed with a non-fat diet among the four genotypes.

Acknowledgments

The authors are grateful to Marina Kang and Jason Kang for editorial assistance.

Author Contributions

Conceived and designed the experiments: JXK. Performed the experiments: VJP BW XYL LW. Analyzed the data: XYL VJP BW JXK. Wrote the paper: JXK.

References

1. Joint WHO/FAO Expert Consultation (2003) Diet, Nutrition and the Prevention of Chronic Diseases (WHO technical report series 916). Geneva: World Health Organization. pp. 81–94.
2. Simopoulos AP (2008) The importance of the omega-6/omega-3 fatty acid ratio in cardiovascular disease and other chronic diseases. Exp Biol Med 233: 674–688.
3. Volpe JJ, Vagelos PR (1976) Mechanisms and regulation of biosynthesis of saturated fatty acids. Physiol Rev 56: 339–417.
4. Paton CM, Ntambi JM (2009) Biochemical and physiological function of stearoyl-CoA desaturase. Am J Physiol Endocrinol Metab 297: E28–37.
5. Leonard AE, Pereira SL, Sprecher H, Huang YS (2004) Elongation of long-chain fatty acids. Prog Lipid Res 43: 36–54.
6. Kang JX, Weylandt KH (2008) Modulation of inflammatory cytokines by omega-3 fatty acids. Subcell Biochem 49: 133–143.
7. Kang JX (2011) The omega-6/omega-3 fatty acid ratio in chronic diseases: animal models and molecular aspects. World Rev Nutr Diet 102: 22–29.
8. Gebauer SK, Psota TL, Harris WS, Kris-Etherton PM (2006) n-3 fatty acid dietary recommendations and food sources to achieve essentiality and cardiovascular benefits. Am J Clin Nutr 83: 1526s–1535s.
9. Watts JL, Browse J (2002) Genetic dissection of polyunsaturated fatty acid synthesis in Caenorhabditis elegans. Proc Natl Acad Sci U S A 99: 5854–5859.
10. Kang JX, Wang J, Wu L, Kang ZB (2004) Transgenic mice: fat-1 mice convert n-6 to n-3 fatty acids. Nature 427: 504.
11. Lai L, Kang JX, Li R, Wang J, Witt WT, et al. (2006) Generation of cloned *fat-1* transgenic pigs rich in omega-3 fatty acids. Nature Biotechnology 24: 435–436.
12. Kang JX (2007) Fat-1 transgenic mice: A new model for omega-3 research. Prostaglandins Leuko Essent Fatty Acids 77: 263–267.

13. Kang JX, Leaf A (2007) Why the omega-3 piggy should go to market. Nature Biotechnology 25: 505–506.
14. Kang JX (2011) Omega-3: a link between global climate change and human health. Biotechnol Adv 29: 388–390.
15. Leaf A, Weber PC (1987) A new era for science in nutrition. Am J Clin Nutr 45: 1048–1053.
16. Cordain L, Eaton SB, Sebastian A, Mann N, Lindeberg S, et al. (2005) Origins and evolution of the Western diet: health implications for the 21st century. Am J Clin Nutr 81: 341–354.
17. Institute of Medicine of the National Academies (2002) Dietary fats: total fat and fatty acids. In: Dietary reference intakes for energy, carbohydrate, fiber, fat, fatty acids, cholesterol, protein, and amino acids (macronutrients). Washington, DC: The National Academy Press, pp. 335–432.
18. Kang JX, Wang J (2005) A simplified method for analysis of polyunsaturated fatty acids. BMC Biochem 6: 5.

9

Synchronization of Developmental Processes and Defense Signaling by Growth Regulating Transcription Factors

Jinyi Liu[1], J. Hollis Rice[1], Nana Chen[1], Thomas J. Baum[2], Tarek Hewezi[1]*

1 Department of Plant Sciences, University of Tennessee, Knoxville, Tennessee, United States of America, 2 Department of Plant Pathology and Microbiology, Iowa State University, Ames, Iowa, United States of America

Abstract

Growth regulating factors (GRFs) are a conserved class of transcription factor in seed plants. GRFs are involved in various aspects of tissue differentiation and organ development. The implication of GRFs in biotic stress response has also been recently reported, suggesting a role of these transcription factors in coordinating the interaction between developmental processes and defense dynamics. However, the molecular mechanisms by which GRFs mediate the overlaps between defense signaling and developmental pathways are elusive. Here, we report large scale identification of putative target candidates of Arabidopsis GRF1 and GRF3 by comparing mRNA profiles of the grf1/grf2/grf3 triple mutant and those of the transgenic plants overexpressing miR396-resistant version of GRF1 or GRF3. We identified 1,098 and 600 genes as putative targets of GRF1 and GRF3, respectively. Functional classification of the potential target candidates revealed that GRF1 and GRF3 contribute to the regulation of various biological processes associated with defense response and disease resistance. GRF1 and GRF3 participate specifically in the regulation of defense-related transcription factors, cell-wall modifications, cytokinin biosynthesis and signaling, and secondary metabolites accumulation. GRF1 and GRF3 seem to fine-tune the crosstalk between miRNA signaling networks by regulating the expression of several miRNA target genes. In addition, our data suggest that GRF1 and GRF3 may function as negative regulators of gene expression through their association with other transcription factors. Collectively, our data provide new insights into how GRF1 and GRF3 might coordinate the interactions between defense signaling and plant growth and developmental pathways.

Editor: Miguel A Blazquez, Instituto de Biología Molecular y Celular de Plantas, Spain

Funding: This work was supported by grants from National Science Foundation (Award #: 1145053) to TJB and TH and by Hewezi Laboratory startup funds from the University of Tennessee, Institute of Agriculture. The funders had no role in study design, data collection and analysis, decision to publish, or preparation of the manuscript.

Competing Interests: The authors have declared that no competing interests exist.

* E-mail: thewezi@utk.edu

Introduction

Plants have evolved complex regulatory mechanisms to defend themselves against a wide range of biotic and abiotic stress factors. In response to pathogen infection plant cells promptly activate defense signaling, which requires considerable metabolic activity, to cope with the infection at the expense of growth-related cellular functions. Accordingly, mutant plants with constitutively activated defense responses frequently exhibit stunted growth and delayed development [1]. The growth-defense trade-off is a well-known phenomenon but the underling molecular mechanisms are elusive. In other words, the cellular factors mediating the overlaps between defense signaling and developmental pathways are unknown. In this context, growth-regulating transcription factors (GRFs) represent exciting targets to investigate the molecular mechanisms that coordinate developmental cell biology changes and defense dynamics. GRFs genes were identified in the genomes of all seed plants examined so far [2–5]. The GRF genes constitute a small gene family containing 9 members in *Arabidopsis thaliana* [3], 12 members in rice (*Oryza sativa*) [4] and 14 members in maize (*Zea mays*) [5]. The GRF gene family is defined by the presence of QLQ and WRC domains in the N-terminal region [3]. The QLQ

domain of GRFs is involved in protein–protein interactions. The WRC domain of the GRFs contains a nuclear localization signal and a DNA-binding motif, which mediates their binding to specific *cis*-acting elements in the promoters of the target genes thereby regulating their expression [6]. It has been shown that Arabidopsis GRF1 and GRF2 act as transcriptional activators and the transactivation activity is mediated by the C-terminal region, which does not contain QLQ or WRC motifs, and through the association with the co-activator GRF-Interacting Factor (GIF) [6]. More recently, Arabidopsis GRF7 was reported to function as transcriptional repressor of osmotic stress–responsive genes by binding to the *cis*-element TGTCAGG [7]. However, the transcriptional repression activity of GRF7 requires the QLQ or WRC motifs. Taken together, these data suggest that GRF proteins can function as transcriptional activators and/or transcriptional repressor, and QLQ-binding cofactors are most likely the major determinants of the transactivation or repression activity.

Several *GRF* genes contain binding sites for microRNA396 (miR396) and thus are post-transcriptionally regulated by the activity of miR396. The induction of miR396 is frequently associated with significant decrease in *GRF* expression levels.

Reduction of the expression of *GRF* genes by overexpressing miR396 suggested a role of GRFs in the development of leaves, and roots [8–11]. For example, miR396 accumulates preferentially in the distal part of young developing leaves and diminishes cell proliferation by inhibiting the activity of GRF2 thereby defining the ultimate number of cells in leaves [9]. Consistent with this finding, a role of GRFs in the establishment of leaf polarity was demonstrated [10]. In addition, the implication of GRFs in coordinating plant response to biotic stress has been recently suggested.

The expression of miR396-regulated *GRF* genes has been shown to be altered in response to various abiotic stress treatments including drought, salinity, low temperature, and UV-B radiation [12,13]. Consistent with a functional role of miR396/GRFs in abiotic stress responses, GRF7 was recently demonstrated to function as a repressor of a wide range of osmotic stress-responsive genes, presumably to prevent growth inhibition under normal conditions [7]. The implication of the miR396/GRFs regulatory system in biotic stress response has been recently reported. For example, miR396 and/or *GRFs* were shown to accumulate in plants treated with the *Pseudomonas syringae* DC3000 *hrcC2* [14] and flg22 [15]. In addition, we recently discovered key functional roles of miR396-targeted *GRF1* and *GRF3* in reprogramming of root cells during cyst nematode parasitism [11,16]. We demonstrated that *GRF1* and *GRF3* are post-transcriptionally regulated by miR396 during cyst nematode infection and that gene expression change of miR396 or its targets *GRF1* and *GRF3* significantly reduced plant susceptibility to nematode infection [16]. More importantly, we found that miR396/GRF1-GRF3 controls about 50% of the gene expression changes described in the syncytium induced by the beet cyst nematode *Heterodera schachtii* in Arabidopsis roots [16]. Collectively, these data point to roles of GRFs in controlling the overlaps between defense signaling and developmental pathways. In this study, we identified a large number of putative targets of GRF1 and GRF3 by comparing gene expression change in transgenic plants overexpressing miRNA396-resitanat version of *GRF1* (*rGRF1*) or *rGRF3* with those of the *grf1*/*grf2*/*grf3* triple mutant. Functional classification of the putative targets revealed that GRF1/3 are involved in a wide range of developmental processes and defense responses. Also, we demonstrate that GRF1/3 control the expression of other miRNA targets and may contribute to the negative regulation of their targets through association with other transcription factors. Together, our data shed lights into possible molecular mechanisms by which GRF1 and GRF3 control various developmental events and coordinate their interactions with defense responses.

Materials and Methods

Identification of putative targets of GRF1 and GRF3

To identify putative target genes of GRF1 and GRF3 we analyzed our recently published microarray data set (accession number GSE31593 in Gene Expression Omnibus at the National Center for Biotechnology Information, http://www.ncbi.nlm.nih.gov/geo/) [16]. In brief, we used Arabidopsis Affymetrix ATH1 GeneChips to compare the mRNA profiles of the *grf1*/*grf2*/*grf3* triple mutant and transgenic plants overexpressing miRNA396-resitanat version of *GRF1* (*rGRF1*) or *rGRF3* with those of the corresponding wild-type (Colombia-0 [Col-0] or Wassilewskija [WS]). The experiment was conducted in a completely randomized design with three independent biological replications for each of the plant types, Col-0, WS, *grf1*/*grf2*/*grf3*, *rGRF1*, and *rGRF3*. A linear model analysis of the normalized expression values was conducted for each gene across the five genetic materials and the

differential expression between Col-0 and *rGRF1* or *rGRF3* and between WS and the triple mutant was determined using a false discovery rate of less than 5% and *P* value <0.05 as described in [16]. Genes showing significant reciprocal expression patterns between overexpression lines and *grf1*/*grf2*/*grf3* mutant were chosen as putative targets.

Biological pathway identification

Biological pathway search for the putative targets of GRF1 and GRF3 was performed using NCBI/BioSystems database (http://www.ncbi.nlm.nih.gov/biosystems), which contains records from several databases including KEGG, WikiPathways, BioCyc, Reactome, the National Cancer Institute's Pathway Interaction Database and Gene Ontology (GO). We conducted the analysis to include only Arabidopsis-specific pathways. The statistical significance of gene set enrichment in each pathway was determined using Chi-square test ($P<0.05$).

Cluster analysis and identification of tissue-specific genes

To identify tissue-specific expression of the putative targets of GRF1 and GRF3, we analyzed microarray data from the AtGenExpress expression atlas (http://www.weigelworld.org/resources/microarray/AtGenExpress) [17] and the Arabidopsis eFP Browser (http://bbc.botany.utoronto.ca/efp/cgi-bin/efpWeb.cgi) [18]. The AtGenExpress expression atlas contains gene expression data for 79 samples covering several tissues and developmental stages, while the Arabidopsis eFP Browser contains gene expression data for more than 1000 microarray data sets. The signal intensity of each probe was retrieved and logarithmically transformed (base 10) and then used to generate the heat map using McV (Multiple Experiment Viewer) software, version 4.9 (http://www.tm4.org/mev.html).

Cis-element identification in the promoter region of GRF1/3 regulated genes

The promoter region, 1,500 bp upstream of the translation initiation codon, of all GRF1/3 putative targets were retrieved from TAIR (http://www.arabidopsis.org/tools/bulk/sequences/index.jsp) and used to search for known transcription factor *cis*-regulatory elements using PLANTPAN software [19]. The frequency of each *cis*-regulatory element was determined in the positively and negatively regulated subsets of GRF1 and GRF3 putative targets. Statistical significance of the differences in the frequency of *cis* elements between the positively and negatively regulated targets was determined using χ^2 test.

RNA isolation and qRT-PCR analysis

For quantification of the expression levels of *GRF1* and *GRF3* in the cytokinin mutants, Wild-type Arabidopsis (ecotypes Col-0), the *ahk2 ahk3* double mutant [20] *ahp1,2,3* triple mutant [21], *type-A arr3,4,5,6* quadruple mutant [22], and *type-B arr1,12* double mutant [23] were grown on MS medium at 26°C under 16-h-light/8-h-dark conditions. Two-week-old plants were collected for RNA isolation using the method described in [24]. DNase treatment of total RNA was performed using DNase I (Invitrogen). Twenty nanograms of DNase-treated RNA were used for cDNA synthesis and PCR amplification using the Verso SYBR Green One-Step qRT-PCR Kit (Thermo Scientific) according to the manufacturer's protocol. The PCR reactions were run in an ABI 7900HT Fast Real-Time PCR System (Applied Biosystems) using the following program: 50°C for 15 min, 95°C for 15 min, and 40 cycles of 95°C for 15 s, 60°C for 30 s and 72°C for 20 s. After PCR amplification, the reactions were subjected to a temperature ramp to generate the dissociation curve to detect the nonspecific amplification products. The dissociation program was 95°C for 15 s, 50°C for 15 s, followed by a slow ramp from 50°C to 95°C. The constitutively expressed gene *Actin8* (AT1G49240) was used as an internal control to normalize gene expression levels. Quantification of the relative changes in gene expression was performed using the $2^{-\Delta\Delta CT}$ method [25].

For quantification of the expression level of miR169, miR172, miR393, miR395, miR844, miR846, and miR857 in the P35S:rGRF1 and P35S:rGRF3 transgenic plants [16], total RNA was extracted from two-week-old plants with TRIzol reagent (Invitrogen) according to the manufacturer's instructions. Total RNA (5 μg) was polyadenylated and reverse transcribed using the Mir-X miRNA First-Strand Synthesis Kit (Clontech) according the manufacturer's protocol. The synthesized cDNAs then were diluted to a concentration equivalent to 40 ng total RNA μL−1 and used as a template in qPCR reactions to quantify mature miRNA expression. PCR was performed using a universal reverse primer (mRQ; supplied with the Mir-X miRNA First-Strand Synthesis Kit), complementary to the poly(T) and the mature miRNA sequences as forward primers. The miRNA-specific forward primers were extended by two A residues on the 3′ end to ensure the binding to the poly(T) region of the mature miRNA cDNA and to evade its hybridization on the miRNA precursor cDNA, as recently described [16]. The PCR reactions were run using the following program: 95°C for 3 min, and 40 cycles of 95°C for 30 s, and 60°C for 30 s. The U6 small nuclear RNA was used as an internal control to normalize the expression levels of mature miRNAs. Quantification of the relative changes in gene expression was performed as described above. Gene-specific primers used in the qPCR analysis are provided in Table S1.

Root Length Measurements

Seeds of the transgenic lines overexpressing *rGRF1* (line 6–8) or *rGRF3* (line 11–15) described in [16], as well as wild-type Col-0 were planted vertically on modified Knop's medium supplemented or not with 100 nM N⁶-benzyladenine (BA, a cytokinin), on 4-well culture plates (BD Biosciences). The root length of at least 30 plants per line was measured as the distance between the crown and the tip of the main root in three independent experiments. Statistically significant differences between the transgenic lines and Col-0 lines were determined by unadjusted paired *t* tests ($P < 0.01$).

Results

Identification of potential targets of GRF1 and GRF3 using microarray analysis

Because both GRF1 and GRF3 function as transcription factors, identifying their direct or indirect target genes will elucidate the pathways in which these transcription factors function. Recently, we used *Arabidopsis* Affymetrix ATH1 Gene-Chips to compare the mRNA profiles of root tissues of the grf1/grf2/grf3 triple mutant and transgenic plants overexpressing miRNA396-resistant version of *GRF1* (*rGRF1*) or *rGRF3* with those of the corresponding wild-type (Col-0 or Ws). We identified 3,944, 2,293 and 2,410 genes as differentially expressed in the grf1/grf2/grf3 triple mutant, *rGRF1* and *rGRF3* plants, respectively, at a false discovery rate (FDR) of <5% and a *P* value of <0.05 [16]. In order to mine these expression data for the most likely GRF-dependent target gene candidates, we hypothesized that *bona fide* target genes of GRF1 and GRF3 likely would exhibit opposite expression patterns in the grf1/grf2/grf3 triple mutant and *rGRF1* or *rGRF3* overexpression plants. To this end, we compared the expression patterns of the 1,135 overlapping genes between the

grf1/grf2/grf3 triple mutant and *rGRF1* and identified 1,098 genes as having opposite expression patterns in both lines (Figure 1A and Table S2). Of these 1,098 genes, 507 genes were found to be upregulated in *rGRF1* and downregulated in the grf1/grf2/grf3 triple mutant, and 591 genes were upregulated in the grf1/grf2/grf3 mutant and downregulated in *rGRF1* (Figure 1A and Table S2). Similarly, we compared the expression patterns of the 796 overlapping genes between grf1/grf2/grf3 triple mutant and *rGR31*. We identified 600 genes as having opposite expression patterns in both lines, and of these, 299 genes were found to be upregulated in *rGRF3* and downregulated in the grf1/grf2/grf3 triple mutant; 301 genes were upregulated in the grf1/grf2/grf3 triple mutant and downregulated in *rGRF3* (Figure 1B and Table S3). We considered these 1,098 and 600 genes as putative candidate targets of GRF1 and GRF3, respectively. When we compared these two groups of genes, we identified a set of 264 genes as common putative targets of GRF1 and GRF3, leaving a unique set of 1434 genes as putative targets of GRF1 or GRF3 (Table S4). Of these 1434 potential targets, 682 are positively regulated and 752 are negatively regulated by GRF1 or GRF3, suggesting that GRF1/3 positively and negatively regulate target genes to similar extent.

Mapping the putative targets of GRF1 and GRF3 to biological pathways reveals their function diversity.

In order to identify specific biological pathways in which the putative targets of GRF1 or GRF3 are involved we subjected the 1434 genes to a comprehensive analysis using NCBI/Biosystem database [26]. We successfully mapped 383 genes for 161 organism specific pathways (Table S5). In Figure 2, we included only pathways that are represented by at least 5 genes and significantly enriched in the putative targets gene list compared with the genome. Genes related to flavonoid biosynthesis, degradation of aromatic compounds and capsaicin biosynthesis constitute half of the genes involved in these pathways. Also, genes involved in the biosynthesis of other secondary metabolites such as phenylpropanoid, stilbenoids, terpenoid and cyanoamino acid were also enriched in the putative targets gene list. Putative targets involved in the biosynthesis of lignin and various amino acids constitute a significant portion of these pathways. Furthermore, putative targets of GRF1 or GRF3 involved in the metabolism of glutathione, nitrogen, or sulfur are enriched in these pathways. This analysis clearly indicates the implication of these targets in a wide range of biological processes, specifically the biosynthesis of amino acid and secondary metabolites.

GRF1 and GRF3 may regulate common targets in a tissue-specific fashion

To test whether the putative targets of GRF1 or GRF3 are associated with tissue specific expression patterns, the expression profiles of the 1434 putative targets were scanned across the AtGenExpress expression atlas [17], which contains 79 samples covering several tissues and developmental stages, from embryogenesis to senescence. Out of 1434 genes, we identified 130 and 13 specifically expressed in root and seed tissues, respectively. After this initial screen, the specific expression patterns of these genes were further verified by exploring a larger microarray database, the Arabidopsis eFP Browser [18], which contains more than 1000 microarray data sets. The second analysis yielded 25 and 10 genes as root and seed-specific genes, respectively (Figure 3 and Table S6). Of the 25 root-specific genes, 6 are common putative targets of both GRF1 and GRF3. Similarly, 2 genes were identified as common targets of both GRF1 and GRF3 out of the 10 seed-

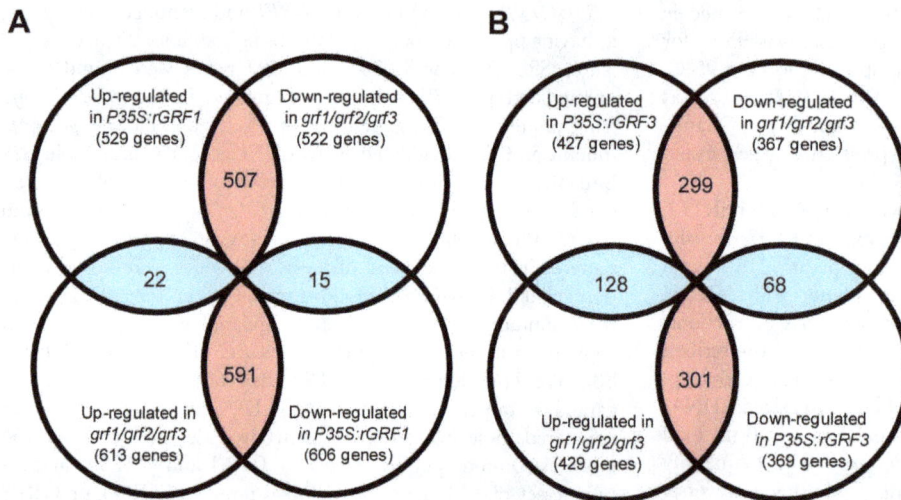

Figure 1. Identification of potential target genes of GRF1 and GRF3. Venn diagram comparing the overlapping differentially expressed genes between *rGRF1* and *grf1/grf2/grf3* (A) or *rGRF3* and *grf1/grf2/grf3* (B). A. Identification of potential target genes of GRF1. Out of the 1,135 overlapping genes between *grf1/grf2/grf3* triple mutant and *rGRF1*, 1,098 genes were identified as having opposite expression patterns in both lines from which 507 genes were found to be upregulated in *rGRF1* and downregulated in *grf1/grf2/grf3* triple mutant, and 591 genes were upregulated in the *grf1/grf2/grf3* mutant and downregulated in *rGRF1*. B. Identification of potential target genes of GRF3. Out of the 796 overlapping genes between *grf1/grf2/grf3* triple mutant and *rGRF3*, 600 genes were identified as having opposite expression patterns in both lines from which 299 genes were found to be upregulated in *rGRF3* and downregulated in *grf1/grf2/grf3* triple mutant, and 301 genes were upregulated in the *grf1/grf2/grf3* mutant and downregulated in *rGRF3*. Numbers in the areas highlighted in red indicate differentially expressed genes that exhibit opposite expression whereas overlapping areas highlighted in blue indicate the number of the differentially expressed genes that exhibited similar expression.

specific genes (Figure 3). These data suggest that GRF1 and GRF3 may regulate common targets in a tissue-specific fashion.

GRF1 and GRF3 regulate the expression of other miRNA targets

To test whether GRF1 or GRF3 regulate other miRNA target genes, we scanned the entire set of the differentially expressed genes in *rGRF1* (2,293 genes) or *rGRF3* (2,410 genes) against all known Arabidopsis miRNAs target genes (205 genes). Interestingly, among the 2,293 genes regulated in *rGRF1*, we identified 19 genes that are post-transcriptionally regulated by 12 different miRNA gene families (Table S7). Also, among the 2,410 genes regulated in *rGRF3*, we identified 19 genes that are targets of 13 different miRNA gene families (Table S7). However, when these comparisons were narrowed to include only the putative direct targets of *GRF1* (1,098 genes) or *GRF3* (600 genes), we identified 15 genes that are targets of 7 miRNA gene families including miR169, miR172, miR393, miR395, miR844, miR846, and miR857 (Table 1). Interestingly, all targets of miR169 (7 genes) are negatively regulated by GRF1 and/or GRF3. This cross regulation seems to be organized in a coordinated manner since three out of the seven targets are co-regulated by both GRF1 and GRF3. Also, we found that GRF1 and GRF3 regulate the expression of miRNA targets in both directions. For example, targets of miR172, miR393, miR846 are positively regulated by GRF1 and/or GRF3. In contrast, targets of miR169, miR395 and miR857 are negatively regulated by GRF1 and/or GRF3.

Recent studies have shown that miRNA expression can be positively or negatively regulated by their targets through negative or positive feedback regulation loops [11,27–31]. Therefore, we tested whether overexpression of *rGRF1* or *rGRF3* affected the expression of 7 miRNA genes (miR169, miR172, miR393, miR395, miR844, miR846, and miR857) whose targets were found to be regulated by GRF1 and/or GRF3. We used qPCR to quantify the abundance of mature miRNAs in the transgenic

plants overexpressing *rGRF1* or *rGRF3* relative to wild-type Col-0. The expression levels of miR169 and miR393 were found to be downregulated both in *rGRF1* and *rGRF3* overexpression plants (Figure 4). In contrast, miR844, miR846 and miR857 showed predominant upregulation in the transgenic plants overexpression *rGRF3*, and to lesser extent in the transgenic plants overexpression *rGRF1* (Figure 4). miR172 and miR395 showed little or no changes in the transgenic plants (Figure 4). These data clearly demonstrate that GRF1 and GRF3 can contribute to the negative or positive regulation of other miRNA genes through altering the expression of their targets.

Because we previously found that *GRF1* and *GRF3* change their expression in the syncytium induced by *H. schachtii* [16], it was of interest to test whether the 15 miRNA targets regulated by GRF1 and/or GRF3 are differentially expressed in the syncytium. Interestingly, these entire target genes were found to be differentially expressed in the syncytium induced by *H. schachtii* according to microarray analysis reported by [32]. However, when these 15 target genes were compared with those reported to be differentially expressed in the giant cells induced by the root-knot nematode *Meloidogyne incognita* [33], none of these genes were found to be overlapped. These data suggest that the regulation of miRNA targets by GRF1/3 is specific to the syncytial cells.

GRF1 and GRF3 regulate cytokinin-responsive genes

Our examination of the GRF-regulated targets for genes involved in hormone biosynthesis pathways led to the identification of a set of genes that are involved in the biosynthesis of cytokinin (6 genes), brassinosteroid (2 genes), auxin (2 genes), gibberellin (2 genes) salicylic acid (2 genes), ethylene (1 gene), and jasmonic acid (1 gene) (Figure 5). The abundance of cytokinin biosynthesis genes in this gene set prompted us to speculate that cytokinin-responsive genes could be also regulated by GRF1/3. To test this hypothesis, the 2,293 genes regulated by GRF1 were compared with the golden list of the cytokinin-responsive genes

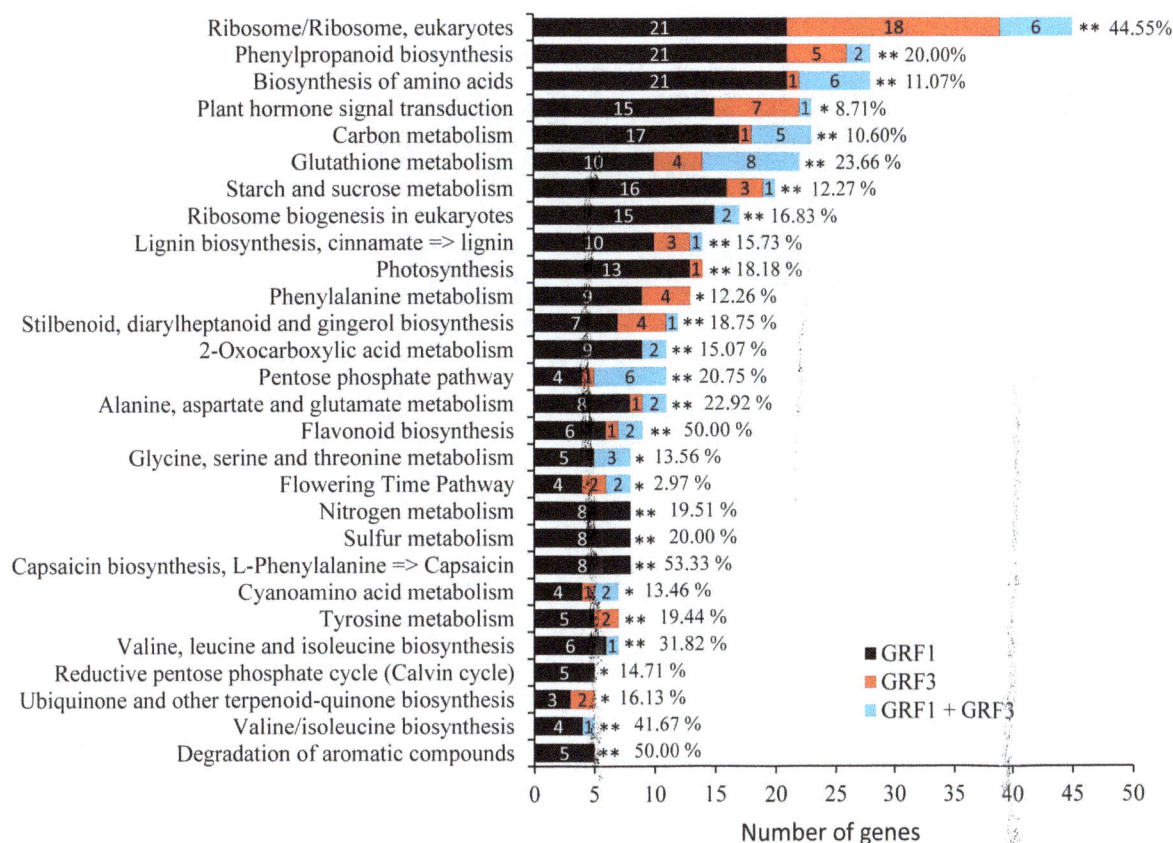

Figure 2. Mapping putative target genes of GRF1 and GRF3 to biological pathways. The 1434 putative target genes of GRF1/3 were subjected to NCBI/Biosystem database to identify specific biological pathways. Out of the 1434 genes, 383 were mapped to 161 organism specific pathways. We included only pathways that are represented by at least 5 genes and significantly enriched in the putative targets gene list compared with the genome. The complete description of the 161 pathways is provided in Table S5.

[34]. Out of the 226 cytokinin-responsive genes, 61 were identified as overlapping with GRF1-regulated genes. Similarly, 43 of the cytokinin-responsive genes overlapped with GRF3-regulated genes. After eliminating duplicates, a total of 92 (41%) cytokinin-responsive genes were identified as overlapping with the GRF1/3-regulated genes (Table S8). When these analyses were conducted to include only the potential targets of GRF1/3 (1434 genes), we identified 48 (21%) of the cytokinin-responsive genes as overlapping (Table 2). These data suggest that GRF1 and GRF3 play major role in controlling gene expression changes of cytokinin-responsive genes.

In plants, cytokinin is perceived through a multi-step phosphor-elay pathway. Based on the current model in Arabidopsis, three histidine Kinases, AHK2, AHK3 and AHK4 have been identified as transmembrane cytokinin receptors. These receptors transfer the signal via Arabidopsis histidine phosphotransfer proteins (AHPs) to the nucleus, activating two types of primary Arabidopsis response regulators (ARRs), known as type-A and type-B response regulators [35]. To provide direct evidence for the connection between GRF1/3 and cytokinin signaling, we measured the expression levels of *GRF1* and *GRF3*, using qPCR, in several cytokinin signaling mutants including the *ahk2 ahk3* double mutant, *ahp1,2,3* triple mutant, type-A *arr3,4,5,6* quadruple mutant and type-B *arr1,12* double mutant. Data from three biological replicates revealed that the expression levels of *GRF1* and *GRF3* are significantly changed in the *ahk2 ahk3* double mutant, showing at least twofold down-regulation in the mutant

relative to wild-type plants (Figure 6A and B). In contrast, the expression levels of *GRF1* and *GRF3* were not significantly altered in the *ahp1,2,3*, type-A *arr3,4,5,6* or type-B *arr1,12* mutant lines (Figure 6A and B). These data support a role for GRF1 and GRF3 in the regulation of cytokinin receptors.

One of the main morphological defects in the transgenic plants overexpressing *rGRF1* or *rGRF3* is the short-root phenotype [16]. Because cytokinin regulates the root meristem activity, root size and overall root length [36], therefore, it was of interest to examine whether the short-root phenotype in the *rGRF1* and *rGRF3* is mediated by cytokinin. To this end, homozygous T3 plants overexpressing *rGRF1* (line 6–8), or *rGRF3* (line 11–15) as well as the wild-type (Col-0) were grown vertically on modified Knop's medium supplemented or not with cytokinin in the form of benzyladenine (BA) at the concentration of 100 nM. Without exogenous application of cytokinin, the transgenic plants overexpressing *rGRF1* or *rGRF3* developed statistically significant shorter roots than the wild-type Col-0 at 9 days after planting (Figure 6C), confirming our previously published data [16]. Because exogenous application of cytokinin reduces root size and growth, we decided to compare the root length of the transgenic plants overexpressing *rGRF1* or *rGRF3* with Col-0 at 9 and 15 days after planting on modified Knop's medium supplemented with 100 nM BA. Interestingly, at both time points, the root lengths of the transgenic plants were found to be very similar to that of the Col-0 and no statistically significant differences were detected (Figure 6C). These results provide further support that GRF1 and GRF3 play key role

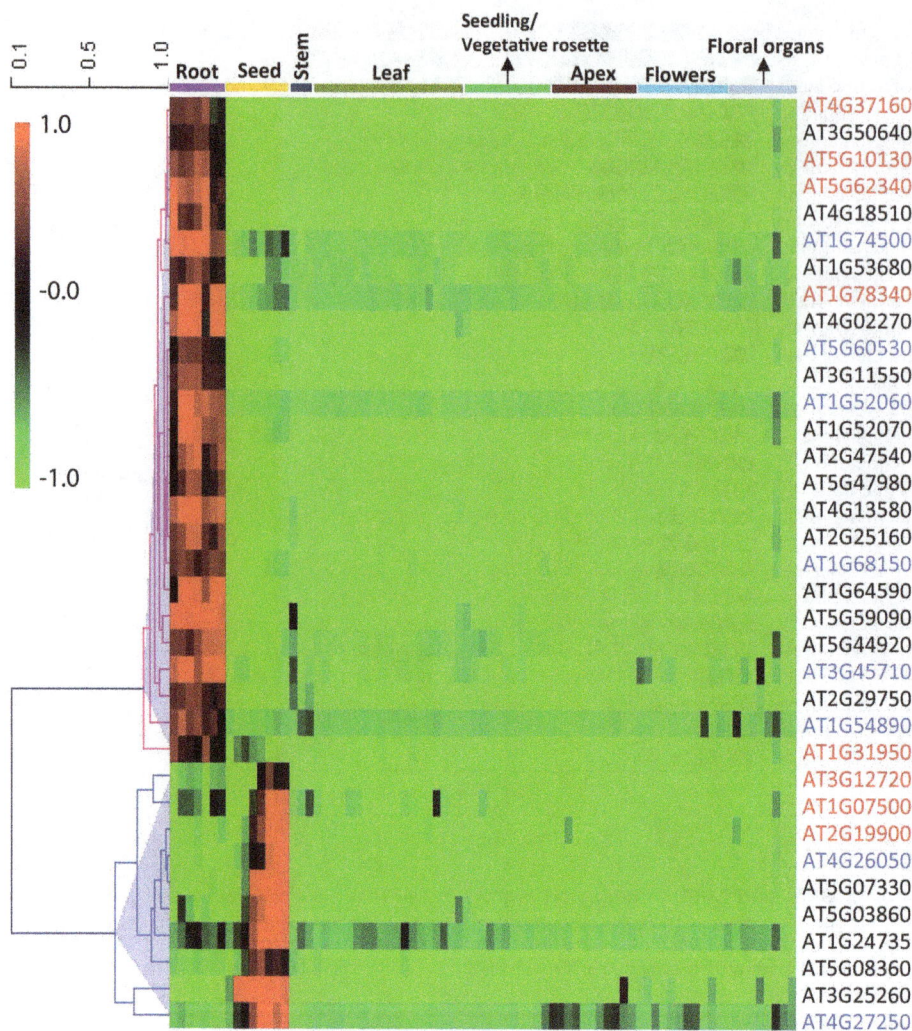

Figure 3. Hierarchical cluster analysis representation of root and seed-specific genes that are putative targets of GRF1 and/or GRF3. The absolute values of gene expression were logarithmically scaled (base 10) and used to generate the heat map using MeV (Multiple Experiment Viewer) software, version 4.9. Genes are represented in lines and different tissues/organs are represented in column. Red and green correspond to transcriptional upregulation and downregulation relative to the average expression level over all tissues included, respectively. Gene IDs highlighted in black, red or blue color indicate putative targets of GRF1, GRF3 or both, respectively.

in regulating gene expression changes of cytokinin-responsive genes.

Several transcription factor gene families are putative targets of GRF1/3

Careful examination of the potential targets of GRF1/3 revealed that high number of these targets code for transcription factors (Figure 7A). Transcription factors of the MYB, ERF NAC, bHLH and NF-YA gene families are highly represented. Interestingly, we identified four bZIP/TGA transcription factor genes (*TGA1, 3, 4* and *7*) that are specifically regulated by GRF1. These genes are members of clade I (*TGA1* [At5g65210] and *TGA4* [At5g10030]) and clade III (*TGA3* [At1g22070] and *TGA7* [At1g77920]). Functional characterization of clade I and III TGA factors has established an essential role in the regulation of pathogenesis-related genes and disease resistance [37–39]. In addition, we identified several MYB transcription factors as potential targets of GRF1 (*MYB58* [AT1G16490], *MYB63* [AT1G79180] and *MYB43* [AT5G16600]), which are involved

in the regulation of secondary cell wall formation [40,41]. Consistent with this finding, genes with cell-wall related functions constitute 10 and 15% of the differentially expressed genes identified in the transgenic plants overexpression *GRF1* or *GRF3*, respectively. Another interesting finding that may connect the function of GRF1 and GRF3 to a wide range of developmental processes and biotic stress tolerance is that several ethylene-responsive element-binding factors (*ERFs*) were identified as putative targets of GRF1 and GRF3. ERFs impact a number of developmental processes and are also function in plant adaptation to biotic and abiotic stresses [42–44].

GRF1 and GRF3 may function as negative regulators of gene expression through their association with other transcription factors

Because GRF1/3 contain the QLQ protein/protein interaction domain, we hypothesized that other transcription factors may form a complex with GRF1/3 and facilitate the binding of GRF1/3 to specific binding motifs in the promoter of their putative

Table 1. Putative targets of GRF1 or GRF3 that are post-transcriptionally regulated by miRNAs.

Gene ID	Annotation	GRF	miRNA
AT1G54160	CCAAT-binding transcription factor	GRF1	miR169
AT3G20910	CCAAT-binding transcription factor	GRF1	miR169
AT5G12840	HAP2A transcription factor	GRF1	miR169
AT1G17590	CCAAT-binding transcription factor	GRF3	miR169
AT1G72830	HAP2C transcription factor	GRF1 + GRF3	miR169
AT3G05690	HAP2B transcription factor	GRF1 + GRF3	miR169
AT5G06510	CCAAT-binding transcription factor	GRF1 + GRF3	miR169
AT3G54990	AP2 domain transcription factor	GRF3	miR172
AT4G03190	Auxin signaling F box protein 1	GRF3	miR393
AT5G10180	Sulfate transporter 68	GRF1	miR395
AT5G51270	Protein kinase family protein	GRF1	miR844
AT1G52070	Jacalin lectin family protein	GRF1	miR846
AT1G52060	Jacalin lectin family protein	GRF3	miR846
AT2G25980	Jacalin lectin family protein	GRF1 + GRF3	miR846
AT3G09220	Laccase 7	GRF3	miR857

targets. Therefore, we searched for known *cis*-elements that would be involved in the transcriptional regulation of all putative target genes of GRF1 and GRF3 in a 1.5 kb promoter region upstream of the translation start codon using PlantPan software [19]. We identified 382 and 361 *cis* elements in the promoters of the putative targets of GRF1 and GRF3, respectively (Table S9). Interestingly, when these *cis* elements were compared to identify common elements, the majority of these elements (357) were found to be common in the promoters of the putative targets of GRF1 and GRF3. These data suggest that both GRF1 and GRF3 may employ similar mechanisms in regulating the expression of their targets, consistent with the redundant function of these two

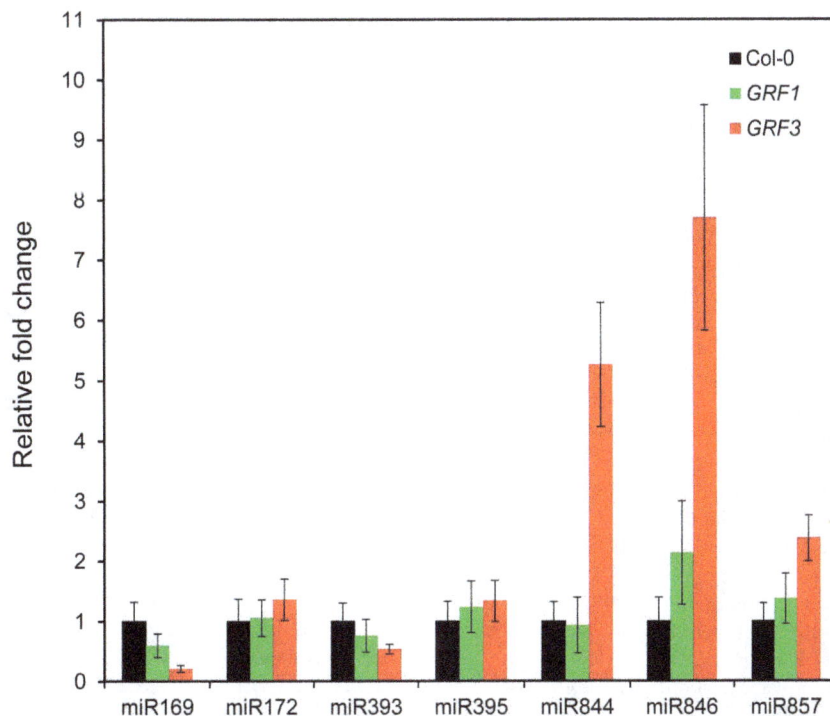

Figure 4. Overexpression of *rGRF1* or *rGRF3* alters the expression of other miRNAs. The expression levels of mature miR169, miR172, miR393, miR395, miR844, miR846, and miR857 were quantified in transgenic plants constitutively expressing the miR396-resistant forms of *GRF1* and *GRF3* (P35S:rGRF1 and P35S:rGRF3) using qPCR. The expression levels of mature miRNAs were normalized using *U6* snRNA as an internal control. The relative fold-change values represent changes of mature miRNA expression levels in the transgenic plants relative to the wild-type control. Data are averages of three biological samples ± SE.

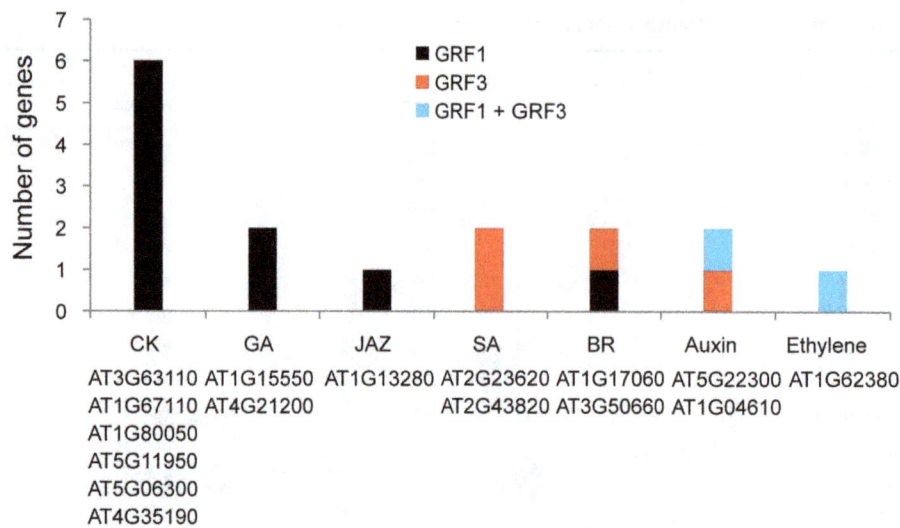

Figure 5. Putative targets of GRF1/3 are involved in hormone biosynthesis pathways. Sixteen potential targets of GRF1/3 are implicated in the biosynthesis of various hormone pathways with cytokinin biosynthesis genes being the most abundant.

transcription factors. In addition, we tested the distribution and frequency of these *cis* elements in the positively and negatively regulated targets of GRF1 (834 genes), GRF3 (336 genes) and both (264 genes). While these *cis* elements are equally distributed between up and downregulated genes, their frequency is much higher in the downregulated genes (Figure 7B), suggesting that GRF1 and GRF3 may function as negative regulators of gene expression through their association with other transcription factors.

Discussion

Despite the efforts to assign the biological processes regulated by GRFs during plant development, very limited number of target genes have been identified and characterized to date [6,7]. One of the most common approaches to identify target genes of the transcription factors involves comparison of the genome-wide transcript profiles of transgenic plants overexpressing transcription factors and the corresponding wild types allowing the identification of genes that are significantly altered as a result of the increased expression of the transcription factors [45,46]. An alternative approach relies on the comparison between the transcriptome of mutants and wild-type plants [47–49]. In the current study, we combined both approaches to identify potential target genes of GRF1 and GRF3. We retained only genes showing opposite expression between *grf1/grf2/grf3* triple mutant and *rGRF1* or *rGRF3* in order to exclude genes whose expression is altered as artifactual effects of the ectopic overexpression and do not reflect authentic roles of the overexpressed transcription factors. Using this approach we identified 1,098 and 600 genes as putative targets of GRF1 and GRF3, respectively. These numbers are relatively low compared with the total number of genes regulated by GRF1 (1,098 genes out of 2293, 47.9%) or GRF3 (600 genes out of 2410, 24.9%), suggesting that the greater part of these genes are indirectly regulated. The indirect regulation of downstream genes could be through the transcription control mediated by transcription factors or proteins with binding activity among those directly regulated by GRF1 or GRF3. Consistent with this interpretation, genes coding for transcription factors or proteins with binding activity represent up to 39% of the GRF1-potential direct target genes and up to 35% of the GRF3- potential

direct targets. The enrichment of transcription factors belonging to Myb, ERF, NAC, bHLH, NY-YA, and C2H2 transcription factor family proteins in GRF1 or GRF3- potential direct target genes suggests key roles of these transcription factors in initiating transcriptional cascades, thereby extending the effects of GRF1 or GRF3 on downstream signaling pathways.

Transcription factors can positively or negatively regulate the expression of their target genes [50]. Our data point to the possibility that GRF1/3 may function as transcriptional repressors since more than half of the GRF1/3 targets are negatively regulated. Initially, members of the GRF gene family have been shown to function as transcriptional activators and this transactivation function involves the C-terminal region [6]. More recently, GRF7 was found to function as transcriptional repressor through its N-terminal QLQ and WRC motifs [7]. Because GRF proteins contain the QLQ protein–protein interaction domain, it is possible that GRF1/3 contribute to the negative regulation of their targets through their association with other transcription factors. This hypothesis is developed based on our data showing that the frequency of known *cis* elements is more abundant in the negatively regulated targets relative to the upregulated targets (Figure 7B). However, we don't rule out the possibility that GRF1/3 may function as transcriptional repressors through their biding to specific *cis* motifs.

Functional classification of the potential targets of GRF1/3 placed these two transcription factors as molecular links connecting defense signaling to plant growth and developmental pathways. Previously, we reported a key role for GRF1/3 in plant response to nematode infection [16]. In the current analysis, the anticipated roles of GRF1/3 in defense responses is further illuminated by identifying crucial factors that are involved in defense response and disease resistance. Four bZIP/TGA transcription factors genes (*TGA1, 3, 4* and *7*) were identified as potential targets of GRF1. *TGA1* and *TGA4*, which belong to clade I are positively regulated, whereas *TGA3* and *TGA7*, which belong to clade III are negatively regulated by GRF1. Characterization of Arabidopsis T-DNA insertion mutants indicated that clade I TGA factors contribute to basal disease resistance and this contribution is most likely independent of NPR1 [39,51,52]. In contrast, NPR1 stimulates the DNA binding of the clade III factors (TGA3 and

Table 2. Cytokinin-responsive genes that are identified as putative targets of GRF1 or GRF3.

Gene ID	Annotation
AT2G01890	PAP8 (PURPLE ACID PHOSPHATASE PRECURSOR)
AT1G13420	sulfotransferase family protein
AT5G63450	CYP94B1 (cytochrome P450, family 94, subfamily B, polypeptide 1)
AT5G10580	Unknown protein
AT5G03380	Heavy-metal-associated domain-containing protein
AT2G17820	HISTIDINE KINASE 1
AT1G59940	ARR3 (RESPONSE REGULATOR 3)
AT5G38020	S-adenosyl-L-methionine:carboxyl methyltransferase family protein
AT1G67110	CYP735A2 (cytochrome P450, family 735, subfamily A, polypeptide 2)
AT1G15550	GA4 (GA REQUIRING 4); gibberellin 3-beta-dioxygenase
AT1G47400	Unknown protein
AT1G14960	Major latex protein-related/MLP-related
AT5G04120	Phosphoglycerate/bisphosphoglycerate mutase family protein
AT3G10960	Xanthine/uracil permease family protein
AT2G17500	Auxin efflux carrier family protein
AT4G21120	AAT1 (CATIONIC AMINO ACID TRANSPORTER 1)
AT1G69040	ACR4 (ACT REPEAT 4); amino acid binding
AT3G57040	ARR9 (RESPONSE REACTOR 4); transcription regulator
AT5G47980	Transferase family protein
AT1G67030	ZFP6 (ZINC FINGER PROTEIN 6)
AT5G05790	Myb family transcription factor
AT4G19030	NLM1 (NOD26-like intrinsic protein 1;1)
AT2G34610	Unknown protein
AT3G15990	SULTR3;4; sulfate transmembrane transporter
AT3G59670	Unknown protein
AT2G23170	GH3.3; indole-3-acetic acid amido synthetase
AT1G64590	Short-chain dehydrogenase/reductase (SDR) family protein
AT3G21670	Nitrate transporter (NTP3)
AT5G60890	ATMYB34
AT2G38750	ANNAT4 (ANNEXIN ARABIDOPSIS 4)
AT4G34950	Nodulin family protein
AT2G46660	CYP78A6 (cytochrome P450, family 78, subfamily A, polypeptide 6)
AT5G01740	Similar to SAG20 (WOUND-INDUCED PROTEIN 12)
AT2G25160	CYP82F1 (cytochrome P450, family 82, subfamily F, polypeptide 1)
AT2G36950	Heavy-metal-associated domain-containing protein
AT4G23750	CRF2 (CYTOKININ RESPONSE FACTOR 2)
AT5G64620	Invertase inhibitors AtC/VIF2
AT3G29250	Oxidoreductase
AT1G49470	Unknown protein
AT5G65210	TGA1
AT5G47990	CYP705A5 (cytochrome P450, family 705, subfamily A, polypeptide 5)
AT4G29700	Type I phosphodiesterase/nucleotide pyrophosphatase family protein
AT1G78000	SULTR1;2 (SULFATE TRANSPORTER 1;2)
AT3G45710	Proton-dependent oligopeptide transport (POT) family protein
AT4G25410	basix helix-loop-helix family protein
AT5G48000	CYP708A2 (cytochrome P450, family 708, subfamily A, polypeptide 2)
AT5G26220	ChaC-like family protein
AT1G66800	Cinnamyl-alcohol dehydrogenase family/CAD family

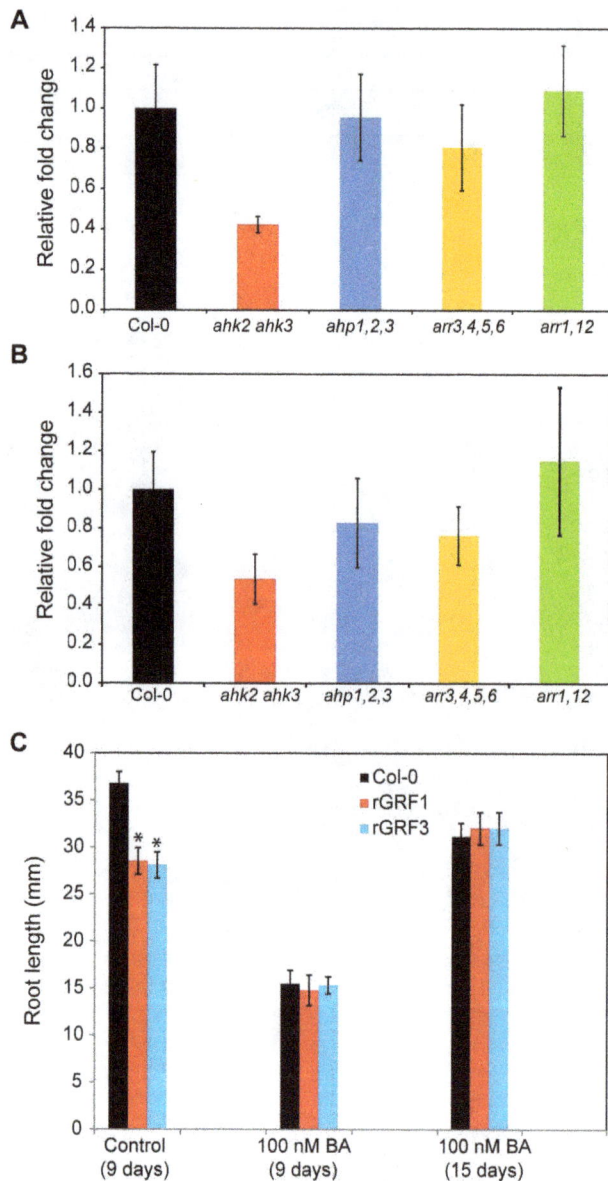

Figure 6. GRF1 and GRF3 regulate cytokinin signaling. A and B, GRF1 and GRF3 may contribute to the activity of cytokinin receptors. The expression levels of *GRF1* (A) and *GRF3* (B) were quantified by qPCR in various cytokinin signaling mutants including the *ahk2 ahk3* double mutant, *ahp1,2,3* triple mutant, type-A *arr3,4,5,6* quadruple mutant and type-B *arr1,12* double mutant. *GRF1* and *GRF3* showed significant downregulation in the *ahk2 ahk3* double mutant. The expression levels of *GRF1* and *GRF3* were normalized using *actin8* as an internal control. The relative fold-change values represent changes of *GRF* expression levels in the mutant lines relative to the wild-type (Col-0). Data are averages of three biological samples ± SE. C, Exogenous application of cytokinin rescued the short-root phenotype of *rGRF1* and *rGRF3* overexpression lines. Homozygous T3 lines overexpressing *rGRF1* (line 6–8), or *rGRF3* (line 11–15) as well as the wild-type Col-0 were grown vertically on modified Knop's medium supplemented or not with 100 nM BA and root lengths were measured 9 and 15 days after planting. Root length values are averages of at least 30 plants ± SE. Mean values significantly different from that of the wild type as determined by unadjusted paired t tests (P<0.01) are denoted by an asterisk.

TGA7) to the promoter of *PR1* in a SA-dependent manner [39,53–56]. It seems that GRF1 regulates the synergistic interactions between clade I and III TGA factors during plant response to pathogen infection by oppositely regulating the expression of genes belonging to both groups. Similar to clade I, clade III factor TGA3 is required for basal resistance [51] as well as for a novel form of cytokinin-induced resistance against virulent *P. syringae* [57]. Cytokinin-induced resistance may be an additional mechanism by which GRF1/3 control pathogen infection. Consistent with this speculation we found that GRF1/3 regulate 92 genes (41%) of the cytokinin-responsive genes from which 48 genes (21%) were identified as putative targets. Our data suggest that the potent control of GRF1/3 over cytokinin-responsive genes could be through targeting these genes directly as well as genes involved in cytokinin biosynthesis and signaling pathways. This suggestion was further supported by our data showing a significant down regulation of *GRF1* and *GRF3* in the cytokinin receptor *ahk2 ahk3* double mutant and that exogenous application of cytokinin rescued the short-root phenotype of the transgenic plants overexpressing *rGRF1* or *rGRF3* (Figure 6). Cytokinins are fundamental hormones for the proper growth and development of the plants [58] and also play critical roles in plant-pathogen interaction as many plant pathogens secrete cytokinins or promote cytokinin accumulation in host plants [57,59–61]. We conclude that targeting cytokinin-responsive and/or biosynthesis genes by GRF1/3 seems to be one of the main mechanisms employed by these two transcription factors to synchronize developmental processes and defense responses during pathogen infection.

Another interesting finding that could explain the coordination between developmental processes and defense responses mediated by GRF1/3 is that several ethylene-responsive element-binding factors (ERFs) are identified as putative targets of GRF1/3. ERFs constitute a plant-specific transcriptional factor superfamily of 147 members in Arabidopsis [62], influence a number of developmental processes, and are also involved in plant response to biotic stress [63–65]. It might be relevant to mention that several ERFs we identified as putative targets of GRF1/3 are implicated in defense responses. For example ERF5 (AT5G47230) plays vital role in phytotoxin-triggered programmed cell death [65] and in regulating both stress tolerance and leaf growth inhibition [66]. In addition, ERF2 (At5g47220) induces high levels of defense gene expression and enhances plant resistance to *Fusarium oxysporum* when overexpressed in Arabidopsis [67,68]. Furthermore, four *ERFs* (AT1G28370, AT2G33710, AT3G50260 and AT5G47220) identified as potential targets of GRF1/3 were found to be highly upregulated in response to chitin, a plant-defense elicitor [69]. These transcription factors may regulate gene expression downstream of chitin-activated defense signaling pathways in association with GRF1/3. Interestingly, WRKY33 was identified as potential direct target of GRF1 and GRF3. WRKY33 is a pathogen-inducible transcription factor, functions downstream of MPK3/MPK6 in controlling the accumulation of camalexin, the major phytoalexin in Arabidopsis. WRKY33 binds directly to the promoter of *PAD3*, which catalyzes the last conversion step of camalexin pathway [70,71]. It is intriguing to find that out of the ten genes known to be involved in the camalexin biosynthetic process, 5 were identified as putative targets of GRF1/3 including *MKK9*, *MPK3*, *PAD3* and *NAC* domain-containing protein 42 in addition to WRKY33. These data suggest that GRF1/3 may contribute significantly to the regulation of camalexin biosynthetic genes and hence defense responses.

Plants respond to invading pathogens by activating various metabolic pathways including induction of an array of secondary metabolites with antimicrobial properties as an integral part of

A

B

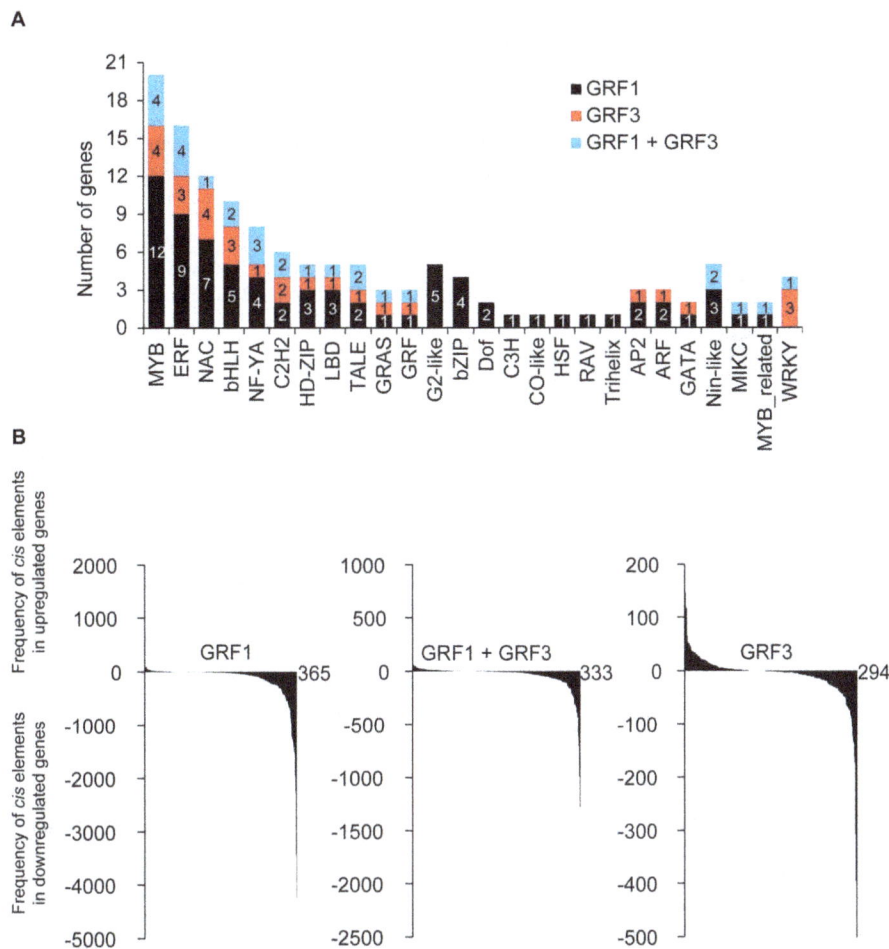

Figure 7. GRF1 and GRF3 may function as negative regulators of gene expression through their association with other transcription factors. A. Histogram showing the number of genes in different transcription factor families that are identified as putative targets of GRF1 or GRF3. B. The frequency of various transcription factor *cis* elements was quantified in the promoters (1,500 bp upstream of the translation start codon) of upregulated putative targets of GRF1, GRF3 or both versus downregulated genes using PlantPan software [19]. For each *cis* element (x axis), the differences in the frequency between upregulated and downregulated targets (y axis) were calculated and used in the plot.

plant disease resistance [72,73]. Regulating the activity of various secondary metabolite pathways appears to be another way by which GRF1/3 regulate defense responses. Our analysis revealed that several genes involved in the biosynthesis of several secondary metabolites including capsaicin, phenylpropanoid, stilbenoids, terpenoid and cyanoamino acid constitute a significant portion of the GRF1/3 putative targets. Unlike primary metabolites, secondary metabolites are not directly involved in the normal growth, development, or reproduction of the plants. However, they frequently play an important role in plant immunity by controlling the entry and/or development of the pathogens into plant cells and tissues as these metabolites can be secreted and delivered directly at the plant-pathogen interface [73,74]. For example, stilbenoids can function as antimicrobial compounds and accumulate as phytoalexins following pathogen infection [73]. Constitutive expression of a grapevine stilbene-synthase gene in alfalfa resulted in increased plant resistance to the leaf spot pathogen *Phoma medicaginis* [75]. Phenylpropanoids serve as precursors for several compounds essential for disease resistance and their association with active defense response are well-known [76–78]. Terpenoids are the biggest and most diverse class of phytochemicals and recent data demonstrate that their accumu-

lation in plant tissues can modify plant interactions with various pathogens [79].

Molecular links between defense and developmental pathways are believed to mediate and control the cross talk between various signaling pathways. This was clearly demonstrated by our data showing that GRF1/3 regulate other miRNA target genes that are involved in various cellular processes including flowering, auxin signaling, and copper and sulfate homeostasis (Table 1). Interestingly, this regulation was extended to include the expression of these miRNAs. As shown in Figure 4, the expression levels of seven miRNAs (miR169, miR172, miR393, miR395, miR844, miR846, and miR857) were altered in the transgenic plants overexpressing *GRF1* or *GRF3*. It is unlikely that GRF1 and GRF3 directly impact the expression of these miRNAs. Most likely, the expression of these miRNAs are altered as a results of positive or negative feedback regulation loops between these miRNAs and their targets that are regulated by GRF1 and/or GRF3. This assigns new and unexpected roles for these transcription factors in regulating the crosstalk between miRNA signaling networks. Our finding that GRF1 and GRF3 regulate the expression of all targets of miR169 (7 genes) from which 3 are co-regulated by both GRF1 and GRF3 suggests that the cross regulation is organized in a coordinated manner. Thus, GRF1/3 may fine tune the expression levels of co-regulated genes and members of multigene families with concomitant biological functions. Consistent with this hypothesis, several genes involved in flowering control (AT3G20910, AT5G12840, AT1G72830, AT1G17590, AT3G05690 and AT3G54990) and negatively regulated by miR169 or miR172 [31,80,81] were identified as putative targets of GRF1/3. Similarly, genes involved in auxin signaling such as auxin response factors, NAC domain-containing proteins, and auxin signaling F box protein1, which are negatively regulated by miR167, miR164 and miR393 [82–84], respectively, are also regulated by GRF1 or GRF3.

It is of interest to find that GRF1 and 3 regulate the expression of their putative targets in a tissue-specific manner. Identifying a subset of putative targets of GRF1/3 that are specifically expressed in roots is consistent with the abundant expression of *GRF1/3* in various root-tissue types and that overexpression of *GRF1* or *GRF3* impacts root growth and development [16]. Also, several recent reports support a role of GRF family members in floral organ development [85–88]. Our identification of several seed-specific genes as putative targets of GRF1/3 in the current study could illuminate the molecular events controlled by GRFs and required for precise floral organ initiation and development.

In conclusion, our data provide new insights into the molecular events by which GRF1/3 directly or indirectly regulate a variety of biological processes to formulate a decisive coordination between plant growth and defense responses. While direct proof is lacking, GRF1/3 may function not only as transcriptional activators or transcriptional repressors but also oppositely regulate genes that share common function or even genes that belong to the same gene family. This bifunctional activity, which reveals an unexpected degree of complexity of GRF1/3 in the regulation of their targets, may count among the main characteristics of key genes linking plant growth and developmental pathways to defense signaling.

Supporting Information

Table S1   Primer sequences used in this study.

Table S2   List of 1,098 differentially expressed genes showing opposite expression in the *grf1/grf2/grf3* triple mutant and *rGRF1* lines.

Table S3   List of 600 differentially expressed genes showing opposite expression in the *grf1/grf2/grf3* triple mutant and *rGRF3* lines.

Table S4   List of 1,434 genes identified as unique putative target genes of GRF1 and GRF3.

Table S5   Biological pathway description of 383 putative targets of GRF1 or GRF3.

Table S6   Putative targets of GRF1 or GRF3 showing root and seed-specific expression.

Table S7   List of miRNA target genes that are identified as differentially expressed in the *rGRF1* or *rGRF3* transgenic plants.

Table S8   List of cytokinin-responsive genes that are identified as differentially expressed in the *rGRF1* or *rGRF3* transgenic plants.

Table S9   List of the *cis* elements that are identified in the promoters of the putative targets of GRF1 and GRF3.

Acknowledgments

We thank Dr. Joseph J. Kieber at University of North Carolina for providing us with cytokinin signaling mutants.

Author Contributions

Conceived and designed the experiments: TH TJB. Performed the experiments: TH JL JHR. Analyzed the data: JL NC TJB TH. Wrote the paper: TH.

References

1. Heidel AJ, Clarke JD, Antonovics J, Dong X (2004) Fitness costs of mutations affecting the systemic acquired resistance pathway in *Arabidopsis thaliana*. Genetics 168: 2197–2206.
2. van der Knaap E, Kim JH, Kende H (2000) A novel gibberellin-induced gene from rice and its potential regulatory role in stem growth. Plant Physiol 122: 695–704.
3. Kim JH, Choi D, Kende H (2003) The AtGRF family of putative transcription factors is involved in leaf and cotyledon growth in Arabidopsis. Plant J 36: 94–104.
4. Choi D, Kim JH, Kende H (2004) Whole genome analysis of the OsGRF gene family encoding plant-specific putative transcription activators in rice (Oryza sativa L.). Plant Cell Physiol 45: 897–904.
5. Zhang DF, Li B, Jia GQ, Zhang TF, Dai JR, et al. (2008) Isolation and characterization of genes encoding GRF transcription factors and GIF transcriptional coactivators in Maize (Zea mays L.). Plant Sci 175: 809–817.
6. Kim JH, Kende H (2004) A transcriptional coactivator, AtGIF1, is involved in regulating leaf growth and morphology in Arabidopsis. Proc Natl Acad Sci U S A 101: 13374–13379.
7. Kim JS, Mizoi J, Kidokoro S, Maruyama K, Nakajima J, et al. (2012) Arabidopsis growth-regulating factor7 functions as a transcriptional repressor of abscisic acid- and osmotic stress-responsive genes, including DREB2A. Plant Cell 24: 3393–3405.
8. Liu D, Song Y, Chen Z, Yu D (2009) Ectopic expression of miR396 suppresses GRF target gene expression and alters leaf growth in Arabidopsis. Physiol Plant 136: 223–236.
9. Rodriguez RE, Mecchia MA, Debernardi JM, Schommer C, Weigel D, et al. (2010) Control of cell proliferation in *Arabidopsis thaliana* by microRNA miR396. Development 137: 103–112.
10. Wang L, Gu X, Xu D, Wang W, Wang H, et al. (2011) miR396-targeted AtGRF transcription factors are required for coordination of cell division and differentiation during leaf development in Arabidopsis. J Eep Bot 62: 761–773.
11. Hewezi T, Baum TJ (2012) Complex feedback regulations govern the expression of miRNA396 and its GRF target genes. Plant Signal Behav 7: 749–751.
12. Casadevall R, Rodriguez RE, Debernardi JM, Palatnik JF, Casati P (2013) Repression of Growth Regulating Factors by the MicroRNA396 Inhibits Cell Proliferation by UV-B Radiation in Arabidopsis Leaves. Plant Cell 25: 3570–3583.
13. Liu HH, Tian X, Li YJ, Wu CA, Zheng CC (2008) Microarray-based analysis of stress-regulated microRNAs in *Arabidopsis thaliana*. RNA 14: 836–843.
14. Fahlgren N, Howell MD, Kasschau KD, Chapman EJ, Sullivan CM, et al. (2007) High-throughput sequencing of Arabidopsis microRNAs: evidence for frequent birth and death of MIRNA genes. PloS One 2: e219.
15. Li Y, Zhang Q, Zhang J, Wu L, Qi Y, et al. (2010) Identification of microRNAs involved in pathogen-associated molecular pattern-triggered plant innate immunity. Plant Physiol 152: 2222–2231.
16. Hewezi T, Maier TR, Nettleton D, Baum TJ (2012) The Arabidopsis microRNA396-GRF1/GRF3 regulatory module acts as a developmental regulator in the reprogramming of root cells during cyst nematode infection. Plant Physiol 159: 321–335.
17. Schmid M, Davison TS, Henz SR, Pape UJ, Demar M, et al. (2005) A gene expression map of *Arabidopsis thaliana* development. Nat genet 37: 501–506.

18. Winter D, Vinegar B, Nahal H, Ammar R, Wilson GV, et al. (2007) An "Electronic Fluorescent Pictograph" browser for exploring and analyzing large-scale biological data sets. PloS One 2: e718.

19. Chang WC, Lee TY, Huang HD, Huang HY, Pan RL (2008) PlantPAN: Plant promoter analysis navigator, for identifying combinatorial cis-regulatory elements with distance constraint in plant gene groups. BMC genomics 9: 561.

20. Higuchi M, Pischke MS, Mahonen AP, Miyawaki K, Hashimoto Y, et al. (2004) In planta functions of the Arabidopsis cytokinin receptor family. Proc Natl Acad Sci U S A 101: 8821–8826.

21. Hutchison CE, Li J, Argueso C, Gonzalez M, Lee E, et al. (2006) The Arabidopsis histidine phosphotransfer proteins are redundant positive regulators of cytokinin signaling. Plant Cell 18: 3073–3087.

22. To JPC, Haberer G, Ferreira FJ, Deruere J, Mason MG, et al. (2004) Type-A Arabidopsis response regulators are partially redundant negative regulators of cytokinin signaling. Plant Cell 16: 658–671.

23. Mason MG, Mathews DE, Argyros DA, Maxwell BB, Kieber JJ, et al. (2005) Multiple type-B response regulators mediate cytokinin signal transduction in Arabidopsis. Plant Cell 17: 3007–3018.

24. Verwoerd TC, Dekker BMM, Hoekema A (1989) A Small-Scale Procedure for the Rapid Isolation of Plant Rnas. Nucleic Acids Res 17: 2362–2362.

25. Livak KJ, Schmittgen TD (2001) Analysis of relative gene expression data using real-time quantitative PCR and the $2^{-\Delta\Delta CT}$ method. Methods 25: 402–408.

26. Geer LY, Marchler-Bauer A, Geer RC, Han L, He J, et al. (2010) The NCBI BioSystems database. Nucleic Acids Res 38: D492–496.

27. Gutierrez L, Bussell JD, Pacurar DI, Schwambach J, Pacurar M, et al. (2009) Phenotypic Plasticity of Adventitious Rooting in Arabidopsis Is Controlled by Complex Regulation of AUXIN RESPONSE FACTOR Transcripts and MicroRNA Abundance. Plant Cell 21: 3119–3132.

28. Wu G, Park MY, Conway SR, Wang JW, Weigel D, et al. (2009) The Sequential Action of miR156 and miR172 Regulates Developmental Timing in Arabidopsis. Cell 138: 750–759.

29. Marin E, Jouannet V, Herz A, Lokerse AS, Weijers D, et al. (2010) miR390, Arabidopsis TAS3 tasiRNAs, and Their AUXIN RESPONSE FACTOR Targets Define an Autoregulatory Network Quantitatively Regulating Lateral Root Growth. Plant Cell 22: 1104–1117.

30. Wang JW, Czech B, Weigel D (2009) miR156-Regulated SPL Transcription Factors Define an Endogenous Flowering Pathway in Arabidopsis thaliana. Cell 138: 738–749.

31. Yant L, Mathieu J, Dinh TT, Ott F, Lanz C, et al. (2010) Orchestration of the Floral Transition and Floral Development in Arabidopsis by the Bifunctional Transcription Factor APETALA2. Plant Cell 22: 2156–2170.

32. Szakasits D, Heinen P, Wieczorek K, Hofmann J, Wagner F, et al. (2009) The transcriptome of syncytia induced by the cyst nematode Heterodera schachtii in Arabidopsis roots. Plant J 57: 771–784.

33. Barcala M, Garcia A, Cabrera J, Casson S, Lindsey K, et al. (2010) Early transcriptomic events in microdissected Arabidopsis nematode-induced giant cells. Plant J 61: 698–712.

34. Bhargava A, Clabaugh I, To JP, Maxwell BB, Chiang YH, et al. (2013) Identification of cytokinin-responsive genes using microarray meta-analysis and RNA-Seq in Arabidopsis. Plant Physiol 162: 272–294.

35. Hwang I, Sheen J, Muller B (2012) Cytokinin Signaling Networks. Annu Rev Plant Biol 63: 353–380.

36. Ioio RD, Linhares FS, Scacchi E, Casamitjana-Martinez E, Heidstra R, et al. (2007) Cytokinins determine Arabidopsis root-meristem size by controlling cell differentiation. Curr Biol 17: 678–682.

37. Kesarwani M, Yoo J, Dong X (2007) Genetic interactions of TGA transcription factors in the regulation of pathogenesis-related genes and disease resistance in Arabidopsis. Plant Physiol 144: 336–346.

38. Wang L, Fobert PR (2013) Arabidopsis clade I TGA factors regulate apoplastic defences against the bacterial pathogen Pseudomonas syringae through endoplasmic reticulum-based processes. PloS One 8: e77378.

39. Shearer HL, Cheng YT, Wang L, Liu J, Boyle P, et al. (2012) Arabidopsis clade I TGA transcription factors regulate plant defenses in an NPR1-independent fashion. Mol Plant Microbe Interact 25: 1459–1468.

40. Zhou J, Lee C, Zhong R, Ye ZH (2009) MYB58 and MYB63 are transcriptional activators of the lignin biosynthetic pathway during secondary cell wall formation in Arabidopsis. Plant Cell 21: 248–266.

41. Zhong R, Ye ZH (2012) MYB46 and MYB83 bind to the SMRE sites and directly activate a suite of transcription factors and secondary wall biosynthetic genes. Plant Cell Physiol 53: 368–380.

42. O'Donnell PJ, Calvert C, Atzorn R, Wasternack C, Leyser HMO, et al. (1996) Ethylene as a Signal Mediating the Wound Response of Tomato Plants. Science 274: 1914–1917.

43. Ecker JR (1995) The ethylene signal transduction pathway in plants. Science 268: 667–675.

44. Penninckx IA, Eggermont K, Terras FR, Thomma BP, De Samblanx GW, et al. (1996) Pathogen-induced systemic activation of a plant defensin gene in Arabidopsis follows a salicylic acid-independent pathway. Plant Cell 8: 2309–2323.

45. Ito T, Nagata N, Yoshiba Y, Ohme-Takagi M, Ma H, et al. (2007) Arabidopsis MALE STERILITY1 encodes a PHD-type transcription factor and regulates pollen and tapetum development. Plant Cell 19: 3549–3562.

46. Pu L, Li Q, Fan X, Yang W, Xue Y (2008) The R2R3 MYB transcription factor GhMYB109 is required for cotton fiber development. Genetics 180: 811–820.

47. Aya K, Ueguchi-Tanaka M, Kondo M, Hamada K, Yano K, et al. (2009) Gibberellin modulates anther development in rice via the transcriptional regulation of GAMYB. Plant Cell 21: 1453–1472.

48. Raffaele S, Vailleau F, Leger A, Joubes J, Miersch O, et al. (2008) A MYB transcription factor regulates very-long-chain fatty acid biosynthesis for activation of the hypersensitive cell death response in Arabidopsis. Plant Cell 20: 752–767.

49. Oh E, Kang H, Yamaguchi S, Park J, Lee D, et al. (2009) Genome-Wide Analysis of Genes Targeted by PHYTOCHROME INTERACTING FACTOR 3-LIKE5 during Seed Germination in Arabidopsis. Plant Cell 21: 403–419.

50. Morimoto RI (1992) Transcription factors: positive and negative regulators of cell growth and disease. Curr Opin Cell Biol 4: 480–487.

51. Kesarwani M, Yoo JM, Dong XN (2007) Genetic interactions of TGA transcription factors in the regulation of pathogenesis-related genes and disease resistance in Arabidopsis. Plant Physiol 144: 336–346.

52. Lindermayr C, Sell S, Muller B, Leister D, Durnera J (2010) Redox Regulation of the NPR1-TGA1 System of Arabidopsis thaliana by Nitric Oxide. Plant Cell 22: 2894–2907.

53. Zhang YL, Fan WH, Kinkema M, Li X, Dong XN (1999) Interaction of NPR1 with basic leucine zipper protein transcription factors that bind sequences required for salicylic acid induction of the PR-1 gene. Proc Natl Acad Sci U S A 96: 6523–6528.

54. Despres C, DeLong C, Glaze S, Liu E, Fobert PR (2000) The arabidopsis NPR1/NIM1 protein enhances the DNA binding activity of a subgroup of the TGA family of bZIP transcription factors. Plant Cell 12: 279–290.

55. Zhou JM, Trifa Y, Silva H, Pontier D, Lam E, et al. (2000) NPR1 differentially interacts with members of the TGA/OBF family of transcription factors that bind an element of the PR-1 gene required for induction by salicylic acid. Mol Plant Microbe Interact 13: 191–202.

56. Rochon A, Boyle P, Wignes T, Fobert PR, Despres C (2006) The coactivator function of Arabidopsis NPR1 requires the core of its BTB/POZ domain and the oxidation of C-terminal cysteines. Plant Cell 18: 3670–3685.

57. Choi J, Huh SU, Kojima M, Sakakibara H, Paek KH, et al. (2010) The Cytokinin-Activated Transcription Factor ARR2 Promotes Plant Immunity via TGA3/NPR1-Dependent Salicylic Acid Signaling in Arabidopsis. Dev Cell 19: 284–295.

58. Choi J, Hwang I (2007) Cytokinin: Perception, signal transduction, and role in plant growth and development. J Plant Biol 50: 98–108.

59. Jameson PE, Zhang H, Lewis DH (2000) Cytokinins. Extraction, separation, and analysis. Methods Mol Biol 141: 101–121.

60. Pertry I, Vaclavikova K, Depuydt S, Galuszka P, Spichal L, et al. (2009) Identification of Rhodococcus fascians cytokinins and their modus operandi to reshape the plant. Proc Natl Acad Sci U S A 106: 929–934.

61. Choi J, Choi D, Lee S, Ryu CM, Hwang I (2011) Cytokinins and plant immunity: old foes or new friends? Trends Plant Sci 16: 388–394.

62. Nakano T, Suzuki K, Fujimura T, Shinshi H (2006) Genome-wide analysis of the ERF gene family in Arabidopsis and rice. Plant Physiol 140: 411–432.

63. Berrocal-Lobo M, Molina A, Solano R (2002) Constitutive expression of ETHYLENE-RESPONSE-FACTOR1 in Arabidopsis confers resistance to several necrotrophic fungi. Plant J 29: 23–32.

64. Lu X, Zhang L, Zhang FY, Jiang WM, Shen Q, et al. (2013) AaORA, a trichome-specific AP2/ERF transcription factor of Artemisia annua, is a positive regulator in the artemisinin biosynthetic pathway and in disease resistance to Botrytis cinerea. New Phytol 198: 1191–1202.

65. Mase K, Ishihama N, Mori H, Takahashi H, Kaminaka H, et al. (2013) Ethylene-Responsive AP2/ERF Transcription Factor MACD1 Participates in Phytotoxin-Triggered Programmed Cell Death. Mol Plant Microbe Interact 26: 868–879.

66. Dubois M, Skirycz A, Claeys H, Maleux K, Dhondt S, et al. (2013) ETHYLENE RESPONSE FACTOR6 Acts as a Central Regulator of Leaf Growth under Water-Limiting Conditions in Arabidopsis. Plant Physiol 162: 319–332.

67. McGrath KC, Dombrecht B, Manners JM, Schenk PM, Edgar CI, et al. (2005) Repressor- and activator-type ethylene response factors functioning in jasmonate signaling and disease resistance identified via a genome-wide screen of Arabidopsis transcription factor gene expression. Plant Physiol 139: 949–959.

68. Brown RL, Kazan K, McGrath KC, Maclean DJ, Manners JM (2003) A role for the GCC-box in jasmonate-mediated activation of the PDF1.2 gene of Arabidopsis. Plant Physiol 132: 1020–1032.

69. Libault M, Wan JR, Czechowski T, Udvardi M, Stacey G (2007) Identification of 118 Arabidopsis transcription factor and 30 ubiquitin-ligase genes responding to chitin, a plant-defense elicitor. Mol Plant Microbe Interact 20: 900–911.

70. Qiu JL, Fiil BK, Petersen K, Nielsen HB, Botanga CJ, et al. (2008) Arabidopsis MAP kinase 4 regulates gene expression through transcription factor release in the nucleus. EMBO J 27: 2214–2221.

71. Mao GH, Meng XZ, Liu YD, Zheng ZY, Chen ZX, et al. (2011) Phosphorylation of a WRKY Transcription Factor by Two Pathogen-Responsive MAPKs Drives Phytoalexin Biosynthesis in Arabidopsis. Plant Cell 23: 1639–1653.

72. Hammerschmidt R (1999) PHYTOALEXINS: What have we learned after 60 years? Annu Rev Phytopathol 37: 285–306.

73. Dixon RA (2001) Natural products and plant disease resistance. Nature 411: 843–847.

74. Bednarek P (2012) Chemical warfare or modulators of defence responses - the function of secondary metabolites in plant immunity. Curr Opin Plant Biol 15: 407–414.

75. Hipskind JD, Paiva NL (2000) Constitutive accumulation of a resveratrol-glucoside in transgenic alfalfa increases resistance to *Phoma medicaginis*. Mol Plant Microbe Interact 13: 551–562.

76. Nicholson RL, Hammerschmidt R (1992) Phenolic-Compounds and Their Role in Disease Resistance. Annu Rev Phytopathol 30: 369–389.

77. Noel JP, Austin MB, Bomati EK (2005) Structure-function relationships in plant phenylpropanoid biosynthesis. Curr Opin Plant Biol 8: 249–253.

78. Cheynier V, Comte G, Davies KM, Lattanzio V, Martens S (2013) Plant phenolics: Recent advances on their biosynthesis, genetics, and ecophysiology. Plant Physiol Biochem 72: 1–20.

79. Szucs I, Escobar M, Grodzinski B (2011) Emerging Roles for Plant Terpenoids. In: Moo-Young M, editor. Comprehensive biotechnology second edition: Agriculture and related biotechnologies. pp. 273–286.

80. Zhu QH, Helliwell CA (2011) Regulation of flowering time and floral patterning by miR172. J Exp Bot 62: 487–495.

81. Mathieu J, Yant LJ, Murdter F, Kuttner F, Schmid M (2009) Repression of Flowering by the miR172 Target SMZ. Plos Biol 7: e1000148.

82. Raman S, Greb T, Peaucelle A, Blein T, Laufs P, et al. (2008) Interplay of miR164, CUP-SHAPED COTYLEDON genes and LATERAL SUPPRES-SOR controls axillary meristem formation in *Arabidopsis thaliana*. Plant J 55: 65–76.

83. Chen ZH, Bao ML, Sun YZ, Yang YJ, Xu XH, et al. (2011) Regulation of auxin response by miR393-targeted transport inhibitor response protein 1 is involved in normal development in Arabidopsis. Plant Mol Biol 77: 619–629.

84. Wu MF, Tian Q, Reed JW (2006) Arabidopsis microRNA167 controls patterns of ARF6 and ARF8 expression, and regulates both female and male reproduction. Development 133: 4211–4218.

85. Yang FX, Liang G, Liu DM, Yu DQ (2009) Arabidopsis MiR396 Mediates the Development of Leaves and Flowers in Transgenic Tobacco. J Plant Biol 52: 475–481.

86. Wynn AN, Rueschhoff EE, Franks RG (2011) Transcriptomic Characterization of a Synergistic Genetic Interaction during Carpel Margin Meristem Development in *Arabidopsis thaliana*. PloS One 6: e26231.

87. Baucher M, Moussawi J, Vandeputte OM, Monteyne D, Mol A, et al. (2013) A role for the miR396/GRF network in specification of organ type during flower development, as supported by ectopic expression of Populus trichocarpa miR396c in transgenic tobacco. Plant Biol 15: 892–898.

88. Liang G, He H, Li Y, Wang F, Yu D (2013) Molecular mechanism of miR396 mediating pistil development in *Arabidopsis thaliana*. Plant Physiol 164: 249–258.

10

Impact of Transgenic Wheat with *wheat yellow mosaic virus* Resistance on Microbial Community Diversity and Enzyme Activity in Rhizosphere Soil

Jirong Wu[1,2,3 ◊], Mingzheng Yu[1,2,3 ◊], Jianhong Xu[1,2,3], Juan Du[1,2,3], Fang Ji[1,2,3], Fei Dong[1,2,3], Xinhai Li[4*], Jianrong Shi[1,2,3*]

1 Institute of Food Safety and Detection, Jiangsu Academy of Agricultural Sciences, Nanjing, China, **2** Key Lab of Food Quality and Safety of Jiangsu Province—State Key Laboratory Breeding Base, Nanjing, China, **3** Jiangsu Center for GMO evaluation and detection, Nanjing, China, **4** Institute of Crop Sciences, Chinese Academy of Agricultural Sciences, Beijing, China

Abstract

The transgenic wheat line N12-1 containing the *WYMV-Nib8* gene was obtained previously through particle bombardment, and it can effectively control the wheat yellow mosaic virus (WYMV) disease transmitted by *Polymyxa graminis* at turngreen stage. Due to insertion of an exogenous gene, the transcriptome of wheat may be altered and affect root exudates. Thus, it is important to investigate the potential environmental risk of transgenic wheat before commercial release because of potential undesirable ecological side effects. Our 2-year study at two different experimental locations was performed to analyze the impact of transgenic wheat N12-1 on bacterial and fungal community diversity in rhizosphere soil using polymerase chain reaction-denaturing gel gradient electrophoresis (PCR-DGGE) at four growth stages (seeding stage, turngreen stage, grain-filling stage, and maturing stage). We also explored the activities of urease, sucrase and dehydrogenase in rhizosphere soil. The results showed that there was little difference in bacterial and fungal community diversity in rhizosphere soil between N12-1 and its recipient Y158 by comparing Shannon's, Simpson's diversity index and evenness (except at one or two growth stages). Regarding enzyme activity, only one significant difference was found during the maturing stage at Xinxiang in 2011 for dehydrogenase. Significant growth stage variation was observed during 2 years at two experimental locations for both soil microbial community diversity and enzyme activity. Analysis of bands from the gel for fungal community diversity showed that the majority of fungi were uncultured. The results of this study suggested that virus-resistant transgenic wheat had no adverse impact on microbial community diversity and enzyme activity in rhizosphere soil during 2 continuous years at two different experimental locations. This study provides a theoretical basis for environmental impact monitoring of transgenic wheat when the introduced gene is derived from a virus.

Editor: Newton C M Gomes, University of Aveiro, Portugal

Funding: This work was supported by the National Special Transgenic Project (2014ZX08011-003), Natural Science Foundation of Jiangsu Province, China (BK20130721) and Jiangsu Agriculture Science and Technology Innovation Fund [cx(11)4064]. The funders had no role in study design, data collection and analysis, decision to publish, or preparation of the manuscript.

Competing Interests: The authors have declared that no competing interests exist.

* E-mail: lixinhai@caas.cn (XL); shiji@jaas.ac.cn (JS)

◊ These authors contributed equally to this work.

Introduction

Since the first successful genetically engineered (GE) plant was reported in 1983 [1], the planting area of transgenic crops has increased rapidly [2]. The global area cultivated commercially with transgenic crops has increased from 1.7 million ha in 1996 to 170.3 million ha in 2012 [3]. With the continued release and use of transgenic crops, there is a growing concern about their impact on the biota and soil microbial processes, such as nutrient cycling, and the potential risk of gene transfer from transgenic crops to indigenous soil microbes [4–5]. The microbes in rhizosphere soil play an important role in plant growth and development [6–7]. Transgenic crops planted in soil will inevitably interact with microorganisms such as bacteria, fungi, and actinomycetes [8–10]. Thus, transgenic crops may affect soil microbial population structure and quantity [11–13]. Additionally, root exudates have marked effects on soil microbial diversity and spatial distribution

[14–15]. At this time, most studies of environmental risk assessment focused on transgenic Bt crops such as transgenic cotton, rice and maize containing the *Bt* gene [16–18]; these studies provided basic methods for environmental risk assessment for other crops.

Enzymes in the rhizosphere soil derived from animal, plant roots and soil microbial cell secretion and decomposition of residues are an important component of the soil ecosystem [19]. They play an important role in soil biochemical processes and directly affect soil fertility [19]. Urease is associated with nitrogen transformation in the soil, while sucrase is associated with soil organic matter, nitrogen and phosphorus contents, and dehydrogenase is associated with the redox ability of the soil [19]. Previous studies showed that transgenic plants might affect enzyme activities in rhizosphere soil [11,20–21]. Therefore, it is important to investigate the impact of transgenic crops on rhizosphere soil enzyme activity when performing environmental safety risk assessments.

The first report of transgenic plants with virus resistance, expressing the coat protein of the *tobacco mosaic virus* (TMV) and delaying the development of disease, appeared in 1986 [22]. The same strategy was subsequently used to create resistance to a range of other viruses [23–24]. The exogenous genes of the transgenic virus-resistant crops are generally derived from the virus itself, including genes encoding coat protein and replicase [22–24]. Sequences derived from the genomes of plant viruses have been used to generate viral resistance in transgenic crop plants, but potential safety issues have been raised due to the environmental risks of transgenic plants with virus resistance, including hetero-encapsidation, virus recombination, gene flow, synergism and effects on non-target organisms [25,26].

Wheat yellow mosaic disease, caused by the wheat yellow mosaic virus (WYMV) at turngreen stage, is a serious illness affecting wheat in the middle and lower reaches of the Yangtze River region in China [27–28]. Disease-resistant variety breeding is one of the most cost-effective ways to control this disease through conventional wheat breeding. In recent years, conventional wheat breeding in combination with genetic engineering techniques has been applied to address wheat yellow mosaic disease, and some disease-resistant wheat lines have been cultivated. Using the particle bombardment method, genes from WYMV encoding replicase WYMV-Nib8 were transferred to the disease-sensitive variety Yangmai158 (Y158), and the disease-resistant transgenic wheat line named N12-1 was obtained by successive backcross with Y158 [29]. N12-1 showed stable and effective resistance to wheat yellow mosaic disease in a previous study [30].

Considering the above risks, transgenic virus-resistant wheat may affect the microbial community diversity in rhizosphere soil and change the population structure. Exogenous insertion of genes may also cause changes in the metabolic pathways of genetically modified crops and alter the composition of root exudates, resulting in changes in soil enzyme activity [31]. Thus, further studies on the impact on soil microbial community diversity and enzyme activities should be performed. In this study, environmental risk assessment of N12-1 was performed during 2 consecutive years of wheat cultivation under field conditions at two different experimental stations. The research involved primarily: (i) differences in soil microbial (bacterial and fungal) diversity in rhizosphere soil between N12-1 and Y158 using polymerase chain reaction–denaturing gradient gel electrophoresis (PCR-DGGE) and (ii) the activity of enzymes (urease, sucrase and dehydrogenase) in rhizosphere soil. In this report, we provide a theoretical basis for environmental transgenic wheat monitoring.

Materials and Methods

Ethics statement

In our study, the research samples were rhizosphere soils in the presence of transgenic and non-transgenic wheat. This presented no ethical issue.

Plant materials and field trial

Transgenic wheat line N12-1 and its recipient Yangmai158 (Y158) provided by the Chinese Academy of Agricultural Sciences (CAAS) were applied in this study. N12-1, which contains the *WYMV-Nib8* gene from wheat yellow mosaic virus, can effectively control the WYMV disease transmitted by *Polymyxa graminis* at turngreen stage. Y158 was one of the most popular varieties in the middle and lower reaches of the Yangtze River region in China.

However, it is sensitive to WYMV disease and the yield decreased significantly due to effects of this severe disease.

This study was performed at Luhe experimental station for transgenic crop, Jiangsu Academy of Agricultural Sciences (Luhe) and Xinxiang experimental station for transgenic crop, Henan Academy of Agricultural Sciences (Xinxiang). The physical and chemical properties of the soil are provided in Table 1. pH value, water content, available nitrogen, phosphorus potassium and organic matter content were determined by potentiometry method, alkali solution diffusion method, sodium bicarbonate method, ammonium acetate extraction method, potassium dichromate method, respectively [32]. The experiment was conducted in two successive growth seasons of wheat (October 2010-June 2011 and October 2011-June 2012) in the same field in which transgenic crops had never been planted. Each variety (line) had four blocks, each of which was 10×6 m. The materials were planted in a row with a row length of 6 m and row spacing of 0.3 m. Distance between plants was 3 cm within a row. Completely random design was applied to arrange the experiment performed in the field, and the wheat was subjected to conventional field management, that was 375 kg/ha of compound fertilizer (N:P_2O_5:K_2O = 1:0.4:1) as base fertilizer and 225 kg/ha of urea as topdressing at seedling stage.

Soil sampling

Rhizosphere soil samples were collected in both years at Luhe and Xinxiang at four growth stages [seeding stage (SS), turngreen stage (TS), grainfilling stage (GS), maturing stage (MS)]. Rhizosphere soil was defined as the soil still attached to the roots after the roots were shaken by hand. For each sampling site, five wheat plants were selected to collect rhizosphere soil and each block contains five sampling site. Rhizosphere soil from the five sampling sites per block was mixed as a composite rhizosphere soil sample. The soil samples were sieved using a 20-mesh sieve and then stored at 4°C until further use, usually within one month before DNA extraction.

Soil DNA extraction

Total community DNA was extracted from 0.5 g of rhizosphere soil using an UltraClean Soil DNA Isolation Kit (MoBio Lab, USA). DNA extraction was performed according to the manufacturer's protocol.

PCR amplification of 16S and18S rDNA fragments for DGGE analysis

The 16S rDNA fragments of bacteria were amplified by using the primer pair GC338f (5′-<u>CGCCGCGCGCGGCGGGGCG-GGGCGGGGGGCACGGGGGG</u>ACTCCTACGGGAGGCAGC-AG-3′, the sequence underlined was the GC clamp) and 518r (5′-ATTACCGCGGCTGCTGG -3′) as described by Bakke et al. [33]. High fidelity polymerase of KOD-Plus-Neo (Toyobo, Japan) was applied to perform PCR amplification and avoid mutations in the PCR product. Briefly, the reaction mixture consisted of 1 µl of template DNA (1–5 ng), 5 µl 10×PCR Buffer, 5 µl of 2 mM dNTPs, 3 µl of 25 mM $MgSO_4$, 0.5 µl of 10 µM forward primer, 0.5 µl of 10 µM reverse primer, and 1 U of DNA polymerase, after which ddH_2O was added to a final volume of 50 µl. The thermal cycling program was performed with an initial denaturation at 94°C for 5 min, followed by 35 cycles at 95°C for 15 sec, 58°C for 15 sec, and 68°C for 30 sec before the final extension at 68°C for 10 min. Products were checked by electrophoresis in 1% (wt/vol) agarose gels followed by ethidium bromide staining.

Table 1. Main physical and chemical properties of the soil from two experiment locations before planting.

Experiment station	Physical and chemical properties					
	pH	water (%)	available nitrogen (mg/kg)	available phosphorus (mg/kg)	available potassium (mg/kg)	organic matter (%)
Luhe	5.8	20.55	110.16	90.81	857.99	1.44
Xinxiang	8.5	4.92	70.39	28.26	863.69	0.68

The 18S rDNA fragments of fungi were amplified by using the primer pair (GC-Fungi: 5′-<u>CGCCCGCCGCGCCCCGCGCCCGGCCCGCCGCCCCCGCCCC</u>ATTCCCCGTTACCCGTT-G-3′; NS1: 5′- GTAGTCATATGCTTGTCTC -3′, the sequence underlined was the GC clamp) as described by Das et al. [34]. The protocol for PCR amplification was similar as above. All products were purified before electrophoresis using a Cycle Pure Kit (Omega, USA).

PCR-DGGE

DGGE analysis for 16S rDNA and 18S rDNA products was performed with the DCode System (Bio-Rad, USA). Polyacrylamide gels were composed of a denaturing gradient of 50–65% (bacteria) and 30–38% (fungi) urea, 0.17% (vol/vol) TEMED, 0.047% (wt/vol) ammonium persulfate, 6% acrylamide-N,N_-methylenebisacrylamide (37.5:1) and 1×TAE. PCR products (up to 50 μl) were applied to the gel. DGGE was performed at 50 V in 1×TAE at 60°C for 12 h (bacteria) and at 50 V in 1×TAE at 60°C for 20 h (fungi), respectively. A silver staining method was used for the detection of DNA in DGGE gels.

Migration and intensity of DGGE bands were analyzed using Quantity One according to the manual. The bands that shared identical migration positions were considered to be the same species. Shannon's diversity index (H) of bacterial and fungal DGGE profiles was calculated with the following formula [35]:

$$H = - \sum_{i=1}^{S} \frac{n_i}{N} \ln \frac{n_i}{N}$$

Simpson's diversity index (D) was calculated with the following formula:

$$D = \sum_{i=1}^{S} \left[\frac{n_i}{N} \right]^2$$

Evenness (E) was calculated with the following formula:

$$E = \frac{H}{\ln S}$$

Figure 2. Shannon's index of bacterial communities at different growth stages. Error bars indicate standard errors (n = 4). Different letters above bars denote a statistically significant difference between the means of the fields. A: Luhe; B: Xinxiang. SS: seeding stage; TS: turngreen stage; GS: grainfilling stage; MS: maturing stage.

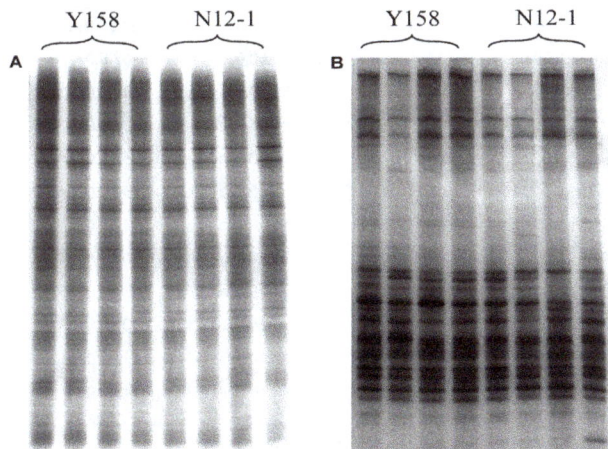

Figure 1. DGGE profiles of 16S rDNA and 18S rDNA fragments amplified from DNA extracted from rhizosphere soil of N12-1 and Y 158 at turngreen stage from Luhe experiment station in 2011. A: bacteria; B: fungus.

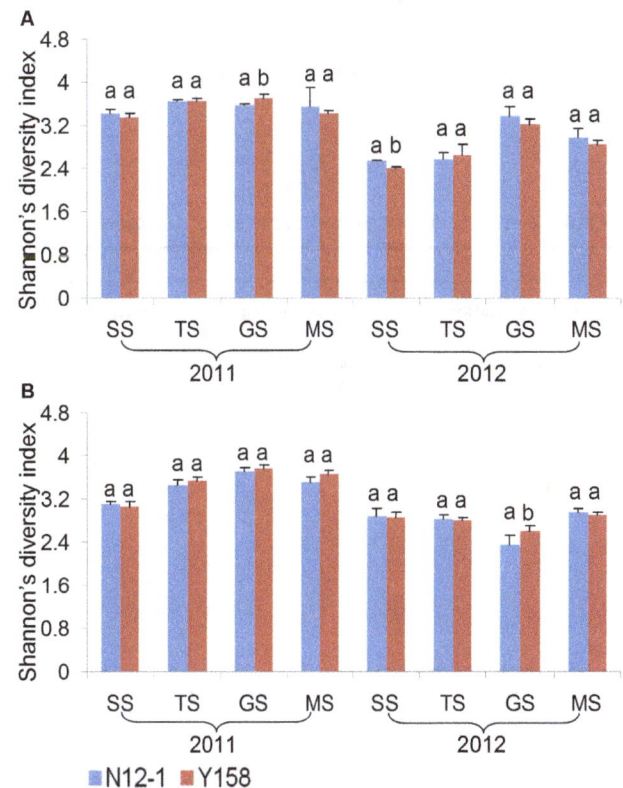

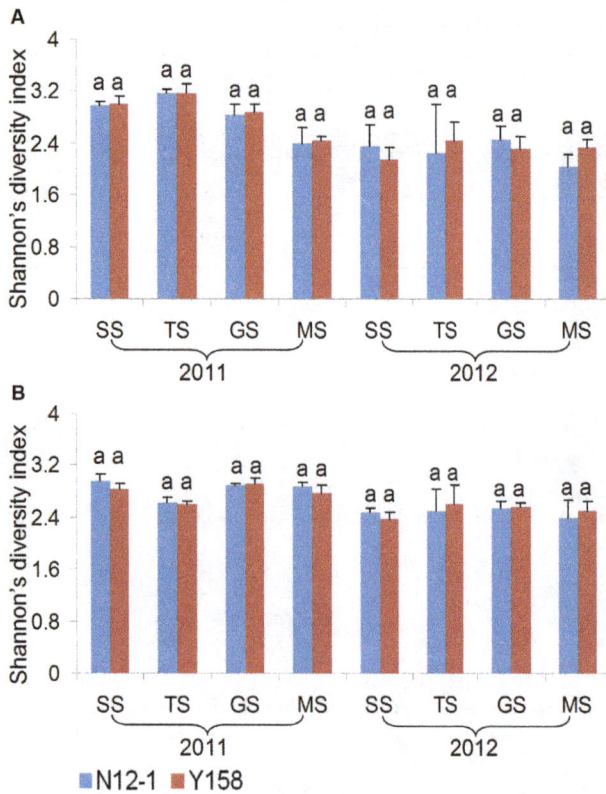

Figure 3. Shannon's index of fungi communities at different growth stages. Error bars indicate standard errors (n = 4). Different letters above bars denote a statistically significant difference between the means of the fields. A: Luhe; B: Xinxiang. SS: seeding stage; TS: turngreen stage; GS: grainfilling stage; MS: maturing stage.

n_i represented the square of individual peaks detected by Quantity One; N represented the square of all peaks in the same lane; S represented the number of bands in the same lane.

Band sequencing

Visible bands in the fungi DGGE gel were picked with sterile tips and transferred into a 200 µl tube. Sterile ddH$_2$O (50 µl) was added to the tube and the gel was pounded to pieces. The tubes with broken gels were incubated at room temperature overnight and then centrifuged at 12000 rpm for 5 min. The supernatant solution was used as the template for PCR, which was performed as described above or for fungi using primers without GC-clamps. The PCR products were purified (Omega, USA), ligated into pM19-T vector (Takara, Japan) and transformed into competent cells (*E coli* DH5α, Takara, Japan) according to the instructions of the manufactures and plated on LB solid medium with ampicillin. Positive clones were selected by PCR with primer pair of NS1 and Fungi (GC-Fungi without GC-clamps) and plasmids were extracted for sequencing (Invitrogen, Shanghai). All the sequences that have been sequenced successfully were submitted to GenBank (Accession numbers: KJ755390-KJ755404).

Enzyme activity analysis

Activities of urease, sucrose and dehydrogenase were analyzed in this study. Urease and sucrose activities in soil were assayed using the method of Guan [36]: urease activity was determined by measuring the release of NH$_3$ as mg.(g.d)$^{-1}$, and sucrose activity was determined based on 3,5-dinitrosalicylic acid colorimetry as mg.(g.d)$^{-1}$. Dehydrogenase activity was determined based on the reduction of triphenyltetrazolium chloride (TTC) to triphenylformazan (TPF), as described by Serra-Wittling et al. [37] with minor modifications, which was expressed as µg.(g.d)$^{-1}$. The data were subjected to analysis of variance, and the means and standard deviations of four replicates were calculated.

Table 2. Simpson's index and Evenness of bacterial community.

Experiment station	Growth stage	Variety(line)	Simpson's index 2011	2012	Evenness 2011	2012
Luhe	SS	N12-1	0.04±0.00a	0.10±0.00a	0.95±0.01a	0.88±0.01a
		Y158	0.04±0.00a	0.12±0.00b	0.95±0.01a	0.86±0.02a
	TS	N12-1	0.03±0.00a	0.09±0.01a	0.84±0.01a	0.92±0.03a
		Y158	0.03±0.00a	0.08±0.02a	0.95±0.01a	0.93±0.01a
	GS	N12-1	0.03±0.00a	0.04±0.01a	0.95±0.00a	0.93±0.01a
		Y158	0.03±0.00b	0.05±0.01a	0.96±0.01a	0.93±0.02a
	MS	N12-1	0.04±0.00a	0.06±0.02a	0.99±0.10a	0.93±0.03a
		Y158	0.04±0.00a	0.07±0.00a	0.95±0.01a	0.90±0.02a
Xinxiang	SS	N12-1	0.05±0.01a	0.06±0.01a	0.93±0.03a	0.98±0.01a
		Y158	0.05±0.01a	0.06±0.01a	0.95±0.01a	1.00±0.01a
	TS	N12-1	0.04±0.01a	0.07±0.01a	0.93±0.02a	0.93±0.02a
		Y158	0.04±0.00a	0.07±0.01a	0.93±0.00a	0.94±0.02a
	GS	N12-1	0.03±0.00a	0.11±0.02a	0.95±0.01a	0.92±0.03a
		Y158	0.03±0.00a	0.08±0.01b	0.95±0.01a	0.96±0.01a
	MS	N12-1	0.04±0.01a	0.07±0.00a	0.93±0.02a	0.90±0.01a
		Y158	0.03±0.00a	0.07±0.01a	0.96±0.01b	0.89±0.02a

SS: seeding stage; TS: turngreen stage; GS: grainfilling stage; MS: maturing stage. The alphabets after the value represented the significance level of the index.

Table 3. Simpson's index and Evenness of fungi community.

Experiment station	Growth stage	variety(line)	Simpson's index		Evenness	
			2011	2012	2011	2012
Luhe	SS	N12-1	0.06±0.01a	0.11±0.03a	0.92±0.02a	0.94±0.02a
		Y158	0.06±0.01a	0.14±0.02a	0.94±0.01a	0.91±0.03a
	TS	N12-1	0.05±0.00a	0.04±0.06a	0.92±0.00a	0.78±0.23a
		Y158	0.05±0.01a	0.11±0.05a	0.91±0.01a	0.97±0.08a
	GS	N12-1	0.07±0.01a	0.10±0.02a	0.93±0.01a	0.95±0.02a
		Y158	0.07±0.01a	0.11±0.03a	0.93±0.01a	0.93±0.02a
	MS	N12-1	0.12±0.02a	0.16±0.03a	0.88±0.03a	0.86±0.04a
		Y158	0.11±0.01a	0.12±0.02b	0.89±0.04a	0.92±0.04a
Xinxiang	SS	N12-1	0.06±0.01a	0.10±0.01a	0.93±0.03a	0.87±0.02a
		Y158	0.07±0.00b	0.10±0.01a	0.93±0.01a	0.86±0.03a
	TS	N12-1	0.08±0.01a	0.09±0.03a	0.95±0.01a	0.97±0.02a
		Y158	0.08±0.00a	0.08±0.02a	0.95±0.01a	0.97±0.00a
	GS	N12-1	0.07±0.00a	0.08±0.01a	0.94±0.02a	0.96±0.01a
		Y158	0.07±0.01a	0.08±0.01a	0.93±0.01a	0.96±0.01a
	MS	N12-1	0.07±0.01a	0.12±0.04a	0.89±0.03a	0.88±0.04a
		Y158	0.07±0.01a	0.10±0.02a	0.91±0.02a	0.89±0.03a

SS: seeding stage; TS: turngreen stage; GS: grainfilling stage; MS: maturing stage. The alphabets after the value represented the significance level of the index.

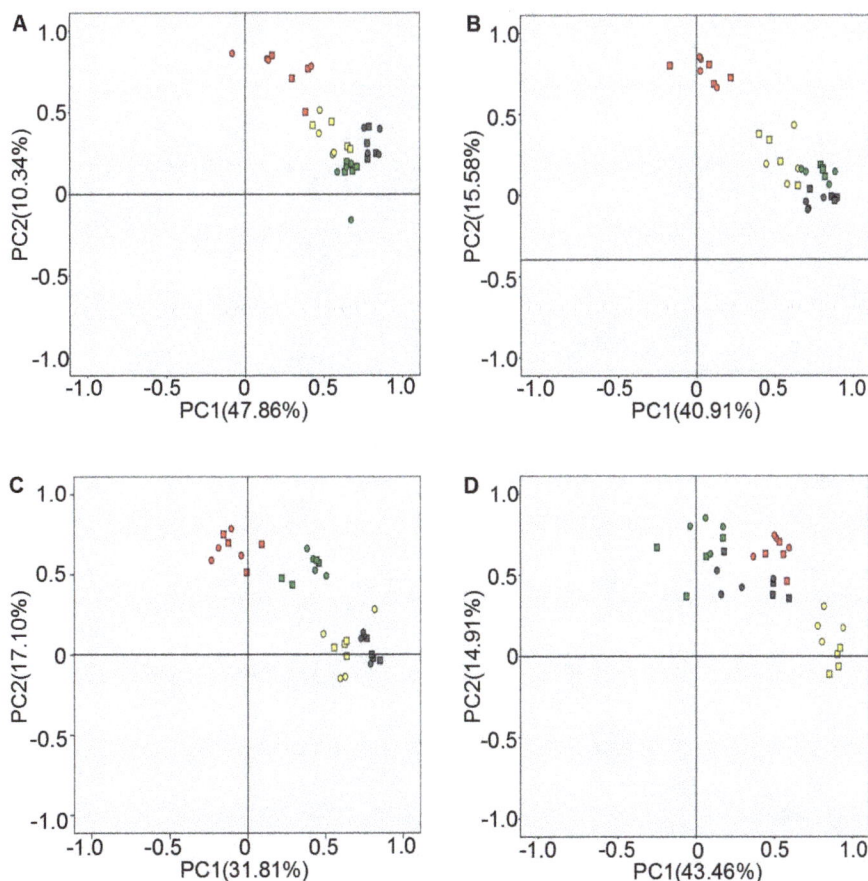

Figure 4. Principal component analysis of bacterial community diversities in rhizosphere soil. A: Luhe in 2011; B: Luhe in 2012; C: Xinxiang in 2011; D: Xinxiang in 2012. Square: N12-1; Round: Y158. Gray: seeding stage; Green: turngreen stage; Red: grainfilling stage; Yellow: maturing stage. Band position and presence (presence/absence) were used to carry out PCA analyses.

Statistical analysis

SPSS 16.0 was applied to determine whether the indices and enzyme activities differed between years, varieties and growth stages by ANOVA. PCA analyses were carried out based on band position and presence (presence/absence), and then the correlation matrix principal component analysis was performed by SPSS 16.0 [35]. Microsoft Excel 2003 was used to construct column diagrams.

Results

Impact of transgenic wheat on bacterial and fungal community diversity

One of the DGGE profiles of 16S rDNA and 18S rDNA fragments amplified from DNA extracted from rhizosphere soil was presented as figure 1. Three diversity indices (Shannon's, Simpson's, evenness) were used to analyze the bacterial and fungal DGGE profiles of the soil samples from Luhe and Xinxiang at four different growth stages in 2011 and 2012. For bacteria, the effect of wheat line on DGGE diversity indices was insignificant, except GS stage in 2011, SS in 2012 at Luhe and GS stage in 2012 at Xinxiang for Shannon's diversity index (Fig. 2). The Simpson's diversity index showed the same results as Shannon's diversity index (Table 2). For evenness, only one difference was found at the MS stage at Xinxiang in 2011 (Table 2). For fungi, the effect of wheat line on DGGE diversity indices was insignificant, except for

SS in 2011 at Xinxiang and MS in 2012 at Luhe for Simpson's index (Fig. 3; Table 3).

Principal component analysis of bacterial community diversity

Principal components analysis (PCA) using both band position and presence/absence as parameters were performed to further analyze DGGE fingerprint profiles. For experiments conducted at Luhe, the contribution rates of the two principal components were 47.86% and 10.34% in 2011 (Fig. 4A) and 40.91% and 15.58% in 2012 (Fig. 4B), respectively. Different growth stages showed a distinct separation along the principal components axes, whereas different replications of experimental materials formed a cluster at the same growth stage. This was consistent with the result of Shannon's diversity analysis. In 2011, the first principal component axis clearly separated the GS and SS stage (Fig. 4A), but separated the GS and MS stage in 2012 (Fig. 4B). The second principal component axis clearly distinguished the GS stage in 2011 (Fig. 4A) and the GS stage in 2012 (Fig. 4B).

For experiments conducted at Xinxiang, the contribution rates of the two principal components were 31.81% and 17.10% in 2011 (Fig. 4C) and 43.46% and 14.91% in 2012 (Fig. 4D), respectively. Different growth stages also showed a distinct separation along the principal components axes, whereas different replications of experimental materials clustered together at the same growth stage. In 2011, the first principal component axis

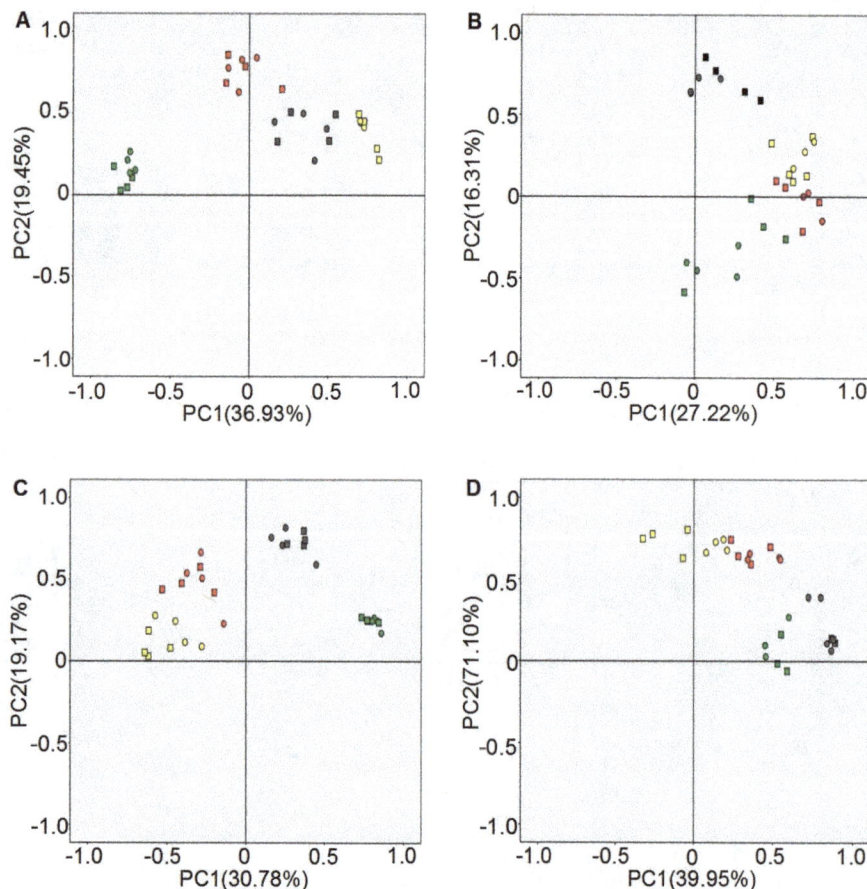

Figure 5. Principal component analysis of fungi communities diversity in rhizosphere soil. A: Luhe in 2011; B: Luhe in 2012; C: Xinxiang in 2011; D: Xinxiang in 2012. Square: N12-1; Round: Y158. Gray: seeding stage; Green: turngreen stage; Red: grainfilling stage; Yellow: maturing stage. Band position and presence (presence/absence) were used to carry out PCA analyses.

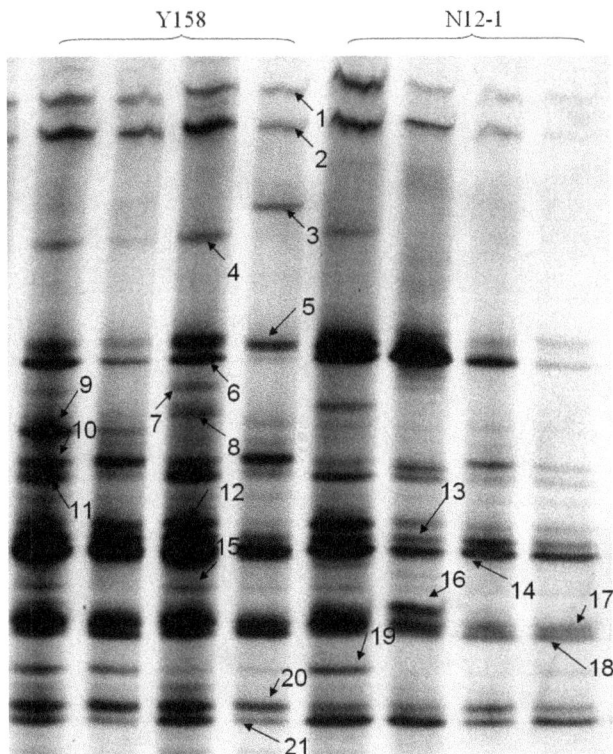

Figure 6. PCR-DGGE gel profile of fungi communities used for band sequencing. The numbers means different bands picked for sequencing.

clearly separated the four growth stages (Fig. 4C), but separated the MS stage from the other three stages in 2012 (Fig. 4D). The second principal component axis clearly distinguished the TS and GS stages from the SS and MS stages in 2011 (Fig. 4C), and distinguished the MS stage in 2012 (Fig. 4D).

These PCA analysis results showed that growth stage played an important role in bacterial community diversity, rather than the presence of transgenic and non-transgenic wheat.

Principal component analysis of fungal community diversity

For experiments at Luhe, the contribution rates of the two principal components were 36.93% and 19.45% in 2011 (Fig. 5A) and 27.22% and 16.31% in 2012 (Fig. 5B), respectively. Different sampling times showed a distinct separation along the principal components axes, whereas different replications of experiment materials formed a cluster at the same sampling time. In 2011, the first principal component axis clearly separated the four growth stages (Fig. 5A), but separated SS and TS from GS and MS in 2012 (Fig. 5B). The second principal component axis clearly distinguished the TS stage in 2011 (Fig. 5A), and the SS and TS stage in 2012 (Fig. 5B).

For experiments at Xinxiang, the contribution rates of the two principal components were 30.78% and 19.17% in 2011 (Fig. 5C) and 39.95% and 17.10% in 2012 (Fig. 5D). Different sampling times also showed a distinct separation along the principal components axes, whereas different replications of experimental materials formed a cluster at the same sampling time. In 2011, the first principal component axis clearly separated the SS and TS stages (Fig. 5C), but separated the SS and MS stages in 2012 (Fig. 5D). The second principal component axis could not clearly

Table 4. Blast results of the bands from the DGGE gels of fungl community analysis.

No. of bands	Accession No.	Blast result	identity
1	GU214699.1	*Septoria dysentericae* strain CPC 12328 18S ribosomal RNA gene	100%
2	GQ330624.1	Uncultured *Mucorales* clone PR3 4E 28 18S ribosomal RNA gene	95%
3		Cannot be amplified	
4	AJ515922.1	Uncultured soil ascomycete partial 18S rDNA gene	100%
5	EU120944.1	Uncultured *Cystofilobasidiales* (aff. Guehomyces) clone Y9 18S ribosomal RNA gene	100%
6	AJ515941.1	Uncultured soil ascomycete partial 18S rDNA gene	99%
7		Cannot be amplified	
8	AY789390.1	*Peziza varia* strain ZW-Geo94-Clark 18S small subunit ribosomal RNA gene	99%
9	FJ176814.1	*Saccobolus dilutellus* isolate AFTOL-ID 1299 18S small subunit ribosomal RNA gene	97%
10	FO181499.1	Balen uncultured eukaryote partial 18S ribosomal RNA	80%
11	AY771600.1	Polyozellus multiplex isolate AFTOL-ID 677 18S small subunit ribosomal RNA gene	99%
12	GU190186.1	*Cochliobolus* sp. Enrichment culture clone NJ-F5 18S small subunit ribosomal RNA gene	100%
13		Cannot be amplified	
14	AJ515948.1	Uncultured soil ascomycete partial 18S rDNA gene	99%
15		Cannot be amplified	
16	AJ301992.1	*Myrothecium leucotrichym* 18S RNA gene	99%
17	JX159444.1	Uncultured *Filobasidium* clone Cegs 957 18S ribosomal RNA gene	99%
18	KC171701.1	Uncultured fungus isolate DGGE gel band f10 18S ribosomal RNA gene	100%
19		Cannot be amplified	
20		Cannot be amplified	
21	EU120947.1	Uncultured *Ascobolus* clone Y12 18S ribosomal RNA gene	99%

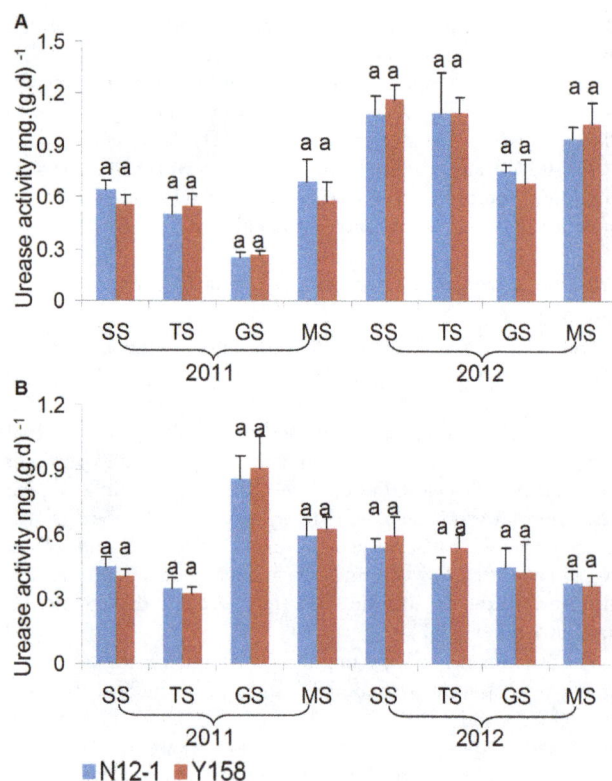

Figure 7. Urease activity in rhizosphere soil at different growth stages. Error bars indicate standard errors (n = 4). Different letters above bars denote a statistically significant difference between the means of the fields. A: Luhe; B: Xinxiang. SS: seeding stage; TS: turngreen stage; GS: grainfilling stage; MS: maturing stage.

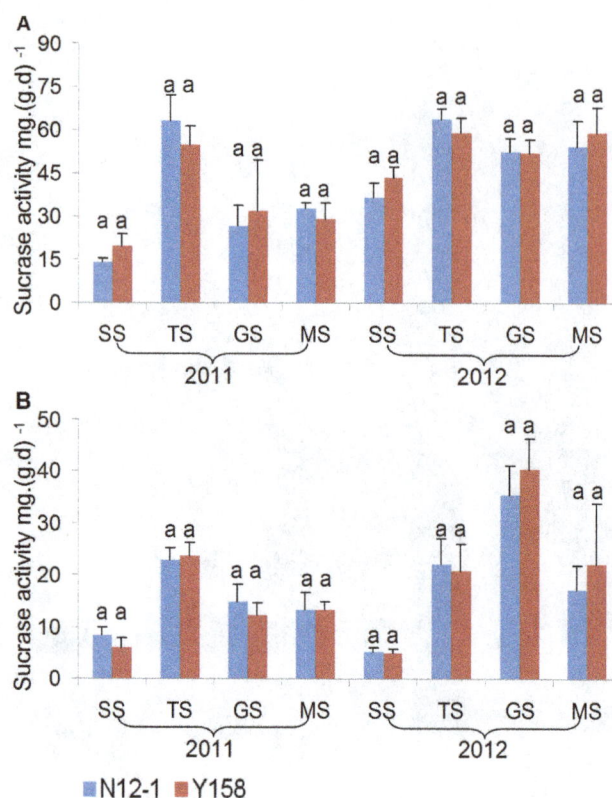

Figure 8. Sucrase activity in rhizosphere soil at different growth stages. Error bars indicate standard errors (n = 4). Different letters above bars denote a statistically significant difference between the means of the fields. A: Luhe; B: Xinxiang. SS: seeding stage; TS: turngreen stage; GS: grainfilling stage; MS: maturing stage.

distinguish any growth stage in 2011 (Fig. 5C), but could distinguish the MS and GS stages from the SS and TS stages in 2012 (Fig. 5D).

These PCA analysis results showed that fungal communities exhibited marked diversity at different growth stages, rather than between the transgenic line and non-transgenic wheat recipient.

Band sequencing

A total of 21 visible bands from the DGGE gel of fungi from Luhe in 2011 were subjected to sequencing (Fig. 6), and 15 were sequenced successfully. Using NCBI BLAST, we found that most of the sequenced bands represented uncultured fungi. Others were partial 18S rRNA sequences of *Septoria dysentericae*, *Peziza varia*, *Saccobolus dilutellus*, *Polyozellus*, *Cochliobolus*, and *Myrothecium leuco-trichym* (Table 4).

Enzyme activity analysis

Urease, sucrase, and dehydrogenase activities in rhizosphere soil were applied as indicators for environmental risk assessment of transgenic wheat N12-1 in this study.

In general, there was no consistent significant difference in the enzyme activity between soils of transgenic wheat N12-1 and its recipient Y158 within the same growth stage during the 2 years. Only one significant difference in activity was observed; for dehydrogenase at the MS stage at Xinxiang in 2011. In 2011, the dehydrogenase activity in soil of N12-1 was significantly ($p < 0.05$) higher than in soil of its recipient Y158 (Figs. 7–9). Significant differences were observed between years ($p < 0.01$) and among

growth stages ($p < 0.001$) at both Luhe and Xinxiang, with the exception of dehydrogenase among growth stages at Xinxiang ($p < 0.25$) (Table 5). These results showed that N12-1 had a minor impact on soil enzyme activities.

Discussion

With the cultivation of more varieties of virus-resistant transgenic plants and large-scale planting, environmental impact monitoring after commercial release has attracted increasing attention from the scientific community and public [38,39]. In soil, there are high microbial population densities and large numbers of microbial species that interact with the plants and surrounding environment and have an effect on the function of the soil ecosystem, such as the enzyme activity and physicochemical properties.

Soil microbial analysis has been used widely to evaluate the impact of various exogenous chemical or environmental pollutants (such as herbicides, fertilizers, heavy metals, et al.) on soil fertility and crop yields [7]. Therefore, monitoring changes in soil microbial populations will increase our understanding of the potential risks of introduction of exogenous genes to soil [4,7]. In our study, two years and two locations of field research was performed to compare the impact of transgenic wheat with genes encoding replicase from WYMV on microbial population diversity in agricultural systems. One of the major outcomes was that transgenic insertion did not significantly alter bacterial or fungal population diversity at each growth stage; however, growth stage and planting year had important effects on microbial diversity.

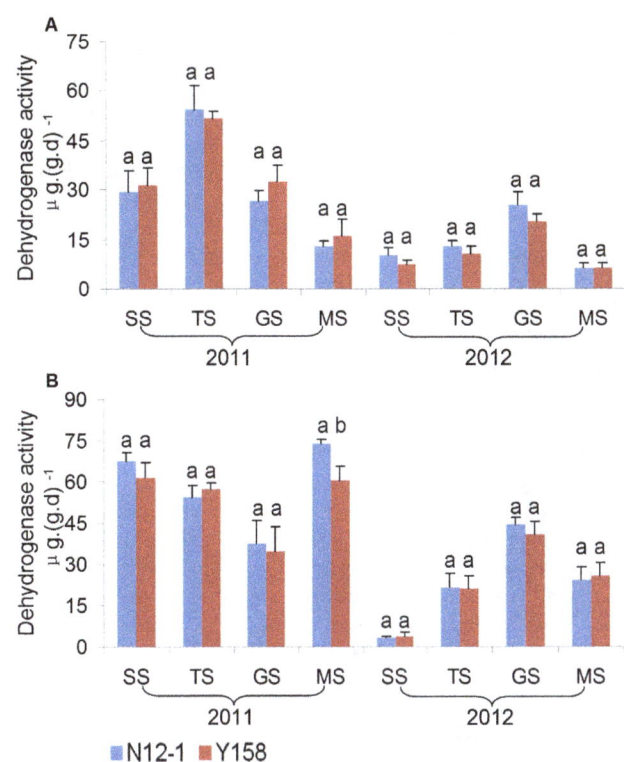

Figure 9. Dehydrogenase activity in rhizosphere soil at different growth stages. Error bars indicate standard errors (n = 4). Different letters above bars denote a statistically significant difference between the means of the fields. A: Luhe; B: Xinxiang. SS: seeding stage; TS: turngreen stage; GS: grainfilling stage; MS: maturing stage.

This result was similar to the findings of Meyer et al. [40]. In that study, the authors found that the effects of GM wheat on plant-beneficial root-colonizing microorganisms are minor and not of ecological importance. Lupwayi et al. reported that glyphosate-resistant wheat–canola rotations under low-disturbance direct seeding and conventional tillage did not affect the functional diversity of rhizosphere soil bacteria in 18 of 20 site-years [20]. The observation that certain growth stages (mainly SS and GS) showed differences between transgenic and non-transgenic wheat may be due to inconsistencies in the soil at seeding time, and at later growth stages the temperature and humidity increased rapidly. The differences between years and growth stages indicated that the diversity of bacteria and fungi might be affected by various environmental factors, such as temperature, humidity, and light. In studies of GM crops against virus, there was no significant difference between microbial communities with transgenic or non-transgenic watermelon resistant or cucumber green mottle mosaic virus (CGMMV), but significant changes in the microbial community were observed during the growing season [41]. Transgenic tomato resistant to cucumber mosaic virus (CMV) had no effect on the variation of soil microbial communities, in which soil position and environmental factors played more dominant roles [42]. Also, non-environmental factor, such as root exudates, may also play an important role for diversity changes of bacteria and fungi between years and growth stages [14–15]. Fang et al. thought that bacterial communities differed due to changes in root exudates quantity and composition by developing corn plant, which select different bacterial groups during root colonization [43]; Donegan et al. have speculated that the reason for the different in the communities of genetically modified plants is due to differences in the root exudates patterns of these plants [44]. However, Wei et al. reported the opposite result [45]. In a transgenic alfalfa study performed using the cultivation-dependent plating method, statistically significant differences in densities of rhizospheric bacteria between transgenic and non-transgenic

Table 5. Generalized Linear Mixed Model results for overall effects on enzyme activity.

Location	Enzyme	Effect	F Value	p Value
Luhe	Urease	Years	282.21	0.00
		Growth stage	36.89	0.00
		Variety (Line)	0.01	0.91
	Sucrase	Years	74.66	0.00
		Growth stage	33.54	0.00
		Variety (Line)	0.05	0.83
	Dehydrogenase	Years	82.20	0.00
		Growth stage	16.23	0.00
		Variety (Line)	0.03	0.87
Xinxiang	Urease	Years	6.38	0.01
		Growth stage	7.97	0.00
		Variety (Line)	0.103	0.75
	Sucrase	Years	11.45	0.00
		Growth stage	23.67	0.00
		Variety (Line)	0.10	0.75
	Dehydrogenase	Years	73.88	0.00
		Growth stage	1.40	0.25
		Variety (Line)	0.11	0.74

alfalfa clones were observed for ammonifying bacteria, cellulolytic bacteria, rhizobial bacteria, denitrifying bacteria and *Azotobacter* spp. [46]. These results indicated that transgenic crops containing a viral gene conferring resistance to viral disease had little effect on soil microbial diversity (excluding a small number of studies) compared with non-transgenic crops. Transgenic wheat also had no adverse effects on soil biological indicators, such as *Folsomia candida* [47] and earthworm [48]. Duc et al. found that GM wheat with race-specific antifungal resistance against powdery mildew (Pm3b), and two with nonspecific antifungal resistance, had no impact on the soil fauna community (mites, springtails, annelids, and diptera). However, sampling date and location significantly influenced the soil fauna community and decomposition processes [49].

Soil enzymes in the soil nutrient cycle and energy transfer play an important role in soil ecology, and are derived mainly from soil microbial populations. Many studies have used soil enzymes as indicators of soil microbial activity and fertility [50–52]. In our study, urease, sucrase and dehydrogenase were used as indicators of the impact of transgenic wheat on soil quality. The results showed no significant difference in enzyme activity in rhizosphere soil between transgenic and non-transgenic wheat at each growth stage at two locations in 2 years, excluding dehydrogenase during the maturing stage at Xinxiang in 2011. In other studies of transgenic crops, there was no consistent significant difference in soil enzymes between transgenic and non-transgenic plants, but there were differences among seasons and crop varieties [53–54]. These results are consistent with our study. In other studies, some enzymes showed significant differences between transgenic and non-transgenic plants [11,20–21]. There have been no previous studies of soil enzyme activities of transgenic wheat. Thus, our results should be confirmed in future studies and at more experimental locations. Additionally, other types of transgenic wheat, such as insect-resistant and stress-tolerant varieties, should be used to perform risk assessments.

Due to the complexity of DGGE profiles, several bands can be difficult to identify visually, and different bands represent different microbes. These issues make it difficult to compare varieties. Thus, the combination of DGGE and cloning sequencing methods is often used to investigate the impact of transgenic plants on microorganisms in rhizosphere soil [55,56]. In our study, most of the bands from fungi DGGE gels represented uncultured fungal taxa. This is in agreement with the fact that only ~1% of microbes in soil can be artificially cultured and identified [57].

With the development of sequencing technology, the way we study microbial communities has been changed. Traditionally, the study of genes from natural environments included cloning DNA into a vector, inserting that vector into a host, screening, and Sanger sequencing. Sequence-by-synthesis methods provide faster, cheaper, and simpler methods for (meta)genome sequence that bypass the PCR amplification bias, cloning bias and labor-intensive Sanger method [58]. Currently, massively parallel high-throughput pyrosequencing methods can process hundreds of thousands of sequences simultaneously [59]. Fierer et al. used metagenomic and small subunit rRNA analyses to study the genetic diversity of bacteria, archaea, fungi, and viruses in soil [60]; Uroz et al. used functional assays and metagenomic analyses to reveal difference between the microbial communities [61]. Li et al. analyzed the impact on bacterial community in midguts of the asian corn borer larvae by transgenic *Trichoderma* strain overexpressing a heterologous *chit42* gene with chitin-binding domain by using 16s rRNA library. All above studies have used the next generation sequencing technology [62]. Now, this technology is being adopted to study the microbial community in rhizosphere soil of transgenic plants gradually [62].

In conclusion, our study has produced weak evidence for the effect of virus-resistant transgenic wheat on soil microbial community diversity and enzyme activities. The community structure was markedly affected by natural variations in the environment related to wheat growth stage and planting year. Little difference was observed in bacterial and fungal communities in the presence of the wild-type Y158 or the transgenic line N12-1. This requires further investigation using extended field observations involving more varieties for more years. Based on this information, we can determine whether the altered composition is attributable to the presence of transgenic crops, or is simply part of the variation driven by the presence of different genotypes [63]. These studies should also involve more soil types and longer-term monitoring to account for the variability of the natural environment.

Acknowledgments

We thank everyone that helped with the fieldwork and Dr. YZ from CAAS for providing seeds of transgenic and non-transgenic wheat.

Author Contributions

Conceived and designed the experiments: JW MY JX XL JS. Performed the experiments: JW MY JX JD FJ FD. Analyzed the data: JW MY. Wrote the paper: JW MY JS.

References

1. Horsch RB, Fraley RT, Rogers SG, Sanders PR, Lloyd A, et al. (1984) Inheritance of functional foreign genes in plants. Science 223: 496–498

2. Vauramo S, Pasonen HL, Pappinen A, Setälä H (2006) Decomposition of leaf litter form chitinase transgenic birch (*Betula pendula*) and effects on decomposer populations in a field trial. Appl Soil Ecol 32: 338–349.

3. James C (2013) Global Status of Commercialized Biotech/GM Crops: 2012. ISAAA Brief No. 44. ISAAA, Ithaca, New York.

4. McGregor AN, Turner MA (2000) Soil effects of transgenic agriculture: biological processes and ecological consequences. NZ Soil News 48(6):166–169

5. Sengelov G, Kristensen KJ, Sørensen AH, Kroer N, Sørensen SJ (2001) Effect of genomic location on horizontal transfer of a recombinant gene cassette between Pseudomonas strains in the rhizosphere and spermosphere of barley seedlings. Cur Microbiol 42:160–167

6. Gyaneshwar P, Naresh Kumar G, Parekh LJ, Poole PS (2002) Role of soil microorganisms in improving P nutrition of plants. Plant Soil 245: 83–93

7. Kent AD, Triplett EW (2002) Microbial communities and their interactions in soil and rhizosphere ecosystems. Annu Rev Microbiol 56:211–236

8. O'Callaghan M, Glare TR. (2001) Impacts of transgenic plants and microorganisms on soil biota. New Zeal Plant Prot 54: 105–110.

9. Andow DA, Zwahlen C (2006) Assessing environmental risks of transgenic plants. Ecology Letters 9:196–214

10. Liu B, Zeng Q, Yan F, Xu H, Xu C (2005) Effects of transgenic plants on soil microorganisms. Plant Soil 271:1–13

11. Chen ZH, Chen LJ, Zhang YL, Wu ZJ (2011) Microbial properties, enzyme activities and the persistence of exogenous proteins in soil under consecutive cultivation of transgenic cottons (*Gossypium hirsutum* L.). Plant Soil Environ, 57: 67–74

12. Blackwood CB, Buyer JS (2004) Soil microbial communities associated with Bt and non-Bt corn in three soils. J Environ Qual 33:832–836

13. Donegan KK, Palm CJ, Fieland VJ, Porteous LA, Ganio LM, et al. (1995) Changes in levels, species and DNA fingerprints of soil microorganisms associated with cotton expressing the *Bacillus thuringiensis* var. *kurstaki endotoxin*. Appl Soil Ecol 2:111–124

14. Bais HP, Weir TL, Perry LG, Gilroy S, Vivanco JM (2006) The role of root exudates in rhizosphere interactions with plants and other organisms. Annu Rev Plant Biol 57:233–266

15. Saxena D, Flores S, Stotzky G (1999) Insecticidal toxin in root exudates from Bt corn. Nature 402: 480–481

16. Raybould A, Higgins LS, Horak MJ, Layton RJ, Storer NP, et al. (2012) Assessing the ecological risks from the persistence and spread of feral populations of insect-resistant transgenic maize. Transgenic Res 21:655–664

17. Devarc MH, Jones CM, Thies JE (2004) Effect of Cry3Bb transgenic corn and tefluthrin on the soil microbial community: biomass, activity, and diversity. J Environ Qual 33: 837–843

18. Lu H, Wu W, Chen Y, Zhang X, Devare M, et al. (2010) Decomposition of *Bt* transgenic rice residues and response of soil microbial community in rapeseed–rice cropping system. Plant Soil 336: 279–290

19. Burns RG (1982) Enzyme activity in soil: Location and a possible role in microbial ecology. Soil Biol Biochem 14: 423–427

20. Lupwayi NZ, Hanson KG, Harker KN, Clayton GW, Blackshaw RE, et al. (2007) Soil microbial biomass, functional diversity and enzyme activity in glyphosate-resistant wheat–canola rotations under low-disturbance direct seeding and conventional tillage. Soil Biol Biochem 39: 1418–1427

21. Sun CX, Chen LJ, Wu ZJ, Zhou LK, Shimizu H (2007) Soil persistence of *Bacillus thuringiensis* (Bt) toxin from transgenic Bt cotton tissues and its effect on soil enzyme activities. Biol Fertil Soils 43: 617–620

22. Abel PP, Nelson RS, De B, Hoffman N, Rogers SG, et al. (1986) Delay of disease development in transgenic plants that express the tobacco mosaic virus coat protein. Science 232, 738–43.

23. Beachy RN, Loesch-Fries S, Tumer NE (1990) Coat-protein mediated resistance against virus infection. Annu Rev Phytopathol 28, 451–474.

24. Fuchs M, Gonsalves D (2007) Safety of virus-resistant transgenic plants two decades after their introduction: lessons from realistic field risk assessment studies. Annu Rev Phytopathol 45: 173–202

25. Tepfer M (2002) Risk assessment of virus-resistant transgenic plants. Annu Rev Phytopathol 40: 467–491

26. Robinson DJ (1996) Environmental risk assessment of release of transgenic plants containing virus-derived inserts. Transgenic Res 5: 359–362

27. Chen J (1993) Occurrence of fungally transmitted wheat mosaic viruses in China. Ann Appl Biol 123: 55–61

28. Han C, Li D, Xing Y, Zhu K, Tian Z, et al. (2000) *Wheat yellow mosaic virus* widely occurring in wheat (*Triticum aestivum*) in China. Plant Dis 84: 627–630

29. Xu H, Pang J, Ye X, Du L, Li L, et al. (2001) Study on the gene transferring of Nib8 into wheat for It's resistance to the yellow mosaic virus by bombardment. Acta Agronomica Sinica 27: 688–693 (in Chinese with English abstract)

30. Wu H, Zhang B, Gao D, Xu H, Cheng S (2006) Disease resistance test of transgenic wheat lines with *WYMV-Nib8* gene and their application in breeding. J Triticeae Crops 26: 11–14 (in Chinese with English abstract)

31. Conner AJ, Glare TR, Nap J (2003) The release of genetically modified crops into the environment. Part II. Overview of ecological risk assessment. The Plant J 33: 19–46

32. Lu RK (1999) Methods of Soil Agrochemical analysis. Beijing: China Agricultural Science and Technology Press

33. Bakke I, Schryver PD, Boon N, Vadstein O (2011) PCR-based community structure studies of bacteria associated with eukaryotic organisms: a simple PCR strategy to avoid co-amplification of eukaryotic DNA. J Microbiol Meth 84: 349–351

34. Das M, Royer TV, Leff LG (2007) Diversity of fungi, bacteria, and actinomycetes on Leaves decomposing in a stream. Appl Environ Microbiol 73. 756–767

35. Liu W, Lu HH, Wu W, Wei QK, Chen YX, et al. (2008) Transgenic Bt rice does not affect enzyme activities and microbial composition in the rhizosphere during crop development. Soil Biol Biochem 40: 475–486

36. Guan SY (1986) Soil Enzyme and Its Research Methods. Beijing: China Agricultural Press, 62–142

37. Serra-Wittling C, Houot S, Barriuso E (1995) Soil enzymatic response to addition of municipal solid-waste compost. Biol Fert Soils 20: 226–236

38. Graef F, Züghart W, Hommel B, Heinrich U, Stachow U, et al. (2005) Methodological scheme for designing the monitoring of genetically modified crops at the regional scale. Environ Monit Assess 111: 1–26

39. Züghart W, Benzler A, Berhorn F, Sukopp U, Graef F (2008) Determining indicators, methods and sites for monitoring potential adverse effects of genetically modified plants to the environment: the legal and conceptional framework for implementation. Euphytica 164: 845–852.

40. Meyer JB, Song-Wilson Y, Foetzki A, Luginbühl C, Winzeler M, et al. (2013) Does wheat genetically modified for disease resistance affect root-colonizing pseudomonads and arbuscular mycorrhizal fungi? PLoS ONE, 8(1): e53825. doi:10.1371/journal.pone.0053825

41. Yi H, Kin H, Kim C, Harn CH, Kim HM, et al. (2009) Using T-RFLP to assess the impact on soil microbial communities by transgenic lines of watermelon rootstock resistant to cucumber green mottle mosaic virus (CGMMV). J Plant Biol 52: 577–584

42. Lin C, Pan T (2010) PCR-denaturing gradient gel electrophoresis analysis to assess the effects of a genetically modified cucumber mosaic virus-resistant tomato plant on soil microbial communities. Applied Environ Microbiol 76: 3370–3373

43. Fang M, Kremer RJ, Motavalli PP, Davis G (2005) Bacterial diversity in rhizospheres of nontransgenic and transgenic corn. Appl Environ Microb 71:4132–4136

44. Donegan KK, Seidler RJ, Doyle JD, Porteous LA, Digiovanni G, et al. (1999) A field study with genetically engineered alfalfa inoculated with recombinant *Sinorhizobium meliloti*: effects on the soil ecosystem. J Appl Ecol 36:920–936

45. Wei XD, Zou HL, Chu LM, Liao B, Ye CM, et al. (2006) Field released transgenic papaya affects microbial communities and enzyme activities in soil. Plant Soil 285:347–358

46. Faragova N, Gottwaldova K, Farago J (2011) Effect of transgenic alfalfa plants with introduced gene for alfalfa mosaic virus coat protein on rhizosphere microbial community composition and physiological profile. Biologia 66: 768–777

47. Romeis J, Battini M, Bigler F (2003) Transgenic wheat with enhanced fungal resistance causes no effects on *Folsomia candida* (Collembola: Isotomidae). Pedobiologia 47: 141–147

48. Lindfield A, Nentwig W (2012) Genetically engineered antifungal wheat has no detrimental effects on the key soil species *Lumbricus terrestris*. The Open Ecology J 5:45–52

49. Duc C, Nentwig W, Lindfeld A (2011) No adverse effect of genetically modified antifungal wheat on decomposition dynamics and the soil fauna community – a field study. PLoS ONE 6(10): e25014. doi:10.1371/journal.pone.0025014

50. Weaver RW, Angle JS, Bottomiey PS (1994) Methods of soil analysis. Part 2. Microbiological and Biochemical properties, No. 5. Soil Sci Soc Am, Madison

51. Alef K, Nannipieri P (1995) Methods in Applied Soil Microbiology and Biochemistry. San Diego, CA: Academic Press

52. Dick RP, Breakwel DP, Tureo RF (1996) Soil enzyme activities and biodiversity measurements as integrating microbiological indicators. In: Doran JW, Jones AJ. ed. Methods for Assessing Soil Quality. Soil Sci Soc Am, Madison, WI, 247–272

53. Icoz I, Saxena D, Andow DA, Zwahlen C, Stotzky G (2008) Microbial populations and enzyme activities in soil In Situ under transgenic corn expressing Cry proteins from *Bacillus thuringiensis*. J. Environ. Qual 37: 647–662

54. Shen RF, Cai H, Gong WH (2006) Transgenic Bt cotton has no apparent effect on enzymatic activities of functional diversity of microbial communities in rhizosphere soil. Plant Soil 285:149–159

55. Tan F, Wang J, Feng Y, Chi G, Kong H, et al. (2010) Bt corn plants and their straw have no apparent impact on soil microbial communities. Plant Soil 329: 349–364

56. Weiner N, Meincke R, Gottwald C, Radl V, Dong X, Schloter M, et al. (2010) Effects of genetically modified potatoes with increased zeaxanthin content on the abundance and diversity of rhizobacteria with in vitro antagonistic activity do not exceed natural variability among cultivars. Plant Soil 326:437–452

57. Kowalchuk GA, Bruinsma M, Van Veen JA (2003) Assessing responses of soil microorganisms to GM plants. Trends Ecol and Evol 18: 403–410

58. Cardenas E, Tiedje JM (2008) New tools for discovering and characterizing microbial diversity. Curr Opin Biotech 19: 544–549

59. Hirsch PR, Mauchline TH, Clark IM (2010) Culture-independent molecular techniques for soil microbial ecology. Soil Bio Biochem 42: 878–887

60. Fierer N, Breitbart M, Nulton J, Peter S, Lozupone C, et al. (2007) Metagenomic and small-subunit rRNA analyses reveal the genetic diversity of bacteria, archaea, fungi, and viruses in soil. Appl Environ Microbiol 73: 7059–7066

61. Uroz S, Ioannidis P, Lengelle J, Cébron A, Morin E, et al. (2013) Functional assays and metagenomic analyses reveals differences between the microbial communities inhabiting the soil horizons of a Norway spruce plantation. PLoS ONE 8(2): e55929. doi:10.1371/journal.pone.0055929

62. Li Y, Fu K, Gao S, Wu Q, Fan L, et al. (2013) Impact on bacterial community in midguts of the asian corn borer larvae by transgenic *Trichoderma* Strain overexpressing a heterologous chit42 gene with chitin-binding domain. PLoS ONE 8(2): e55555. doi:10.1371/journal.pone.0055555

63. Chun YJ, Kim DY, Kim H, Park KW, Jeong S, et al. (2011) Do transgenic chili pepper plants producing viral coat protein affect the structure of a soil microbial community? Applied Soil Ecol 51: 130–1

11

Transgenic Tobacco Overexpressing *Brassica juncea* HMG-CoA Synthase 1 Shows Increased Plant Growth, Pod Size and Seed Yield

Pan Liao[1], Hui Wang[1¤], Mingfu Wang[1], An-Shan Hsiao[1], Thomas J. Bach[2], Mee-Len Chye[1]*

1 School of Biological Sciences, The University of Hong Kong, Hong Kong, China, 2 Centre National de la Recherche Scientifique, UPR 2357, Institut de Biologie Moléculaire des Plantes, Strasbourg, France

Abstract

Seeds are very important not only in the life cycle of the plant but they represent food sources for man and animals. We report herein a mutant of 3-hydroxy-3-methylglutaryl-coenzyme A synthase (HMGS), the second enzyme in the mevalonate (MVA) pathway that can improve seed yield when overexpressed in a phylogenetically distant species. In *Brassica juncea,* the characterisation of four isogenes encoding HMGS has been previously reported. Enzyme kinetics on recombinant wild-type (wt) and mutant BjHMGS1 had revealed that S359A displayed a 10-fold higher enzyme activity. The overexpression of wt and mutant (S359A) BjHMGS1 in *Arabidopsis* had up-regulated several genes in sterol biosynthesis, increasing sterol content. To quickly assess the effects of BjHMGS1 overexpression in a phylogenetically more distant species beyond the Brassicaceae, wt and mutant (S359A) BjHMGS1 were expressed in tobacco (*Nicotiana tabacum* L. cv. Xanthi) of the family Solanaceae. New observations on tobacco OEs not previously reported for *Arabidopsis* OEs included: (i) phenotypic changes in enhanced plant growth, pod size and seed yield (more significant in OE-S359A than OE-wtBjHMGS1) in comparison to vector-transformed tobacco, (ii) higher *NtSQS* expression and sterol content in OE-S359A than OE-wtBjHMGS1 corresponding to greater increase in growth and seed yield, and (iii) induction of *NtIPPI2* and *NtGGPPS2* and downregulation of *NtIPPI1, NtGGPPS1, NtGGPPS3* and *NtGGPPS4.* Resembling *Arabidopsis* HMGS-OEs, tobacco HMGS-OEs displayed an enhanced expression of *NtHMGR1, NtSMT1-2, NtSMT2-1, NtSMT2-2* and *NtCYP85A1.* Overall, increased growth, pod size and seed yield in tobacco HMGS-OEs were attributed to the up-regulation of native *NtHMGR1, NtIPPI2, NtSQS, NtSMT1-2, NtSMT2-1, NtSMT2-2* and *NtCYP85A1.* Hence, S359A has potential in agriculture not only in improving phytosterol content but also seed yield, which may be desirable in food crops. This work further demonstrates HMGS function in plant reproduction that is reminiscent to reduced fertility of *hmgs* RNAi lines in *let-7* mutants of *Caenorhabditis elegans.*

Editor: Joshua L Heazlewood, Lawrence Berkeley National Laboratory, United States of America

Funding: This work was supported by the Wilson and Amelia Wong Endowment Fund and the University of Hong Kong (CRCG 10400945, CRCG 104001061, University Postgraduate Fellowship (PL) and studentships (HW and ASH)). The funders had no role in study design, data collection and analysis, decision to publish, or preparation of the manuscript.

Competing Interests: The authors have declared that no competing interests exist.

* E-mail: mlchye@hkucc.hku.hk

¤ Current address: Key Laboratory of Microorganism and Genetic Engineering, College of Life Sciences, Shenzhen University, Shenzhen, China

Introduction

Isoprenoids form a large and diverse group of natural products, which have promising pharmacological applications including anti-cancer, antibacterial and anti-malarial properties [1–4]. Some isoprenoids including gibberellic acids, abscisic acid, cytokinins, sterols and brassinosteroids (BRs) play significant roles in plant growth and development [4–6]. Furthermore, carotenoids and chlorophylls are involved in photosynthesis [7]. Phytosterols are important in regulating growth and mediating stress tolerance in plants [4,8] and their nutritional value and health benefits in the human diet has been recognized [9–11].

In higher plants, two pathways generate isopentenyl diphosphate (IPP), which constitutes the universal precursor of all isoprenoids: the mevalonate (MVA) pathway in the cytosol, and the non-MVA, methylerythritol phosphate (MEP) pathway in plastids [1,3,12](and references cited therein), with some crosstalk between them [13,14] (Figure 1). Sterols and BRs are synthesized

in the cytoplasm and thereby derive from MVA, while gibberellic acids and abscisic acid precursors, active cytokinins, carotenoids and chlorophylls are produced in plastids [1,15–21] and thus depend on the MEP pathway (Figure 1).

In agriculture, it is desirable to increase seed yield because grains represent significant sources of food, and the relevant key genes must be identified. Plant isoprenoids including sterols and BRs are essential in plant growth and reproduction [6,22–24] and genes from the BR-specific biosynthetic pathway, including *DWF4* and *DWF5*, affect seed production [22–24]. Transgenic *Arabidopsis* overexpressing *DWF4* showed better vegetative growth and seed yield [23], while the *Arabidopsis dwf5* mutant demonstrated a dwarf phenotype accompanied by abnormal seeds [22]. The genes in the first and third steps of the MVA pathway also affect plant growth and development. RNAi lines of *Arabidopsis* downregulated for cytoplasmic *ACETOACETYL-COA THIOLASE2 (AACT2)* displayed reduction in apical dominance, seed yield and root length, accompanied by sterility and dwarfing [25]. Also, the *Arabidopsis*

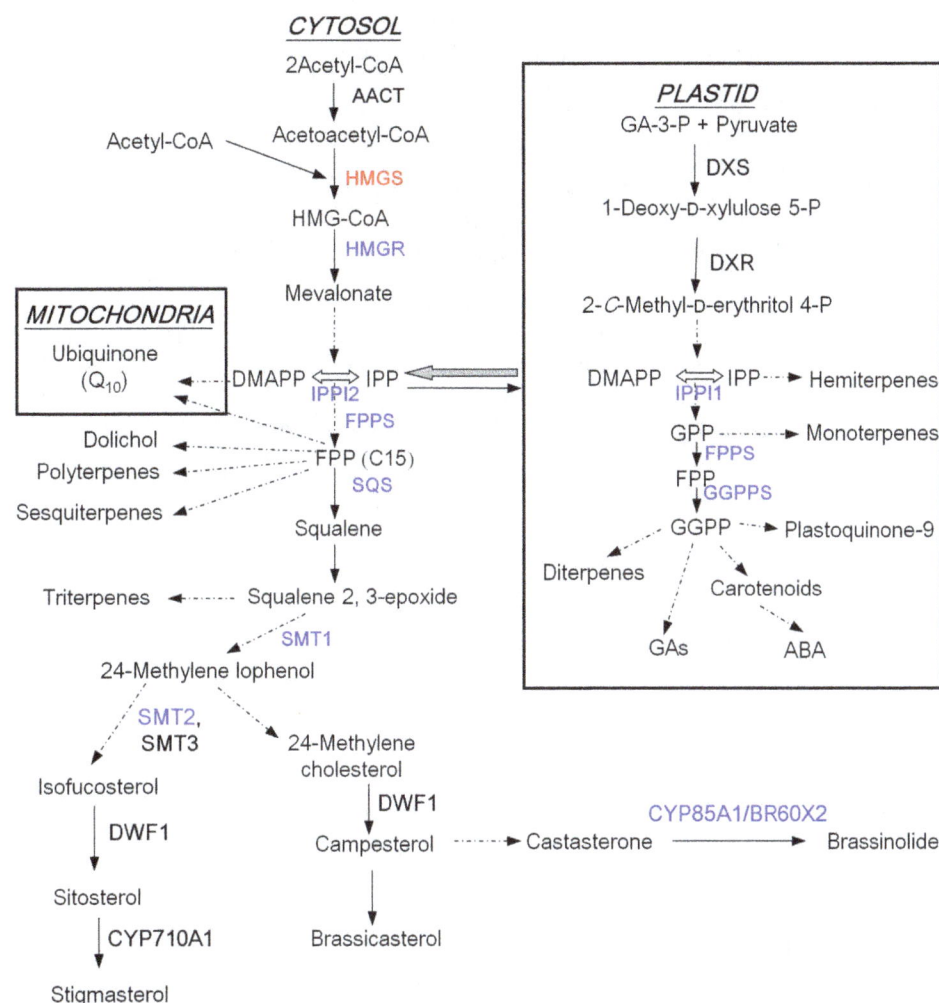

Figure 1. Outline of isoprenoid biosynthesis pathways in plants. Enzymes are shown in bold. Pathway inside the mitochondria and plastid are boxed. Arrows between cytosolic and plastid compartments represent metabolic flow between them (greater arrow for more flux). Abbreviations: ABA, abscisic acid; AACT, acetoacetyl-CoA thiolase; BR6OX2, brassinosteroid-6-oxidase 2; CYP710A1, sterol C-22 desaturase; CYP85A1, cytochrome P450 monooxygenase; DMAPP, dimethylallyl diphosphate; DWF1, delta-24 sterol reductase; DXR, 1-deoxy-D-xylulose 5-phosphate reductoisomerase; DXS, 1-deoxy-D-xylulose 5-phosphate synthase; FPP, farnesyl diphosphate; GA-3-P, glyceraldehyde-3-phosphate; FPPS, farnesyl diphosphate synthase; GAs, gibberellins; GGPP, geranylgeranyl diphosphate; GGPPS, geranylgeranyl diphosphate synthase; GPP, geranyl diphosphate; HMG-CoA, 3-hydroxy-3-methylglutaryl-CoA; HMGS, 3-hydroxy-3-methylglutaryl-CoA synthase; HMGR, 3-hydroxy-3-methylglutaryl-CoA reductase; IPP, isopentenyl diphosphate; IPPI, isopentenyl/dimethylallyl diphosphate isomerase; Q_{10}, coenzyme Q_{10}; SMT, sterol methyltransferase; SQS, squalene synthase. HMGS is marked in red colour. The expression levels of enzymes analysed in this work are marked in blue colour.

hmgr1 mutant is dwarf-like and male sterile, and has a lower sterol content [26].

3-Hydroxy-3-methylglutaryl-coenzyme A synthase (HMGS) is the second enzyme in the MVA pathway [27–31]. Besides 3-hydroxy-3-methylglutaryl-coenzyme A reductase (HMGR), HMGS is a key enzyme in cholesterol biosynthesis in mammals and cytoplasmic isoprenoid biosynthesis in plants [3,4,32–36]. Four genes designated *BjHMGS1* to *BjHMGS4* encode HMGS in *Brassica juncea* [34] and investigations revealed that BjHMGS1 is cytosolic. The expression of recombinant BjHMGS1 led to the elucidation of its kinetic and physiological properties [37,38] and of its crystal structure [39]. Enzyme kinetics of recombinant wild-type (wt) and mutant BjHMGS1 had revealed that H188N showed 8-fold lower enzyme activity and loss of acetoacetyl-CoA inhibition, while S359A displayed a 10-fold higher enzyme activity [37]. Given these interesting results, mutant (H188N, S359A and H188N/S359A) and wt BjHMGS1 were overexpressed in

Arabidopsis, which like *Brassica*, belongs to the family Brassicaceae [4]. *BjHMGS1* overexpression in transgenic *Arabidopsis* up-regulated several genes in sterol biosynthesis (cf. Figure 1), for instance those encoding HMGR, SMT2 (sterol methyltransferase 2), DWF1 (sterol C-24-reductase), CYP710A1 (sterol C-22 desaturase) and BR6OX2 (brassinosteroid-6-oxidase 2), increasing sterol content and thereby enhancing stress tolerance [4]. Analysis of the *Arabidopsis hmgs* mutant demonstrated the role of HMGS in tapetal development and pollen fertility [35].

To quickly assess the effects of BjHMGS1 overexpression in a more distant species, the overexpression of BjHMGS1 was carried out on a plant outside the Brassicaceae family. Hence, tobacco (*Nicotiana tabacum* L. cv. Xanthi), another model plant from the family of Solanaceae was selected, also because of the easiness of its genetic transformation. Subsequently, the genes downstream of *HMGS* that were tested encode enzymes that produce intermediates in phytosterol and BR biosynthesis, for instance

N. tabacum 3-hydroxy-3-methylglutaryl-CoA reductase (NtH-MGR1 and NtHMGR2), isopentenyl diphosphate isomerase (NtIPPI1 and NtIPPI2), farnesyl diphosphate synthase (NtFPPS), squalene synthase (NtSQS), sterol methyltransferases (NtSMT1-2, NtSMT2-1 and NtSMT2-2) and cytochrome P450 monooxygenase (NtCYP85A1). In addition, we examined the expression of genes encoding geranylgeranyl diphosphate synthases (NtGGPPS1, NtGGPPS2, NtGGPPS3 and NtGGPPS4), enzymes that are not implied in the formation of an intermediate in the sterol pathway. Resultant transgenic tobacco (OE-wtBjHMGS1 and OE-S359A) not only showed an increased sterol content but also displayed enhanced plant growth, pod size and seed yield that were not previously observed in transgenic *Arabidopsis* HMGS-OEs. Furthermore, OE-S359A conferred better plant growth and seed production than OE-wtBjHMGS1, and this was attributed to higher *NtSQS* expression and total sterol content, realizing the potential application of *BjHMGS1* in being quite active in phylogenetically distant species.

Materials and Methods

Plant materials and growth conditions

Wt tobacco (*N. tabacum* L. cv. Xanthi) obtained from the Institute of Molecular and Cell Biology (Singapore) was used in this study. Tobacco plants were grown at 25°C (16 h light)/22°C (8 h dark). Tobacco seedlings were cultured in Murashige and Skoog (MS) medium [40].

Generation of transgenic plants overexpressing HMGS

Plasmids pBj134 (wtBjHMGS1) and pBj136 (S359A) were used in *Agrobacterium*-mediated leaf disc transformation of *N. tabacum* [4,41]. The binary vector pSa13 [42] was used as vector control in transformation. T_1 transgenic tobacco seeds were selected on MS containing kanamycin (50 μg ml^{-1}) and verified using PCR and DNA sequence [4]. T_2 homozygous plants with a single-copy transgene were compared in mRNA expression, metabolite composition, plant growth and seed yield.

Western blot analysis

Total protein was extracted [43] from 21-d-old tobacco leaves. Protein concentration was determined using the Bio-Rad Protein Assay Kit I (Bio-Rad). Protein (20 μg per well) separated on 12% SDS-PAGE was transferred onto Hybond-ECL membrane (Amersham) using a Trans-Blot® cell (Bio-Rad). Antibodies raised against the synthetic peptide (DESYQSRDLEKVSQQ) corresponding to BjHMGS1 amino acids 290 to 304 were used in western blot analyses [4,44]. Cross-reacting bands were detected using the ECL^{TM} Western Blotting Detection Kit (Amersham).

Northern blot analysis

Tobacco total RNA was extracted from 21-d-old tobacco leaves using TRIzol reagent (Invitrogen). RNA (20 μg per well), separated on 1.3% agarose gels containing 6% formaldehyde, was transferred to Hybond-N membrane (Amersham) for northern blot analysis [45]. Digoxigenin-labelled probes were synthesized using the PCR Digoxigenin Probe Synthesis (Roche) with primer pairs ML276 and ML860 for *BjHMGS1*. Primers are listed in Table S1.

Southern blot analysis

Genomic DNA (40 μg) from 4-week-old tobacco leaves prepared by the CTAB method [46] was digested by *Eco*RI and separated on 0.7% agarose gel by electrophoresis, together with a 1-kb plus DNA standard ladder (Invitrogen). DNA was transferred from the agarose gel onto Hybond-N membrane (Amersham) by capillary transfer [47]. Southern blot analysis of tobacco using a ^{32}P-labelled full-length of *BjHMGS1* cDNA probe with primer pair ML264 and ML860 was performed [4]. Primers are listed in Table S1.

Extraction and quantitative analysis of sterols

For sterol profiling, freeze-dried materials from 20 mg of 60-d-old soil-grown tobacco leaves and 10 mg of 20-d-old MS plate-cultured tobacco seedlings were used. Extraction and quantitative analysis of sterols were carried out as described [4,48]. GC-MS analysis (GC: Hewlett Packard 6890 with an HP-5MS capillary column: 30 m long, 0.25 mm i.d., film thickness 0.25 μm; MS: Hewlett Packard 5973 mass selective detector, 70 eV) was used to determine sterol content, with He as the carrier gas (1 ml/min). The column temperature program used included a fast rise from 60°C to 220°C (30°C/min) and a slow rise from 220°C to 300°C (5°C/min), then kept at 300°C for 10 min. The inlet temperature was 280°C. Compounds were identified using the National Institute of Standards and Technology (NIST) libraries of peptide tandem mass spectra (Agilent, USA). The sterol masses were determined by comparison of the peak area of each compound with that of the internal standard (lupenyl-3,28-diacetate). Two independent lines for each OE genotype were analysed. Five independent repeats (samples) for each independent line were used for sterol extraction. Each sample was injected twice in GC-MS analyses and an average of the sterol mass was taken. Sitosterol, campesterol and stigmasterol contents in transgenic tobacco HMGS-OEs were compared to those in vector (pSa13)-transformed plants following previous reports [4,17].

Seed germination assay

Tobacco seeds collected simultaneously from vector (pSa13)-transformed control and HMGS-OE lines were sterilized in 20% bleach, 70% ethanol and then spread on MS medium agar plates supplemented with kanamycin (50 mg/l). About 30 tobacco seeds were sown on one plate. Five duplicate plates were used for each independent line [4]. All the plates were incubated at 4°C for 4 days and transferred to a culture room for 2 days under a photoperiod of 22°C 8-h dark and 23°C 16-h light. Subsequently, the number of germinated seeds was counted every 12 h for 60 h using a dissecting microscope. The emergence of the radicle was defined as germination [4]. The germination rates were calculated and compared using the Student's *t*-test. Two independent lines of OE-wtBjHMGS1 ("401" and "402") and two independent lines of OE-S359A ("603" and "606") were tested in seed germination assays. The experiment to measure seed germination was repeated twice.

Growth rate measurements

Growth rate was measured according to previous reports [49–53]. Four-d-old seedlings were transferred onto fresh MS plates placed vertically for a further 10-d growth. The dry weight of 14-d-old seedlings was then measured. Five seedlings were grouped for weight measurements and a total of 30 groups were analysed per individual line.

For greenhouse plants, 7-d-old tobacco seedlings of similar size were transferred from MS medium to soil for further growth rate measurements. The height of 80-, 98- and 210-d-old tobacco were measured. As 80-d-old plants did not have flowers, the height measurement did not include the inflorescence. However, 98- and 210-d-old plants were flowering and the height measurement included the inflorescence. For 98-d-old tobacco, measurements of leaf fresh weight, length and width of the four bottom-most leaves were also analysed for the vector-transformed control, OE-wtBjHMGS1 and OE-S359A. Two independent lines from each OE construct were analysed for 80-d-old tobacco plants and three independent lines from each OE construct were analysed for 98- and 210-d-old tobacco plants. For each line, six plants were used.

Comparison in tobacco seed yield

Seed yield was measured [49,51,52,54] to test the differences between HMGS-OEs (OE-wtBjHMGS1 and OE-S359A) and the vector-transformed control. Ten plants each from two independent lines from each OE construct were examined and T_2 homozygous seeds of each line were germinated on MS. Fourteen-d-old seedlings were transferred to soil in a greenhouse. Pods (30 per group) were harvested at maturity from each of 10 plants per line to determine total dry pod weight, average dry pod weight, total dry seed weight and total seed number. The experiment to measure seed yield was repeated twice (2–3 groups were analysed for each repeat).

To further determine if increase in seed size occurred, the dry weight of 100 seeds from each line was measured and 29 repeats were carried out per line. The average dry weight was calculated from 30 measurements of 100 seeds per line.

RNA analysis

Total RNA (5 µg) of 20-d-old tobacco seedlings and 14-d-old Arabidopsis were extracted using RNeasy Plant Mini Kit (Qiagen) and were reverse-transcribed into first-strand cDNA using the SuperScript First-Strand Synthesis System (Invitrogen). Quantitative Reverse Transcription-PCR (qRT-PCR) was carried out with a StepOne Plus Real-time PCR System (Applied Biosystems) and FastStart Universal SYBR Green Mater (Roche). The conditions for qRT-PCR were as follows: denaturation at 95°C for 10 min, followed by 40 cycles of 95°C for 15 s and 60°C for 1 min. Three experimental replicates for each reaction were carried out using gene-specific primers and tobacco ACTIN and Arabidopsis ACTIN2 were used as internal controls. The relative changes in expression from three independent experiments were analysed [55]. Primers for qRT-PCR are listed in Table S1.

Accession numbers

Sequence data included herein can be found in the GenBank/EMBL data libraries under accession numbers AF148847 (BjHMGS1), AY140008 (AtHMGS), U60452 (NtHMGR1), AF004232 (NtHMGR2), AB049815 (NtIPI1), AB049816 (NtIPI2), GQ410573 (NtFPPS), U60057 (NtSQS), GQ911583 (NtGGPPS1), GQ911584 (NtGGPPS2), AF053766 (NtSMT1-2), U71108 (NtSMT2-1), U71107.1 (NtSMT2-2), DQ649022 (NtCYP85A1), U60489 (NtACTIN), BT003419 (AtSQS) and AY096381 (AtACTIN2).

Statistical analysis

Analyses of data in this work was carried out using the Student's t-test to determine any significant differences between means.

Results

Molecular analyses of transgenic tobacco HMGS-OEs

The presence of wt and mutant BjHMGS1 in transgenic tobacco was verified by PCR (Figure S1A-B) followed by DNA sequence analysis of the PCR product. Putative tobacco HMGS-OEs were designated as OE-wtBjHMGS1 (lines "401", "402" and "404") and OE-S359A (lines "602", "603" and "606"). PCR-positive HMGS-OE lines were confirmed by western blot analysis (Figure 2A). As the peptide used to generate anti-BjHMGS1

antibodies shows 100% homology to tobacco HMGS (GenBank accession number EF636813), a faint band was detected in the vector (pSa13)-transformed control (Figure 2A). Northern blot analyses revealed that transgenic lines verified by western blot analysis expressed BjHMGS1 mRNA (Figure 2B). Single-insertional lines identified by Southern blot analyses (Figure S2) were selected for further experiments.

Tobacco HMGS-OEs accumulate sterols in both seedlings and leaves

The contents of the three major sterols (campesterol, stigmasterol and sitosterol) in 20-d-old tobacco HMGS-OE seedlings and

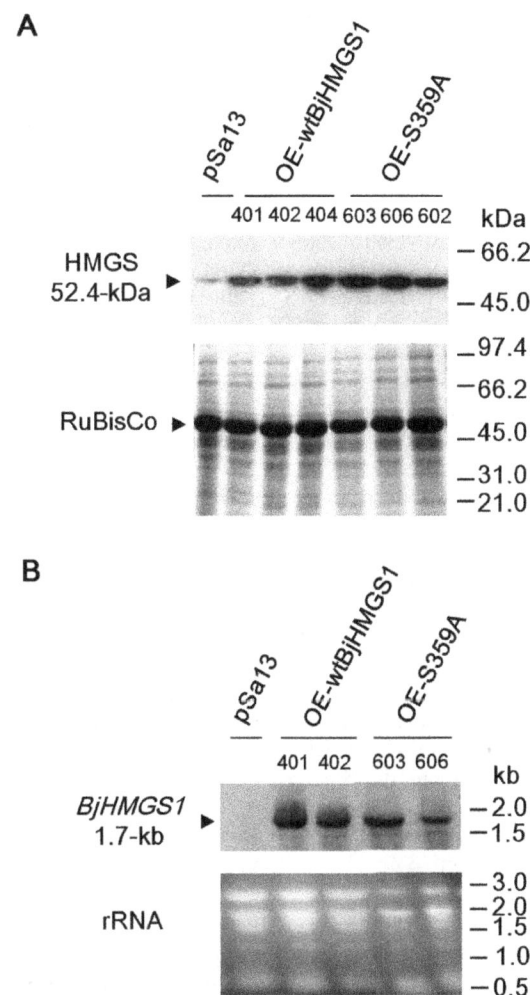

Figure 2. Molecular analysis of representative transgenic tobacco HMGS-OEs. (A) Western blot analysis using antibodies against BjHMGS1 to verify the expression of BjHMGS1 (52.4-kDa) in representative vector (pSa13)-transformed control and HMGS-OEs (OE-wtBjHMGS1 and OE-S359A). Putative tobacco HMGS-OEs were designated as OE-wtBjHMGS1 (lines "401", "402" and "404") and OE-S359A (lines "602", "603" and "606"). Bottom, Coomassie Blue-stained gel of total protein loaded (20 µg per well). Three independent lines per construct were analysed. (B) Northern blot analysis of BjHMGS1 in representative vector (pSa13)-transformed control and HMGS-OEs. The expected 1.7-kb BjHMGS1 band is marked with an arrowhead. Bottom gels show rRNA (20 µg per lane). Two independent lines per construct are shown. The two independent lines of OE-wtBjHMGS1 plants labelled "401" and "402", and two independent lines of OE-S359A plants labelled "603" and "606" used in further tests are underlined.

60-d-old leaves were analysed. GC-MS results of changes represented in µg per mg dry weight showed that the average campesterol, stigmasterol, sitosterol and total sterol contents of the OE-S359A seedlings were significantly higher than the vector (pSa13)-transformed control and OE-wtBjHMGS1 (Table 1). In particular, the average elevations over the vector (pSa13)-transformed control in OE-S359A seedlings were noted for campesterol (31.7%), stigmasterol (24.0%), sitosterol (25%) and total sterol (25.7%) (Table 2) and average elevations over OE-wtBjHMGS1 for campesterol (25.4%), stigmasterol (19.0%), sitosterol (20%) and total sterol (20.4%) (Table 2). However, OE-wtBjHMGS1 seedlings did not show significant changes from the vector-transformed control and increases were merely ~4–5% for each sterol (Table 2).

In leaves, except for stigmasterol, the average amounts of campesterol, sitosterol and total sterol were significantly higher in OE-wtBjHMGS1 than the vector (pSa13)-transformed control (Table 1): campesterol (12.9%), sitosterol (42.9%) and total sterol (12.1%) (Table 2). Furthermore, the average amounts of stigmasterol and total sterol in OE-S359A leaves were significantly higher (31.8% and 19.0%, respectively) over the vector (pSa13)-transformed control (Table 2). The differences between OE-wtBjHMGS1 and OE-S359A leaves were not significant and OE-S359A average stigmasterol and total sterol contents were only slightly higher than OE-wtBjHMGS1 (Table 1).

The % increase of sterols between transgenic tobacco (observed herein) and transgenic Arabidopsis (OE-wtBjHMGS1 and OE-S359A) [4] were also compared (Table 2 and S2). A similar trend was observed in transgenic Arabidopsis and tobacco seedlings; OE-S359A transformants displayed higher increase than the OE-wtBjHMGS1 not only in each sterol (campesterol, stigmasterol and sitosterol) but also in total sterol (Table S2). OE-S359A transformants also showed similar increase over the OE-wtBjHMGS1 in both Arabidopsis and tobacco leaves for stigmasterol and total sterol (Table 2 and S2).

Tobacco HMGS-OE seeds germinated earlier

As seeds from Arabidopsis HMGS-OEs were observed to germinate earlier than the vector (pSa13)-transformed control [4], the germination of tobacco HMGS-OE seeds was investigated. Tobacco seeds of OE-wtBjHMGS1 and OE-S359A not only germinated earlier but also displayed significantly higher germination rates than the control at 60 to 120 h post-germination (Figure S3). Also, OE-S359A germinated faster than OE-wtBjHMGS1 (Figure S3).

Tobacco HMGS-OE plants show increased growth

As sterols or steroid plant hormones have been reported to regulate plant growth [8,56], phenotyping was carried out on 14-d-old seedlings and 80-d-old plants. In 14-d-old HMGS-OE (OE-wtBjHMGS1 and OE-S359A) seedlings, root length (Figure 3A–B) and dry weight (Figure 3C) were significantly greater than the vector (pSa13)-transformed controls. Although the root length of 14-d-old seedlings in OE-S359A was not significantly greater than the OE-wtBjHMGS1 (Figure 3B), their dry weight was significantly heavier than OE-wtBjHMGS1 (Figure 3C). Consistently, 80-d-old tobacco HMGS-OE greenhouse plants grew better than the vector-transformed control (Figure 3D). HMGS-OEs (OE-wtBjHMGS1 and OE-S359A) were taller at 80-d than the control (Figure 3E). More interestingly, 80-d-old OE-S359A displayed significantly greater height than the OE-wtBjHMGS1 (Figure 3E).

Growth differences in height (Figure 4A–B) and leaf size (Figure 4C–D) between 98-d-old HMGS-OEs (OE-wtBjHMGS1 and OE-S359A) and vector (pSa13)-transformed plants were also

Table 1. Sterol profiles of tobacco HMGS-OE seedlings and leaves (µg/mg dry weight).

Construct	Sterol content of 20-d-old seedlings				Sterol content of 60-d-old leaves			
	Campesterol	Stigmasterol	Sitosterol	Total sterol	Campesterol	Stigmasterol	Sitosterol	Total sterol
pSa13	0.60±0.08	1.21±0.15	0.48±0.08	2.49±0.29	0.85±0.06	0.66±0.05	0.14±0.02	1.74±0.10
401	0.64±0.05	1.25±0.04	0.51±0.02	2.59±0.07	**0.98±0.05^a**	0.68±0.04	**0.21±0.01^a**	**1.99±0.08^a**
402	0.63±0.03	1.26±0.06	0.49±0.02	2.61±0.11	**0.93±0.04^a**	0.73±0.02	**0.18±0.01^a**	**1.91±0.05^a**
603	**0.79±0.04^{a,b}**	**1.50±0.04^{a,b}**	**0.59±0.02^{a,b}**	**3.14±0.10^{a,b}**	0.87±0.03	**0.91±0.05^a**	0.16±0.02	**2.12±0.06^a**
606	**0.79±0.03^{a,b}**	**1.49±0.05^{a,b}**	**0.61±0.02^{a,b}**	**3.12±0.09^{a,b}**	0.89±0.05	**0.82±0.04^a**	0.16±0.01	**2.01±0.04^a**

Two independent lines for each OE genotype were analysed. For OE-wtBjHMGS1, transformants "401" and "402" were tested. For OE-S359A, transformants "603" and "606" were tested. a indicates significant difference between HMGS-OE and the vector (pSa13)-transformed control; b indicates significant difference between OE-wtBjHMGS1 and OE-S359A. Bold font indicates significant higher sterol content than vector (pSa13)-transformed control and/or the OE-wtBjHMGS1 (P<0.01 by the Student's t-test). Values are mean ±SD, n=5.

Table 2. Increase (%) of sterol composition in tobacco HMGS-OE seedlings and leaves in comparison to vector (pSa13)-transformed control and elevation of OE-S359A over OE-wtBjHMGS1.

Construct	Compared to	Elevation (%) in 20-d-old seedlings				Elevation (%) in 60-d-old leaves			
		Campesterol	Stigmasterol	Sitosterol	Total sterol	Campesterol	Stigmasterol	Sitosterol	Total sterol
401	pSa13	6.7	3.3	6.3	4.0	15.3	3.0	50.0	14.4
402	pSa13	5.0	4.1	2.1	4.8	9.4	10.6	28.6	9.8
603	pSa13	**31.7**	**24.0**	**22.9**	**26.1**	2.4	**37.9**	14.3	**21.8**
606	pSa13	**31.7**	**23.1**	**27.1**	**25.3**	4.7	**24.2**	14.3	**15.5**
OE-S359A	OE-wtBjHMGS1	25.4	19.0	20.0	20.4	−8.3	24.3	−20.0	6.2

Two independent lines for each OE genotype were analysed. For tobacco OE-wtBjHMGS1, transformants "401" and "402" were tested. For tobacco OE-S359A, transformants "603" and "606" were tested. Values = [(mean$_{OEs}$ − mean$_{pSa13}$)/mean$_{pSa13}$]*100. The data presented for OE-S359A in comparison to OE-wtBjHMGS1 was calculated from an average of two transformants (average of "603" and "606" for OE-S359A in comparison to average of "401" and "402" for OE-wtBjHMGS1). Bold font indicates % increase value in OE-S359A which was higher than the corresponding OE-wtBjHMGS1.

evident (Figure 4). Both OE-wtBjHMGS1 and OE-S359A had a significant increase (91% and 97%, respectively) in height over the vector-transformed control (Figure 4B). Leaf fresh weight and size (length and width) (Figure 4C–D) in some of the OE-wtBjHMGS1 lines and all three OE-S359A lines were significantly heavier and bigger, respectively, than the control at similar age (Figure 4D).

Furthermore, growth differences in height between 210-d-old HMGS-OEs (OE-wtBjHMGS1 and OE-S359A) and vector-transformed plants were also observed (Figure 5). OE-wtBjHMGS1 showed a significant increase (21%) in height over the control, while OE-S359A displayed an even higher increase (45%) (Figure 5B).

Tobacco HMGS-OEs produce an enhanced seed yield

Comparison in seed yield by seed weight measurement between HMGS-OEs (OE-wtBjHMGS1 and OE-S359A) and the vector (pSa13)-transformed control indicated that both OE-wtBjHMGS1 and OE-S359A were higher than the control (Figure 6A–D); seed yield of OE-wtBjHMGS1 increased by 21 to 32% ($P<0.05$) (Figure 6D–F), while OE-S359A showed a 55 to 80% rise ($P<0.01$) (Figure 6D–F). OE-S359A (lines "603" and "606") showed an average of 32% increase over OE-wtBjHMGS1 (lines "401" and "402") by the Student's t-test ($P<0.05$) (Figure 6D–F). No significant difference in dry seed weight of 100 seeds was noted between the vector-transformed control and HMGS-OEs (Figure 6G), suggesting that seed size was not affected. Hence, HMGS-OE increase in seed yield was attributed to increase in pod size and seed number rather than seed size (Figure 6).

Change in expression of isoprenoid biosynthesis genes in tobacco HMGS-OEs

qRT-PCR was performed to check the effect of *BjHMGS1* overexpression on the expression of genes downstream of *HMGS* in tobacco HMGS-OE seedlings and to explore possible molecular mechanism of HMGS function in plant growth and seed production. The results from qRT-PCR revealed that the expression of *NtHMGR1*, *NtIPPI2*, *NtSQS*, *NtSMT1-2*, *NtSMT2-1*, *NtSMT2-2* and *NtCYP85A1* was significantly higher than in the vector (pSa13)-transformed control for both OE-wtBjHMGS1 and OE-S359A tobacco seedlings with the exception of *NtSQS*, *NtSMT1-2*, *NtSMT2-2* and *NtCYP85A1* in one OE-wtBjHMGS1 line (401) ($P<0.01$) (Figure 7). However, there was no difference in the expression of *NtHMGR2* between all the HMGS-OE lines and the vector-transformed control (Figure 7). For the expression of *NtFPPS*, there was no disparity amongst the two lines of OE-wtBjHMGS1 (401 and 402) and the vector-transformed control, while the expression of *NtFPPS* in another OE-wtBjHMGS1 line (404) and in two OE-S359A lines (602 and 606) was slightly higher than the control ($P<0.05$) (Figure 7). Conversely, the expression of *NtIPPI1*, *NtGGPPS1*, *NtGGPPS3* and *NtGGPPS4* were down-regulated in tobacco HMGS-OE seedlings ($P<0.01$) (Figures 7–8) while the expression of *NtGGPPS2* was higher than the control ($P<0.05$) in two OE-wtBjHMGS1 lines (402 and 404) and two OE-S359A lines (602 and 606) (Figure 8). Observations that (i) *NtSQS* expression in all three OE-S359A lines was higher than all three OE-wtBjHMGS1 lines, (ii) *NtHMGR1* and *NtCYP85A1* expression in all three OE-S359A lines were higher than two ("401" and "402") of three OE-wtBjHMGS1 lines, and (iii) *NtSMT2-1* expression in two ("602" and "603") of three OE-S359A lines was higher than two ("401" and "402") of three OE-wtBjHMGS1 lines suggest that the differences in expression levels of *NtSQS*, *NtHMGR1*, *NtSMT2-1* and *NtCYP85A1* in OE-wtBjHMGS1 and OE-S359A do correspond to the expected

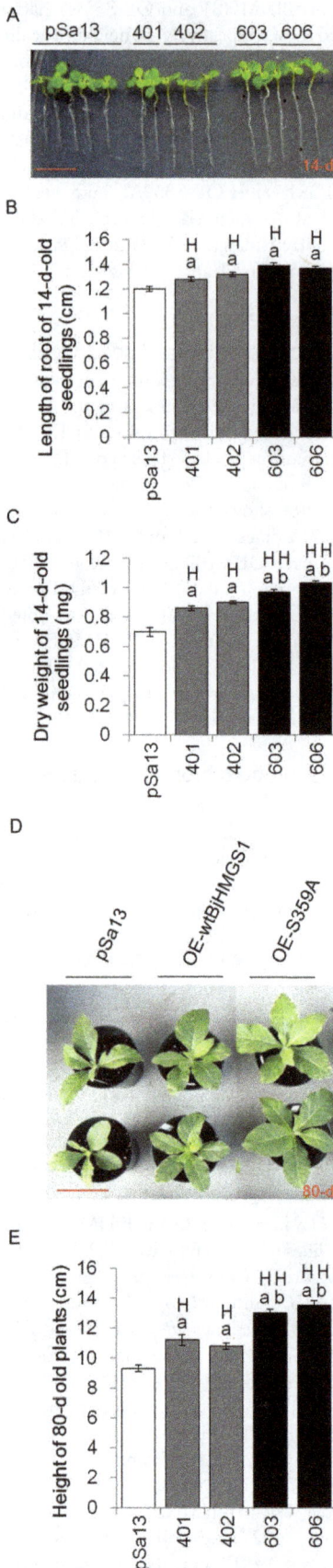

Figure 3. Comparison in growth between tobacco HMGS-OE seedlings/plants and vector-transformed control. (A) Seedlings 14-d post-germination. The vector-transformed control is labelled "pSa13", two independent lines of OE-wtBjHMGS1 plants are labelled "401" (two representative seedlings of this OE construct were shown) and "402" (three representative seedlings of this OE construct were shown) and two independent lines of OE-S359A plants are labelled "603" (two representative seedlings of this OE construct were shown) and "606" (three representative seedlings of this OE construct were shown). Bar = 1 cm. (B) Root length measurements of 14-d-old seedlings showed that tobacco HMGS-OE roots grow faster than the vector (pSa13)-transformed control. Values are mean ±SD (n = 30); Bars are SD. (C) Dry weight determination of 14-d-old seedlings shows that tobacco HMGS-OEs possess a higher mass than the vector-transformed control. Values are mean ± SD (n = 30); Bars are SD. (D) Representative greenhouse-grown plants photographed 80-d after germination. OE plants are labelled OE-wtBjHMGS1 and OE-S359A. Two independent lines of OE-wtBjHMGS1 plants, "401" (upper) and "402" (lower) and two independent lines of OE-S359A plants, "603" (upper) and "606" (lower) are shown. Bar = 10 cm. (E) Statistical analysis on height of 80-d-old transgenic plants. Values are mean ±SD (n = 6); Bars are SD; H, higher than control; a indicates significant difference between HMGS-OE and the vector (pSa13)-transformed control (P<0.01 by the Student's t-test); b indicates significant difference between OE-wtBjHMGS1 and OE-S359A (P<0.01 by the Student's t-test). pSa13, vector-transformed control; two independent lines of OE-wtBjHMGS1 ("401" and "402") and two independent lines of OE-S359A ("603" and "606") were used for growth rate measurement.

differences in enzyme activities between recombinant wtBjHMGS1 and S359A [37].

Discussion

New observations from tobacco HMGS-OEs

Our investigations on the overexpression of HMGS in transgenic tobacco revealed new observations not previously evident in *Arabidopsis* HMGS-OEs including the upregulation of *NtIPP12*, *NtSQS* and *NtGGPPS2* and downregulation of *NtIPP11*, *NtGGPPS1*, *NtGGPPS3* and *NtGGPPS4* (Figures 7–8). However, similar to findings from *Arabidopsis* HMGS-OEs, enhanced *NtHMGR1*, *NtSMT1-2*, *NtSMT2-1*, *NtSMT2-2* and *NtCYP85A1* expression in tobacco HMGS-OEs was seen (Figure 7). Other new findings from tobacco HMGS-OEs included growth stimulation in the tobacco HMGS-OE lines, confirming the positive role of HMGS overexpression in plant growth. Furthermore, tobacco HMGS-OEs show increased pod size and seed yield (Figure 6), indicative of a specific HMGS function in seed production. Improved growth, pod size and seed yield of OE-S359A in comparison to OE-wtBjHNMGS1 may be attributed to the higher *NtSQS* expression (Figure 7) and sterol content in OE-S359A transformants (Table 1).

Function of HMGS in reproduction and development

In plants, the floral organs are involved in reproduction. HMGS has been shown to play a crucial role in floral development [4,34,35,37]. In *Arabidopsis*, higher *AtHMGS* expression had been observed in flowers than seedlings or leaves from RT-PCR analysis [4]. Using mutants in *HMGS*, *AtHMGS* was demonstrated essential for pollen fertility and proper development of tapetum-specific organelles in *Arabidopsis* [35]. In *B. juncea*, northern blot analysis had previously revealed that *BjHMGS1* mRNA was highly expressed in flowers and seedling hypocotyls [34] and *in situ* hybridization analysis had shown that *HMGS* mRNA was predominantly localized in the stigmata and ovules of flower buds and in the piths of seedling hypocotyls [37]. *BjHMGS1* and *BjHMGS2*, but not *BjHMGS3* and *BjHMGS4* expression was

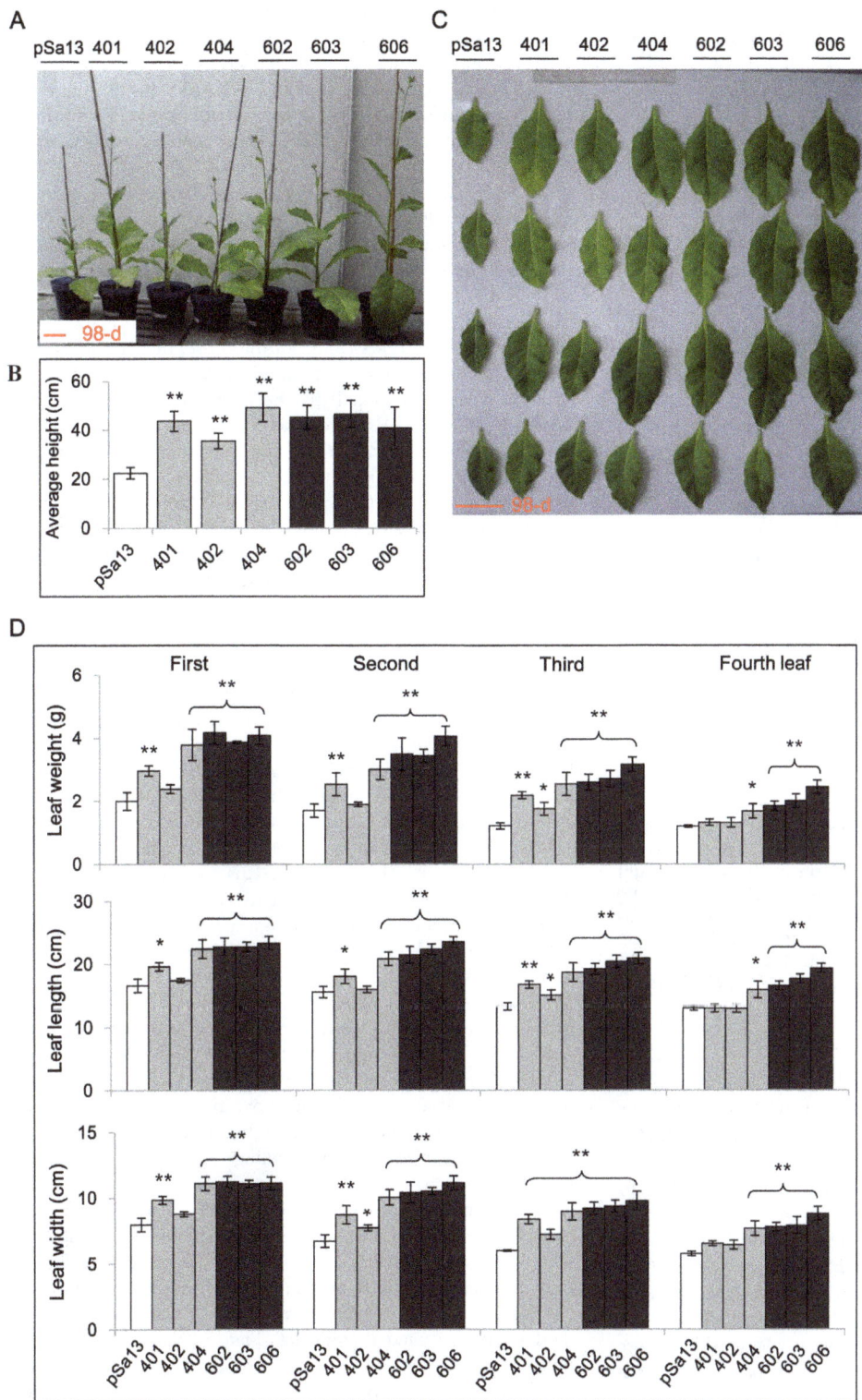

Figure 4. Comparison in plant growth between 98-d-old greenhouse-grown HMGS-OEs and vector-transformed tobacco. (A) Representative plants photographed 98-d after germination show differences in growth between HMGS-OE tobacco plants and vector-transformed control. Bar = 10 cm. (B) Analysis on height of 98-d-old transgenic plants. (C) Representative tobacco leaves photographed 98-d after germination with growth differences between HMGS-OE and vector-transformed tobacco. Bar = 10 cm. (D) Analysis on fresh weight, length and width of bottom-most four leaves from a 98-d-old tobacco plant. Values are mean ± SD (n = 6); Bars are SD; **, $P < 0.01$; *, $P < 0.05$; ** and *, significantly higher than control, by the Student's t-test. The vector-transformed control is labelled "pSa13", three independent lines of OE-wtBjHMGS1 plants are labelled "401", "402" and "404", and three independent lines of OE-S359A plants are labelled "602", "603" and "606".

A

B

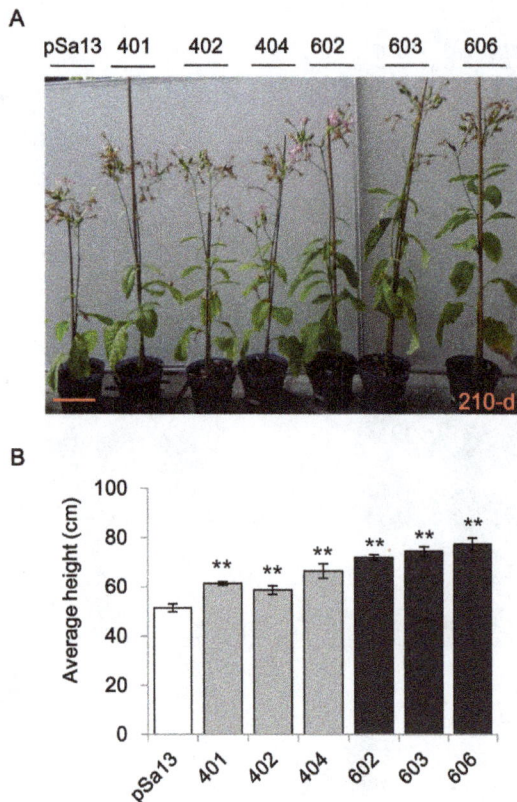

Figure 5. Comparison in plant growth between 210-d-old greenhouse-grown HMGS-OEs and vector-transformed tobacco. (A) Representative plants photographed 210-d after germination show differences in growth between HMGS-OE tobacco plants and the vector (pSa13)-transformed control. Bar = 10 cm. (B) Analysis on height of 210-d-old transgenic plants. Values are mean ± SD (n = 6); Bars are SD; **, $P<0.01$; *, $P<0.05$; ** and *, significantly higher than control, by the Student's t-test. The vector-transformed control is labelled "pSa13", three independent lines of OE-wtBjHMGS1 plants are labelled "401", "402" and "404", and three independent lines of OE-S359A plants are labelled "602", "603" and "606".

detected in the floral buds as examined by RT-PCR analysis [37]. The effect on the overexpression of BjHMGS1 in transgenic tobacco observed herein further extends the significance of HMGS in reproduction related to seed production as well as to whole plant development (Figures 3–6). More interestingly, OE-S359A lines were found to display greater effect in growth, pod size and seed yield than OE-wtBjHNMGS1 (Figures 3–6). OE-S359A, which was expected to possess higher HMGS activity than OE-wtBjHMGS1, caused higher expression of tobacco native genes downstream of HMGS such as *NtSQS*, *NtHMGR1*, *NtSMT2-1* and *NtCYP85A1* (Figure 7), and increased sterol levels, which more effectively enhanced seed production in comparison to OE-wtBjHMGS1.

Besides HMGS, other enzymes in the early steps of the MVA pathway are important in these development processes. It has been observed that both *hmg1/hmg1* and *HMG1/hmg1 hmg2/hmg2* *Arabidopsis* mutants deficient in HMGR activity are male sterile [26,57]. The *hmg1hmg2* male gametophytes in the *HMG1/hmg1 hmg2/hmg2* mutant were lethal [57]. Furthermore, the characterization of *Arabidopsis AACT1* and *AACT2* led to suggest a specific role of AACT2 in catalyzing the first step of the MVA pathway [58], while AACT1 is rather involved in the peroxisomal fatty acid

degradation process, like in tobacco seedlings [59]. *Arabidopsis* *AACT2 RNAi* lines further showed reduction in apical dominance, seed yield and root length accompanied by sterility and dwarfing [25]. These studies using the *AACT* RNAi lines, and mutants in *HMGS* and *HMGR* together with observations herein confirm the significance of the MVA pathway in plant reproduction and development.

Recently, two genes were cloned and characterized from two miRNA-action deficient (*MAD*) mutants; *MAD3* encodes the MVA pathway enzyme HMGR1, while *MAD4* encodes sterol C-8 isomerase in dedicated sterol biosynthesis [60]. Their results showed that the lack in HMGR1 catalytic activity is sufficient to inhibit miRNA activity and that sterol is essential for the normal activity of plant miRNAs [60]. Furthermore, their results implied that besides sterols, other isoprenoids may also affect the normal function of miRNA [60]. It has been reported that *Caenorhabditis elegans* HMGS1 (CeHMGS1) plays an important role in the miRNA pathway; *CeHMGS1* regulates the function of many, if not all, miRNAs at multiple tissues and stages during *C. elegans* development [61]. Furthermore, *CeHMGS1* affects the fertility of *C. elegans* in the miRNA defective *let-7* worms [61]. This effect on fertility is reminiscent of our observations on tobacco HMGS-OEs herein on seed production which represents fertility in plants.

Effects of HMGS in regulating isoprenoid biosynthesis genes in tobacco HMGS-OEs

In transgenic *Arabidopsis*, the overexpression of wt and mutant (H188N, S359A and H188N/S359A) BjHMGS1 caused a feed-forward effect in the upregulation of several genes in sterol biosynthesis including *HMGR*, *SMT2*, *DWF1*, *CYP710A1* and *BR6OX2* [4]. This study using tobacco HMGS-OEs demonstrated that some differences exist between tobacco and *Arabidopsis* HMGS-OEs in the expression of genes encoding HMGR and SMT (cf. Figure 1). Although HMGR is considered to be the rate-limiting enzyme in the MVA pathway in plants [62], only *NtHMGR1* but not *NtHMGR2* was upregulated in tobacco HMGS-OEs (Figure 7). This can perhaps be attributed to some differences in the localization and function of NtHMGR1 and NtHMGR2 [63,64]. *NtHMGR1* is a house-keeping gene that likely participates in sterol biosynthesis, plant growth and development, while *NtHMGR2* is stress-inducible [63,64]. Also elicitor-inducible HMGR activity is known to be associated with defence-related sesquiterpenoid accumulation in tobacco cell suspension cultures [65]. Thus it was not surprising that rather than *NtHMGR2*, *NtHMGR1* was upregulated in seedlings undergoing rapid growth and development.

Isopentenyl diphosphate isomerase (IPPI) catalyses the interconversion of IPP and its allyl isomer dimethylallyl diphosphate (DMAPP) and provides the first key intermediate for the biosynthesis of all kinds of isoprenoids including sterols in the MVA pathway and carotenoids in the MEP pathway [1,3,12,66] (and references cited therein) (cf. Figure 1). IPP is most likely involved in cross-talk between the cytosolic MVA pathway and the plastidial MEP pathway [13,14]. *AtIPPI1* and *AtIPPI2* have been reported to be critical to sterol biosynthesis in the MVA pathway and *Arabidopsis* development [67]. Analysis of the expression of the two *NtIPPI* genes in tobacco HMGS-OE seedlings revealed that *NtIPPI1* was downregulated, while *NtIPPI2* was upregulated (Figure 7). Their corresponding proteins are apparently differentially localized in tobacco [68]. NtIPPI1 is targeted to the chloroplast, while NtIPPI2 is cytosolic, similar to BjHMGS1 [37,68]. Possibly, upregulation of *BjHMGS1* and *NtIPPI2* in the cytosol of tobacco HMGS-OE seedlings promoted cross-talk between the MVA and MEP pathways. The MEP pathway

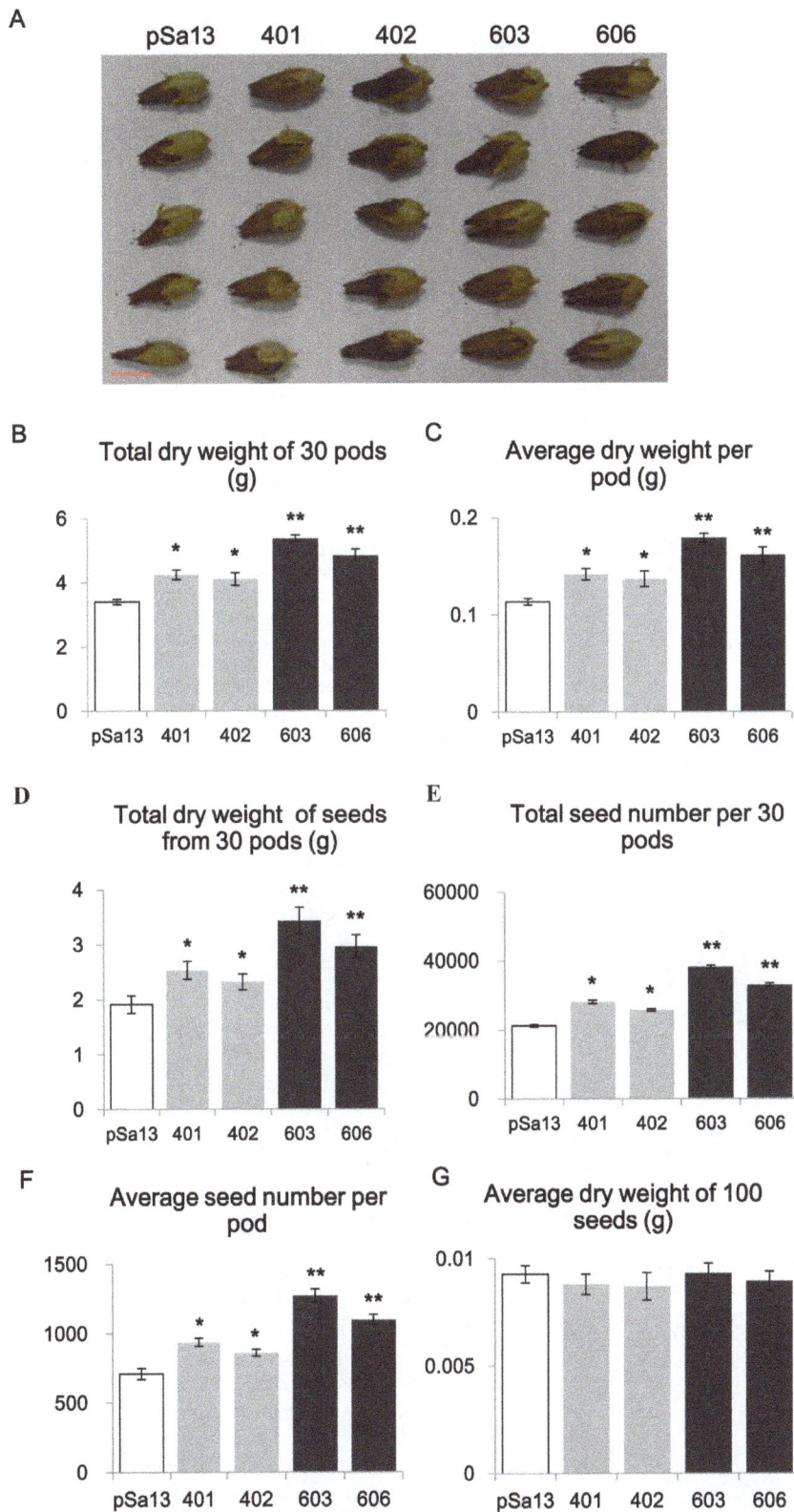

Figure 6. Tobacco HMGS-OEs show increased seed yield. (A) Phenotype of tobacco pods. pSa13, vector-transformed control; "401" and "402", two independent lines of OE-wtBjHMGS1 and "603" and "606", two independent lines of OE-S359A. Scale bar = 1 cm. (B) Total dry weight of 30 tobacco pods. (C) Average dry weight per pod. (D) Total dry weight of seeds from 30 pods. (E) Total seed number per 30 pods. (F) Average seed number per pod. (G) Average dry weight of 100 seeds in control and HMGS-OEs. Thirty independent readings were taken for each line. Values are means ± SD, n = 30. a indicates significant difference between HMGS-OE and the vector (pSa13)-transformed control; b indicates significant difference between OE-wtBjHMGS1 and OE-S359A. H, value higher than the control (P<0.05 or 0.01 by the Student's t-test).

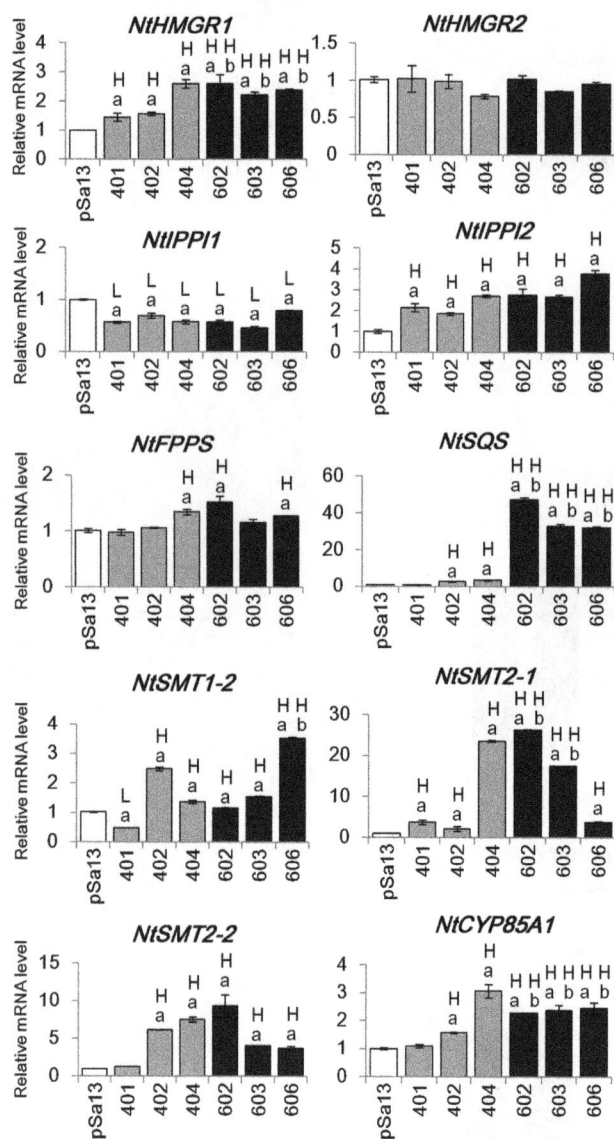

Figure 7. Expression of HMGS downstream genes by qRT-PCR in 20-d-old tobacco seedlings of HMGS-OEs. Total RNA was extracted from 20-d-old tobacco seedlings of vector (pSa13)-transformed control, three independent lines of OE-wtBjHMGS1 (lines "401", "402" and "404") and three independent lines of OE-S359A (lines "602", "603" and "606"). H, value higher than the control ($P<0.05$, Student's t-test); L, value lower than the control ($P<0.05$, Student's t-test). Values are means $\pm$SD (n = 3). a indicates significant difference between HMGS-OE and the vector (pSa13)-transformed control for at least two independent lines from three independent lines; b indicates significant difference between OE-wtBjHMGS1 and OE-S359A for at least two independent lines from three independent lines.

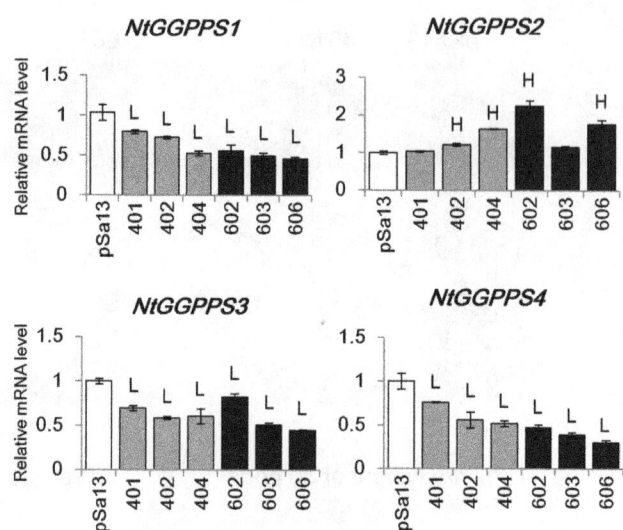

Figure 8. Expression of plastidial *GGPPSs* determined by qRT-PCR in 20-d-old tobacco seedlings of HMGS-OEs. Total RNA was extracted from 20-d-old tobacco seedlings of vector (pSa13)-transformed control, three independent lines of OE-wtBjHMGS1 (lines "401", "402" and "404") and three independent lines of OE-S359A (lines "602", "603" and "606"). H, value higher than the control ($P<0.05$, Student's t-test); L, value lower than the control ($P<0.05$, Student's t-test). Values are means $\pm$ SD (n = 3).

dolichols and sterols [69,70]. In plants, FPPS isozymes that are encoded by a small gene family, exert differential roles, based on their subcellular localisation [69,71]. *NtFPPS* expression was slightly elevated in seedlings of only one OE-wtBjHMGS1 line (Figure 7). Given that NtFPPS functions as the key provider of the universal product FPP in the biosynthesis of many C-15 related products, a moderate change in *NtFPPS* mRNA in the HMGS-OE lines may not be significant enough to affect sterol accumulation. Also, other *NtFPPS* isogenes or post-translational regulation may be involved [72–74].

SQS catalyses the biosynthesis of squalene by the reductive dimerization of two FPP molecules (cf. Figure 1), and represents the first committed step in the biosynthesis of sterols, BRs and triterpenes [75–79]. The change in *NtSQS* expression in seedlings was the most dramatic, with a 2.1-fold increase in two lines of OE-wtBjHMGS1 and 36.5-fold in OE-S359A, in comparison to the vector-transformed control (Figure 7). The increase of *NtSQS* mRNA in OE-S359A seedlings was also much higher (11.1-fold) than OE-wtBjHMGS1 (Figure 7). Interestingly, *NtSQS* expression and NtSQS activity have been detected predominantly at the shoot apical meristem (SAM) rather than leaves or roots, implying that sterol biosynthesis occurs especially in the SAM [77]. Furthermore, the SAM is critical in plant growth and development, and stem cells from the SAM continuously generate all the aerial organs and tissues of a plant [80]. Results from qRT-PCR (Figure 7) herein support a role for *NtSQS* in *HMGS*-associated sterol accumulation related to growth and seed yield. Also, enhanced sterol accumulation, growth and seed yield in OE-S359A, over OE-wtBjHMGS1 (Figure 7), corresponded to higher *NtSQS* expression (Figure 7). Consistently, Arabidopsis *SQS* (*AtSQS*) displayed higher expression in HMGS-OEs than the vector-transformed control; and *AtSQS* expression in OE-S359A was higher than OE-wtBjHMGS1 (Figure S4). However the elevation of *NtSQS* in tobacco OE-S359A over OE-wtBjHMGS1 (Figure 7)

produces simultaneously IPP and DMAPP, and plastidial NtIPPI1 is possibly needed to adjust the ratio of starter DMAPP to elongation units IPP for longer prenyl chains. If IPP is imported from the cytosol because of "overproduction", then plastidial *NtIPPI1* would be downregulated.

FPPS catalyses the condensation of two molecules of IPP with DMAPP to form farnesyl diphosphate (FPP) (C_{15}) (cf. Figure 1), which provides the key precursor for the biosynthesis of essential isoprenoids such as sesquiterpenes, ubiquinones, polyterpenes, was greater in comparison to *AtSQS* in Arabidopsis OE-S359A (Figure S4). Furthermore, our results correspond well to a recent study on the overexpression of *Glycine max* SQS1 (GmSQS1) in Arabidopsis that yielded a 50% increase of seed sterol content [81]. An enhanced flux of MVA to FPP might present some risk as phosphatases always being present might liberate farnesol, which can be quite toxic to cells [82]. Thus SQS could remove a potentially dangerous intermediate and get it channelled into the synthesis and accumulation of chemically inert sterols and their derivatives.

In the MEP pathway, GGPPS catalyses the consecutive condensation of three molecules of IPP and one DMAPP to generate the 20-carbon geranylgeranyl diphosphate (GGPP) (cf. Figure 1), which is the universal key intermediate for the biosynthesis of carotenoids and of abscisic acid as derivative, of gibberellins, chlorophylls, tocopherols, phylloquinone, plastoquinone, dolichols, polyprenols and oligoprenols [12,83]. Although four GGPPS-like cDNAs have been reported from tobacco [84], only *NtGGPPS2* was upregulated in two lines of OE-wtBjHMGS1 and all three lines of OE-S359A, while *NtGGPPS1*, *NtGGPPS3* and *NtGGPPS4* were observed to be downregulated in all the HMGS-OE seedlings (Figure 8), implying that HMGS overexpression had a positive effect on *NtGGPPS2* expression and a negative role on *NtGGPPS1*, *NtGGPPS3* and *NtGGPPS4* expression. However, it cannot be discounted that NtGGPPS1, NtGGPPS3 and NtGGPPS4 may be subject to other modes of regulation such as post-translational modification that has been reported for AtGGPPS3, AtGGPPS7, AtGGPPS9 and AtGGPPS10 [85,86]. Most recently, a new relationship between the MVA pathway and the MEP pathway has been proposed in which the monoterpene *S*-carvone inhibited the production of MVA-derived capsidiol, a cellulose-induced sesquiterpenoid phytoalexin in tobacco by down-regulation of MEP-pathway dependent protein isoprenylation [87].

The overexpression of NtSMT1(cf. Figure 1), which catalyses the conversion of cycloartenol to 24-methylene cycloartanol, considered as the first methylation step in phytosterol biosynthesis, resulted in a higher total sterol content in tobacco seeds [88–90]. Transgenic tobacco overexpressing AtSMT2/NtSMT2, which converts 24-methylene lophenol to 24-ethylidene lophenol, showed an increase in sitosterol but not total sterol content [91–94]. HMGS overexpression in tobacco upregulated both *NtSMT1* and *NtSMT2* expression in seedlings of all three OE-S359A lines and two OE-wtBjHMGS1 lines with the exception of OE-wtBjHMGS1 line 401 (Figure 7). The upregulation of *SMT2* was also observed in 21-d-old rosette leaves of transgenic *Arabidopsis* overexpressing *BjHMGS1* [4]. Our results suggest that *NtSMT1* affects *HMGS*-associated sterol accumulation, which had not been previously observed in transgenic *Arabidopsis* HMGS-OEs [4].

BR is a steroid hormone essential for plant growth and development [95]. Several mutants in BR biosynthesis affect seed yield [6,22–24]. The cytochrome P-450 monooxygenases (CYP85A family) are involved in the last several oxidative reactions in the BR pathway [96]. In *Arabidopsis*, two members of CYP85A exist: AtCYP85A1 (brassinosteroid-6-oxidase 1, BR60X1) that catalyses several reactions in the biosynthesis of castasterone [97], and AtCYP85A2 (BR60X2) in the conversion of castasterone to brassinolide [96]. The *Arabidopsis* cyp85a1 mutant showed a semi-sterile phenotype and the *cyp85a2* mutant exhibited dwarfness and reduced fertility [96–97]. The *cyp85a1/cyp85a2* double mutants displayed severe dwarfism [96]. To test the effect in HMGS overexpression on BR biosynthesis, *NtCYP85A1* (cf. Figure 1) mRNA was measured in tobacco HMGS-OE seedlings

and was observed to significantly increase in all three OE-S359A lines and two OE-wtBjHMGS1 lines with the exception of OE-wtBjHMGS1 line 401 (Figure 7). Although OE-wtBjHMGS1 line 401 did not show higher expression in *NtCYP85A1*, as well as in *NtSQS*, *NtSMT1-2*, *NtSMT2-2* and *NtGGPPS2*, the expression of all these genes were maintained a level similar to the control (Figures 7–8). Furthermore, *NtHMGR1*, *NtIPPI2* and *NtSMT2-1* displayed significantly higher expression in this line than the control (Figure 7), implying that they positively affected plant growth and seed yield. Taken together with observations on a general up-regulation of *AtCYP85A2* (*BR60X2*) in 21-d-old rosette leaves of transgenic *Arabidopsis* overexpressing BjHMGS1 [4], our studies reinforce that HMGS overexpression likely leads to upregulation of BR synthesis, and thereby promotes growth and seed production.

Supporting Information

Figure S1　The BjHMGS1 constructs used in tobacco transformation and resultant PCR analysis on transgenic tobacco lines. (A) Schematic map of transformation vector indicating primer location. *BjHMGS1* wild-type and mutant inserts were derived from plasmids, pBj134 (WT *BjHMGS1*) and pBj136 (S359A) [4]. *CaMV35S*: Cauliflower Mosaic Virus *35S* promoter; *NOSpro*: nopaline synthase (*NOS*) promoter; *NOSter*: *NOS* terminator; *NPTII*: gene encoding neomycin phosphotransferase II conferring resistance to kanamycin; RB: right border of T-DNA; LB: left border of T-DNA. *35S*: *35S* promoter 3'-end forward primer; ML264: *BjHMGS1*-specific 3'-end reverse primer. (B) Agarose gel showing the expected 1.65-kb *BjHMGS1* cDNA band (arrowed) from transgenic tobacco following PCR using primer pair 35S/ML264; representative lines are shown here. OE-wtBjHMGS1 (lanes 1–3); OE-S359A (lanes 4–6); positive control (PC) (lane 7, PCR template plasmid pBj134); blank control (BC) (lane 8, no DNA band after PCR). Putative tobacco HMGS-OEs were designated as OE-wtBjHMGS1 (lines "401", "402" and "404") and OE-S359A (lines "602", "603" and "606").

Figure S2　Southern blot analysis on transgenic tobacco plants. (A) Schematic map of transformation vector indicating *Eco*RI (E) sites. *BjHMGS1* wild-type and mutant inserts were derived from plasmids pBj134 (wt*BjHMGS1*) and pBj136 (S359A). *CaMV35S*: Cauliflower Mosaic Virus *35S* promoter; *NOSpro*: nopaline synthase (*NOS*) promoter; *NOSter*: *NOS* terminator; *NPTII*: gene encoding neomycin phosphotransferase II conferring resistance to kanamycin; RB: right border of T-DNA; LB: left border of T-DNA. Dotted lines denote position of nucleotide on vector. (B) Southern blot analysis of genomic DNA digested by restrictive endonuclease *Eco*RI and probed with ^{32}P-labelled *BjHMGS1* full-length cDNA in representative blots. Arrowheads indicate hybridizing bands. OE-wtBjHMGS1 transformants (lanes 1–2), OE-S359A transformants (lanes 3–5). Representative single insertion lines (transformants "401" and "402" for OE-wtBjHMGS1 and "603" and "606" for OE-S359A) are underlined. Transformant "601" likely has a more than one inserts and was not included in further analysis.

Figure S3　Comparison in seed germination of tobacco HMGS-OEs. Statistical data on seed germination rates recorded at 60, 72, 84, 96, 108 and 120 h after incubation at 23°C indicates (a) significant difference (P<0.01 by the Student's *t*-test) between HMGS-OE and the vector (pSa13)-transformed control; (b) indicates significant difference (P<0.01 by the Student's *t*-test) between OE-wtBjHMGS1 and OE-S359A. Values are mean ±SD (n = 5); bars represent SD. pSa13, vector-transformed control; the two independent lines of OE-wtBjHMGS1 ("401" and "402") and two independent lines of OE-S359A ("603" and "606") were tested in seed germination assays. The data represents the average from two transformants.

Figure S4 Expression of Arabidopsis *SQS* by qRT-PCR in 14-d-old HMGS-OE seedlings.

Total RNA was extracted from 14-d-old Arabidopsis seedlings of vector (pSa13)-transformed control, two independent lines of OE-wtBjHMGS1 (lines "134-L1" and "134-L2") and two independent lines of OE-S359A (lines "136-L1" and "136-L2") previously generated [4]. H, value higher than the control (*P*<0.01, Student's *t*-test). Values are means ± SD (n = 3). a indicates significant difference between HMGS-OE and the vector (pSa13)-transformed control;

b indicates significant difference between OE-wtBjHMGS1 and OE-S359A.

Table S1 Oligonucleotide primers used in this study.
Restriction sites are underlined.

Table S2 Increase (%) of sterol composition in Arabidopsis HMGS-OE seedlings and leaves in comparison to vector (pSa13)-transformed control.

Author Contributions

Conceived and designed the experiments: PL HW TJB MLC. Performed the experiments: PL HW MFW ASH TJB. Analyzed the data: PL HW TJB MLC. Contributed reagents/materials/analysis tools: MLC TJB. Wrote the paper: PL HW TJB MLC. Coordinated the project: TJB MLC.

References

1. Bach TJ (1995) Some new aspects of isoprenoid biosynthesis in plants - a review. Lipids 30: 191–202.
2. Briskin DP (2000) Medicinal plants and phytomedicines. Linking plant biochemistry and physiology to human health. Plant Physiol 124: 507–514.
3. Hemmerlin A, Harwood JL, Bach TJ (2012) A *raison d'etre* for *two distinct* pathways in the early steps of plant isoprenoid biosynthesis? Prog Lipid Res 51: 95–148.
4. Wang H, Nagegowda DA, Rawat R, Bouvier-Navé P, Guo D, et al. (2012) Overexpression of *Brassica juncea* wild-type and mutant HMG-CoA synthase 1 in Arabidopsis up-regulates genes in sterol biosynthesis and enhances sterol production and stress tolerance. Plant Biotechnol J 10: 31–42.
5. Shani E, Ben-Gera H, Shleizer-Burko S, Burko Y, Weiss D, et al. (2010) Cytokinin Regulates compound leaf development in tomato. Plant Cell 22: 3206–3217.
6. Vriet C, Russinova E, Reuzeau C (2012) Boosting crop yields with plant steroids. Plant Cell 24: 842–857.
7. Demmig-Adams B, Adams WW III (1996) The role of xanthophyll cycle carotenoids in the protection of photosynthesis. Trends Plant Sci 1: 21–26.
8. He JX, Fujioka S, Li TC, Kang SG, Seto H, et al. (2003) Sterols regulate development and gene expression in Arabidopsis. Plant Physiol 131: 1258–1269.
9. Bradford PG, Awad AB (2007) Phytosterols as anticancer compounds. Mol Nutr Food Res 51: 161–170.
10. Moreau RA, Whitaker BD, Hicks KB (2002) Phytosterols, phytostanols, and their conjugates in foods: structural diversity, quantitative analysis, and health-promoting uses. Prog Lipid Res 41: 457–500.
11. Woyengo TA, Ramprasath VR, Jones PJ (2009) Anticancer effects of phytosterols. Eur J Clin Nutr 63: 813–820.
12. Rohmer M (1999) The discovery of a mevalonate-independent pathway for isoprenoid biosynthesis in bacteria, algae and higher plants. Nat Prod Rep 16: 565–574.
13. Hemmerlin A, Hoeffler JF, Meyer O, Tritsch D, Kagan IA, et al. (2003) Cross-talk between the cytosolic mevalonate and the plastidial methylerythritol phosphate pathways in tobacco bright yellow-2 cells. J Biol Chem 278: 26666–26676.
14. Laule O, Fürholz A, Chang HS, Zhu T, Wang X, et al. (2003) Crosstalk between cytosolic and plastidial pathways of isoprenoid biosynthesis in *Arabidopsis thaliana*. Proc Natl Acad Sci USA 100: 6866–6871.
15. Bush PB, Grunwald C (1972) Sterol changes during germination of *Nicotiana tabacum* seeds. Plant Physiol 51: 69–72.
16. Bach TJ, Lichtenthaler HK (1983) Inhibition by mevinolin of plant growth, sterol formation and pigment accumulation. Physiol Plantarum 59: 50–60.
17. Schaller H, Grausem B, Bouvier-Navé P, Chye ML, Tan CT, et al. (1995) Expression of the *Hevea brasiliensis* (H.B.K) Mull. Arg. 3-hydroxy-3-methylglutaryl-coenzyme A reductase 1 in tobacco results in sterol overproduction. Plant Physiol 109: 761–770.
18. Hedden P, Kamiya Y (1997) Gibberellin biosynthesis: enzymes, genes and their regulation. Annu Rev Plant Physiol Plant Mol Biol 48: 431–460.
19. Clouse SD, Sasse JM (1998) Brassinosteroids: essential regulators of plant growth and development. Annu Rev Plant Physiol Plant Mol Biol 49: 427–451.
20. Eisenreich W, Rohdich F, Bacher A (2001) Deoxyxylulose phosphate pathway to terpenoids. Trends Plant Sci 6: 78–84.
21. Montoya T, Nomura T, Yokota T, Farrar K, Harrison K, et al. (2005) Patterns of Dwarf expression and brassinosteroid accumulation in tomato reveal the importance of brassinosteroid synthesis during fruit development. Plant J 42: 262–269.
22. Choe S, Tanaka A, Noguchi T, Fujioka S, Takatsuto S, et al. (2000) Lesions in the sterol Δ^7 reductase gene of Arabidopsis cause dwarfism due to a block in brassinosteroid biosynthesis. Plant J 21: 431–443.
23. Choe S, Fujioka S, Noguchi T, Takatsuto S, Yoshida S, et al. (2001) Overexpression of *DWARF4* in the brassinosteroid biosynthetic pathway results in increased vegetative growth and seed yield in *Arabidopsis*. Plant J 26: 573–582.
24. Li FL, Asami T, Wu XZ, Tsang EWT, Cutler AJ (2007) A putative hydroxysteroid dehydrogenase involved in regulating plant growth and development. Plant Physiol 145: 87–97.
25. Jin H, Song Z, Nikolau BJ (2012) Reverse genetic characterization of two paralogous acetoacetyl CoA thiolase genes in Arabidopsis reveals their importance in plant growth and development. Plant J 70: 1015–1032.
26. Suzuki M, Kamide Y, Nagata N, Seki H, Ohyama K, et al. (2004) Loss of function of *3-hydroxy-3-methylglutaryl coenzyme A reductase 1 (HMG1)* in *Arabidopsis* leads to dwarfing, early senescence and male sterility, and reduced sterol levels. Plant J 37: 750–761.
27. Balasubramaniam S, Goldstein JL, Brown MS (1977) Regulation of cholesterol synthesis in rat adrenal gland through coordinate control of 3-hydroxy-3-methylglutaryl coenzyme A synthase and reductase activities. Proc Natl Acad Sci USA 74: 1421–1425.
28. Ferguson JJ Jr, Rudney H (1959) The biosynthesis of β-hydroxy-β-methylglutaryl coenzyme A in yeast. I. Identification and purification of the hydroxymethylglutaryl coenzyme-condensing enzyme. J Biol Chem 234: 1072–1075.
29. Lynen F (1967) Biosynthetic pathways from acetate to natural products. Pure Appl Chem 14: 137–167.
30. Rudney H, Ferguson JJ Jr (1959) The biosynthesis of β-hydroxy-β-methylglutaryl coenzyme A in yeast. II. The formation of hydroxymethylglutaryl coenzyme A via the condensation of acetyl coenzyme A and acetoacetyl coenzyme A. J Biol Chem 234: 1076–1080.
31. Stewart PR, Rudney H (1966) The biosynthesis of β-hydroxy-β-methylglutaryl coenzyme A in yeast. IV. The origin of the thioester bond of β-hydroxy-β-methylglutaryl coenzyme A. J Biol Chem 241: 1222–1225.
32. Bach TJ (1986) Hydroxymethylglutaryl-CoA reductase, a key enzyme in phytosterol synthesis? Lipids 21: 82–88.
33. Dooley KA, Millinder S, Osborne TF (1998) Sterol regulation of 3-hydroxy-3-methylglutaryl-coenzyme A synthase gene through a direct interaction between sterol regulatory element binding protein and the trimeric CCAAT-binding factor/nuclear factor Y. J Biol Chem 273: 1349–1356.
34. Alex D, Bach TJ, Chye ML (2000) Expression of *Brassica juncea* 3-hydroxy-3-methylglutaryl CoA synthase is developmentally regulated and stress-responsive. Plant J 22: 415–426.
35. Ishiguro S, Nishimori Y, Yamada M, Saito H, Suzuki T, et al. (2010) The Arabidopsis *FLAKY POLLEN1* gene encodes a 3-hydroxy-3-methylglutaryl-coenzyme A synthase required for development of tapetum-specific organelles and fertility of pollen grains. Plant Cell Physiol 51: 896–911.
36. Suwanmanee P, Sirinupong N, Suvachittanont W (2013) Regulation of 3-hydroxy-3-methylglutaryl-CoA synthase and 3-hydroxy-3-methylglutaryl-CoA reductase and rubber biosynthesis of *Hevea brasiliensis* (B.H.K.) Mull. Arg. In: Bach TJ, Rohmer M editors. Isoprenoid synthesis in plants and microorganisms: new concepts and experimental approaches, New York: Springer. pp. 315–327.
37. Nagegowda DA, Bach TJ, Chye ML (2004) *Brassica juncea* 3-hydroxy-3-methylglutaryl (HMG)-CoA synthase 1: expression and characterization of recombinant wild-type and mutant enzymes. Biochem J 383: 517–527.
38. Nagegowda DA, Ramalingam S, Hemmerlin A, Bach TJ, Chye ML (2005) *Brassica juncea* HMG-CoA synthase: localization of mRNA and protein. Planta 221: 844–856.
39. Pojer F, Ferrer JL, Richard SB, Nagegowda DA, Chye ML, et al. (2006) Structural basis for the design of potent and species-specific inhibitors of 3-hydroxy-3-methylglutaryl CoA synthases. Proc Natl Acad Sci USA 103: 11491–11496.

40. Murashige T, Skoog F (1962) A revised medium for rapid growth and bio assays with tobacco tissue cultures. Physiol Plant 15: 473–497.

41. Horsch R, Fry J, Hoffmann N, Neidermeyer J, Rogers S, et al. (1989) Leaf disc transformation. In: Gelvin S, Schilperoort R, editors. Plant Molecular Biology Manual, A5. Netherlands: Springer. pp. 63–71.

42. Xiao S, Li HY, Zhang JP, Chan SW, Chye ML (2008) Arabidopsis acyl-CoA-binding proteins ACBP4 and ACBP5 are subcellularly localized to the cytosol and ACBP4 deletion affects membrane lipid composition. Plant Mol Biol 68: 574–583.

43. Chye ML, Huang BQ, Zee SY (1999) Isolation of a gene encoding *Arabidopsis* membrane-associated acyl-CoA binding protein and immunolocalization of its gene product. Plant J 18: 205–214.

44. Xiao S, Gao W, Chen QF, Chan SW, Zheng SX, et al. (2010) Overexpression of *Arabidopsis* acyl-CoA binding protein ACBP3 promotes starvation-induced and age-dependent leaf senescence. Plant Cell 22: 1463–1482.

45. Chen QF, Xiao S, Chye ML (2008) Overexpression of the Arabidopsis 10-kilodalton acyl-coenzyme A-binding protein ACBP6 enhances freezing tolerance. Plant Physiol 148: 304–315.

46. Rogers SO, Bendich AJ (1985) Extraction of DNA from milligram amounts of fresh, herbarium and mummified plant tissues. Plant Mol Biol 5: 69–76.

47. Southern E (2006) Southern blotting. Nat Protoc 1: 518–525.

48. Babiychuk E, Bouvier-Navé P, Compagnon V, Suzuki M, Muranaka T, et al. (2008) Allelic mutant series reveal distinct functions for *Arabidopsis* cycloartenol synthase 1 in cell viability and plastid biogenesis. Proc Natl Acad Sci USA 105: 3163–3168.

49. Fang Z, Xia K, Yang X, Grotemeyer MS, Meier S, et al. (2013) Altered expression of the *PTR/NRT1* homologue *OsPTR9* affects nitrogen utilization efficiency, growth and grain yield in rice. Plant Biotechnol J 11: 446–458.

50. Gévaudant F, Duby G, von Stedingk E, Zhao R, Morsomme P, et al. (2007) Expression of a constitutively activated plasma membrane H+-ATPase alters plant development and increases salt tolerance. Plant Physiol 144: 1763–1776.

51. Sun F, Suen PK, Zhang Y, Liang C, Carrie C, et al. (2012) A dual-targeted purple acid phosphatase in *Arabidopsis thaliana* moderates carbon metabolism and its overexpression leads to faster plant growth and higher seed yield. New Phytol 194: 206–219.

52. Zhang Y, Yu L, Yung KF, Leung DY, Sun F, et al. (2012) Over-expression of AtPAP2 in *Camelina sativa* leads to faster plant growth and higher seed yield. Biotechnol Biofuels 5: 19.

53. Bae H, Choi SM, Yang SW, Pai HS, Kim WT (2009) Suppression of the ER-localized AAA ATPase NgCDC48 inhibits tobacco growth and development. Mol Cells 28: 57–65.

54. Li D, Wang L, Wang M, Xu YY, Luo W, et al. (2009). Engineering *OsBAK1* gene as a molecular tool to improve rice architecture for high yield. Plant Biotechnol J 7: 791–806.

55. Schmittgen TD, Livak KJ (2008) Analyzing real-time PCR data by the comparative C_T method. Nat Protoc 3: 1101–1108.

56. Grove MD, Spencer GF, Rohwedder WK, Mandava N, Worley JF, et al. (1979) Brassinolide, a plant growth-promoting steroid isolated from *Brassica Napus* pollen. Nature 281: 216–217.

57. Suzuki M, Nakagawa S, Kamide Y, Kobayashi K, Ohyama K, et al. (2009) Complete blockage of the mevalonate pathway results in male gametophyte lethality. J Exp Bot 60: 2055–2064.

58. Ahumada I, Cairó A, Hemmerlin A, González V, Paterakí I, et al. (2008) Characterisation of the gene family encoding acetoacetyl-CoA thiolase in *Arabidopsis*. Funct Plant Biol 35: 1100–1111.

59. Wentzinger L, Gerber E, Bach TJ, Hartmann MA (2013) Occurrence of two acetoacetyl-coenzyme A thiolases with distinct expression patterns and subcellular localization in tobacco. In: Bach TJ, Rohmer M editors. Isoprenoid synthesis in plants and microorganisms: New concepts and experimental approaches. New York: Springer. pp. 347–365.

60. Brodersen P, Sakvarelidze-Achard L, Schaller H, Khafif M, Schott G, et al. (2012) Isoprenoid biosynthesis is required for miRNA function and affects membrane association of ARGONAUTE 1 in *Arabidopsis*. Proc Natl Acad Sci USA 109: 1778–1783.

61. Shi Z, Ruvkun G (2012) The mevalonate pathway regulates microRNA activity in *Caenorhabditis elegans*. Proc Natl Acad Sci USA 109: 4568–4573.

62. Chappell J, Wolf F, Proulx J, Cuellar R, Saunders C (1995) Is the reaction catalyzed by 3-hydroxy-3-methylglutaryl coenzyme A reductase a rate-limiting step for isoprenoid biosynthesis in plants? Plant Physiol 109: 1337–1343.

63. Hemmerlin A, Gerber E, Feldtrauer JF, Wentzinger L, Hartmann MA, et al. (2004) A review of tobacco BY-2 cells as an excellent system to study the synthesis and function of sterols and other isoprenoids. Lipids 39: 723–735.

64. Merret R, Cirioni J, Bach TJ, Hemmerlin A (2007) A serine involved in actin-dependent subcellular localization of a stress-induced tobacco BY-2 hydroxymethylglutaryl-CoA reductase isoform. FEBS Lett 581: 5295–5299.

65. Chappell J, Vonlanken C, Vögeli U (1991) Elicitor-inducible 3-hydroxy-3-methylglutaryl coenzyme A reductase activity is required for sesquiterpene accumulation in tobacco cell suspension cultures. Plant Physiol 97: 693–698.

66. Sacchettini JC, Poulter CD (1997) Creating isoprenoid diversity. Science 277: 1788–1789.

67. Okada K, Kasahara H, Yamaguchi S, Kawaide H, Kamiya Y, et al. (2008) Genetic evidence for the role of isopentenyl diphosphate isomerases in the mevalonate pathway and plant development in *Arabidopsis*. Plant Cell Physiol 49: 604–616.

68. Nakamura A, Shimada H, Masuda T, Ohta H, Takamiya K (2001) Two distinct isopentenyl diphosphate isomerases in cytosol and plastid are differentially induced by environmental stresses in tobacco. FEBS Lett 506: 61–64.

69. Hemmerlin A, Rivera SB, Erickson HK, Poulter CD (2003) Enzymes encoded by the farnesyl diphosphate synthase gene family in the big sagebrush *Artemisia tridentata* ssp. spiciformis. J Biol Chem 278: 32132–32140.

70. Dudareva N, Klempien A, Muhlemann JK, Kaplan I (2013) Biosynthesis, function and metabolic engineering of plant volatile organic compounds. New Phytol 198: 16–32.

71. Closa M, Vranová E, Bortolotti C, Bigler L, Arró M, et al. (2010) The *Arabidopsis thaliana* FPP synthase isozymes have overlapping and specific functions in isoprenoid biosynthesis, and complete loss of FPP synthase activity causes early developmental arrest. Plant J 63: 512–525.

72. Cunillera N, Arró M, Delourme D, Karst F, Boronat A, et al. (1996) *Arabidopsis thaliana* contains two differentially expressed farnesyl-diphosphate synthase genes. J Biol Chem 271: 7774–7780.

73. Masferrer A, Arró M, Manzano D, Schaller H, Fernández-Busquets X, et al. (2002) Overexpression of *Arabidopsis thaliana* farnesyl diphosphate synthase (FPS1S) in transgenic *Arabidopsis* induces a cell death/senescence-like response and reduced cytokinin levels. Plant J 30: 123–132.

74. Hemmerlin A (2013) Post-translational events and modifications regulating plant enzymes involved in isoprenoid precursor biosynthesis. Plant Sci 203-204: 41–54.

75. Abe I, Rohmer M, Prestwich GD (1993) Enzymatic cyclization of squalene and oxidosqualene to sterols and triterpenes. Chem Rev 93: 2189–2206.

76. Devarenne TP, Shin DH, Back K, Yin SH, Chappell J (1998) Molecular characterization of tobacco squalene synthase and regulation in response to fungal elicitor. Arch Biochem Biophys 349: 205–215.

77. Devarenne TP, Ghosh A, Chappell J (2002) Regulation of squalene synthase, a key enzyme of sterol biosynthesis, in tobacco. Plant Physiol 129: 1095–1106.

78. Lee MH, Jeong JH, Seo JW, Shin CG, Kim YS, et al. (2004) Enhanced triterpene and phytosterol biosynthesis in *Panax ginseng* overexpressing squalene synthase gene. Plant Cell Physiol 45: 976–984.

79. Seo JW, Jeong JH, Shin CG, Lo SC, Han SS, et al. (2005) Overexpression of squalene synthase in *Eleutherococcus senticosus* increases phytosterol and triterpene accumulation. Phytochemistry 66: 869–877.

80. Murray JA, Jones A, Godin C, Traas J (2012) Systems analysis of shoot apical meristem growth and development: integrating hormonal and mechanical signaling. Plant Cell 24: 3907–3919.

81. Nguyen HTM, Neelakadan AK, Quach TN, Valliyodan B, Kumar R, et al. (2013) Molecular characterization of Glycine max squalene synthase genes in seed phytosterol biosynthesis. Plant Physiol Biochem. 73: 23–32.

82. Hemmerlin A, Bach TJ (2000) Farnesol-induced cell death and stimulation of 3-hydroxy-3-methylglutaryl-coenzyme A reductase activity in tobacco cv bright yellow-2 cells. Plant Physiol 123: 1257-1268.

83. Lichtenthaler HK (1999) The 1-deoxy-D-xylulose-5-phosphate pathway of isoprenoid biosynthesis in plants. Annu Rev Plant Physiol Plant Mol Biol 50: 47–65.

84. Orlova I, Nagegowda DA, Kish CM, Gutensohn M, Maeda H, et al. (2009) The small subunit of snapdragon geranyl diphosphate synthase modifies the chain length specificity of tobacco geranylgeranyl diphosphate synthase in planta. Plant Cell 21: 4002–4017.

85. Durek P, Schmidt R, Heazlewood JL, Jones A, MacLean D, et al. (2010) PhosPhAt: the *Arabidopsis thaliana* phosphorylation site database. An update. Nucleic Acids Res 38: D828–D834.

86. Gnad F, Gunawardena J, Mann M (2011) PHOSIDA 2011: the posttranslational modification database. Nucleic Acids Res 39: D253–D260.

87. Huchelmann A, Gastaldo C, Veinante M, Zeng Y, Heintz D, et al. (2014) S-Carvone suppresses cellulase-induced capsidiol production in *Nicotiana tabacum* by interfering with protein isoprenylation. Plant Physiol 164: 935–950.

88. Bouvier-Navé P, Husselstein T, Benveniste P (1998) Two families of sterol methyltransferases are involved in the first and the second methylation steps of plant sterol biosynthesis. Eur J Biochem 256: 88–96.

89. Holmberg N, Harker M, Gibbard CL, Wallace AD, Clayton JC, et al. (2002) Sterol C-24 methyltransferase type 1 controls the flux of carbon into sterol biosynthesis in tobacco seed. Plant Physiol 130: 303–311.

90. Holmberg N, Harker M, Wallace AD, Clayton JC, Gibbard CL, et al. (2003) Co-expression of N-terminal truncated 3-hydroxy-3-methylglutaryl CoA reductase and C24-sterol methyltransferase type 1 in transgenic tobacco enhances carbon flux towards end-product sterols. Plant J 36: 12–20.

91. Fonteneau P, Hartmann-Bouillon MA, Benveniste P (1977) A 24-methylene lophenol C-28 methyltransferase from suspension cultures of bramble cells. Plant Sci Lett 10: 147–155.

92. Bouvier-Navé P, Husselstein T, Desprez T, Benveniste P (1997) Identification of cDNAs encoding sterol methyl-transferases involved in the second methylation step of plant sterol biosynthesis. Eur J Biochem 246: 518–529.

93. Schaller H, Bouvier-Navé P, Benveniste P (1998) Overexpression of an Arabidopsis cDNA encoding a sterol-C24^1-methyltransferase in tobacco modifies the ratio of 24-methyl cholesterol to sitosterol and is associated with growth reduction. Plant Physiol 118: 461–469.

94. Sitbon F, Jonsson L (2001) Sterol composition and growth of transgenic tobacco plants expressing type-1 and type-2 sterol methyltransferases. Planta 212: 568–572.

95. Li JM, Nagpal P, Vitart V, McMorris TC, Chory J (1996) A role for brassinosteroids in light-dependent development of *Arabidopsis*. Science 272: 398–401.

96. Nomura T, Kushiro T, Yokota T, Kamiya Y, Bishop GJ, et al. (2005) The last reaction producing brassinolide is catalyzed by cytochrome P-450s, CYP85A3 in tomato and CYP85A2 in *Arabidopsis*. J Biol Chem 280: 17873–17879.

97. Pérez-España VH, Sánchez-León N, Vielle-Calzada JP (2011) *CYP85A1* is required for the initiation of female gametogenesis in *Arabidopsis thaliana*. Plant Signal Behav 6: 321–326.

12

Identification of Novel SHOX Target Genes in the Developing Limb Using a Transgenic Mouse Model

Katja U. Beiser[1], Anne Glaser[1], Kerstin Kleinschmidt[2], Isabell Scholl[1], Ralph Röth[1], Li Li[3], Norbert Gretz[3], Gunhild Mechtersheimer[4], Marcel Karperien[5], Antonio Marchini[1,6], Wiltrud Richter[2], Gudrun A. Rappold[1]*

1 Department of Human Molecular Genetics, Heidelberg University Hospital, Heidelberg, Germany, 2 Division of Experimental Orthopaedics, Orthopaedic University Hospital, Heidelberg, Germany, 3 Medical Research Center (ZMF), Medical Faculty Mannheim at Heidelberg University, Mannheim, Germany, 4 Institute of Pathology, Heidelberg University Hospital, Heidelberg, Germany, 5 Department of Developmental Bioengineering, University of Twente, Enschede, The Netherlands, 6 German Cancer Research Center (DKFZ), Heidelberg, Germany

Abstract

Deficiency of the human short stature homeobox-containing gene (SHOX) has been identified in several disorders characterized by reduced height and skeletal anomalies such as Turner syndrome, Léri-Weill dyschondrosteosis and Langer mesomelic dysplasia as well as isolated short stature. SHOX acts as a transcription factor during limb development and is expressed in chondrocytes of the growth plates. Although highly conserved in vertebrates, rodents lack a SHOX orthologue. This offers the unique opportunity to analyze the effects of human SHOX expression in transgenic mice. We have generated a mouse expressing the human SHOXa cDNA under the control of a murine Col2a1 promoter and enhancer (Tg(Col2a1-SHOX)). SHOX and marker gene expression as well as skeletal phenotypes were characterized in two transgenic lines. No significant skeletal anomalies were found in transgenic compared to wildtype mice. Quantitative and in situ hybridization analyses revealed that Tg(Col2a1-SHOX), however, affected extracellular matrix gene expression during early limb development, suggesting a role for SHOX in growth plate assembly and extracellular matrix composition during long bone development. For instance, we could show that the connective tissue growth factor gene Ctgf, a gene involved in chondrogenic and angiogenic differentiation, is transcriptionally regulated by SHOX in transgenic mice. This finding was confirmed in human NHDF and U2OS cells and chicken micromass culture, demonstrating the value of the SHOX-transgenic mouse for the characterization of SHOX-dependent genes and pathways in early limb development.

Editor: Andre van Wijnen, University of Massachusetts Medical, United States of America

Funding: This work was supported by the Deutsche Forschungsgemeinschaft (grant number RA 380/13-2; URL: www.dfg.de) and the Baden-Wurttemberg Foundation (grant number 1.1601.07; URL: www.bwstiftung.de). The funders had no role in study design, data collection and analysis, decision to publish, or preparation of the manuscript.

Competing Interests: The authors have declared that no competing interests exist.

* E-mail: gudrun.rappold@med.uni-heidelberg.de

Introduction

Height is a complex trait defined by multiple biological and environmental factors that are involved in bone formation and growth. The development of the long bones is characterized by coordinated gene expression from early embryonic stages until adulthood. Disturbances in bone development can affect growth and lead to clinical consequences. The homeodomain transcription factor SHOX is involved in different human short stature syndromes (Turner syndrome, Léri-Weill dyschondrosteosis LWD [MIM 127300] and Langer mesomelic dysplasia [MIM 249700]) and isolated (idiopathic) short stature [MIM 300582] [1,2,3,4,5,6,7]. Mutations and deletions of the SHOX gene and its enhancers have been identified as etiologic for the short stature and skeletal anomalies in these disorders [8,9,10,11,12,13]. Comprehensive case studies have shown that SHOX defects have also been identified in the more common nonsyndromic (isolated) forms of short stature with a prevalence of 5–17% in geographically different populations [6,12,14]. An overdosage of SHOX as in patients with Triple-X or Klinefelter syndrome results in tall stature [15].

Phenotypic characteristics are variable in SHOX-deficient patients and include disproportional (mesomelic) short stature, shortening of the forearms as well as Madelung deformity, a skeletal abnormality of the wrist characteristic for LWD [4,16]. Histopathological evaluation of LWD growth plates revealed a variable disruption of the architecture and an irregular chondrocyte stacking [17], and the SHOX protein was mainly detected in prehypertrophic and hypertrophic chondrocytes of fetal and childhood growth plates by immunohistochemistry [18,19,20]. Since clinical studies have demonstrated that growth hormone (somatropin) therapy before the onset of puberty effectively ameliorates the short stature in SHOX-deficient patients [21], a somatropin-based therapy is proposed in affected individuals.

Despite the high clinical relevance of SHOX mutations, surprisingly little is known about the molecular mechanisms that are governed by SHOX deficiency. This is mainly due to the limited availability of patient tissue samples (growth plate material) and the lack of cellular systems that reliably express SHOX endogenously at sufficiently high levels [22]. Mice do not have a SHOX orthologue, thus a knock-out model cannot be generated. Since the vast majority of genes that govern early developmental processes are highly conserved between human and mouse [23], characterization of genes that are divergent between the two species has not attracted much attention. SHOX has been shown to act as both a transcriptional activator and repressor of target genes [8,20,24,25,26]. Functional studies have also shown that overexpression of the SHOX protein can induce growth arrest and apoptosis, suggesting that SHOX may regulate chondrocyte hypertrophy by inducing apoptosis [19].

The clinical relevance of *SHOX* in short stature prompted us to generate a transgenic mouse to study the effect of the human *SHOX* gene during early chondrogenesis. While the phenotypic features are sparse in these animals, we demonstrate that *Ctgf*, among other genes, is regulated by SHOX in transgenic mice as well as in human and chicken cell cultures. In addition, microarray and molecular analyses revealed that the *SHOX*-transgene can effectively regulate genes important in early processes during limb formation.

Materials and Methods

Animals and genotyping

All animal experiments were conducted according to German animal protection laws and approved by the regional board of Baden Württemberg (permission No. 35–9185.81/G–64/05 and A-30/09). To express *SHOX* (genomic coordinates according to GRCh37: X:585,078-620,145) in mouse limbs, the *SHOXa* cDNA (CCDS14107.1) was cloned into the murine expression vector p1757 including the rat *Col2a1* promoter (1 kb), a Globin splicing sequence (640 bp) and the *Col2a1* enhancer (1.4 kb) [27,28,29] and a SV40 polyadenylation signal from pGL3 Basic (Promega). The construct (p1757 SHOX) was linearized with *AgeI* and microinjected into pronuclei of fertilized C57BL/6 x DBA/2 hybrid eggs to generated transgenic mice. Founders were identified by extraction of genomic DNA from tails followed by PCR using primers SHOX1 and XHO_REV (1-409 of the *SHOXa* cDNA) and SHOX_ECORI_FOR and LUMI-OSHOXCTER_REV (242-TGA of the *SHOXa* cDNA). Southern Blot was carried out according to standard procedures with a probe spanning nucleotides 1-409 to confirm the integration of the transgene at a single locus. Primer sequences are included in the Table S2 in File S1.

Limb preparation and RNA samples

Limbs of wildtype and transgenic littermates at E10.5-E14.5 were dissected and frozen in liquid nitrogen. RNA was isolated using the RNeasy Kit (Qiagen), following homogenization using a PT1300 D polytron (Kinematica). DNA was hydrolyzed using the RNAse-free DNAse Kit (Qiagen). RNA yield was measured using a NanoDrop 2000 spectrophotometer (Nanodrop technologies) and quality-checked on agarose gels. For microarray analysis, RNA from 2-4 E12.5 wildtype and transgenic littermates was pooled and the quality-checked on a 2100 Bioanalyzer (Agilent).

In vitro transcription and quantitative RT-PCR

In vitro transcription of 1 µg RNA was performed using the Superscript II First Strand Synthesis System for RT-PCR (Invitrogen). qRT-PCR was carried out using the Applied Biosystems 7500 Real-Time PCR System and Absolute SYBR Green ROX Mix (Abgene). Each sample and the housekeeping genes were run in duplicates. Relative mRNA levels were calculated according to the delta-delta Ct method [30] by normalization to mRNA expression of the housekeeping genes *Sdha* and *Adam9*. Primer sequences are included in Table S2 in File S1.

µCT imaging and analysis

Transgenic and wildtype littermates were anesthetized by i.p. injection of Ketamin (75 mg/kg) and Domitor (1 mg/kg) at the age of 4 (P28–30), 12 (P84–86) and 24 weeks (P168–170). Microcomputed tomography analyses on tibiae and femora of narcotized mice was performed using a Skyscan 1076 *in vivo* scanner (Skyscan, Antwerp, Belgium) at a resolution of 17.7 µm/pixel with an 0.5 mm aluminium filter. A source voltage of 48 kV, current of 200 µA, exposure time of 320 ms and a rotation step of 0.6 degree were used. Reconstructions (NRecon, Skyscan, Antwerp, Belgium) were made using an under-sampling factor of 1, a threshold for defect pixel mask of 30%, a beam hardening correction factor of 100%, minimum of 0.0061 and maximum of 0.0674 for CS to image conversion. Length of long bones and cortical thickness were measured manually using ruler tool function (CTAn, Skyscan, Antwerp, Belgium). Equal anatomical bone markers were used for reproducibility. For quantitative analysis of bone volume (BV) and bone mineral density (BMD) a region of interest was chosen that included the total bone and thresholds of 68-255 were used for binarisation. For BMD measurement mice were euthanized at the age of 24 weeks, legs were prepared and scanned again in water. Phantoms with known densities of 0.25 and 0.75 g/cm^3 and water were scanned for houndsfield unit calibration. Statistical analyses were carried out using Student's t-test and GraphPad Prism 5 software.

Microarray analysis

Gene expression profiling was performed using GeneChip Mouse Genome 430.2 from Affymetrix (Santa Clara, CA, USA). Duplicate Arrays were done for each genotype (transgene or wildtype). cDNA, cRNA synthesis and hybridization to arrays were performed according to the recommendations of the manufacturer. Microarray data were submitted to NCBI GEO, sample number GSE47902. Microarray data was analyzed based on ANOVA using the software package JMP Genomics, version 4.0 (SAS Institute, Cary, NC, USA). Values of perfect-matches were log transformed, quantile normalized and fitted with log-linear mixed models, with probe_ID and genotype considered to be constant and the sample ID random. Custom CDF version 13 with Entrez gene based gene/transcript definitions (http://brainarray.mbni.med.umich.edu/Brainarray/Database/CustomCDF/genomic_curated_CDF.asp) different from the original Affymetrix probe set definitions were used to annotate the arrays. Gene Set Enrichment analysis (GSEA 2.0) was applied to reveal biological pathways modulated between sample groups. Genes were ranked according to the expression change between genotypes. All Gene Ontology terms were examined using 1000 rounds of permutation of gene sets. Pathways with absolute NES (normalized enrichment score) more than 1.7 and NP (normalized p-value) <0.02 were considered to be differentially modulated.

The nCounter system assay

Assays were performed using 100 ng of total RNA plus reporter and capture probes for 10 genes (nanostring codeset). After overnight hybridization, sample purification and nCounter digital reading, counts for each RNA species were extracted and analyzed using a home-made Excel macro. Codesets include positive controls (spiked RNA at various concentrations) as well as negative controls (alien probes for background calculation). Background correction consisted of the subtraction of negative control average plus two SD from the raw counts. To avoid negative values, signals lower than one after correction were thresholded to one. The positive controls were used as a quality assessment. For each sample, the ratio between sample-related positive control average and the smallest positive control average was accepted when lower

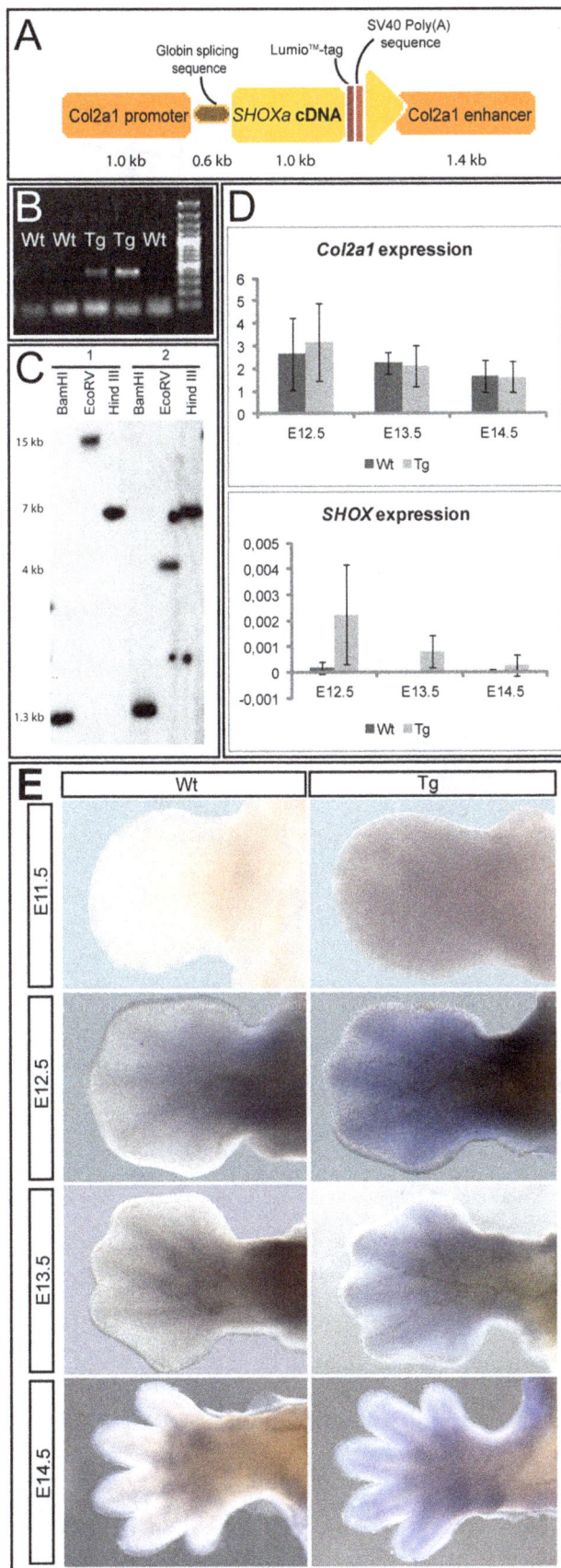

Figure 1. Generation and expression analysis of *SHOX*-transgenic mice. (A): The *SHOXa* cDNA was tagged with a Lumio and SV40 Poly(A) sequence and cloned under the control of a murine *Col2a1* promotor/enhancer expression cassette. (B): Genotyping was performed using specific primers spanning the first 409 nucleotides of the *SHOXa* cDNA. No PCR product was detected in wildtype animals. (C):-Southern Blot analysis of the two transgenic lines (1 and 2) used for our investigations. Genomic DNA was digested with *BamHI*, *EcoRV* and *Hind III*. *BamHI* digestion results in a 1.3 kb fragment that corresponds to the Lumio/SV40-tagged *SHOX* cDNA, which was flanked by *BamHI* sites. The presence of only one signal per lane indicates a single integration site of the transgene. (D): Relative quantitative expression of *Col2a1* and *SHOXa* transcripts in limbs of wildtype and transgenic littermates (N = 5–8 per litter) at E12.5, E13.5 and E14.5. The expression of the transgene corresponds to the expression dynamics of *Col2a1*. SHOX levels are generally low with highest expression at E12.5. Values are variable among individual animals as indicated by the standard deviation (SD). (E): WISH of wildtype (Wt) and transgenic (Tg) embryonic limbs from E11.5-E14.5 (N = 20 for each stage). The transgene is weakly expressed in the developing limb at E11.5 and becomes defined around the cartilaginous anlagen at E12.5. From E13.5 onwards, the expression is mainly seen in the mesenchyme around the developing cartilage and in the perichondrium and decreases during later stages.

than 3. To select adequate normalization genes from series of candidates included in the CodeSet, the geNorm method (5) was implemented. Therefore, the geometric mean of the selected normalization genes according to geNorm was calculated and used as normalization factor. Normalized values were then compared between samples. Probe sequences are included in Table S2 in File S1.

In situ hybridization

Whole-mount *in situ* hybridization using embryos fixed in 4% paraformaldehyde was performed according to standard procedures. Section *in situ* hybridisation was performed on 12 μm paraffin sections using standard protocols. Antisense riboprobe for *Ctgf* was cloned using the pSTBlue-1 AccepTor vector Kit (Novagen) with the primers Ctgf_ISH_FOR: AAA TGC TGC GAG GAG TGG GTG and Ctgf_ISH_REV: GTG CGT TCT GGC ACT GTG CGC. Antisense riboprobe for SHOX was generated from a *Bam/XhoI* fragment of pBSK SHOX, *Shox2* riboprobe was used as reported [31]. Templates for antisense *in vitro* transcription were digested and digoxigenin-labelled antisense RNA was synthesized using MEGAscript ® Kit (Ambion) as follows: SHOX: *KpnI*/Sp6; Ctgf: *BamHI*/Sp6; Ihh: *XbaI*/T7; Col10a1: *XhoI*/T3; Col2a1: *EcoRI*/T7; Fgfr3: *NdeI,*/T7; Shh: *HindIII*/T3; Runx2: *SpeI*/T7; Shox2: *SacI*/T7; Ogn: *XhoI*/T7.

Cell culture, transfections and luciferase assays

Cells were cultivated and transfections as well as reporter gene assays were carried out as reported before [26]. Primers used for the cloning of the reporter construct are included in Table S2 in File S1.

Electrophoretic Mobility Shift Assays (EMSA)

EMSA were carried out as described [10] using the probes sequences included in the Table S2 in File S1.

Immunohistochemistry

Immunohistochemistry was performed on growth plate sections from a pubertal 12 years old boy (tibial growth plate) as described [19] using anti SHOX- and anti-CTGF (clone L20, Santa Cruz) antibodies at the dilution of 1:25 and 1:100, respectively.

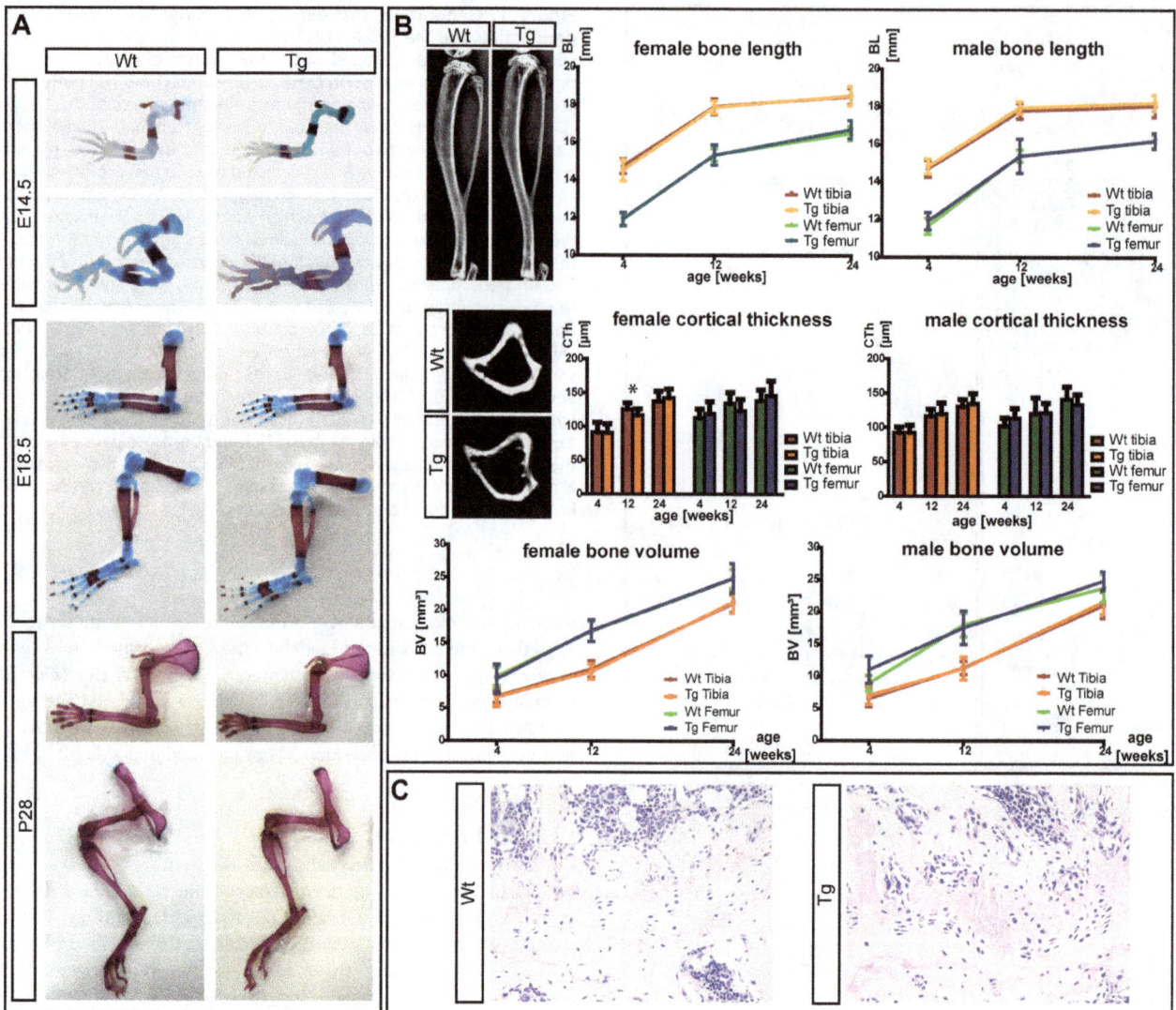

Figure 2. Analysis of postnatal bone parameters of *Col2a1-SHOX*-transgenic mice. (A): Alcian Blue/Alizarin Red S staining at different developmental (E14.5, E18.5) and postnatal (P28) stages does not reveal apparent differences between transgenic and wildtype skeletal elements. (B): Postnatal *in vivo* time-course analysis of bone growth in 65 animals of two transgenic lines by µ-CT analysis. Tibiae and femora of wildtype and *Tg(Col2a1-SHOX)* littermates at the age of 4, 12 and 24 weeks were scanned, female and male individuals were evaluated separately. Total bone length, cortical bone thickness and bone volume do not show significant differences between wildtype and transgenic females or males. Some transgenic animals presented longer bones and weaker structures of the cortical bone in the subcartilaginous region (indicated in the µ-CT images). Other micromorphological parameters (bone mineral density (BMD), trabecular volume and thickness) showed no significant differences. Statistical analyses were performed using student's t-test. (C): hematoxilin and eosin (H&E) stainings of the growth plate in wildtype and transgenic tibiae. Consistent differences between wildtype and *Tg(Col2a1-SHOX)* adult growth plates (24 weeks of age) did not exist (N = 8), but some transgenic tibiae showed a buckling, and the columns of chondrocytes became shorter and were not strictly oriented in a parallel assembly compared to the wildtype (right image).

Histology

For histological examination of growth plates, femora and tibiae of wildtype and transgenic mice (24 weeks of age) were fixed in 4% formalin and decalcified in 10% EDTA. The femora and tibiae were then bisected in the middle, and paraffin embedded. Subsequently, paraffin sections were cut at 4 µm intervals in the plane of the physis. The sections were stained with hematoxylin and eosin (H&E), periodic acid-Schiff (PAS) and Masson's trichrome (MT) by standard protocols.

Results

Generation and expression studies of *Col2a1-SHOX*-transgenic mice

To generate transgenic mice expressing the human *SHOX* gene, the *SHOXa* coding sequence was cloned into a murine transgene expression vector harbouring the rat *Collagen type II (Col2a1)* promoter and enhancer sequence (Fig. 1A). This system was previously used to drive the expression of transgenic constructs in proliferating chondrocytes [27,28,29]. Transgenic founders were identified by the presence of the construct *Tg(Col2a1-SHOX)* using

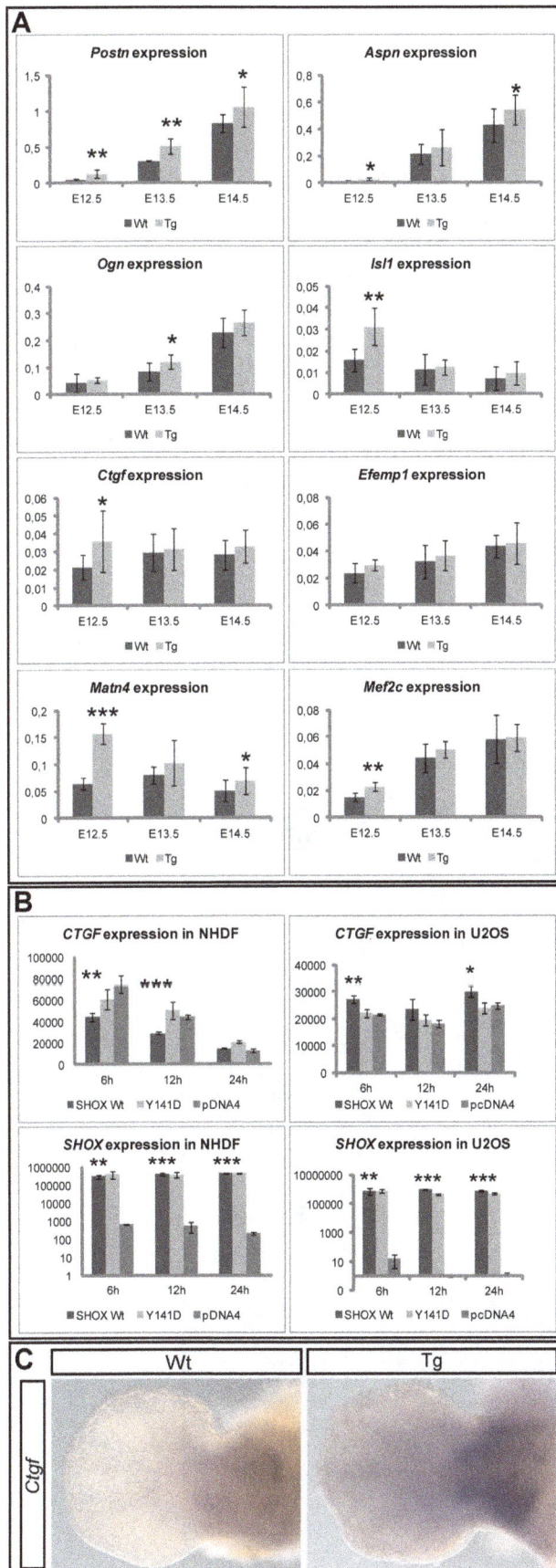

Figure 3. Regulated genes in transgenic mice and validation of Ctgf as a target. (A): qRT-PCR using limb RNA (E12.5-E14.5) from wildtype (Wt) and transgenic littermates (Tg) (N = 8–10 for each stage). Measurements were carried out individually, in duplicates, and normalized to *Adam9* and *Sdha*. Relative normalized values are presented on the y-axis. Significances are indicated in each diagram by asterisks (*: $p \leq 0.05$, **: $p \leq 0.01$, ***: $p \leq 0.001$). Variations are indicated by the standard deviation (SD). In 7/8 candidates an upregulation was confirmed as significant in at least one embryonic stage. (B): nCounter analysis of *CTGF* and *SHOX* expression in NHDF and U2OS cells after transient transfections of *SHOX* and *p.Y141D*. *CTGF* is significantly downregulated in NHDF cells, whereas it is significantly upregulated in U2OS cells. Values on y-axis represent absolute counts of mRNA, normalized to *ADAM9*, *HPRT1* and *SDHA*. Significancies are indicated by asterisks. (C): *In situ* hybridization using a *Ctgf* antisense riboprobe on embryonic limbs from wildtype and *SHOX*-transgenic littermates (N = 8) at stage E12.5. In transgenic embryos, enhanced and distalized expression of *Ctgf* was detected in the middle part of the developing limbs.

PCR and were mated with C57Bl/6 mice (Fig. 1B). Two independent heterozygous transgenic lines were investigated in more detail. Southern blot analysis using genomic DNA from animals of the two transgenic lines showed a single integration locus of the transgenic DNA (Fig. 1C). All transgenic animals were viable and fertile, and the *Tg(Col2a1-SHOX)* allele was transmitted according to Mendelian ratios.

Transgenic expression was analyzed by quantitative RT-PCR and whole mount *in situ* hybridization (WISH), demonstrating that *Tg(Col2a1-SHOX)* was expressed in the developing limbs (Fig. 1D–E). The expression started from E11.5 onwards (Fig. 1E) with a variable expression level among different transgenic mice. Following the expression dynamics of the endogenous *Col2a1*, *Tg(Col2a1-SHOX)* quantities were highest at around E12.5 and gradually decreased during later stages of embryonic development (Fig. 1D). The expression pattern of *Tg(Col2a1-SHOX)* at E12.5 resembled *Col2a1* expression which is transcribed at high levels in chondrogenic tissues [32] (Fig. 1E). During later embryonic stages (e.g. E14.5), transgenic expression was confined to the region around the developing cartilage including the perichondrium (Fig. 1E). Thus, the detected expression pattern of the *SHOX*-transgene was comparable to the endogenous *SHOX* expression domains reported in the developing limbs of human and chick embryos [33,34].

Analysis of skeletal parameters in Col2a1-SHOX-transgenic mice

Transgenic animals showed no obvious difference compared to their wildtype littermates. To investigate whether the *Col2a1-SHOX*-transgene has an effect on embryonic cartilage and bone development, E14.5 and E18.5 embryos were stained with Alcian Blue/Alizarin Red S (Fig. 2A). The transgenic embryos were indistinguishable from wildtype littermates at these stages, indicating that bone formation was grossly normal. As some phenotypic features in patients with SHOX deficiency (e.g. Madelung deformity) are sometimes not detectable before the onset of puberty [4], we also investigated the skeletal elements at postnatal stage P28. Again, no striking phenotype was detected in the transgenic animals (Fig. 2A).

To determine if bone length is increased in transgenic animals, we measured the postnatal bone length in 65 animals of two transgenic lines *in vivo* using micro-computed tomography (µ-CT), which enabled the analysis of different bone-specific parameters simultaneously (Fig. 2B). Tibiae and femora of anaesthetized wildtype and *SHOX*-transgenic mice were scanned *in vivo* in a time-

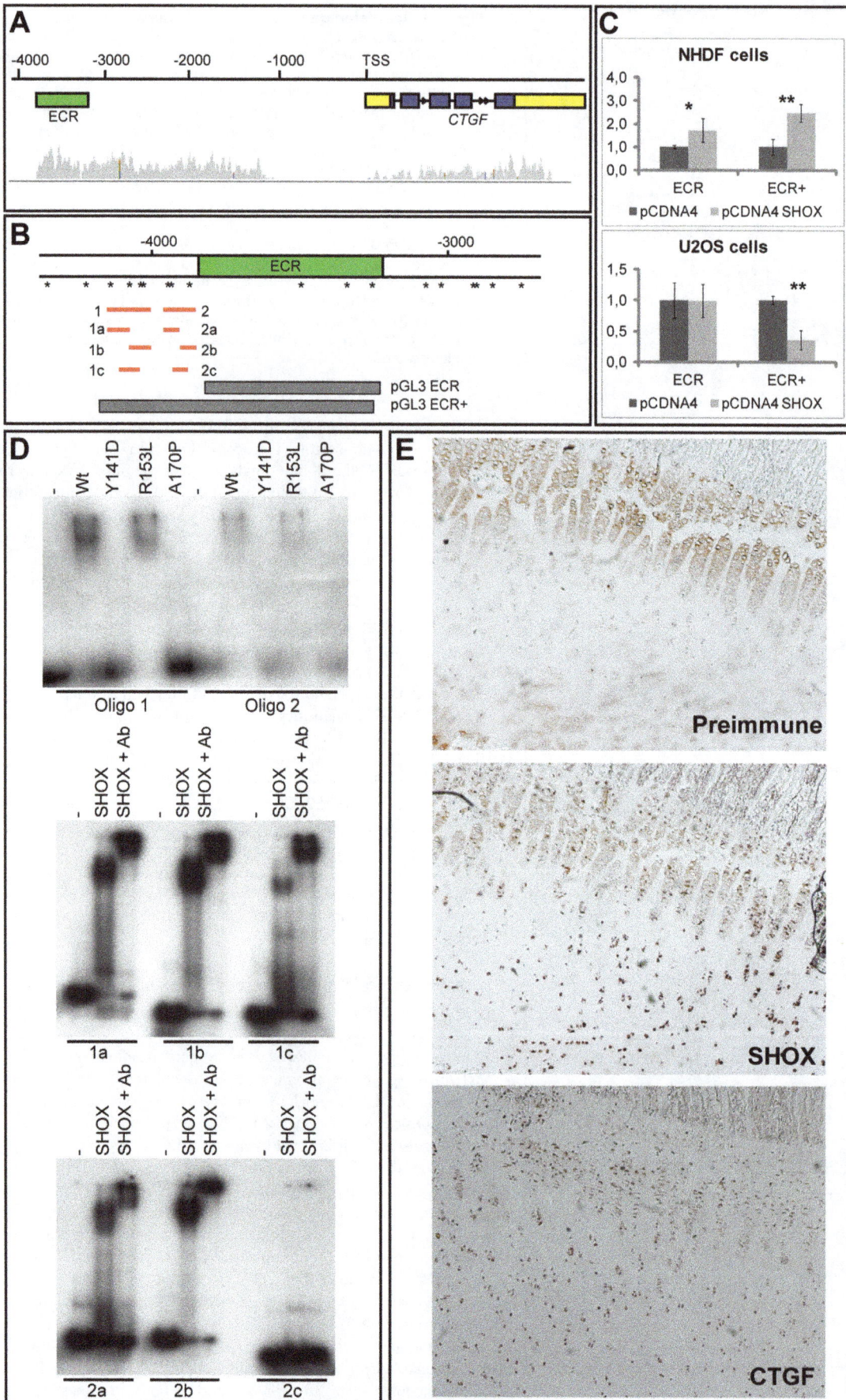

Figure 4. Analysis of *CTGF* as a direct transcriptional target of SHOX. (A): Genomic structure of the human *CTGF* region. ChIP-Seq analysis in ChMM cultures revealed an accumulation of Shox binding in the *Ctgf* promoter region (grey peaks), especially in a region 3–4 kb from the transcriptional start site (TSS) where an evolutionary conserved sequence (ECR) of 597 bp (human chr6:132317086-132318077) was identified (green bar). (B): Location of the pGL3 ECR and pGL3 ECR+ reporter constructs (grey bars) within the *CTGF* upstream region. The ECR+ construct encompasses the ECR and an upstream region including ATTA/TAAT motifs and palindromes. SHOX binding motifs (ATTA/TAAT sites and palindromes) in the *CTGF* 5′ region around the ECR are indicated by asterisks. Red bars represent the location of the generated oligonucleotides for EMSA. (C): Luciferase reporter gene assays in NHDF and U2OS cells. pcDNA4/TO *SHOX* was cotransfected with a luciferase reporter vector harbouring either the ECR or the ECR+ sequence. Transfections and measurements were carried out in triplicates. A significant activation in the luciferase activity was observed 24 h after *SHOX* transfection in NHDF cells using both reporter constructs (1.7-fold/2.5-fold with $p = 0.02/0.007$ for ECR/ECR+). In U2OS cells, an alteration was not observed for the ECR reporter, but a significant reduction was demonstrated for the ECR+ reporter construct (1.0-fold/2.8-fold with $p = 0.1/0.003$ for ECR/ECR+). (D): EMSA. The SHOX wildtype (Wt) and the mutant p.R153L proteins bind to oligonucleotides 1 and 2, whereas the defective proteins p.Y141D and p.A170P cannot. All fragments of oligonucleotides 1 and 2 containing an ATTA/TAAT site are sensitive to SHOX binding (1a–c, 2a–b). The fragment lacking this motif does not bind (oligonucleotide 2c). Using the SHOX-3 antibody (Ab), we demonstrate that the binding is SHOX-specific. (E): Immunohistochemistry performed on pubertal tibial growth plates. Staining was performed using preimmune serum as a negative control, SHOX antibody [19] and a CTGF-specific antibody. Both the SHOX and CTGF proteins were detected in growth plate chondrocytes.

course until 24 weeks of age. Data from female and male mice were analyzed separately to eliminate gender-specific effects. Even though we observed increases in bone length in some transgenic animals, these were not significant (Student's t-test). Significant differences in bone volume and bone mineral density were not found either, indicating that long bone development was largely normal upon *Tg(Col2a1-SHOX)* expression. A statistically significant decrease of the cortical bone thickness (CTh) was identified in 12 weeks old female transgenic mice, but not in males or at any other time points. Since the assessment of the growth plate in patients with LWD previously demonstrated a normal to disorganized morphology including abnormal chondrocyte stacking [17], we analyzed the femoral and tibial growth plate morphology of transgenic and wildtype mice (24 weeks of age) using hematoxylin and eosin (H&E), periodic acid-Schiff (PAS) and Masson's trichrome (MT) stainings. In some cases, a buckling of the growth plate was observed, and the columns of chondrocytes became shorter and were not strictly oriented in a parallel assembly (Fig 2C). However, these alterations were not consistently found in all transgenic samples.

Target gene expression and microarray analyses in *Col2a1-SHOX*-transgenic mice

We performed expression analysis of cartilage- and bone-specific markers from E11.5 to E14.5 using whole mount *in situ* hybridization (WISH) to identify whether limb specific markers show aberrant expression in the *Tg(Col2-SHOX)* embryos (Fig. S1A). We found that early genes such as *Shh* were not altered in the transgenic embryos, indicating that limb initiation and limb bud outgrowth were grossly normal. The expression of *Col2a1*, *Shox2*, *Runx2*, *Ihh* as well as *Col10a1* was similar in transgenic and wildtype embryos, suggesting that chondrocyte proliferation and maturation were largely unaffected. The expression levels of these marker genes were also quantified by qRT-PCR, but no significant differences in the amount of the respective transcripts could be detected.

A regulatory effect of *SHOX* on *FGFR3*, *AGC1* and *NPPB* (*BNP*) was recently reported using human cell lines [20,24,26]. We therefore analyzed whether the *SHOX*-transgene was able to alter the expression of the mouse *Fgfr3*, *Agc1* and *Nppb* genes. By using reversely transcribed RNA from E12.5-E14.5 wildtype and transgenic limbs, we detected no effect on *Fgfr3*, but an increasing effect on *Agc1* (in all three tested stages) and *Nppb* (at E13.5) (Fig. S1B). The finding that *Fgfr3* did not respond to *SHOX*-transgenic expression in mouse is consistent with the fact that the relevant SHOX-regulatory elements in the human *FGFR3* promoter do not exist in mouse, while they are present in *Agc1* and *Nppb*.

The altered expression of two known SHOX target genes in transgenic mice prompted us to perform microarray analyses of wildtype and transgenic limb RNA. Prior to hybridization, *Tg(Col2a1-SHOX)* expression was confirmed by qRT-PCR and pooled whole limb RNA of either E12.5 wildtype or transgenic littermates were hybridized to microarrays. Selection of differentially regulated genes was carried out using a significant change of expression in both experiments ($\log 2f > 0.2$ or < -0.2 and $p < 0.05$). According to these criteria, 189 genes (83%) were upregulated and 40 genes (17%) were downregulated, suggesting that the *Col2a1*-driven *SHOX*-transgene mainly exerted activating effects. A categorization of differentially expressed genes was performed by gene ontology-based pathway analysis and the most significantly regulated genes were identified in biological pathways associated with either the extracellular matrix or skeletal muscle. The eight most significantly upregulated candidate genes (*Postn*, *Aspn*, *Ogn*, *Isl1*, *Ctgf*, *Efemp1*, *Matn4*, *Mef2c*) that were either known to be involved in limb development, extracellular matrix or skeletal muscle pathways are summarized on Table S1 in File S1. qRT-PCR of the candidate genes was carried out using RNA from wildtype and transgenic limbs of stages E12.5-E14.5. An increase in expression of all candidate target genes including *CTGF* was detected in the transgenic embryos (Fig. 3A).

To further confirm the regulatory effects of SHOX on these genes, we carried out transient transfections of wildtype *SHOX* and a *SHOX* mutant (Y141D) in human U2OS and NHDF cell lines which have been previously used for the characterization of target genes [20,24,26]. The p.Y141D variant was identified in two short stature patients and functionally characterized as a defective SHOX protein [10]. For subsequent expression analysis, we applied the nCounter technology that allows direct RNA quantification without reverse transcription into cDNA, resulting in sensitive and reliable detection of mRNA expressed at low abundance. Since the effect of SHOX on validated genes differed between U2OS and NHDF cells, we concluded that the SHOX transcriptional regulation is strongly cell type-dependent (Fig. S2). Most strongly and significantly regulated was the chondrogenic matrix gene *CTGF*, which showed a reduced expression upon *SHOX*-transfections in NHDF and an increased expression in U2OS cells (Fig. 3B). *In situ* hybridization of *Ctgf* on wildtype and *Tg(Col2a1-SHOX)* embryonic limbs showed an increased and a more distal expression in the transgenic limbs (Fig. 3C).

The connective tissue growth factor gene *CTGF* represents a target of SHOX transactivating functions

Analyses of the *SHOX*-transgenic mouse and human cell lines overexpressing *SHOX* have demonstrated a regulatory effect of SHOX on *Ctgf/CTGF* expression. In addition, previous ChIP-Seq data on chicken micromass cultures transduced with RCAS-Shox [26] suggest *Ctgf* as a putative cell target of SHOX with several binding sites identified within the 5′ region of the gene. Computational analyses of the human *CTGF* upstream region (5 kb) identified more than 40 binding motifs of the ATTA/TAAT type which have been reported to be the target sites of SHOX [8,26]. Of these, eight motifs were arranged as palindromes. Furthermore, the region with the highest ChIP-Seq reads in the chicken *Ctgf* locus includes an ECR (evolutionary conserved region) that is also present in the human *CTGF* upstream sequence (Fig. 4A). To demonstrate that *CTGF* could be directly targeted by SHOX, we performed luciferase reporter gene assays in NHDF and U2OS cells. We used two constructs: the smaller one included the human ECR sequence (ECR) and the larger construct included the ECR as well as putative SHOX binding sites (ECR+) (Fig. 4B). As shown in Fig. 4C, significant regulatory effects of SHOX on the ECR+ reporter constructs were observed in both NHDF and U2OS cell lines, whereas for the ECR reporter construct a significant regulation could only be demonstrated in NHDF cells. To confirm a direct binding of SHOX to the *CTGF* upstream region, electrophoretic mobility shift assays (EMSA) were carried out using two oligonucleotide sequences of the ECR+ construct (Oligo 1 and Oligo 2) encompassing the ATTA/TAAT motifs (Fig. 4B). As controls, mutant SHOX proteins (p.Y141D, p.R153L and p.A170P; previously detected in patients with short stature) were used [10]. While the wildtype SHOX and p.R153L proteins bound to the tested sequences, p.Y141D and p.A170P did not (Fig. 4D). Further subdivision of oligonucleotides 1 and 2 narrowed down SHOX binding to all fragments where ATTA/TAAT sites were present (Fig. 4B and 4D). To demonstrate physiological relevance of these data, immunohistological staining on sections from human pubertal growth plate specimen were carried out. Using CTGF and SHOX specific antibodies, coexpression was detected in hypertrophic chondrocytes (Fig. 4E).

Discussion

Generation and expression studies of *Col2a1-SHOX*-transgenic mice

For a small number of human protein-coding genes, a mouse ortholog does not exist [35]. One approach to learn more about the biology of these human genes is to introduce them into mice. We have generated transgenic mice that express the human *SHOX* cDNA in embryonic limbs under the control of the murine *Col2a1* promoter/enhancer. Expression of the *SHOX*-transgene was detected between E12.5 and E14.5. Compared to *Col2a1*, a highly abundant major structural component of the extracellular matrix, the expression of the transcription factor *SHOX* was very weak and differed between animals. The generation of a transgenic mouse using a different promoter and/or enhancer may eventually yield in higher *SHOX* expression levels. However, low expression levels are characteristic for *SHOX* and have been found in all tissues and cell lines tested [22], suggesting that SHOX functions do not rely on high mRNA or protein abundance in the cell.

Analysis of skeletal parameters

Phenotypic analyses of the developing limbs in transgenic mice did not reveal significant differences compared to wildtype (with the exception of cortical thickness in female tibiae at 12 weeks and almost significant differences in female femora). Thus, there may be gender-specific effects in the transgenic mice during postnatal growth, however, to address this question, more detailed experiments would be necessary. Phenotypic clinical features have been previously assessed in patients with isolated SHOX

deficiency and LWD [6], but not much data on cortical bone structures, bone volume or mineral density is available. Patients with Turner syndrome (45,X) suffer from a high fracture risk and have reduced cortical bone structures and bone mineral density [36], but whether this is due to reduced *SHOX* expression is not known. Disorganization of the growth plate has been noted in some of our *SHOX*-transgenic mice, but is not a consistent feature. Disturbed growth plate morphology has been described in patients with LWD [17], but no data is available on patients with additional *SHOX* copies.

Gene expression and microarray analyses

To determine if the critical stages in endochondral ossification were altered in the transgenic mice, we carried out expression analysis of embryonic limb marker genes and could demonstrate that expression of these genes remained intact. A key question also concerned the extent to which the human gene is correctly read by the mouse transcriptional machinery. We therefore tested expression of all three known SHOX target genes [20,24,26] and obtained elevated mean expression levels for *Agc1* and *Nbbp* as expected, probably due to the conserved SHOX-sensitive binding sites in the *Agc1* and *Nppb* enhancer and promoter regions, while the human SHOX-sensitive binding sites in the *Fgfr3* promoter do not exist in mouse.

To further search for effects of the *SHOX*-transgene, we carried out microarray analysis and identified many regulated genes belonging to the extracellular matrix and skeletal muscle pathways. It is interesting that several of these genes, including *Postn* and *Matn4*, have been previously also identified as targets in *Shox2*-deficient mice and thus may represent targets for both SHOX and Shox2 [37]. The mouse Shox2 protein is 79% identical to human SHOX and their 60 amino acid binding domains (the homeodomain) are identical [33]. *In situ* analysis have demonstrated a more proximal expression domain of the *SHOX* paralog *SHOX2* in human and also in chick embryonic limbs [33,34], and conditional deletion of *Shox2* in the developing mouse limbs dramatically impairs the formation of the proximal limb elements [38,39]. A substitution of the *Shox2* locus by human *SHOX* in mouse has demonstrated that *SHOX* is able to ameliorate but not to fully rescue *Shox2*-deficient limb anomalies, suggesting only partial functional redundancy [25].

We have selected eight putatively regulated genes (*Postn, Aspn, Ogn, Isl1, Ctgf, Efemp1, Matn4, Mef2c*) for further analysis and could demonstrate a significant deregulation in E12.5-E14.5 *SHOX*-transgenic limbs compared to wildtype in seven of the eight genes. To further validate these candidates, we also tested them in NHDF and U2OS cells and 5/8 (NHDF) and 4/8 (U2OS) were shown to be significantly regulated in these human cells. Taken together, our data demonstrate that the identified target genes of *Tg(Col2a1-SHOX)* are SHOX-specific and do not represent transgenic artifacts. It is also reassuring that the human *SHOX* is expressed in the appropriate stage- and cell-type specific manner in mouse and we confirm previous data that SHOX can act both as an activator and repressor of target genes in a cell-type specific fashion [25,26].

CTGF represents a direct SHOX target gene

Quantitative analyses in mouse and human cells identified *CTGF/Ctgf* as the most consistently regulated candidate target gene. Enhanced and slightly distalized expression of *Ctgf* was also seen at E12.5 (the stage of highest *SHOX* expression) in transgenic mouse limbs using WISH. Further evidence for *Ctgf* as a target of Shox was derived from ChIP-Seq data in chicken which identified several Shox binding sites in the *Ctgf* upstream region. Multiple

SHOX binding motifs and an ECR were identified, and by luciferase and EMSA experiments, we could show that the extended ECR region (ECR+) is responsive to SHOX in human cells. The finding that the *CTGF* mRNA was either down- (NHDF) or upregulated (U2OS) indicates a complex transcriptional regulation. The remarkable accumulation of Shox-mediated reads (respective binding sites) identified in the chicken *Ctgf* upstream region using ChIP-Seq (Fig. 4A) suggests that additional response elements outside the ECR may also be sensitive to SHOX, and these, together with a spatio-temporal composition of cofactors, may contribute to the fine regulation of *CTGF* expression in a given cellular environment. Physiological relevance of the SHOX-*Ctgf*/*CTGF* relationship is suggested by the coexpression of both, SHOX and CTGF proteins in hypertrophic chondrocytes of the human growth plates.

According to its expression pattern, *SHOX* deficiency results in shortening and deformation of radii/ulnae and tibiae/fibulae. Comparable to *SHOX* deficiency, the skeletal defects in *Ctgf* null mice are also specific for radii/ulnae and tibiae/fibulae and not for the proximal elements of the limbs [40]. Interestingly, the phenotypes of *SHOX*- as well as *Ctgf*-transgenic mice [41] are less severe than the loss-of-function phenotypes and strongly dependent on the expression level of the transgenes. Even though *Ctgf*-transgenic mice show more stigmata than *SHOX*-transgenic individuals, phenotypic differences were reported only at postnatal stages and also include cortical thickness and *Agc1* expression [41]. Since *Agc1* has been found to be reduced in *Ctgf* mutant mice [40] and to be regulated by SHOX in human cells [20], the demonstrated regulation may be indirect and mediated through *Ctgf*. This is also supported by our finding that the response of *CTGF* is an immediate consequence following *SHOX* overexpression, whereas the regulation of *AGC1* occurs at a later time point (Fig. S2). *Ctgf* null mice suffer from multiple defects, such as failure in growth plate chondrogenesis, angiogenesis, extracellular matrix production and bone formation/mineralization [40]. A role of SHOX during angiogenesis has been speculated, since *Shox* expression was detected in the vasculature of the developing chicken limbs [34]. However, a contribution of SHOX in other *CTGF*-associated conditions such as fibrotic disease, inflammation and cancer [42,43,44] is not known.

In summary, we have established a transgenic mouse model expressing *SHOX* under the control of the *Col2a1* promoter and enhancer. By combining data from mouse and chicken micromass cultures and human cell culture experiments, we could identify activating or repressing effects of SHOX on target genes, depending on spatio-temporal conditions and cell types. We have also demonstrated a direct regulatory effect on *CTGF* which may take place in the hypertrophic zone of the human growth plate. We have shown a direct binding of the SHOX protein to a highly conserved upstream region of the *CTGF* gene, identified by ChIP-Seq, resulting in regulatory effects in reporter gene assays in human cell lines. Since CTGF is involved in various biological processes, the effect of SHOX on *CTGF* expression in these different processes can now be investigated.

Supporting Information

Figure S1 Marker and target gene analysis during embryonic development. (A): WISH of limb marker genes from E11.5 to

E14.5. At E11.5, when *Tg(Col2a1-SHOX)* expression was first detected in the developing limb, limb buds in transgenic animals were indistinguishable from the wildtype. Expression of the *Shh* morphogen as a marker gene during limb initiation and outgrowth was normal. Also at E12.5 when *Tg(Col2a1-SHOX)* is most prominently expressed, chondrocyte proliferation in the transgenic animals appeared normal, as represented by *Col2a1* expression comparable to the wildtype. Also, the *SHOX*-homologue *Shox2* and its downstream gene *Runx2* were normally expressed in *SHOX*-transgenic animals at E12.5. *Runx2* is known to regulate chondrocyte maturation and *Ihh* expression, which was also unaffected in *Tg(Col2a1-SHOX)* limbs at E13.5. Following chondrocyte proliferation at E14.5 in both wildtype and transgenic embryos, a specific *Col10a1* pattern is detected which defines chondrocyte hypertrophy. (B): Quantitative RT-PCR on embryonic limb RNA of stages E12.5–E14.5 using primers for the SHOX target genes *Fgfr3*, *Agc1* and *Nppb*. cDNA of wildtype and transgenic littermates of each stage (N = 8–12) were measured individually and in duplicates. Measurements were normalized to *Adam9* and *Sdha*; values on y-axis represent relative normalized expression. The expression of *Fgfr3* was unaltered in transgenic limbs. Mean *Agc1* expression was increased during E12.5 and E13.5, a trend which did, however not reach significance (E12.5: 2.0-fold, $p = 0.068$; E13.5: 2.6-fold, $p = 0.092$; E14.5: 1.3-fold, $p = 0.377$). *Nppb* expression levels were weakly increased at E13.5 (1.7-fold, $p = 0.104$).

Figure S2 nCounter analysis of eight selected candidate genes in NHDF and U2OS cells. RNA was isolated 6 h, 12 h and 24 h after transfection of expression constructs for SHOX, SHOX Y141D (a defective SHOX variant (1)) and a control (pCDNA4). Measurements were carried out in triplicates and normalized to *ADAM9*, *HPRT1* and *SDHA*. As a control, *SHOX* expression upon its target gene *AGC1* was analyzed. Upon strong increase of *SHOX*, *AGC1* was significantly activated 12 hours after *SHOX*-tranfection. Values on y-axis represent absolute counts of mRNA. Significancies of the *SHOX*-transfected samples are indicated in each diagram by asterisks. *: $p \leq 0.05$, **: $p \leq 0.01$, ***: $p \leq 0.001$.

File S1 Contains Table S1, Genes, gene characterization, fold regulation and p-values of eight selected upregulated genes in the microarray. Table S2, Primers, Probes and Oligonucleotides.

Acknowledgments

We thank Andrea Vortkamp for providing p1757 vector and *Col2* and *Shh* probe vectors, Jochen Hecht for providing *Runx2* and *Runx3* probe vectors and Nenja Krüger for experimental support. We thank Roland Knopf for animal care. This publication is dedicated to Ruediger J. Blaschke.

Author Contributions

Conceived and designed the experiments: KUB GAR. Performed the experiments: KUB AG IS RR KK AM GM. Analyzed the data: KUB KK WR LL NG GAR. Contributed reagents/materials/analysis tools: MK. Wrote the paper: KUB GR.

References

1. Rao E, Weiss B, Fukami M, Rump A, Niesler B, et al. (1997) Pseudoautosomal deletions encompassing a novel homeobox gene cause growth failure in idiopathic short stature and Turner syndrome. Nat Genet 16: 54–63.

2. Shears DJ, Vassal HJ, Goodman FR, Palmer RW, Reardon W, et al. (1998) Mutation and deletion of the pseudoautosomal gene SHOX cause Leri-Weill dyschondrosteosis. Nat Genet 19: 70–73.

3. Ross JL, Scott C, Jr., Marttila P, Kowal K, Nass A, et al. (2001) Phenotypes Associated with SHOX Deficiency. J Clin Endocrinol Metab 86: 5674–5680.
4. Rappold GA, Ross JL, Blaschke RJ, Blum W (2002) Understanding *SHOX* deficiency and its role in growth disorders. A reference guide. Oxfordshire, UK: TMG Healthcare Communications Ltd.
5. Shears DJ, Guillen-Navarro E, Sempere-Miralles M, Domingo-Jimenez R, Scambler PJ, et al. (2002) Pseudodominant inheritance of Langer mesomelic dysplasia caused by a SHOX homeobox missense mutation. Am J Med Genet 110: 153–157.
6. Rappold G, Blum WF, Shavrikova EP, Crowe BJ, Roeth R, et al. (2007) Genotypes and phenotypes in children with short stature: clinical indicators of SHOX haploinsufficiency. J Med Genet 44: 306–313.
7. Belin V, Cusin V, Viot G, Girlich D, Toutain A, et al. (1998) SHOX mutations in dyschondrosteosis (Leri-Weill syndrome). Nat Genet 19: 67–69.
8. Rao E, Blaschke RJ, Marchini A, Niesler B, Burnett M, et al. (2001) The Leri-Weill and Turner syndrome homeobox gene SHOX encodes a cell-type specific transcriptional activator. Hum Mol Genet 10: 3083–3091.
9. Sabherwal N, Schneider KU, Blaschke RJ, Marchini A, Rappold G (2004) Impairment of SHOX nuclear localization as a cause for Leri-Weill syndrome. J Cell Sci 117: 3041–3048.
10. Schneider KU, Marchini A, Sabherwal N, Roth R, Niesler B, et al. (2005) Alteration of DNA binding, dimerization, and nuclear translocation of SHOX homeodomain mutations identified in idiopathic short stature and Leri-Weill dyschondrosteosis. Hum Mutat 26: 44–52.
11. Schneider KU, Sabherwal N, Jantz K, Roth R, Muncke N, et al. (2005) Identification of a major recombination hotspot in patients with short stature and SHOX deficiency. Am J Hum Genet 77: 89–96.
12. Chen J, Wildhardt G, Zhong Z, Roth R, Weiss B, et al. (2009) Enhancer deletions of the SHOX gene as a frequent cause of short stature: the essential role of a 250 kb downstream regulatory domain. Journal of Medical Genetics 46: 834–839.
13. Benito-Sanz S, Aza-Carmona M, Rodriguez-Estevez A, Rica-Etxebarria I, Gracia R, et al. (2012) Identification of the first PAR1 deletion encompassing upstream SHOX enhancers in a family with idiopathic short stature. Eur J Hum Genet 20: 125–127.
14. Rosilio M, Huber-Lequesne C, Sapin H, Carel JC, Blum WF, et al. (2012) Genotypes and phenotypes of children with SHOX deficiency in France. J Clin Endocrinol Metab 97: E1257–1265.
15. Kanaka-Gantenbein C, Kitsiou S, Mavrou A, Stamoyannou L, Kolialexi A, et al. (2004) Tall stature, insulin resistance, and disturbed behavior in a girl with the triple X syndrome harboring three SHOX genes: offspring of a father with mosaic Klinefelter syndrome but with two maternal X chromosomes. Horm Res 61: 205–210.
16. Jorge AA, Souza SC, Nishi MY, Billerbeck AE, Liborio DC, et al. (2007) SHOX mutations in idiopathic short stature and Leri-Weill dyschondrosteosis: frequency and phenotypic variability. Clin Endocrinol (Oxf) 66: 130–135.
17. Munns CF, Glass IA, LaBrom R, Hayes M, Flanagan S, et al. (2001) Histopathological analysis of Leri-Weill dyschondrosteosis: disordered growth plate. Hand Surg 6: 13–23.
18. Munns CJ, Haase HR, Crowther LM, Hayes MT, Blaschke R, et al. (2004) Expression of SHOX in human fetal and childhood growth plate. J Clin Endocrinol Metab 89: 4130–4135.
19. Marchini A, Marttila T, Winter A, Caldeira S, Malanchi I, et al. (2004) The short stature homeodomain protein SHOX induces cellular growth arrest and apoptosis and is expressed in human growth plate chondrocytes. J Biol Chem 279: 37103–37114.
20. Aza-Carmona M, Shears DJ, Yuste-Checa P, Barca-Tierno V, Hisado-Oliva A, et al. (2011) SHOX interacts with the chondrogenic transcription factors SOX5 and SOX6 to activate the aggrecan enhancer. Hum Mol Genet 20: 1547–1559.
21. Blum WF, Crowe BJ, Quigley CA, Jung H, Cao D, et al. (2007) Growth hormone is effective in treatment of short stature associated with short stature homeobox-containing gene deficiency: Two-year results of a randomized, controlled, multicenter trial. J Clin Endocrinol Metab 92: 219–228.
22. Durand C, Roeth R, Dweep H, Vlatkovic I, Decker E, et al. (2011) Alternative splicing and nonsense-mediated RNA decay contribute to the regulation of SHOX expression. PLoS One 6: e18115.
23. Waterston RH, Lindblad-Toh K, Birney E, Rogers J, Abril JF, et al. (2002) Initial sequencing and comparative analysis of the mouse genome. Nature 420: 520–562.
24. Marchini A, Haecker B, Marttila T, Hesse V, Emons J, et al. (2007) BNP is a transcriptional target of the short stature homeobox gene SHOX. Hum Mol Genet 16: 3081–3087.
25. Liu H, Chen CH, Espinoza-Lewis RA, Jiao Z, Sheu I, et al. (2011) Functional redundancy between human SHOX and mouse Shox2 genes in the regulation of sinoatrial node formation and pacemaking function. J Biol Chem 286: 17029–17038.
26. Decker E, Durand C, Bender S, Rodelsperger C, Glaser A, et al. (2011) FGFR3 is a target of the homeobox transcription factor SHOX in limb development. Hum Mol Genet 20: 1524–1535.
27. Long F, Schipani E, Asahara H, Kronenberg H, Montminy M (2001) The CREB family of activators is required for endochondral bone development. Development 128: 541–550.
28. Minina E, Wenzel HM, Kreschel C, Karp S, Gaffield W, et al. (2001) BMP and Ihh/PTHrP signaling interact to coordinate chondrocyte proliferation and differentiation. Development 128: 4523–4534.
29. Yang Y, Topol L, Lee H, Wu J (2003) Wnt5a and Wnt5b exhibit distinct activities in coordinating chondrocyte proliferation and differentiation. Development 130: 1003–1015.
30. Pfaffl MW (2001) A new mathematical model for relative quantification in real-time RT-PCR. Nucleic Acids Res 29: e45.
31. Blaschke RJ, Monaghan AP, Schiller S, Schechinger B, Rao E, et al. (1998) SHOT, a SHOX-related homeobox gene, is implicated in craniofacial, brain, heart, and limb development. Proc Natl Acad Sci U S A 95: 2406–2411.
32. Cheah KS, Lau ET, Au PK, Tam PP (1991) Expression of the mouse alpha 1(II) collagen gene is not restricted to cartilage during development. Development 111: 945–953.
33. Clement-Jones M, Schiller S, Rao E, Blaschke RJ, Zuniga A, et al. (2000) The short stature homeobox gene SHOX is involved in skeletal abnormalities in Turner syndrome. Hum Mol Genet 9: 695–702.
34. Tiecke E, Bangs F, Blaschke R, Farrell ER, Rappold G, et al. (2006) Expression of the short stature homeobox gene Shox is restricted by proximal and distal signals in chick limb buds and affects the length of skeletal elements. Dev Biol 298: 585–596.
35. Stahl PD, Wainszelbaum MJ (2009) Human-specific genes may offer a unique window into human cell signaling. Sci Signal 2: pe59.
36. Soucek O, Lebl J, Snajderova M, Kolouskova S, Rocek M, et al. (2011) Bone geometry and volumetric bone mineral density in girls with Turner syndrome of different pubertal stages. Clinical Endocrinology 74: 445–452.
37. Vickerman L, Neufeld S, Cobb J (2011) Shox2 function couples neural, muscular and skeletal development in the proximal forelimb. Dev Biol 350: 323–336.
38. Cobb J, Dierich A, Huss-Garcia Y, Duboule D (2006) A mouse model for human short-stature syndromes identifies Shox2 as an upstream regulator of Runx2 during long-bone development. Proc Natl Acad Sci U S A 103: 4511–4515.
39. Yu L, Liu H, Yan M, Yang J, Long F, et al. (2007) Shox2 is required for chondrocyte proliferation and maturation in proximal limb skeleton. Dev Biol 306: 549–559.
40. Ivkovic S, Yoon BS, Popoff SN, Safadi FF, Libuda DE, et al. (2003) Connective tissue growth factor coordinates chondrogenesis and angiogenesis during skeletal development. Development 130: 2779–2791.
41. Tomita N, Hattori T, Itoh S, Aoyama E, Yao M, et al. (2013) Cartilage-specific over-expression of CCN family member 2/connective tissue growth factor (CCN2/CTGF) stimulates insulin-like growth factor expression and bone growth. PLoS One 8: e59226.
42. Dhar A, Ray A (2010) The CCN family proteins in carcinogenesis. Exp Oncol 32: 2–9.
43. Cicha I, Goppelt-Struebe M (2009) Connective tissue growth factor: context-dependent functions and mechanisms of regulation. Biofactors 35: 200–208.
44. Leask A, Holmes A, Abraham DJ (2002) Connective tissue growth factor: a new and important player in the pathogenesis of fibrosis. Curr Rheumatol Rep 4: 136–142.

13

Genotype-Phenotype Correlations in a Mountain Population Community with High Prevalence of Wilson's Disease: Genetic and Clinical Homogeneity

Relu Cocoş[1,3], **Alina Şendroiu**[5], **Sorina Schipor**[4], **Laurenţiu Camil Bohîlţea**[1,6], **Ionuţ Şendroiu**[5], **Florina Raicu**[1,2]*

1 Chair of Medical Genetics, "Carol Davila" University of Medicine and Pharmacy, Bucharest, Romania, **2** Francisc I. Rainer Anthropological Research Institute, Romanian Academy, Bucharest, Romania, **3** Genome Life Research Centre, Bucharest, Romania, **4** National Institute of Endocrinology "C. I. Parhon", Bucharest, Romania, **5** Family Medical Centre, Rucar, Romania, **6** Sf. Pantelimon Clinical Emergency Hospital, Bucharest, Romania

Abstract

Wilson's disease is an autosomal recessive disorder caused by more than 500 mutations in ATP7B gene presenting considerably clinical manifestations heterogeneity even in patients with a particular mutation. Previous findings suggested a potential role of additional genetic modifiers and environment factors on phenotypic expression among the affected patients. We conducted clinical and genetic investigations to perform genotype-phenotype correlation in two large families living in a socio-culturally isolated community with the highest prevalence of Wilson's disease ever reported of 1:1130. Sequencing of ATP7B gene in seven affected individuals and 43 family members identified a common compound heterozygous genotype, H1069Q/M769H-fs, in five symptomatic and two asymptomatic patients and detected the presence of two out of seven identified single nucleotide polymorphisms in all affected patients. Symptomatic patients had similar clinical phenotype and age at onset (18±1 years) showing dysarthria and dysphagia as common clinical features at the time of diagnosis. Moreover, all symptomatic patients presented Kayser-Fleischer rings and lack of dystonia accompanied by unfavourable clinical outcomes. Our findings add value for understanding of genotype-phenotype correlations in Wilson's disease based on a multifamily study in an isolated population with high extent of genetic and environmental homogeneity as opposed to majority of reports. We observed an equal influence of presumed other genetic modifiers and environmental factors on clinical presentation and age at onset of Wilson's disease in patients with a particular genotype. These data provide valuable inferences that could be applied for predicting clinical management in asymptomatic patients in such communities.

Editor: Bart Dermaut, Pasteur Institute of Lille, France

Funding: This work was financially supported by two grants of the Romanian National Authority for Scientific Research, CNCS-UEFISCDI, project numbers 2005-CNCSIS-27677 and PN-II-ID-PCCE 2011-2-0013. www.cncs-nrc.ro and http://uefiscdi.gov.ro/. The funders had no role in study design, data collection and analysis, decision to publish, or preparation of the manuscript.

Competing Interests: The authors have declared that no competing interests exist.

* E-mail: florina_raicu@yahoo.com

Introduction

Wilson's Disease (WD, OMIM #277900) is an autosomal recessive disorder of copper metabolism caused by mutations in the responsible gene, ATP7B, that codes for a membrane-bound copper-transporting P-type ATPase [1,2,3]. The ATP7B gene is located on chromosome 13 and has 21 exons spanning a DNA region of about 100 kb [4,5,6]. Over 500 mutations within the ATP gene have been identified along the whole length of the entire coding region and also in promoter and intronic regions (http://www.wilsondisease.med.ualberta.ca/database.asp). The worldwide prevalence of WD is estimated at one in 30000 and one in 100000 in most populations [7,8], with a carrier frequency of 1 in 90 to 122 [9,7]. The highest prevalence of WD was reported in the Sardinian (1:7000) and Gran Canaria Island (1:2600) populations due to inbreeding and founder effects [10,11]. The frequency and distribution of ATP7B mutations in Romanian WD patients are not known precisely [12,13].

The diagnosis of WD is made by clinical symptomatology in conjunction with biochemical, histological, imagistic data as established by Scheinber and Sternlieb [14,15,16] and/or genetic testing. Phenotypic classification could be realized using the WD classification scheme proposed by Ferenci et al [17]. Due to the wide range of clinical and biochemical features, Wilson's disease is difficult to characterize clinically.

The clinical architecture of Wilson's disease results from interactions between ATP7B and a spectrum of other genetic modifiers, environmental or lifestyle and stochastic factors that could have a degree of population and geographic specificity.

Copper accumulation can affect many organs especially brain or liver function generating diverse clinical presentations. Hepatic manifestations can range from asymptomatic liver and spleen enlargement to acute liver failure and cirrhosis, while neuropsychiatric manifestations can range from tremor, dysarthria, dystonia and cognitive dysfunction with or without hepatic presentation.

No definite genotype-phenotype correlations have been established so far due to the allelic heterogeneity and the rareness of the disease. However, few papers have suggested possible relationships between age at onset or type of presentation and a specific genotype [18,19,20,21].

Genealogic investigation allowed us to cluster two large families in a common multigenerational pedigree in a socio-culturally isolated mountain community with the highest WD prevalence ever reported. As outlined by other reports, an isolated community is a powerful resource for genetic studies as a consequence of limited genetic heterogeneity of their inhabitants who are more likely to share additional common genetic modifier factors and have similar eating habits that could increase the chances of finding subjects with the same ATP7B genotype and performing genotype-phenotype correlations [22,23].

Here, we conducted a genetic analysis of seven WD patients and 43 family members in two large families spanning six generations and focused on a detailed evaluation of genotype-phenotype correlation. Our results could support the hypothesis of equal effect on WD clinical presentation and age at onset of genetic modifier factors in such populations.

Subjects and Methods

We studied two large families, which spanned six generations consisting of 50 living members, of which 7 were affected by WD. We were able to link these two families based on information provided by relatives of patients in an extensive multigenerational pedigree sharing 4 unique family names (Figure 1). No consanguinity was recorded among parents and the families are interconnected only through the last three generations. The proband was a 59-year-old male who underwent clinical assessment and was diagnosed with WD at the age of 19. The pedigree has been modified to protect the anonymity of the families. Informed written consent was obtained in accordance with protocols approved by the "Carol Davila" University of Medicine and Pharmacy's ethical committee. These families are located in Rucar in a mountain region having a current population size estimated at 6200 with a possible high level of consanguinity in the past.

Clinical Diagnosis

We initially based the diagnosis of affected members on neurological clinical symptomatology like the presence of a Kayser-Fleischer (K-F) ring using slit-lamp examination, typical neurological symptoms, and hepatic features including symptoms of acute, chronic cirrhosis and fulminant liver failure, and the presence of conventional biochemical markers like low serum ceruloplasmin (<20 mg/dL) and elevated baseline 24-hour urinary copper excretion (baseline levels >100 µg/24 h), (Table 1). Due to parental refusal or religious convictions and since most of the patients were neurologic, liver biopsy was not carried out as a diagnostic measure. We performed a second complete clinical evaluation complemented with liver biochemical and genetic analysis in this study for all affected and asymptomatic subjects.

Mutation Sequence Analysis

We extracted genomic DNA from whole blood in EDTA using PureLink Genomic DNA Mini kit (Invitrogen, USA). We performed mutation analysis on PCR amplified DNA for the entire 21 coding exons, their exon-intron boundaries and 600 base pair of promoter with primers previously reported [24,25] and other primers that are available on request using AmpliTaq Gold polymerase (Applied Biosystems, USA) according standard protocols. We purified PCR products with QIAquick PCR purification kit (Hilden, Germany) and sequenced with the Big Dye Terminator v3.1 Cycle Sequencing kit using ABI PRISM 310 and ABI 3130XL Genetic Analyzers (Applied Biosystems, USA). Analyses of ATP7B DNA sequencing data were performed using the ABI PRISM DNA Sequencing Analysis Software, Version 3.7. The sequences were aligned and compared with the revised Cambridge Reference Sequence rCRS (NM_000053, NCB), using the SeqScape Software Version 2.5. We confirmed the detected mutations and SNPs on both sequencing platforms in forward and reverse directions and compared with the revised Cambridge Reference Sequence rCRS using the SeqScape Software, Version 2.5 and MEGA5.

Results

Sequencing Results

Sequencing results revealed two mutations, c.3207C>A (p.His1069Gln) and c.2304insC (p.Met769His-fs), and seven additional single nucleotide polymorphisms (SNPs) in exons/introns: exon 2, c.1216 T>G (p.Ser406Ala); exon 3, c.1366G>C (p.Val456Leu); exon 10, c.2495A>G (p.Lys832Arg); exon 12, c.2855G>A (p.Arg952Lys); intron 13, c.2866–13G>C; exon 16, c.3419C>T (p.Val1140Ala) and intron 18, c.3903+6C>T in seven members affected by WD outlined in a common pedigree. Of these two mutations, one is a missense mutation, H1069Q, located in exon 14 and the other is a frameshift mutation, M769H-fs, lying in exon 8. The mutations described herein, p.H1069Q and p.M769H-fs, alter ATP loop and Tm 4 domain, respectively.

These two mutations are represented as compound heterozygous in the proband (V.20) and other four symptomatic patients in two families outlined in a large apparently non-consanguineous pedigree in a socio-culturally isolated community with high prevalence of WD (Figure 1). By direct sequencing of ATP7B gene we identified the same compound heterozygous genotype, H1069Q/M769H-fs, in both asymptomatic children (VI.3 and VI.4) of parents (V.8 and V.9). The mutation M769H-fs is represented as heterozygous in 13 unaffected members and the mutation H1069Q is represented as heterozygous in 8 unaffected members in the pedigree.

We screened a total of 43 family members for the identified mutations and SNPs. The discovered SNPs, previously reported [26,27,28,29,30,31,32] were tested in 102 healthy controls with their frequencies presented in Table 2.

In the pedigree, two or more SNPs occurred simultaneously at different haplotype combinations among affected patients and healthy members. Two of the seven SNPs, c.2495A>G and c.3419C>T, were present in all symptomatic and asymptomatic patients.

Clinical Data

Patient V.20, the proband of pedigree, was initially diagnosed with WD at the age of 20 in 1974, although he first exhibited neurological signs including dysarthria in one hand, mild dysphagia and malaise at the age of 19 without seeking medical advice. He was started on D-penicillamine treatment at age of 21 years due to the unavailability of the drug in Romania at the time of his diagnosis. The patient responded to the treatment and his conditions improved slightly. Ophthalmologic examination revealed bilateral Kayser-Fleischer rings. After the onset of the

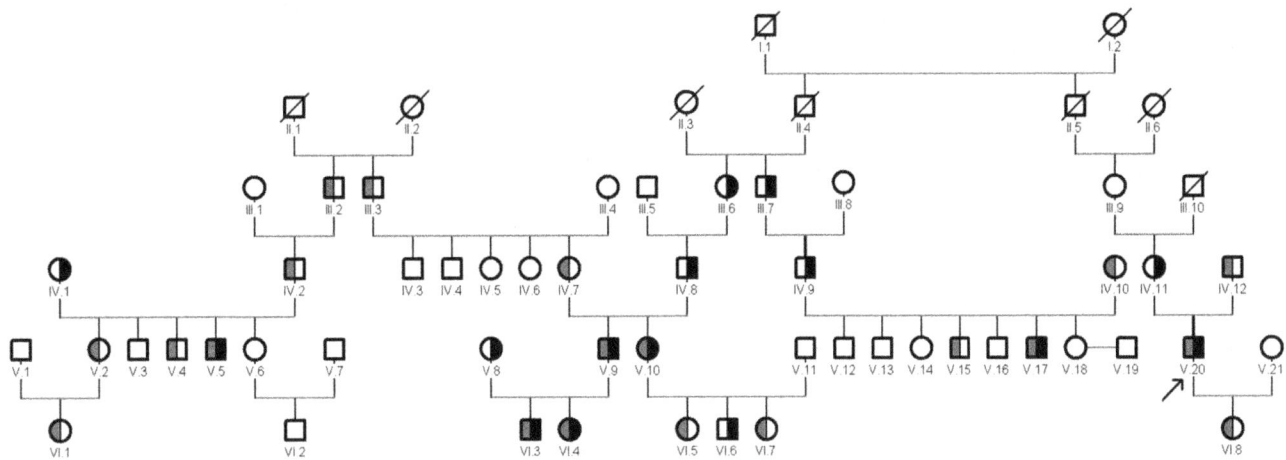

Figure 1. Pedigree and genetic analyses of the two large families. Genetic analyses were performed on all individuals indicated by filled, divided and open symbols. Pedigree symbols: slashed symbol, deceased individual; open symbol, unaffected individual; divided gray symbol, carrier for M769H-fs mutation; divided black symbol, carrier for H1069Q mutation; filled symbol, affected individual with compound heterozygous genotype. A filled arrowhead indicates proband. Roman numbers indicate generations.

disease, he gradually developed neuropsychiatric complications including advanced dysarthria, drooling, postural tremor and instability, chorea, parkinsonism and cognitive manifestations like slowness of thinking and executive dysfunction. Decompensated liver cirrhosis was present with Child-Pugh score B at the age of 36 years with mild ascites, jaundice, coagulopathy, hyperbilirubinemia and hypoalbuminemia.

Patient V.5 was initially admitted to hospitalization at the age of 17 years in 2000 with signs of liver failure including subfebrile temperature, jaundice, fatigue and vomiting. The pathological findings were initially interpreted as a presumed hepatic viral infection but investigation of his nervous system revealed clinical symptoms including dysarthria, mild dysphagia and malaises. His serum ceruloplasmin and his 24-hours urinary copper levels were in pathological range (Table 1). In addition, the presence of the pathognomonic clinical sign, presence of the Kayser-Fleischer rings in both eyes, was confirmed by ophthalmoscopic examination by slit lamp examination. He was started on low D-penicillamine (250 mg/day) dose and continued with slowly increasing doses that resulted in a modest improvement but was diagnosed 2 years later with compensated liver cirrhosis with Child-Pugh score A. In 2013, he was diagnosed with acute myeloid leukemia.

Patient V.17 was diagnosed with WD in 1987 at the age of 19, presenting with mild dysphagia, mild dysarthria and headache. Kayser-Fleischer rings initially unobserved were subsequently detected bilaterally by slit lamp examination. His serum ceruloplasmin and 24-hours urinary copper levels are presented in Table 1. He has been treated with D-pencillamine with incremental doses ever since, with initial slight improvement in his condition, but subsequent worsening the disease. Signs of mild hepatic manifestations without jaundice and haemolytic anemia presented later. The patient has developed major neurological features including pronounced dysphagia and dysarthria, drooling, unsteady gait and psychiatric manifestations like agoraphobia, slowness of thinking and mood disturbances.

Patient V.9 was diagnosed with WD at the age of 18 presenting with progressive neurological symptoms, including dysarthria, dysphagia without impairment of motor skills. His ceruloplasmin levels were low (<2.6 mg/dL) and his 24-hours urinary copper high (>1019 µg/day). Ophthalmologic examination revealed

bilateral Kayser-Fleischer rings. Response was initially positive to D-penicillamine therapy. Five years later, abdominal ultrasonography examination detected a mild echogenicity in the liver. Dysarthria became pronounced over the years with subsequently developed body bradykinesia, resting and postural tremor.

Both children of this patient (VI.3 and VI.4), were clinically asymptomatic, but were diagnosed with WD by biochemical tests at the age of 6 and 7, respectively. An apparent pseudo-dominat inheritance with two consecutive generations affected by Wilson's disease could be seen in parents and their children in this family.

The children's neurological tests were normal, however laboratory findings were clearly abnormal consisting of increased 24 h urinary copper values, decreased serum ceruloplasmin levels and high levels of aspartate transaminase (AST) and alanine transaminase (ALT), (Table 1). Preventive treatment with D-pencillamine was started for both children once they were diagnosed.

For patient V.10, the sister of V.9 (affected), symptoms began when she was 19 years old. She indicated that the symptoms started insidiously when she was 18. At the time of diagnosis, signs of neuropsychiatric disturbances without liver presentation were present including dysphagia, nystagmus, and dysarthria, lacking signs of tremor and postural instability, and mild cognitive impairment like difficulties in school performance. Kayser-Fleischer rings were present on both eyes. The response to D-penicillamine therapy was adequate, showing slight improvement, but was discontinued during her pregnancies. In a very short period of time, she suffered a gradual psychiatric deterioration resulting in depression, personality changes and behavioral disturbances associated with advanced neurological clinical features including body bradykinesia, tremor and postural instability.

Unaffected family members of all WD subjects had a full evaluation and were found normal. Their physical examination, serum ceruloplasmin, serum copper, ALT, AST and 24 h urine copper levels were normal.

Severe neurological deterioration was observed in all symptomatic patients without relevant side effects. Three of the symptomatic patients with initial neurological signs, patients V10, V.7 and V.15, developed a parallel deterioration of hepatic function while under treatment with evidence of cirrhosis, either compensated in

Table 1. Clinical and laboratory findings of WD patients.

Patient No	Age at onset (y), sex	Age at diagnosis (y)	Clinical presentation at diagnosis			Laboratory Findings at diagnosis			
			Hepatic	Neurological	K-F ring	Serum CP (mg/dL)	Urinary Cu (µg/day)	ALT (U/L)	AST (U/L)
V.5	17, M	17	–	+	+	3.5	340	35	40
V.10	18, M	18	–	+	+	0.9	612	20	19
V.9	18, F	19	–	+	+	2.6	1019	27	20
VI.3	6, M	6	–	–	–	0.4	70.5	278	133
VI.4	7, F	7	–	–	–	0.1	210	339	143
V.17	19, M	19	–	+	+	1.2	413	NA	NA
V.20	19, M	20	–	+	+	2.3	507	NA	NA

Abbreviation and Notes: M, male; F, female; y, years; NA, Not available; K-F, Kayser-Fleischer; CP, Ceruloplasmin; Cu, Copper; ALT, Alanine Transaminase; ASP, Aspartate Transaminase.
Serum CP was measured by immunoturbimetric test. Serum CP normal values are 20–60 mg/dL.
Normal urinary copper values is less than 100 µg/day.
Normal ranges of liver enzymes are: ALT (10–45 Units/L) and AST (15–47 Units/L).
Patient numbering is represented as indicated in the pedigree.

patient V.10 or decompensated in patient V.20, whereas the other two remained only neurologically symptomatic with stable liver function. Both asymptomatic patients showed no normalization of biochemical tests under drug therapy.

Discussion

The extensive variation in hepatic and neurological presentations in affected patients carrying an identical genotype in different families or within the same family is one of the most intriguing aspects of Wilson's disease [2,19,33]. It is unknown why a particular genotype is not associated to a specific behavior of the disease, albeit some authors have tried to establish a correlation between the type of presentation, age at onset or clinical course and the presence of a specific mutation in heterozygous or homozygous state [20,21,34,35].

Since the Middle Ages for more than 600 years wealthy families of shepherds in Rucar region practiced marriages between members of the same clan as a way of protecting the family inheritance. This marriage pattern changed only at the beginning of 20th century. As a result of this socio-cultural practice, a period of genetic isolation could have occurred definitely affecting the actual genetic structure of this population. Although the actual prevalence of consanguineous marriages is very low in Romanian population, a substantial level of consanguinity would have been inevitable in scattered mountain rural communities in the past. Isolation could still play a role in this region as a result of the absence of immigration generated by poor economic resources.

On the basis of the number of Wilson's disease patients and births recorded between 1975–2012 we calculated the prevalence of the disease to be 1:1130. This is the highest prevalence ever reported [10,11,36].

As summarized in our paper, there were significant similarities at the time of diagnosis with respect to clinical features and ages at onset. Pedigree analysis revealed an apparent pseudo-dominant inheritance case in which two consecutive generations presented family members with Wilson's disease. A similar situation was reported by other studies in consanguineous or distant consanguineous families [24,37,38].

In most published papers, authors compared patients from distinct families that are homozygous for a particular mutation, while others compared homozygotes for the same type of mutation [19,39].

As a result of identical age at onset and similar clinical presentation among all our symptomatic patients, we suggest a dominance effect of frameshift mutation p.M769H-fs over missense mutation p.H1069Q. Although the presence of an intermediate effect as previously suggested by Gromadzka et al [34] or the influence of other genetic factors could not be excluded.

The presence of a coexisting frameshift mutation, p.H1069Q, in compound heterozygous state in our patients was associated with lower age at onset that is in agreement with previously reported results regarding the ages at onset for p.H1069Q homozygotes and p.H1069Q/missense patients [39,40].

Møller et al [41] classified mutations either as severe or moderate based on whether they cause clinical symptoms before or after the age of 20, assuming the disease severity defined by the age of onset is determined by the less severe of two mutations. In our compound heterozygote patients carrying two severe mutations according to Møller classification, p.H1069Q and p.M769H-fs, the age at presentation (18±1 years) fits exactly with the proposed algorithm. In contrast, Gupta et al [33] obtained contradictory data suggesting that the age of onset is established by the most severe from the two mutations.

Table 2. The SNPs of the ATP7B gene in healthy control group.

Exon/intron	Nucleotide	Amino acid	Protein domain	Type	SNP	Allele frequency (%)
2	c.1216 T>G	p.Ser406Ala	Cu$_4$	Missense	Known	39
3	c.1366G>C	p.Val456Leu	Cu$_5$	Missense	Known	42
10	c.2495A>G	p.Lys832Arg	A-domain/Td	Missense	Known	65
12	c.2855G>A	p.Arg952Lys	TM5	Missense	Known	16
Intron 13	c.2866–13G>C	-	-	-	Known	46
16	c3419C>T	p.Val1140Ala	ATP loop	Missense	Known	31
Intron 18	c.3903+6C>T	-	-	-	Known	26

Abbreviation and Notes: SNP, single nucleotide polymorphism. Nucleotide numbering refers to the cDNA according GenBank Accession number NM000053, where the first nucleotide of ATG translation codon is considered nt +1. Total number of alleles was 204.

Dysarthria and dysphagia, either mild or advanced, were the first common signs observed for all symptomatic WD patients, except for two asymptomatic children that could follow an identical clinical course without medication. Moreover, in all symptomatic patients Kayser-Fleischer rings were present without showing a very common neurological sign, namely, dystonia.

Several familial studies have shown that despite phenotypic variation, siblings present an identical clinical type or age at onset [42,43] while a few authors observed no genotype-phenotype association even among the same homozygote or compound heterozygote genotype siblings or monozygotic twins [33,44,45].

In a recent paper, Chabik et al [43] reported results similar to our findings demonstrating a high intra-familial concordance of WD patients with a less predictability for neurological presentation. Furthermore, our study indicated a great clinical predictability even for neurological presentation by the presence of the same set of clinical features at the time of diagnosis and identical ages at onset (Table 3).

For our symptomatic patients long-term follow-up revealed unfavourable outcomes with respect to the course of neuropsychiatric symptoms. Subsequently, occurrence of other clinical features, neuropsychiatric and/or hepatic, in addition to initial common neurological signs and overall progressive clinical picture could be especially explained by the failure of medication, the time from diagnosis to treatment or periods of drug therapy discontinuation in some of patients. However, the implication of other presumed genetic factors could not be completely excluded. Development of acute myeloid leukemia, a very rare clinical feature in WD, was attributed by a single study to toxicity of D-penicillamine [46]. It seems unlikely that the occurrence of acute myeloid leukemia can be explained by the toxic effect of D-penicillamine for our patient repeatedly discontinued his medication use.

One of our most notable findings was the accelerated rate of disease progression of all symptomatic patients while under treatment with D-penicillamine. The progression of neuropsychiatric symptoms for all our patients while under treatment could be in concordance with Lee et al [47] finding that indicates a less favourable outcome for patients with neurological presentation compared to patients showing hepatic presentation.

The involvement of other presumed genetic modifiers factors such as ATOX1, COMMD1 and/or environmental factors could complicate the clear prediction of a specific phenotypic expression but their influence remains contentious as was suggested by other reports [47,48,49,50]. Clinical heterogeneity in compound heterozygous patients influenced by the same environmental factors could not be entirely explained by the differing severity of particular alleles or other supplementary genetic modifiers but also by the additive effect of SNPs in ATP7B gene. Thus, two of the five exonic SNPs, c.2495A>G and c.3419C>T, were found present in all affected patients suggesting an identical additive effect of SNPs on phenotypic expression.

Table 3. Genotype-phenotype correlations found in patients with Wilson's disease.

Patient No	Mean age at onset 18±1 (y)	ATP7B Genotype H1069Q/M769H-fs	Clinical symptoms at diagnosis		K-F ring
			Neurological presentation	Clinical findings	
V.5	+	+	+	Dysarthria, dysphagia	+
V.10	+	+	+	Dysarthria, dysphagia, nystagmus	+
V.9	+	+	+	Dysarthria, dysphagia	+
V.17	+	+	+	Mild dysarthria, mild dysphagia	+
V.20	+	+	+	Dysarthria, mild dysphagia	+
VI.3	A	+	-	−	-
VI.4	A	+	-	−	-

Abbreviation and Notes: "−", negative; "+", positive; A, asymptomatic; y, years; K-F, Kayser-Fleischer;
These results demonstrate that the H1069Q/M769H-fs genotype is associated with common neurological symptoms at the time of diagnosis (dysarthria, dysphagia and K-F rings) and similar ages of onset, except for the two asymptomatic children that can have an identical clinical course without treatment. Patient numbering is represented as indicated in the pedigree.

In conclusion, according to our results additional genetic modifiers and environmental factors would be expected to exert an equal influence on clinical picture and age at onset of WD in patients with a given genotype within the same or different families in relatively small isolated communities. Whereas a diverse effect would be expected some patients from diverse regions as a result of environmental and genetic heterogeneity as was demonstrated by other research.

Our patients offered a rare opportunity for assessing genotype-phenotype correlations considering the reduced worldwide availability of WD patients with a particular genotype living in isolated populations even if we could not draw definite conclusions. Our results suggest the use of genotypes to predict clinical manifestation and age at onset in asymptomatic patients in such communities.

Acknowledgments

We acknowledge and thank all participants for their cooperation and sample contributions.

Author Contributions

Conceived and designed the experiments: FR RC. Performed the experiments: RC FR SS IS AS. Analyzed the data: FR RC. Contributed reagents/materials/analysis tools: RC FR LCB. Wrote the paper: RC FR SS. Discussed the results and commented on the manuscript: RC FR SS.

References

1. Wilson SAK (1912) Progressive lenticular degeneration: a familial nervous disease associated with cirrhosis of the liver. Brain 34: 295–507.
2. Ala A, Walker AP, Ashkan K, Dooley JS, Schilsky ML (2007) Wilson's disease. Lancet 369: 397–408.
3. Gitlin JD (2003) Wilson disease. Gastroenterology 125: 1868–1877.
4. Frydman F, Bonne-Tamir B, Farrer LA, Conneally PM, Magazanik A, et al. (1985) Assignment of the gene for Wilson disease to chromosome 13: linkage to esterase D locus. Proc Natl Acad Sci USA 82: 1819–1821.
5. Bull PC, Thomas GR, Rommens JM, Forbes JR, Cox DW (1993) The Wilson disease gene is a putative copper transporting P-type ATPase similar to the Menkes gene. Nat Genet 5: 327–337.
6. Yamaguchi Y, Heiny ME, Gitlin JD (1993) Isolation and characterization of a human liver cDNA as a candidate gene for Wilson disease. Biochem Biophys Res Commun 197: 271–277.
7. Reilly M, Daly L, Hutchinson M (1993) An epidemiological study of Wilson's disease in the Republic of Ireland. J Neurol Neurosurg Psychiatry 56: 298–300.
8. Roberts EA, Schilsky ML (2008) Diagnosis and treatment of Wilson disease: an update. Hepatology 47: 2089–2111.
9. Figus A, Angius A, Loudianos G, Bertini C, Dessi V, et al. (1995) Molecular Pathology and Haplotype Analysis of Wilson Disease in Mediterranean Populations. Am J Hum Genet 57: 1318–1324.
10. Loudianos G, Dessi V, Lovicu M, Angius A, Figus A, et al. (1999) Molecular characterization of Wilson disease in the Sardinian population-evidence of a founder effect. Hum Mutat 14: 294–303.
11. García-Villarreal L, Daniels S, Shaw SH, Cotton D, Galvin M, et al. (2000) High prevalence of the very rare Wilson disease gene mutation Leu708Pro in the Island of Gran Canaria (Canary Islands, Spain): a genetic and clinical study. Hepatology 32: 1329–36.
12. Iacob R, Iacob S, Nastase A, Vagu C, Ene AM, et al. (2012) The His1069Gln mutation in the ATP7B gene in Romanian patients with Wilson's disease referred to a tertiary gastroenterology centre. Gastrointestin Liver Dis 21: 181–185.
13. Lepori MB, Zappu A, Incollu S, Dessì V, Mameli E, et al. (2012) Mutation analysis of the ATP7B gene in a new group of Wilson's disease patients: contribution to diagnosis. Mol Cell Probes 26: 147–150.
14. Scheinberg IH, Gitlin D (1952) Deficiency of ceruloplasmin in patients with hepatolenticular degeneration (Wilson's disease). Science 116: 484–485.
15. Scheinberg IH, Sternlieb I (1984) Wilson's disease. Philadelphia: WB Saunders. 23–25.
16. Sternlieb I (1990) Perspectives on Wilson's disease. Hepatology 12: 1234–1239.
17. Ferenci P, Caca K, Loudianos G, Mieli-Vergani G, Tanner S, et al. (2003) Diagnosis and phenotypic classification of Wilson disease. Liver Int 23: 139–142.
18. Stapelbroek JM, Bollen CW, van Amstel JK, van Erpecum KJ, van Hattum J, et al. (2004) The H1069Q mutation in ATP7B is associated with late and neurologic presentation in Wilson disease: results of a meta-analysis. J Hepatol 41: 758–763.
19. Nicastro E, Loudianos G, Zancan L, D'Antiga L, Maggiore G, et al. (2009) Genotype-phenotype correlation in Italian children with Wilson's disease. J Hepatol 50: 555–561.
20. Barada K, Nemer G, ElHajj II, Touma J, Cortas N, et al. (2007) Early and severe liver disease associated with homozygosity for an exon 7 mutation, G691R, in Wilson's disease. Clinical Genetics 72: 264–267.
21. Merle U, Weiss KH, Eisenbach C, Tuma S, Ferenci P, et al. (2010) Truncating mutations in the Wilson disease gene ATP7B are associated with very low serum ceruloplasmin oxidase activity and an early onset of Wilson disease. BMC Gastroenterol 18: 10–18.
22. Fraga MF, Ballestar E, Paz MF, Ropero S, Setien F, et al. (2005) Epigenetic differences arise during the lifetime of monozygotic twins. Proc Natl Acad Sci U S A 102: 10604–10609.
23. Bittles AH, Black ML (2010) Consanguinity, human evolution, and complex diseases. Proc Natl Acad Sci U S A 107 (suppl 1): 1779–1786.
24. Coffey AJ, Durkie M, Hague S, McLay K, Emmerson J, et al. (2013) A genetic study of Wilson's disease in the United Kingdom. Brain 136: 1476–1487.
25. Vrabelova S, Letocha O, Borsky M, Kozak L (2005) Mutation analysis of the ATP7B gene and genotype/phenotype correlation in 227 patients with Wilson disease. Mol Genet Metab 86: 277–285.
26. Gupta A, Maulik M, Nasipuri P, Chattopadhyay I, Das SK, et al. (2007) Molecular Diagnosis of Wilson Disease Using Prevalent Mutations and Informative Single-Nucleotide Polymorphism Markers. Clin Chem 53: 1601–1608.
27. Olsson C, Waldenström E, Westermark K, Landegre U, Syvänen AC (2000) Determination of the frequencies of ten allelic variants of the Wilson disease gene (ATP7B), in pooled DNA samples. Eur J Hum Genet 8: 933–938.
28. Duc HH, Hefter H, Stremmel W, Castañeda-Guillot C, Hernández HA, et al. (1998) His1069Gln and six novel Wilson disease mutations: analysis of relevance for early diagnosis and phenotype. Eur J Hum Genet 6: 616–623.
29. Wang LH, Huang YQ, Shang X, Su QX, Xiong F, et al. (2011) Mutation analysis of 73 southern Chinese Wilson's disease patients: identification of 10 novel mutations and its clinical correlation. J Hum Genet 56: 660–665.
30. Ye S, Gong L, Shui QX, Zhou LF (2007) Wilson disease: identification of two novel mutations and clinical correlation in Eastern Chinese patients. World J Gastroenterol 13: 5147–5150.
31. Thomas GR, Forbes JR, Roberts EA, Walshe JM, Cox DW (1995) The Wilson disease gene: spectrum of mutations and their consequences. Nat Genet 9: 210–217.
32. Cox DW, Prat L, Walshe JM, Heathcote J, Gaffney D (2005) Twenty-four novel mutations in Wilson disease patients of predominantly European ancestry. Hum Mutat 26: 280.
33. Gupta A, Aikath D, Neogi R, Datta S, Basu K, et al. (2005) Molecular pathogenesis of Wilson disease: haplotype analysis, detection of prevalent mutations and genotype-phenotype correlation in Indian patients. Hum Genet 118: 49–57.
34. Gromadzka G, Schmidt HH, Genschel J, Bochow B, Rodo M, et al. (2005) Frameshift and nonsense mutations in the gene for ATPase7B are associated with severe impairment of copper metabolism and with an early clinical manifestation of Wilson's disease. Clin Genet 68: 524–532.
35. Usta J, Abu DH, Halawi H, Al-Shareef I, El-Rifai O, et al. (2012) Homozygosity for Non-H1069Q Missense Mutations in ATP7B Gene and Early Severe Liver Disease: Report of Two Families and a Meta-analysis. JIMD Rep 4: 129–137.
36. Dedoussis GV, Genschel J, Sialvera TE, Bochow B, Manolaki N, et al. (2005) Wilson disease: high prevalence in a mountainous area of Crete. Ann Hum Genet 69: 268–274.
37. Firneisz G, Szonyi L, Ferenci P, Gorog D, Nemes B, et al. (2001) Wilson disease in two consecutive generations: an exceptional family. Am J Gastroenterol 96: 2269–2271.
38. Dziezyc K, Gromadzka G, Czlonkowska A (2011) Wilson's disease in consecutive generations of one family. Parkinsonism Relat Disord 17: 577–578.
39. Gromadzka G, Schmidt HH, Genschel J, Bochow B, Rodo M, et al. (2006) p.H1069Q mutation in ATP7B and biochemical parameters of copper metabolism and clinical manifestation of Wilson's disease. Mov Disord 21: 245–248.

40. Panagiotakaki E, Tzetis M, Manolaki N, Loudianos G, Papatheodorou A, et al. (2004) Genotype–phenotype correlations for a wide spectrum of mutations in the Wilson disease gene (ATP7B). Am J Med Genet A 131: 168–173.

41. Møller LB, Horn N, Jeppesen TD, Vissing J, Wibrand F, et al. (2011) Clinical presentation and mutations in Danish patients with Wilson disease. Eur J Hum Genet 19: 935–941.

42. Santhosh S, Shaji RV, Eapen CE, Jayanthi V, Malathi S, et al. (2008) Genotype phenotype correlation in Wilson's disease within families-a report on four south Indian families. World J Gastroenterol 14: 4672–4676.

43. Chabik G, Litwin T, Członkowska A (2014) Concordance rates of Wilson's disease phenotype among siblings. J Inherit Metab Dis 37: 131–135.

44. Takeshita Y, Shimizu N, Yamaguchi Y, Nakazono H, Saitou M, et al. (2002) Two families with Wilson disease in which siblings showed different phenotypes. J Hum Genet 47: 543–547.

45. Czlonkowska A, Gromadzka G, Chabik G (2009) Monozygotic female twins discordant for phenotype of Wilson's disease. Mov Disord 24: 1066–1069.

46. Gilman PA, Holtzman NA (1982) Acute lymphoblastic leukemia in a patient receiving penicillamine for Wilson's disease. JAMA 248: 467–468.

47. Lee BH, Kim JH, Lee SY, Jin HY, Kim KJ, et al. (2011) Distinct clinical courses according to presenting phenotypes and their correlations to ATP7B mutations in a large Wilson's disease cohort. Liver Int 31: 831–839.

48. Brage A, Tomé S, García A, Carracedo A, Salas A (2007) Clinical and molecular characterization of Wilson disease in Spanish patients. Hepatol Res 37: 18–26.

49. Simon I, Schaefer M, Reichert J, Stremmel W (2008) Analysis of the human Atox 1 homologue in Wilson patients. World J Gastroenterol 14: 2383–2387.

50. Stuehler B, Reichert J, Stremmel W, Schaefer M (2004) Analysis of the human homologue of the canine copper toxicosis gene MURR1 in Wilson disease patients. J Mol Med 82: 629–634.

14

Development of Genetically Stable *Escherichia coli* Strains for Poly(3-Hydroxypropionate) Production

Yongqiang Gao[1,2]**, Changshui Liu**[1,2]**, Yamei Ding**[3]**, Chao Sun**[1]**, Rubing Zhang**[1]**, Mo Xian**[1,4]*****, **Guang Zhao**[1,4]*****

1 CAS Key Laboratory of Biobased Materials, Qingdao Institute of Bioenergy and Bioprocess Technology, Chinese Academy of Sciences, Qingdao, China, **2** University of Chinese Academy of Sciences, Beijing, China, **3** Institute of Oceanology, Chinese Academy of Sciences, Qingdao, China, **4** Collaborative Innovation Center for Marine Biomass Fibers, Materials and Textiles of Shandong Province, Qingdao, China

Abstract

Poly(3-hydroxypropionate) (P3HP) is a biodegradable and biocompatible thermoplastic. In our previous study, a pathway for P3HP production was constructed in recombinant *Esecherichia coli*. Seven exogenous genes in P3HP synthesis pathway were carried by two plasmid vectors. However, the P3HP production was severely suppressed by strain instability due to plasmid loss. In this paper, two strategies, chromosomal gene integration and plasmid addiction system (PAS) based on amino acid anabolism, were applied to construct a genetically stable strain. Finally, a combination of those two methods resulted in the best results. The resultant strain carried a portion of P3HP synthesis genes on chromosome and the others on plasmid, and also brought a tyrosine-auxotrophy based PAS. In aerobic fed-batch fermentation, this strain produced 25.7 g/L P3HP from glycerol, about 2.5-time higher than the previous strain with two plasmids. To the best of our knowledge, this is the highest P3HP production from inexpensive carbon sources.

Editor: John R. Battista, Louisiana State University and A & M College, United States of America

Funding: This research was financially supported by the 100-Talent Project of CAS (for GZ), Natural Science Foundation of Shandong Province (ZR2013EMZ002), National Natural Science Foundation of China (21206185, 21376255), and Qingdao Sci-Tech Development Programme (12-1-4-9-(3)-jch). The funders had no role in study design, data collection and analysis, decision to publish, or preparation of the manuscript.

Competing Interests: The authors have declared that no competing interests exist.

* E-mail: xianmo@qibebt.ac.cn (MX); zhaoguang@qibebt.ac.cn (GZ)

Introduction

Escherichia coli strains are widely used as hosts for microbial production of valuable compounds, like biofuels, chemicals, polymers, and proteins, and the production processes often depend on expression of heterologous genes carried by plasmid vectors [1]. Plasmids have been regarded as important tools for microbial genetic modifications. However, plasmids are separate genetic elements and autonomously replicated, and the redundant DNA carried by plasmids may cause metabolic burden in host strains, which could result in plasmid loss [2–4]. For plasmid maintenance, the cloning and expression vectors harbor antibiotic resistance genes and require the addition of antibiotics to the medium. Though it is feasible at the laboratory scale, the use of antibiotics at industrial scale will increase the production cost and raise the ecological issues. Furthermore, plasmid loss can even occur with presence of antibiotics during cultivation [5].

Plasmid addiction system (PAS) is an efficient strategy to prevent the survival of plasmid-free cells due to selective killing. Up to now, three major groups of PAS have been described: (1) toxin/antitoxin-based systems, (2) metabolism-based systems, and (3) operator repressor titration systems [1]. PASs have been successfully used in metabolically engineered strains to increase the product yield by stabilizing the plasmid in the cells, which carries the genes associated with product synthesis pathway and an addiction system. For instance, an antibiotic-free plasmid selection system based on glycine auxotrophy was constructed and used for overproduction of recombinant protein [6]. To maintenance the plasmid carrying cyanophycin synthesis gene *cphA*, Kroll et al. [7] established a novel anabolism-based addiction system. This system consisted of two components: an *E. coli ispH* mutant that cannot synthesize isopentenyl pyrophosphate (IPP), an essential precursor for isoprenoid biosynthesis, and a synthetic plasmid harboring *cphA* gene and the relevant genes of a foreign IPP-producing mevalonate pathway. The resultant strain revealed a plasmid stability of 100% and improved cyanophycin production.

Chromosomal gene integration (CGI) is another strategy to stabilize foreign genes. For example, pyruvate decarboxylase gene *pdc* and alcohol dehydrogenase II gene *adhB* from *Zymomonas mobilis* were integrated into the *E. coli* chromosome for ethanol biosynthesis, and the integration improved the stability of the *Z. mobilis* genes in *E. coli* and ethanol production [8]. Recently, a method was developed to insert multiple desired genes into target loci on *E. coli* chromosome and up to six copies of *lacZ* gene were simultaneously integrated into different loci. The β-galactosidase activity increased corresponding to the copy number of inserted *lacZ* genes [9]. To a certain extent, multiple insertions have resolved the main problem of CGI, low expression level of recombinant protein due to a low copy number.

Poly(3-hydroxypropionate) (P3HP) is a biodegradable and biocompatible plastic exhibiting high rigidity, ductility, and exceptional tensile strength in drawn films, and was regarded as one of the alternatives to petrochemical-derived plastic [10]. The biosynthesis of P3HP and 3HP-containing copolymers was

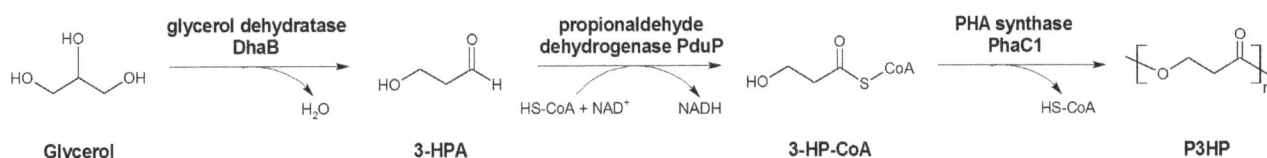

Figure 1. P3HP synthesis pathway used in this study. 3-HPA, 3-hydroxypropionaldehyde; 3-HP-CoA, 3-hydroxypropionyl coenzyme A.

previously dependent on structurally related precursors, such as 3HP, acrylate and 1,3-propanediol [11–14]. However the addition of these expensive precursors increased P3HP production cost. To solve this problem, we constructed a recombinant *E. coli* strain to synthesize P3HP using inexpensive carbon source glycerol (Figure 1) [15]. The genes involved in P3HP synthesis were cloned into two plasmids: the glycerol dehydratase and its reactivating factor genes, *dhaB123* and *gdrAB*, from *Klebsiella pneumoniae* were inserted in the expression vector pACYCDuet-1 to generate plasmid pWQ04, and the propionaldehyde dehydrogenase gene *pduP* from *Salmonella typhimurium* and polyhydroxyalkanoate synthase gene *phaC1* from *Cupriavidus necator* were carried by pWQ02. Under the optimized culture conditions, the recombinant *E. coli* strain accumulated 10.1 g/L P3HP (representing 46.4% of the cell dry weight) in a fed-batch fermentation.

To optimize the P3HP production strain, 5 strategies (Figure 2) were designed and tested in this study. Strategy I used the previously constructed plasmids pWQ02 and pWQ04. Strategy II was designed to construct a phenylalanine/tyrosine-auxotrophy based PAS and all the genes associated with P3HP synthesis were integrated in *E. coli* chromosome in Strategy III. Strategy IV strains carried portion of genes involved in P3HP synthesis on chromosome and others on plasmid, which was further developed using a tyrosine-auxotrophy based PAS to improve the plasmid stability in Strategy V. As a result, the Strategy V strain Q1738 produced 25.7 g/L P3HP from glycerol in aerobic fed-batch fermentation, 2.5-time higher than the previous report.

Materials and Methods

Bacterial Strains and Growth Conditions

All strains and plasmids used in this study are listed in Table 1. *E. coli* DH5α was used as the host to construct and store all recombinant plasmids, *E. coli* χ7213 strain was used for preparation of all suicide vectors, and *E. coli* BL21(DE3) strain was used for protein expression and P3HP production. Bacteria were grown at 37°C in Luria-Bertani (LB) broth unless specified. Diaminopimelic acid (DAP) (50 µg/ml) was used for the growth of χ7213 strain. When necessary, antibiotics were added at final concentration of 100 µg/ml for ampicillin and 34 µg/ml for chloramphenicol. LB agar containing 10% sucrose was used for *sacB* gene-based counter selection in allelic exchange experiments.

Construction of Recombinant Plasmids

The primers used are listed in Table S1 in File S1. The plasmid pG01 was constructed using PCR fragments containing the *tyrA* coding region generated with primers 366 and 367 and BL21(DE3) chromosomal DNA as a template, which were digested with *Xho*I and then ligated with pWQ02 digested by the same enzyme. The plasmid pG02 was constructed using PCR fragments containing *pheA* coding region generated with primers 364 and 365 and BL21(DE3) chromosomal DNA as template, which were digested with *Afl*II and *Hind*III and then ligated with pWQ04 digested by the same enzyme.

Construction of BL21(DE3) Strains with Chromosomal Mutations

The primers used are listed in Table S1in File S1. The mutations were constructed using suicide vector pRE112 as previously described [16]. For the *pheA tyrA* deletion, two pairs of primers, 369/370 and 371/372, were used to amplify approximately 500-bp fragments upstream and downstream of these genes from BL21(DE3) chromosome, respectively. The two fragments were then joined by PCR using primers 369 and 372. The PCR product was digested with *Sac*I and *Xba*I and then ligated between the *Sac*I and *Xba*I sites of vector pRE112 to generate plasmid pG03. The *pheA tyrA* mutation was introduced into BL21(DE3) by allelic exchange using suicide vectors pG03. A similar strategy was used to construct *tyrA* mutation using suicide vector pG04 constructed with primers 569/570 and 571/572.

To integrate the *dhaB123* and *gdrAB* genes into *prpR* locus on BL21(DE3) chromosome, two pairs of primers, 268/269 and 270/271, were used to amplify approximately 500-bp fragments upstream and downstream of *prpR* gene from BL21(DE3) chromosome, respectively. The two fragments were then joined by PCR using primers 268 and 271. The PCR product was digested with *Sac*I and *Nhe*I and then ligated between the *Sac*I and *Xba*I sites of vector pRE112 to generate plasmid pRE112-Δ*prpR*. Then the *Kpn*I-*Xba*I fragment containing *dhaB123 gdrAB* genes from plasmid pWQ04 was inserted into the corresponding sites of plasmid pRE112-Δ*prpR* to generate suicide vector pLC01, which was used to mediate the allelic exchange to generate Δ*prpR::lacI* P_{T7} *gdrAB* P_{T7} *dhaB123* strain. A similar strategy was used to construct suicide vectors pG05, pG06, pG07, and pG08 to generate the chromosomal gene integration of Δ*ascF::*P_{T7} *phaC pduP*, Δ*mtlA::*P_{T7} *phaC pduP*, Δ*ebgR::*P_{T7} *phaC pduP*, and Δ*melR::*P_{T7} *phaC pduP*, respectively.

Shake Flask Cultivation

The strain was inoculated into 500 ml baffled Erlenmeyer flasks containing 100 mL of minimal medium, which contains 3 g/L glucose, 20 g/L glycerol, 1.5 g/L KH_2PO_4, 3 g/L $(NH_4)_2SO_4$, 1 g/L citric acid, 1 g/L citrate sodium, 1.9 g/L KCl, 3 g/L $MgSO_4$, 0.138 g/L $FeSO_4 \cdot 7H_2O$, 4.5 mg/L vitamin B_1, and 1 ml of trace element solution. The trace element solution contained (per liter): 0.37 g $(NH_4)_6Mo_7O_{24} \cdot 4H_2O$, 2.47 g H_3BO_4, 1.58 g $MnCl_2 \cdot 4H_2O$, 0.29 g $ZnSO_4 \cdot 7H_2O$, 0.25 g $CuSO_4 \cdot 5H_2O$. The culture broth was inoculated with the overnight culture and incubated in a gyratory shaker incubator at 37°C and 200 rpm. The cells were induced at OD_{600}~0.6 with 0.05 mM IPTG and further incubated at 30°C. 5 µM of vitamin B_{12} (VB_{12}) and appropriate antibiotic were added every 12 h. The cell dry weight (CDW) and P3HP yield were determined after 48 h culturing. All shake flask experiments were carried out in triplicates.

Fed-batch Fermentation

Fed-batch cultures were carried out in a Biostat B plus MO5L fermentor (Sartorius Stedim Biotech GmbH, Germany) containing

Strategy I, double plasmids

Q1359

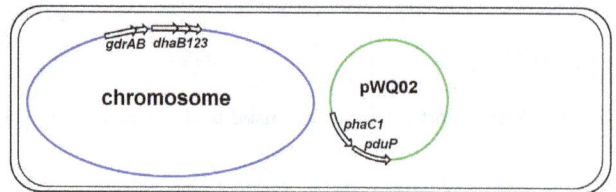

Strategy II, plasmid addiction system (PAS)

Q1509

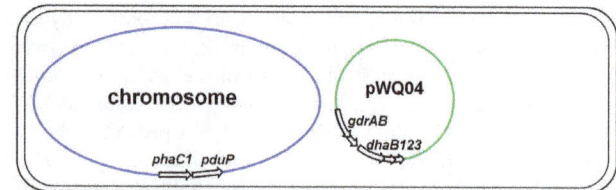

Strategy III, chromosomal gene integration (CGI)

Q1599

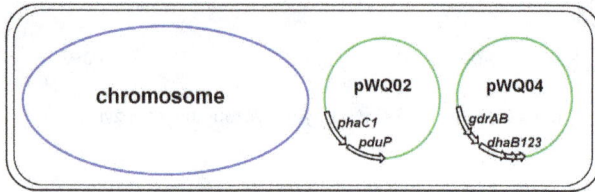

Strategy IV, CGI + plasmid

Q1638

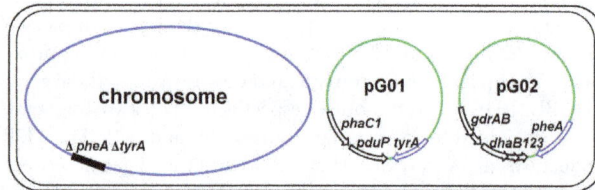

Strategy IV, CGI + plasmid

Q1802

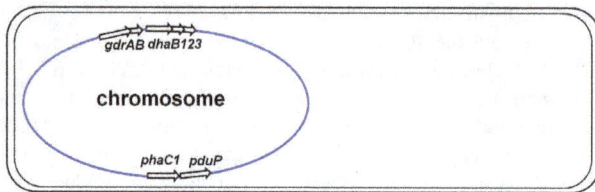

Strategy V, CGI + PAS

Q1738

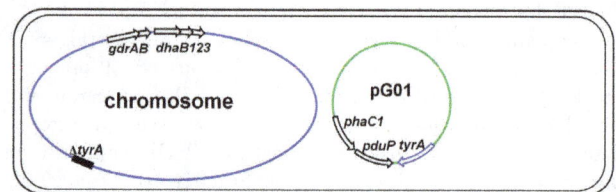

Figure 2. Schematic representation of the strains and strategies for P3HP production used in this study.

3 L of minimal medium as described above. During the fermentation process, pH was controlled at 7.0 via automated addition of 5 M KOH and antifoam 204 was used for foam control. The dissolved oxygen (DO) concentration was maintained at 5% saturation by associating with agitation from 400 rmp to 800 rpm and aeration with the airflow rate of 2 liters per min. After the initial carbon sources were nearly exhausted, fed-batch mode was commenced by feeding a solution containing 10 M glycerol at 0.5 mL/min. The expression of exogenous genes was initiated at an OD_{600} of 12 by adding 0.05 mM IPTG and 5 µM VB_{12}. IPTG, VB_{12} and appropriate antibiotic were added every 12 h.

Cell Harvest, P3HP Extraction and Characterization

Cells of *E. coli* were harvested with centrifugation at 5000×g for 15 min, and washed with distilled water twice. To determine CDW, cell pellets were lyophilized, and the CDW was gravimetrically determined. P3HP was extracted from lyophilized cells with hot chloroform in a Soxhlet apparatus, and precipitated by ice-cold ethanol as described [17]. P3HP structure was confirmed by NMR analysis using an Avance III 600 NMR spectrometer (Bruker, Switzerland) as described previously [15].

Analysis of Antibiotic Concentration

Samples were withdrawn from the cultivation flask and centrifuged at 5000×g for 10 min. The supernatant was filtered using 0.22-µm filter, and concentrations of ampicillin and chloramphenicol were determined using Agilent 1200 HPLC System as described previously [18,19].

Analysis of Plasmid Stability

To determine plasmid stability, samples were withdrawn from the cultivation flask at assigned time, diluted and spread onto LB agar plates with or without antibiotics supplemented. Ampicillin and chloramphenicol were used for plasmids pWQ02/pG01 and pWQ04/pG02, respectively. The plates were incubated for 16 h at 37°C, and the colony-forming units (CFU) were determined and analyzed by comparing the CFU on LB agar plates containing antibiotics and CFU on LB agar plates without antibiotics.

SDS-PAGE. The strain Q1599 and *E. coli* BL21(DE3) were grown in MM and induced by 0.05 mM IPTG. The cells were harvested by centrifugation and lysed by sonication. The whole-cell lysate was used for SDS-PAGE. Protein concentration was determined using the Bradford Protein Assay Kit (Tiangen, China). Proteins were separated in 12% acrylamide gels and visualized with Coomassie brilliant blue R250.

Results

P3HP Production and Plasmid Stability of Strategy I Strain

In our previous study, the genes involved in P3HP synthesis pathway were carried by two plasmids pWQ02 and pWQ04 [15]. An *E. coli* BL21(DE3) strain carrying pWQ02 and pWQ04, named as Q1359, was used to test the P3HP production and plasmid stability with the presence and absence of antibiotics. Without the addition of antibiotics, strain Q1359 accumulated 0.34 g/L P3HP representing 15.5% of the CDW under shaking flask condition. When ampicillin and chloramphenicol were added into the culture media, P3HP production and content increased to 0.52 g/L and

Table 1. Bacteria strains and plasmids used in this study.

Strains or plasmid	Description	Source
E. coli strains		
DH5α	F⁻ supE44 ΔlacU169 (φ80 lacZ ΔM15) hsdR17 recA1 endA1 gyrA96 thi-1 relA1	lab collection
BL21(DE3)	F⁻ ompT gal dcm lon hsdSB (rB⁻ mB⁻) λ(DE3),	lab collection
χ7213	thi-1 thr-1 leuB6 glnV44 fhuA21 lacY1 recA1 RP4-2-Tc::Mu λpir ΔasdA4 Δzhf-2::Tn10	[29]
Q1475	ΔpheA ΔtyrA	BL21(DE3)
Q1463	ΔprpR::lacI P$_{T7}$ gdrAB P$_{T7}$ dhaB123	BL21(DE3)
Q1599	ΔprpR::lacI P$_{T7}$ gdrAB P$_{T7}$ dhaB123 ΔmtlA::P$_{T7}$ phaC pduP	BL21(DE3)
Q1633	ΔprpR::lacI P$_{T7}$ gdrAB P$_{T7}$ dhaB123 ΔascF::P$_{T7}$ phaC pduP ΔmtlA::P$_{T7}$ phaC pduP	BL21(DE3)
Q1693	ΔprpR::lacI P$_{T7}$ gdrAB P$_{T7}$ dhaB123 ΔascF::P$_{T7}$ phaC pduP ΔmtlA::P$_{T7}$ phaC pduP ΔebgR::P$_{T7}$ phaC pduP	BL21(DE3)
Q1736	ΔprpR::lacI P$_{T7}$ gdrAB P$_{T7}$ dhaB123 ΔascF::P$_{T7}$ phaC pduP ΔmtlA::P$_{T7}$ phaC pduP ΔebgR::P$_{T7}$ phaC pduP ΔmelR::P$_{T7}$ phaC pduP	BL21(DE3)
Q1779	ΔmtlA::P$_{T7}$ phaC pduP	BL21(DE3)
Q1734	ΔtyrA ΔprpR::lacI P$_{T7}$ gdrAB P$_{T7}$ dhaB123	BL21(DE3)
Q1359	BL21(DE3) carrying pWQ02 and pWQ04	[15]
Q1509	Q1475 carrying pG01 and pG02	BL21(DE3)
Q1638	Q1463 carrying pWQ02	BL21(DE3)
Q1802	Q1779 carrying pWQ04	BL21(DE3)
Q1738	Q1734 carrying pG01	BL21(DE3)
Recombinant plasmids		
pWQ02	rep$_{pBR322}$ Amp^R lacI P$_{T7}$ phaC pduP	[15]
pWQ04	rep$_{p15A}$ Cm^R lacI P$_{T7}$ gdrAB P$_{T7}$ dhaB123	[15]
pG01	rep$_{pBR322}$ Amp^R lacI P$_{T7}$ phaC pduP TT P$_{lac1-6}$ tyrA	pWQ02
pG02	rep$_{p15A}$ Cm^R lacI P$_{T7}$ gdrAB TT P$_{pheA}$ pheA P$_{T7}$ dhaB123	pWQ04
Suicide plasmids		
pRE112	oriT oriV sacB cat	[16]
pG03	ΔpheA ΔtyrA	pRE112
pG04	ΔtyrA	pRE112
pG05	ΔascF::P$_{T7}$ phaC pduP	pRE112
pG06	ΔmtlA::P$_{T7}$ phaC pduP	pRE112
pG07	ΔebgR::P$_{T7}$ phaC pduP	pRE112
pG08	ΔmelR::P$_{T7}$ phaC pduP	pRE112
pLC01	ΔprpR::lacI P$_{T7}$ gdrAB P$_{T7}$ dhaB123	pRE112

17.3%. After 48 h of cultivation, cultures were appropriately diluted and spread onto LB agar plates with and without the antibiotics to calculate the plasmid stability. Most of the cells lost their plasmids even with antibiotic selection (Table 2), and it was assumed that the segregational plasmid instability was caused by two reasons. First, plasmid duplication increased metabolic burden of the strain. Secondly, the antibiotic in medium was degraded during culturing process. To test this speculation, the ampicillin and chloramphenicol concentration in medium was determined by HPLC. Surprisingly, no antibiotic can be detected after 48-h cultivation even with periodic antibiotic addition every 12 h, indicating that the antibiotic degraded very fast under concentrations used in this study.

Construction and Characterization of Strain with PAS

To improve P3HP production and plasmid stability, a phenylalanine/tyrosine-auxotrophy based PAS was designed and constructed. The biosynthetic pathways of aromatic amino acids phenylalanine and tyrosine share the first step, from chorismate to prephenate catalyzed by bifunctional chorismate mutase/prephenate dehydratase PheA or TyrA. Besides that, PheA also carries out the second step in phenylalanine synthesis, converting prephenate into 2-keto-phenylpyruvate [20], and TyrA is responsible for the formation of 4-hydroxyphenylpyruvate from prephenate in tyrosine synthesis [21]. The pheA and tyrA genes are located next to each other in E. coli chromosome, and transcription of these 2 loci proceeds in opposite direction. In this study, an E. coli ΔpheA ΔtyrA mutant Q1475 was constructed using suicide vector pRE112 [16]. The chromosomal knockout of pheA and tyrA was verified by PCR and DNA sequencing. As shown in Figure 3, strain Q1475 was not able to grow in minimal medium. When phenylalanine and tyrosine were added, growth level and rate was similar to the wild-type E. coli BL21(DE3) strain, confirming the Phe⁻Tyr⁻ phenotype of strain Q1475.

To complement the phenylalanine/tyrosine-auxotrophy phenotype, the tyrA and pheA genes were cloned into pWQ04 and pWQ02, and the resulting plasmids were named as pG01 and pG02, respectively (Fig. S1 in File S1). For pheA gene, the structure

Table 2. P3HP production and plasmid stability of plasmid-containing strains.

Strains	antibiotics	CDW (g/L)	P3HP (g/L)	P3HP content	pWQ02 or pG01 stability	pWQ04 or pG02 stability
Q1359	–	2.27±0.19	0.34±0.04	15.5±1.2%	6.7±1.0%	4.4±1.1%
	+	3.07±0.22	0.52±0.02	17.3±2.1%	12.8±2.3%	8.2±1.7%
Q1509	–	2.33±0.31	0.55±0.03	22.5±1.1%	42.9±3.7%	27.2±0.9%
	+	2.59±0.29	0.84±0.07	31.2±0.5%	59.1±2.4%	36.9±3.1%
Q1638	–	4.67±0.23	2.01±0.09	42.6±1.7%	76.4±4.0%	–
	+	4.54±0.36	2.34±0.05	52.4±4.2%	87.3±2.8%	–
Q1802	–	2.78±0.06	0.56±0.06	20.3±1.9%	–	6.5±0.5%
	+	3.32±0.14	0.83±0.05	25.0±0.9%	–	13.1±.06%
Q1738	–	4.46±0.24	1.99±0.06	45.0±2.9%	80.8±1.7%	–
	+	4.51±0.18	2.40±0.04	53.7±3.4%	90.2±4.0%	–

The experiment was performed under shake flask condition in triplicate.

gene and its own promoter region were amplified from *E. coli* chromosomal DNA. As *tyrA* is the second gene in cluster and does not have its own promoter, a constitutive promoter P_{lac1-6} [22] was fused to the *E. coli* tyrA coding region by PCR. In order to rule out the possible interferer between transcription of *pheA/tyrA* gene and P3HP synthesis associated genes, a bi-directional Rho-independent transcriptional terminator [23] was added behind the *pheA* and *tyrA* structural genes. The resulting plasmids pG01 and pG02 were confirmed by DNA sequencing, and transformed into Q1475 to generate Strategy II strain Q1509. In minimal medium without addition of phenylalanine and tyrosine, strain Q1509 revealed a similar growth with the wild-type strain (Fig. 3).

To verify P3HP accumulation and plasmid stability, strain Q1509 was cultivated as described above. The results showed that the phenylalanine/tyrosine-auxotrophy based PAS greatly increased the stabilities of pG01 and pG02 (Table 2). In absence of antibiotics, Q1509 exhibited plasmid stabilities of 42.9% for pG01 and 27.2% for pG02, more than 6-time higher than Strategy I strain Q1359. When antibiotics were added into the medium, the plasmid stabilities of pG01 and pG02 increased to 59.1% and 36.9%, respectively, about 4.5-time higher than strain Q1359. Unfortunately, the P3HP production did not raise in proportion of the plasmid stabilities. Only 0.55 g/L and 0.84 g/L P3HP were harvested under the condition without and with antibiotics addition, about 1.5-time higher compared with Q1359 (Table 2).

Construction and Characterization of CGI Strains

To insert the P3HP synthesis associated genes into *E. coli* BL21(DE3) chromosome, a set of suicide plasmid was constructed based on the vector pRE112 [16]. For example, the flanking regions of *prpR* gene were amplified and linked up with each other by overlap extension PCR, and the restriction sites of *Kpn*I and *Xho*I were introduced at the connection point by primer design. This fragment was cloned into the vector pRE112 to generate pRE112-Δ*prpR*. Then an *Xho*I-*Kpn*I fragment from pWQ04, encoding transcriptional regulator LacI and glycerol dehydratase system, was inserted into the corresponding site of pRE112-Δ*prpR*, and the resulting plasmid was defined as pLC01, which was used to mediate the allelic exchange. After two rounds of selection based on the positive marker chloramphenicol resistance gene *cat* and negative marker levan-sucrase gene *sacB* from *Bacillus* spp. [24], we obtained the strain Q1463 carrying chromosomal copy of *dhaB123* and *gdrAB* genes (Figure S2 in File S1 and Table 1). The *phaC1* and *pduP* genes were inserted into the *mtlA* locus, mediated

by suicide vector pG06 similarly, to generate Strategy III strain Q1599 (Figure 2, Figure S3 in File S1 and Table 1). The *prpR* gene and *mtlA* gene encode propionate catabolism regulator and mannitol-specific phosphotransferase system (PTS) enzyme IIA component, respectively. Under conditions used in this study, the *prpR* and *mtlA* mutations shouldnot affect the cell metabolism.

Cultivated in minimal medium for 48 h, strain Q1599 accumulated 0.26 g/L P3HP, which represented 6.4% of the CDW and was much lower than the P3HP productions of Q1359 and Q1509. To figure out the reason of low P3HP production in strain Q1599, SDS-PAGE analysis of whole-cell lysate was performed. Compared with the control strain *E. coli* BL21(DE3), DhaB1 and GdrA were observed as distinct bands with the expected molecular weights on SDS-PAGE, however we cannot find the bands for PhaC1 and PduP at appropriate position. It is assumed that the low copy number limited the expression level of *phaC1* and *pduP* genes, so another three copies of these two genes were inserted into the chromosomal loci of *ascF*, *ebgR* and *melR*. These three genes are all involved in degradation of carbon compounds like cellobiose and lactose. As these disaccharide molecules are absent, the inactivation of those genes should not be detrimental to cell metabolism and growth. The result of shake flask cultivation showed that the P3HP production increased with the crescent copy number of *phaC1* and *pduP* genes (Table 3), to the similar level of strains Q1359 and Q1509. Although this strain is stable and antibiotic-free, its P3HP production was still too low.

Combination of CGI Strategy and Plasmid Vector

To further improve the P3HP production, we constructed the Strategy IV strains, in which the CGI strategy and plasmid vector were used simultaneously. The strain Q1638 carries a chromosomal copy of genes encoding glycerol dehydratase and its reactivatase and plasmid-borne *phaC1* and *pduP* genes, while in strain Q1802 the *phaC1* and *pduP* genes were integrated into the *mtlA* locus and the *dhaB* and *gdrAB* genes were carried by plasmid pWQ04 (Figure 2).

The strains Q1638 and Q1802 were inoculated into minimal medium to test the P3HP production and plasmid stability. After 48 h of cultivation in shake flasks, the strain Q1638 accumulated 2.01 g/L and 2.34 g/L P3HP without and with addition of ampicillin, respectively, while the strain Q1802 only produced 0.56 g/L and 0.83 g/L P3HP under the same conditions (Table 2). In respect of plasmid stability, 76.4% of strain Q1638 cells still carried plasmid pWQ02 at the end of cultivation even without

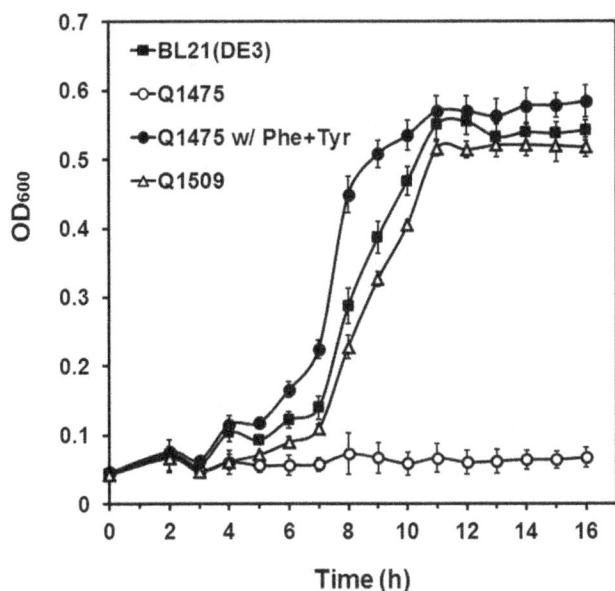

Figure 3. Growth of E. coli strains with and without the phenylalanine/tyrosine- auxotrophy based PAS, tested using minimal medium. The experiment was performed under shake flask condition in triplicate.

addition of ampicillin, however only 13.1% of strain Q1802 cells possessed the expression plasmid pWQ04 with the presence of chloramphenicol (Table 2).

Based on Strategy IV strain Q1638, we developed Strategy V strain Q1738 as following: the chromosomal *tyrA* gene was knocked out and plasmid pWQ02 was replaced by pG01 (Figure 2). In shake flask cultivation, strain Q1738 revealed similar P3HP production and slightly higher plasmid stability than strain Q1638 (Table 2). Compared with original strain Q1359, strains Q1638 and Q1738 presented about 4.5-time higher P3HP production and only required the addition of ampicillin in the growth process. Even without the usage of antibiotics, the P3HP production of strains Q1638 and Q1738 was still about 4-time higher than that of strain Q1359 with presence of ampicillin and chloramphenicol.

Fed-batch Fermentation

To evaluate the P3HP production in a scalable process, fed-batch fermentation of Q1638 and Q1738 was carried out at 5-L scale under aerobic condition. Cell growth and P3HP accumulation were monitored over the course of fermentation. As shown in Figure 4, CDW and P3HP reached the maximum in 36 h. With

presence of ampicillin, the P3HP productions of trains Q1638 and Q1738 were 24.3 g/L (58.1% of CDW) and 25.7 g/L (67.9% of CDW), respectively. Even without antibiotic addition, 15.1 g/L and 16.2 g/L P3HP was accumulated by trains Q1638 and Q1738, respectively, higher than the previously reported P3HP yield from glycerol [15].

Discussion

In this study, we are trying to construct a genetically stable strain for P3HP biosynthesis. As shown in Table 2, the stability of plasmid pWQ02 in stain Q1638 and plasmid pWQ04 in strain Q1802 was significantly improved when compared with strain Q1359. In strains Q1738 and Q1509, similar phenotype of plasmid pG01 was also observed. All these three strains with increased plasmid stability contain only one plasmid, whereas two in strains Q1359 and Q 1509. This phenomenon indicated that the plasmid stability would decrease if two or more types of plasmid exist in the same strain. It was reported that the segregative plasmid stability decreased with the size increasing, and the metabolic burden caused by plasmid duplication is a major reason for plasmid loss [25,26]. Multiple plasmids brought about heavier burden on cell metabolism obviously, and a single plasmid had better stability reasonably.

Another noticeable result is that the CDW and P3HP production of strain Q1638 were much higher than that of Q1359, and the only difference between these strains is the copy number of glycerol dehydratase genes. Besides burden caused by plasmid duplication, the toxic product of glycerol dehydratase was also responsible for the low P3HP content. Glycerol dehydratase converts glycerol into 3-hydroxypropionaldehyde, which is a major component of antimicrobial substance Reuterin and inhibits the growth of some bacteria, yeasts and protozoa [27,28]. The difference between strains Q1509 and Q1738 can be explained in the same way, and the intermediate toxicity was also the reason that the CDW and P3HP production were significantly lower when ampicillin was not supplemented in strains Q1638 and Q1738 (Figure 4). They carry a chromosomal copy of glycerol dehydratase gene, and the other genes involved in P3HP synthesis were borne by plasmids with ampicillin resistance. When ampicillin was absent, the plasmid instability increased, resulting in intracellular accumulation of 3-hydroxypropionaldehyde and growth depression.

In sum, the microbial P3HP production from glycerol was improved greatly by constructing a genetically stable *E. coli* recombinant strain. To overcome the strain instability due to plasmid loss, two strategies, amino acid anabolism based PAS and chromosomal integration, were tested. Finally, a combination of those two methods led to the best result. Our recombinant strain Q1738 produced 25.7 g/L P3HP from glycerol in aerobic

Table 3. P3HP production of strains with various copy numbers of *phaC1* and *pduP* genes.

Strains	copy number of *phaC1* and *pduP* genes	CDW (g/L)	P3HP (g/L)	P3HP content
Q1599	1	4.02±0.26	0.26±0.04	6.4±0.4%
Q1633	2	3.58±0.14	0.38±0.03	10.7±0.8%
Q1693	3	3.37±0.21	0.44±0.01	13.2±1.4%
Q1736	4	2.65±0.26	0.50±0.01	18.9±0.3%

The experiment was performed under shake flask condition in triplicate.

A

B

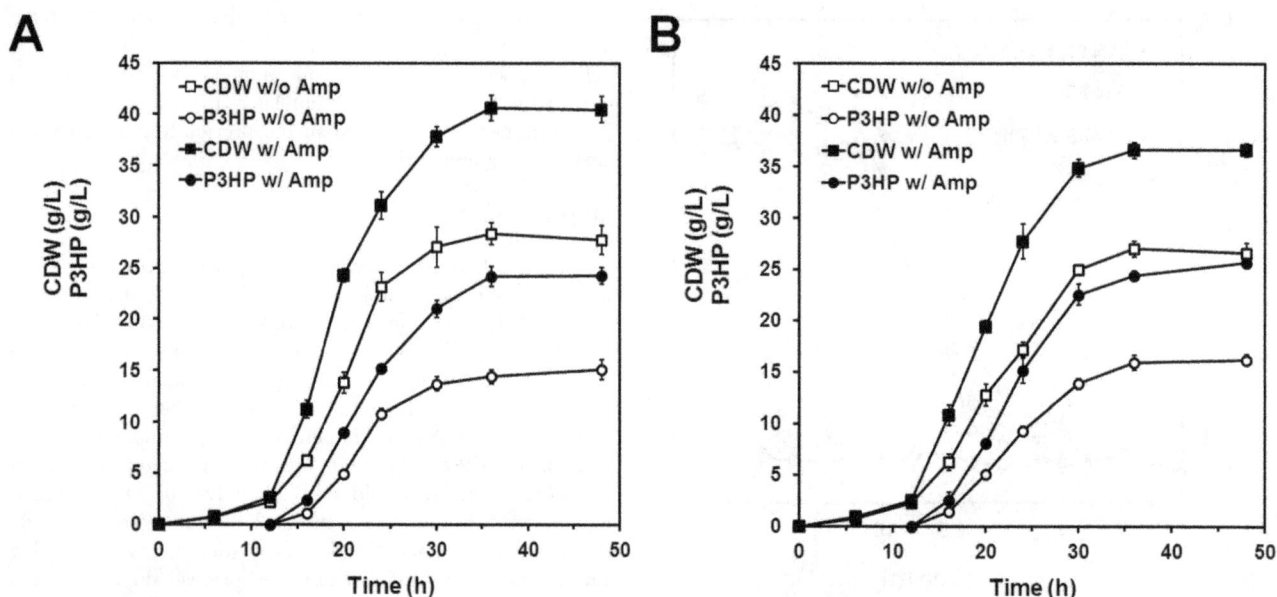

Figure 4. Time profiles for CDW and P3HP production during aerobic fed-batch fermentation of Strain Q1638 (A) and Q1738 (B). The experiment was performed in triplicate.

fed-batch fermentation. To the best of our knowledge, this is the highest P3HP production from inexpensive carbon sources.

Acknowledgments

We acknowledge Dr. Roy Curtiss III (Arizona State University) for supplying the suicide vector pRE112 and strain χ7213.

Author Contributions

Conceived and designed the experiments: MX GZ. Performed the experiments: YG CL CS RZ. Analyzed the data: YG YD MX GZ. Wrote the paper: GZ.

References

1. Kroll J, Klinter S, Schneider C, Voss I, Steinbuchel A (2010) Plasmid addiction systems: perspectives and applications in biotechnology. Microbial Biotechnol 3: 634–657.
2. Peredelchuk MY, Bennett GN (1997) A method for construction of E. coli strains with multiple DNA insertions in the chromosome. Gene 187: 231–238.
3. Jones KL, Keasling JD (1998) Construction and characterization of F plasmid-based expression vectors. Biotechnol Bioeng 59: 659–665.
4. Wang Z, Xiang L, Shao J, Wegrzyn A, Wegrzyn G (2006) Effects of the presence of ColE1 plasmid DNA in Escherichia coli on the host cell metabolism. Microb Cell Fact 5: 34.
5. Zabriskie DW, Arcuri EJ (1986) Factors influencing productivity of fermentations employing recombinant microorganisms. Enzyme Microb Technol 8: 706–717.
6. Vidal L, Pinsach J, Striedner G, Caminal G, Ferrer P (2008) Development of an antibiotic-free plasmid selection system based on glycine auxotrophy for recombinant protein overproduction in Escherichia coli. J Biotechnol 134: 127–136.
7. Kroll J, Steinle A, Reichelt R, Ewering C, Steinbuchel A (2009) Establishment of a novel anabolism-based addiction system with an artificially introduced mevalonate pathway: complete stabilization of plasmids as universal application in white biotechnology. Metab Eng 11: 168–177.
8. Ohta K, Beall DS, Mejia JP, Shanmugam KT, Ingram LO (1990) Genetic improvement of Escherichia coli for ethanol production: chromosomal integration of Zymomonas mobilis genes encoding pyruvate decarboxylase and alcohol dehydrogenase II. J Bacteriol 57: 893–900.
9. Koma D, Yamanaka H, Moriyoshi K, Ohmoto T, Sakai K (2012) A convenient method for multiple insertions of desired genes into target loci on the Escherichia coli chromosome. Appl Microbiol Biot 93: 815–829.
10. Andreeben B, Steinbuchel A (2010) Biosynthesis and biodegradation of 3-hydroxypropionate-containing polyesters. Appl Environ Microbiol 76: 4919–4925.
11. Ichikawa M, Nakamura K, Yoshie N, Asakawa N, Inoue Y, et al. (1996) Morphological study of bacterial poly(3-hydroxybutyrate-co-3-hydroxypropionate). Macromol Chem Phys 197: 2467–2480.
12. Green PR, Kemper J, Schechtman L, Guo L, Satkowski M, et al. (2002) Formation of short chain length/medium chain length polyhydroxyalkanoate copolymers by fatty acid beta-oxidation inhibited Ralstonia eutropha. Biomacromolecules 3: 208–213.
13. Zhou Q, Shi ZY, Meng DC, Wu Q, Chen JC, et al. (2011) Production of 3-hydroxypropionate homopolymer and poly(3-hydroxypropionate-co-4-hydroxybutyrate) copolymer by recombinant Escherichia coli. Metab Eng 13: 777–785.
14. Meng DC, Shi ZY, Wu LP, Zhou Q, Wu Q, et al. (2012) Production and characterization of poly(3-hydroxypropionate-co-4-hydroxybutyrate) with fully controllable structures by recombinant Escherichia coli containing an engineered pathway. Metab Eng 14: 317–324.
15. Wang Q, Yang P, Liu C, Xue Y, Xian M, et al. (2013) Biosynthesis of poly(3-hydroxypropionate) from glycerol by recombinant Escherichia coli. Bioresour Technol 131: 548–551.
16. Edwards RA, Keller LH, Schifferli DM (1998) Improved allelic exchange vectors and their use to analyze 987P fimbria gene expression. Gene 207: 149–157.
17. Brandl H, Gross RA, Lenz RW, Fuller RC (1988) Pseudomonas oleovorans as a source of poly(β-hydroxyalkanoates) for potential applications as biodegradable polyesters. Appl Environ Microbiol 54: 1977–1982.
18. Burns DT, O'Callaghan M, Smyth WF, Ayling CJ (1991) High-performance liquid chromatographic analysis of ampicillin and cloxacillin and its application to an intramammary veterinary preparation. Fresenius' J Anal Chem 340: 53–56.
19. Shen HY, Jiang HL (2005) Screening, determination and confirmation of chloramphenicol in seafood, meat and honey using ELISA, HPLC–UVD, GC–ECD, GC–MS–EI–SIM and GCMS–NCI–SIM methods. Anal Chim Acta 535: 33–41.
20. Dopheide TA, Crewther P, Davidson BE (1972) Chorismate mutase-prephenate dehydratase from Escherichia coli K-12. II. Kinetic properties. J Biol Chem 247: 4447–4452.

21. Sampathkumar P, Morrison JF (1982) Chorismate mutase-prephenate dehydrogenase from *Escherichia coli*. Purification and properties of the bifunctional enzyme. Biochim Biophys Acta 702: 204–211.

22. Liu M, Tolstorukov M, Zhurkin V, Garges S, Adhya S (2004) A mutant spacer sequence between -35 and -10 elements makes the P$_{lac}$ promoter hyperactive and cAMP receptor protein-independent. Proc Natl Acad Sci USA 101: 6911–6916.

23. Lesnik EA, Sampath R, Levene HB, Henderson TJ, McNeil JA, et al. (2001) Prediction of Rho-independent transcriptional terminators in *Escherichia coli*. Nucleic Acids Res 29: 3583–3594.

24. Gay P, Le Coq D, Steinmetz M, Ferrari E, Hoch JA (1983) Cloning structural gene *sacB*, which codes for exoenzyme levansucrase of *Bacillus subtilis*: expression of the gene in *Escherichia coli*. J Bacteriol 153: 1424–1431.

25. Wojcik K, Wieckiewicz J, Kuczma M, Porwit-Bobr Z (1993) Instability of hybrid plasmids in *Bacillus subtilis*. Acta Microbiol Pol 42: 127–136.

26. Friehs K (2004) Plasmid copy number and plasmid stability. Adv Biochem Engin/Biotechnol 86: 47–82.

27. Vollenweider S, Grassi G, Konig I, Puhan Z (2003) Purification and structural characterization of 3-hydroxypropionaldehyde and its derivatives. J Agric Food Chem 51: 3287–3293.

28. Talarico TL, Dobrogosz WJ (1989) Chemical characterization of an antimicrobial substance produced by *Lactobacillus reuteri*. Antimicrob Agents Chemother 33: 674–679.

29. Roland K, Curtiss R, 3rd, Sizemore D (1999) Construction and evaluation of a D*cya* D*crp Salmonella typhimurium* strain expressing avian pathogenic *Escherichia coli* O78 LPS as a vaccine to prevent airsacculitis in chickens. Avian Dis 43: 429–441.

15

Overexpression of a Defensin Enhances Resistance to a Fruit-Specific Anthracnose Fungus in Pepper

Hyo-Hyoun Seo[1], Sangkyu Park[2], Soomin Park[3], Byung-Jun Oh[4], Kyoungwhan Back[2], Oksoo Han[2], Jeong-Il Kim[2,5]*, Young Soon Kim[2,5]*

1 Medicinal Nanomaterial Institute, BIO-FD&C Co. Ltd., Incheon, Korea, 2 Department of Biotechnology, Chonnam National University, Gwangju, Korea, 3 Experiment Research Institute, National Agricultural Products Quality Management Service, Seoul, Korea, 4 Biological Control Center, Jeonnam Bioindustry Foundation, JeollaNamdo, Korea, 5 Kumho Life Science Laboratory, Chonnam National University, Gwangju, Korea

Abstract

Functional characterization of a defensin, J1-1, was conducted to evaluate its biotechnological potentiality in transgenic pepper plants against the causal agent of anthracnose disease, *Colletotrichum gloeosporioides*. To determine antifungal activity, J1-1 recombinant protein was generated and tested for the activity against *C. gloeosporioides*, resulting in 50% inhibition of fungal growth at a protein concentration of 0.1 mg·mL^{-1}. To develop transgenic pepper plants resistant to anthracnose disease, *J1-1* cDNA under the control of 35S promoter was introduced into pepper via *Agrobacterium*-mediated genetic transformation method. Southern and Northern blot analyses confirmed that a single copy of the transgene in selected transgenic plants was normally expressed and also stably transmitted to subsequent generations. The insertion of T-DNA was further analyzed in three independent homozygous lines using inverse PCR, and confirmed the integration of transgene in non-coding region of genomic DNA. Immunoblot results showed that the level of J1-1 proteins, which was not normally accumulated in unripe fruits, accumulated high in transgenic plants but appeared to differ among transgenic lines. Moreover, the expression of jasmonic acid-biosynthetic genes and pathogenesis-related genes were up-regulated in the transgenic lines, which is co-related with the resistance of J1-1 transgenic plants to anthracnose disease. Consequently, the constitutive expression of J1-1 in transgenic pepper plants provided strong resistance to the anthracnose fungus that was associated with highly reduced lesion formation and fungal colonization. These results implied the significance of the antifungal protein, J1-1, as a useful agronomic trait to control fungal disease.

Editor: Sung-Hwan Yun, Soonchunhyang University, Republic of Korea

Funding: This work was supported by Next-Generation BioGreen 21 Program, Rural Development Administration, Republic of Korea (Grant No. PJ00949101), and by Technology Development Program for Agriculture and Forestry, Ministry for Agriculture, Forestry and Fisheries, Republic of Korea (Grant No. 309017-5). The funders had no role in study design, data collection and analysis, decision to publish, or preparation of the manuscript.

Competing Interests: The authors have the following interests. Hyo-Hyoun Seo is employed by BIO-FD&C Co., Ltd. There are no patents, products in development or marketed products to declare.

* E-mail: kimji@chonnam.ac.kr (JIK); youngskim@chonnam.ac.kr (YSK)

Introduction

Higher plants have innate defense systems to protect themselves against biotic stresses [1–3]. A range of protective molecules, including antimicrobial proteins, are synthesized in the tissues invaded by pathogens or accumulated during normal growth [4–6]. Defensins that belong to antimicrobial peptide superfamily are a large class of small peptides occurring in various living organisms, ranging from microorganisms to plants and mammals [7,8]. On the basis of structural and functional similarity with insect defensin, plant antimicrobial peptide called γ–thionin in wheat and barley grains was renamed as defensin [9]. Plant defensins are composed of three anti-parallel β-strands and one α-helix with a characteristic three-dimensional folding stabilized by four disulfide bonds [10]. The cysteine-stabilized α-helix/β-sheet (CSαβ) motif confers great stability on the peptide to maintain the functional activity [11].

The main biological function of plant defensins was found to inhibit the growth of a broad range of phytopathogenic fungi at micromolar concentrations [12]. Other biological activities of defensins have also been proposed as protein synthesis inhibitors, α-amylase inhibitors, zinc tolerance mediators, and ion channel blockers [13–16]. Although the action mode of plant defensin in fungal growth inhibition has not been clearly understood, the inhibition of fungal growth is followed by initial binding of the defensin on fungal membrane due to electrostatic and/or hydrophobic interactions. Indeed, a higher concentration of defensins causes severe membrane permeabilization, which leads to fungal death [17–20]. However, this arouses a controversy that the peptides could disrupt the integrity of membranes not only in the fungal cells, but also in plant cells. Regarding the localization, plant defensins were generally predicted to be secreted to extracellular space due to the occurrence of signal peptide at their N-terminal. Previously, subcellular localization analysis showed that several plant defensins were deposited to cell wall [9]. Otherwise, a flower defensin, NaD1, has been immunolocalized in the vacuole in *Nicotiana alata* [21], and AhPDF1.1 from *Arabidopsis halleri* was retained in internal compartments while moving to the lytic vacuole [22]. Therefore, further studies are necessary to clarify the action mode in association with the localization of defensins.

Cumulative studies demonstrate that defensins are expressed in various vegetative tissues of plant [23,24]. In addition, reproductive organs, such as floral organs or fruits, express defensins as part of a predetermined developmental program or induce defensins under stressed conditions associated with the invasion of a pathogen [25,26]. A defensin, designated as J1-1, has been previously described in the fruit of bell pepper [27]. Expression of the *J1-1* gene was found to occur during the ripening of the fruit and was also inducible by wound or pathogen. Thus, J1-1 was suggested to play a role in protecting the reproductive organs against biotic and abiotic stresses. *In vitro* antifungal assay has shown that J1-1 protein isolated from ripe pepper fruits effectively suppressed the mycelial growth of *Fusarium oxysporum* and *Botrytis cinerea*. However, further questions about the functional activity of the protein in the infected pepper cells need to be addressed. For example, although the expression of the *J1-1* gene in ripe fruit during an incompatible interaction with *Colletotrichum gloeosporioides* provided indirect evidence for the role of J1-1 in protecting the ripe fruits against pathogen attack [28], genetic and biological studies of J1-1 in relation to disease resistance have been lacking.

Chilli peppers have been cultivated worldwide and are one of the most important vegetable fruits in some areas. Major constraints to the pepper fruit production include pests and diseases, and anthacnose disease is the most notifiable infectious disease caused by *Colletotrichum* species [29,30]. Immature pepper fruits in green color are vulnerable to the pathogen, causing widespread outbreaks of the disease. Despite large-scale breeding efforts to control the anthacnose disease, it remains as an endemic disease, resulting in large reductions of annual yields worldwide [31]. Thus, it is necessary to build up novel genetic resources for the development of anthracnose disease-resistant peppers, and genetic engineering is a feasible approach to generate anthracnose-resistant pepper plants [32]. Although constitutive expression of plant defensins has shown enhanced resistance in tomato, potato, and canola against various pathogens [33–37], anthracnose disease-resistant plants have not yet been developed.

In this study, we investigated the antifungal activity of J1-1 protein against the anthracnose fungus and the localization of the peptide in infected fruits. The characteristics of the pepper defensin prompted us to develop transgenic pepper plants and to assess the contribution of the defensin to pepper resistance against the anthracnose fungal infection using the transgenic pepper plants overexpressing J1-1. The results suggest that J1-1 is an attractive candidate for biotechnological application to provide enhanced resistances in pepper, especially to the anthracnose fungus.

Materials and Methods

Plant Materials and Growth Conditions

Capsicum annuum cv. Nokkwang was used in plant transformation experiments, as previously described [38]. Wild-type and transgenic seedlings were grown in a growth chamber at 25°C and 50% humidity in a light/dark cycle of 16/8 h. Pepper plants were transferred to soil and grown in a greenhouse for further experiments. Samples were collected from 2-month-old plants, except for ripening fruits. Mature fruits were harvested at the following ripening stages: stage I, green fruit; stage II, early breaker fruit; stage III, turning fruit; stage IV, purple fruit; stage V, red fruit.

Fungal Pathogens and Inoculation

The monoconidial isolate KG13 of *C. gloeosporioides*, which is only compatible with unripe pepper fruits, was used to elucidate the functional role of J1-1 protein in infected unripe fruits [39]. Growth and spore harvests of the fungus were performed as described previously [40]. Fungal inoculation was conducted by applying a drop of spore suspension (density of 5×10^5 spores per 1 mL in distilled water) onto mature unripe green fruits. The inoculated fruits were placed in high humidity and dark conditions for 1 day to stimulate infection. Thereafter, the fruits were incubated at 26°C in a growth chamber until harvesting. For analysis of the J1-1 protein, a piece (5×5 mm) of pericarp was taken from the inoculated sites of the fruits at 0, 3, 24, 48, and 72 hours.

The growth of the fungus on infected fruits was analyzed using the non-transformed or transgenic unripe fruits after fungal infection. For microscopic observation, 0.1% toluidine blue was topically applied on the infection area of the fruits at one day after infection and then peeled the skin off to observe the fungus under microscope. At five days after infection, transversal sections of the fruits were stained with lactophenol trypan blue to visualize the fungal hyphae. In addition, the development of anthracnose symptom was monitored until 9 days after infection. Then, disease rate was expressed as percentage of the number of lesions from infected spots. The sporulation was determined by counting the number of spores from a lesion.

In vitro Antifungal Activity Assay with Recombinant J1-1 Protein

The cDNA encoding *J1-1* was cloned into pGEX-6P-1 (Amersham Biosciences, Freiburg, Germany) between *Eco*RI and *Xho*I, creating an in-frame fusion with the sequence encoding glutathione-S-transferase (GST). The primers used were a forward primer (5′-GGAATTCCTTATGGCTGGCTTTTCCAAAG-3′) and a reverse primer (5′-CCCTCGAGGGATTAAGCA-CAGGGCTTCGT-3′). The GST fusion protein was then expressed in *E. coli* strain BL21 and purified according to the manufacturer's instructions. The protein concentration was determined using the Bradford method. Following purification, the antifungal activities of the GST/J1-1 fusion protein were examined against *C. gloeosporioides*. The fungal growth was monitored by microscopic examination on cover glass with 5×10^2 spores in sterile water containing various concentrations of GST/J1-1 recombinant protein or heated protein obtained by incubating at 90°C for 10 min. The spores were treated with the proteins for 24 hours at 26°C, and then counted for germination and appressorium formation in at least five microscopic fields. The experiment was conducted in triplicate.

Plasmid Construction and Pepper Transformation

A full-length cDNA of *J1-1* was amplified by PCR using the primers, 5′-GCTCTAGAGCATGGCTGGCTTTTCCAAAG-3′ (forward) and 5′-CGGATCCGTTAAGCA-CAGGGCTTCGT-3′ (reverse). The resulting fragment was cloned between *Xba*I and *Bam*HI in pBI121. Then, the expression cassette spanning the CaMV 35S promoter to the Nos terminator was transferred into pCAMBIA1300 between *Hin*dIII and *Eco*RI sites. The pCAMBIA1300/*J1-1* was mobilized into *Agrobacterium tumefaciens* GV3101 and used to generate transgenic pepper plants.

The cotyledon and hypocotyl explants were inoculated with *Agrobacterium* suspensions as described previously [38]. Following infection, the regeneration of the primary transformants was accomplished on selection medium containing 20 mg·L^{-1} hygromycin and 300 mg·L^{-1} cefotaxime. Plantlets resistant to hygromycin were then transferred onto rooting medium containing MS basal salts supplemented with 300 mg·L^{-1} cefotaxime. All cultures were incubated at 26°C under a 16/8 hr (light/dark) photoperiod. Plants having well-developed roots were transplanted to pots and grown in a greenhouse until they flowered. Primary transgenic plants (T_0) were self-pollinated and their seeds (T_1) were germinated in MS medium containing 20 mg·L^{-1} hygromycin. The number of green seedlings resistant to hygromycin was counted. The data were then analyzed using the χ^2 test to determine the number of functional *HPT* gene loci on the pepper genome. Self-pollinated T_2 progenies were also tested for hygromycin resistance to identify homozygosity. From these procedures, four transgenic pepper lines were generated and used for further analysis.

Southern and Northern Blot Analyses

To analyze the genomic DNA for integration of the *J1-1* gene, pepper genomic DNA was isolated using a DNeasy Plant Maxi Kit (Qiagen, Hilden, Germany) as described by the manufacturer. For Southern blot analysis, 15 μg of each DNA sample was digested with *Eco*RI and separated on a 1.0% (w/v) agarose gel. The digested DNA was then transferred to a nylon membrane and hybridized with *HPT* or *J1-1* gene probes that was labeled with [$\alpha^{32}P$] dCTP using the Rediprime II Random Prime Labeling System (Amersham Biosciences, UK). After hybridization, the membranes were exposed at $-80°C$ on Kodak XAR-5 film (Kodak, Rochester, NY) using an intensifying screen.

For Northern blot analysis, total RNA was extracted from the pepper fruits using a RNeasy Plant Kit (Qiagen, Hilden, Germany). 10 μg of the total RNA was separated on 1.2% denaturing agarose gels and blotted onto a Hybond N^+ membrane (GE Healthcare, Buchinghamshire, UK). The blots were then hybridized with [$\alpha^{32}P$] dCTP-labeled respective probes that were amplified by PCR. The primers used for probes were shown in Table S1.

Immunoblot Analysis and J1-1 Antibody Production

The samples were homogenized in an extraction buffer (50 mM Tris, pH 8.0, 2 mM EDTA, 2 mM DTT, 0.25 M sucrose and protease inhibitor cocktail (Roche, Mannheim, Germany)) at cold conditions and subjected to centrifugation at 3000 g for 15 min. The supernatant was used as a total protein. For immunoblot analysis, total proteins were separated by 12% SDS-PAGE and transferred onto polyvinylidene fluoride (PVDF) membranes. A dilution of polyclonal anti-J1-1 rabbit antibody was used for immunoblot analysis, which was followed by peroxidase-conjugated anti-rabbit antibody. The anti-J1-1 serum was raised against a KLH-conjugated peptide corresponding to amino acid sequences (26-39 AA; KICEALSGNFKGLCL) of J1-1 as described previously [27].

Immunohistochemical Localization of J1-1 Protein

For immunolocalization, pepper fruits were fixed in 0.1% glutaraldehyde and 4% paraformaldehyde in a 50 mM sodium phosphate buffer (pH 7.0), dehydrated in ethanol, and embedded in paraffin. Tissues were sliced into 5-μm thick transverse sections. The deparaffinized sections were incubated with anti-J1-1 antibody (1:2000) for 4 h at 12°C, followed by incubation with peroxidase-conjugated anti-rabbit antibody according to the manufacturer's instruction (DAKO, Carpinteria, CA). Control experiments using pre-immune serum were not reactive (data not shown).

Inverse PCR Analysis

The genomic DNA (gDNA) sequences flanking a T-DNA insertion were cloned from the transgenic lines using inverse PCR (i-PCR). gDNA was isolated from the leaves of independent T_2 transgenic pepper plants and digested with a restriction enzyme that was chosen in the T-DNA region of the pCAMBIA1300/*J1-1* binary vector with unique restriction sites, such as *Bgl*II, *Eco*RI, or *Hin*dIII. After purification, 1 μg of gDNA was self-ligated in a 250-μL reaction volume using 40 units of T_4 DNA ligase (Promega, Madison, WI). The circularized DNA was purified, and 100 ng of DNA was used as a template for i-PCR reactions. Two sets of primers, which were specifically designed for the sequences of T-DNA and the *J1-1* gene, were sequentially used: IP-F1/IP-R1 and IP-F2/IP-R2 for right border (RB), and IP-F3/IP-R3 and IP-F4/IP-R4 for left border (LB) (Table S1). The PCR condition was 5 min at 94°C, 35 cycles of 94°C for 30 sec, 58°C for 30 sec, and 72°C for 2 min with a 10-min extension period at 72°C. The resulting PCR products were cloned into a TOPO vector (Invitrogen, Carlsbad, CA) and subjected to sequencing. The gDNA sequences were compared using the basic local alignment search tool (BLAST).

Statistical analysis

Experimental data were subjected to analysis of variance (ANOVA) using IBM SPSS statistics 20 software. Significant difference of mean values was compared by the LSD and DMRT at $P < 0.05$. All of the data were represented as the mean ± SD of at least three independent experiments.

Results

Fruit-specific Accumulation of J1-1 Protein upon Ripening and Its Inducibility by Fungus

To understand how the expression of J1-1 is associated with fruit development, immunoblot analysis was used to compare the expression levels in various pepper tissues. J1-1 was not detected in non-fruit tissues such as leaf, stem, root, flower and unripe green fruit, but was detected in the ripe red fruit (Figure 1A and S1). J1-1 protein was gradually increased in the fruits from the early stage of the ripening, indicating the developmental regulation of the fruit-specific expression of J1-1 (Figure S2). Additionally, the presence of a higher band which is detected in the flower suggests the occurrence of another defensin member. Then, the induction of the J1-1 was monitored in the pepper fruits infected with the fruit-specific fungal pathogen, *C. gloeosporioides*. As shown in Figure 1B, J1-1 was not detected in infected unripe fruits, even though expression of the gene at the transcriptional level was previously reported in fruits in response to a pathogen [28]. On the contrary, the level of the J1-1 protein was increased at 3 hours after infection (HAI) in the ripe fruit and was maintained during the period of observation. These results suggest that differential regulation of the J1-1 expression likely occurred at both transcriptional and translational levels in infected fruits during ripening.

Moreover, immunolocalization analysis using J1-1 antibody demonstrated that the consistent expression of the J1-1 protein occurred in healthy ripe fruit (Figure 1C). J1-1 protein was not detected in unripe fruit, even around the infection site undergoing necrotized cell death in the epidermis (Figure 1C-c). In contrast, the protein was evenly expressed in the pericarp of the ripe fruit (Figure 1C-d). In response to fungal attack, a strong positive signal was noted near the position where the fungus had broken into the cell wall of the epidermis of ripe fruit. Figure 1C-f shows that the protein was secreted to the outside of epidermal cells. The protein was also highly detected in the cytoplasm of some cells around the

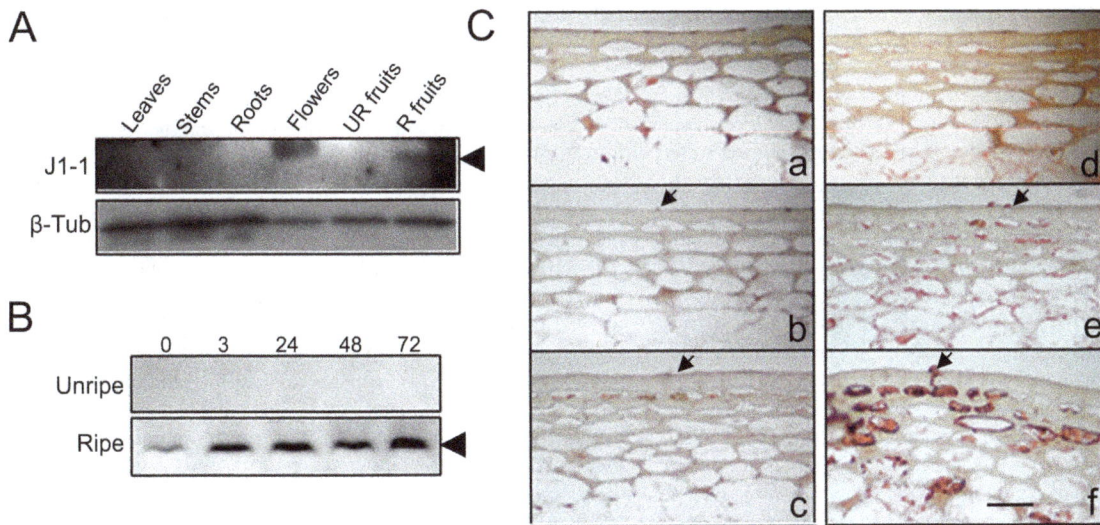

Figure 1. Expression of J1-1 is related to fruit ripening and induced by fungal infection in pepper fruits. A, Organ-specific expression of J1-1 protein in leaves, stems, roots, flowers, unripe (UR) and ripe (R) fruits of pepper. β-tubulin was shown as a loading control. An arrowhead indicates the protein band of J1-1. **B,** Fungal-induced J1-1 accumulation in unripe and ripe pepper fruits infected with *C. gloeosporioides*. Numbers on the top represent hours after infection (HAI). Immunoblot analysis was performed with total soluble proteins from pepper tissues using polyclonal J1-1 antibody. **C,** Immunolocalization of J1-1 in unripe (**a–c**) and ripe (**d–f**) fruits at 0, 24, and 48 h after inoculation. To localize the protein, transverse sections of pepper fruits were incubated with polyclonal J1-1 antibody that was detected with AEC (3-amino-9-ethylcarbazole) chromogen, shown as red. The arrows indicate fungal spores on the surface of the pepper fruits. Bar = 50 μm.

infected area. In the negative control with preimmune serum, no positive signal was detected in the tissues (data not shown). The results suggest that the J1-1 protein is preferentially accumulated in the peripheral cell layers of the ripe fruits and secreted to the invading pathogen, indicating the role in the first line of plant defense against pathogen attack.

Antifungal Activity of J1-1 Recombinant Protein

To assess the possible function of J1-1, the antifungal activity of J1-1 protein was examined against a major fruit pathogen that causes sunken disease in unripe fruits. Its effect was evaluated according to the spore germination and appressorium development of *C. gloeosporioides*. For microscopic observation, 10 μL of spores diluted in sterile water to a density of 5×10^5 mL^{-1} was mixed with GST/J1-1 protein to yield mixtures of 0.001, 0.01, 0.1, and 1 mg·mL^{-1} on cover glass and kept in a humidified chamber at 26°C for 24 hours. The results showed that the 0.1 mg·mL^{-1} mixture of GST/J1-1 protein had a 50% inhibitory effect on appressorium formation from the germination tube of the fungus (Figure 2). However, spore germination was barely affected regardless of the protein concentration (Figure S3). When the protein was heated by incubating at 90°C for 10 min, the growth inhibition was reduced by approximately 25% compared with active J1-1. These results indicate the activity of J1-1 against the pathogenic fungus, *C. gloeosporioides*, and also suggest the potential of J1-1 for plant disease control in economically important pepper cultivation.

Development of Stable Transgenic Pepper Plants Carrying *J1-1*

Based on the proposed function of J1-1, transgenic pepper plants were generated to express the *J1-1* gene under the control of the CaMV 35S promoter and nopaline synthase transcriptional terminator (Figure 3A). Cotyledonary explants were infected with *Agrobacterium* cells harboring pCAMBIA1300/*J1-1* as described

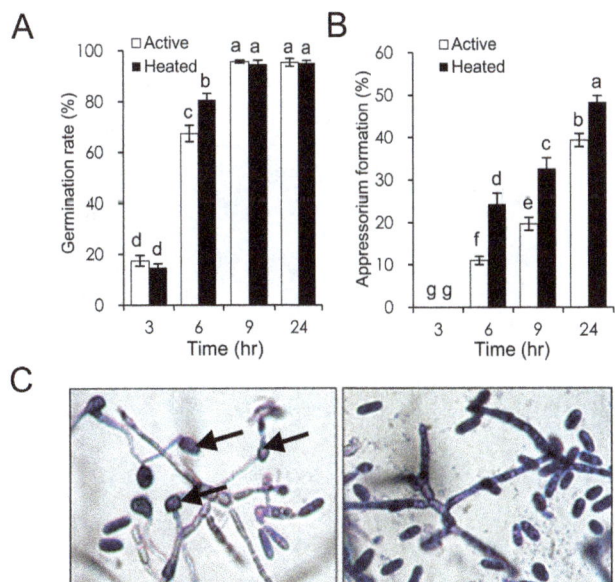

Figure 2. J1-1 recombinant protein shows antifungal activity against *C. gloeosporioides*. A, Spore germination. **B,** Appressorium formation. Spore suspensions were amended with 10 μL of the GST/J1-1 recombinant protein or its heated protein to final concentrations of 0.1 mg·mL^{-1}. The protein was heated by incubating at 90°C for 10 min. A minimum of 100 spores were counted per replicate. Each value represents the mean ± SD of three replicates. Means with different letters in each column are significantly different at P<0.05. **C,** Representative photos of fungi that were treated with 0.1 mg·mL^{-1} of GST/J1-1 recombinant protein for 48 hours (right). Control was treated with distilled water (left). Arrows indicate appressorium.

previously [38]. During subcultures of explants, 20 mg·L^{-1} hygromycin was used for the callus induction and 10 mg·L^{-1} was used at the regeneration stage to select transgenic shoots. Adventitious buds were transferred to rooting media, in which putative transgenic plants were produced (Figure S4). Since these plants displayed normal phenotypes in the pots, their seeds were obtained from self-pollination. Of the nine primary transformants, eight were identified as containing T-DNA using preliminary PCR screening with the combination of a sequence from the 35S CaMV promoter as a forward primer and a sequence from the *J1-1* cDNA as a reverse primer (data not shown).

To determine whether the transgenes were stably inherited to the next generation, seeds harvested from eight primary transgenic pepper lines were evaluated for resistance to 20 mg·L^{-1} hygromycin. Segregation ratios of 2.4~3.8:1 were observed in four lines and χ^2 analysis verified a 3:1 segregation for the *HPT* gene, indicating Mendelian segregation of a single dominant gene (Table S2). The results also suggest that transgenic pepper plants carrying the *J1-1* gene were genetically stable in advanced generations. Thus, after analyzing the antibiotic resistance of T$_2$ seeds, four

homozygous lines carrying a single copy of T-DNA were selected and used in the subsequent experiments.

Molecular Characterization of Transgenic Pepper Plants

Stable integration and expression of the transgenes was further investigated in four selected T$_1$ progenies. Southern blot analysis was conducted with genomic DNA isolated from J15-1, J32-2, J51-4, and J19-7 plants, as well as non-transformed wild-type (WT) plant as a negative control (Figure 3B). The genomic DNAs were digested with *Hind*III and hybridized with a probe composed of the *HPT* gene. While genomic DNA from control plant (WT) showed no hybridization signal to the probe DNA, each primary transgenic line exhibited a single band with a different band pattern, except for J32-2 which exhibited the same band mobility as J15-1. The result indicates that the J15-1/J32-2, J19-7, and J51-4 plants were independent events of transformation. To confirm the integration of *J1-1* in the transgenic plants, the membrane was deprobed and rehybridized with the *J1-1* gene as the probe. The result demonstrated that each transgenic plant showed the same band patterns corresponding to the *HPT* band (Figure 3B). In addition, two endogenous *J1-1* bands were detected in all lanes including the control plant. Since the T-DNA of pCAMBIA1300/ *J1-1* has a unique *Hind*III site, the result indicates that a single copy of the *J1-1* gene along with *HPT* gene was integrated into the pepper genome.

Northern blot analysis was carried out using four transgenic T$_1$ plants to confirm stable expression of the introduced transgene in the transgenic pepper plants. Total RNA was extracted from three homozygous and hemizygous progeny plants from genetically independent T$_1$ plants and hybridized with a *J1-1* cDNA probe. At the mRNA level, the introduced *J1-1* gene was transcriptionally active in the unripe fruits of the transgenic lines, as well as in the wild-type ripe pepper fruit as a positive control, while no signal was detected in the unripe fruits of the non-transformed control plant (Figure 4). In general, homozygous plants showed higher levels of transgene expression, while a hemizygous state led to weak expression in the transgenic plants carrying a single copy of T-DNA. These results indicate that the introduced gene was stably expressed, but the expression level was dependent on the hemizygosity of the transgene in the transgenic progenies.

To understand the correlation between T-DNA integration and its genetic stability, we cloned and sequenced the genomic DNA flanking both the RB and LB of T-DNA in each transgenic line. Two consecutive primers were designed in the vicinity of a unique restriction enzyme, such as *Eco*RI, *Hind*III or *Bgl*II, in the T-DNA region (Figure 3A). Genomic DNA was extracted from three independent transgenic pepper lines and digested with an appropriate enzyme to clone right and left T-DNA/gDNA junctions. After self-ligation of the gDNA, the ligated DNA was used as a template for i-PCR with a pair of primers; IP F1/IP R1 and IP F3/IP R3 were typically used for the RB and LB, respectively. As summarized in Table 1, five events from either side of the T-DNA revealed that deletion of a short DNA fragment ranging from 52 bp occurred in the border sequence of the T-DNA integrated in the pepper genome. There were two instances of more extensive deletion on the left border of T-DNA: 344 bp in J15 and 289 bp in J19 transgenic plants (Table S3). In transgenic line J15, the deleted fragment at the LB included 28 bp in the C-terminal of the *hygromycin phosphotransferase* (*HPT1*) gene, which resulted in an impairment of the original stop codon that caused an additional tail with 20 amino acids. However, the J15 transgenic progenies were capable of retaining antibiotic resistance against hygromycin. The results indicate that the genetic stability of the *HPT* gene was maintained in all transgenic pepper lines in

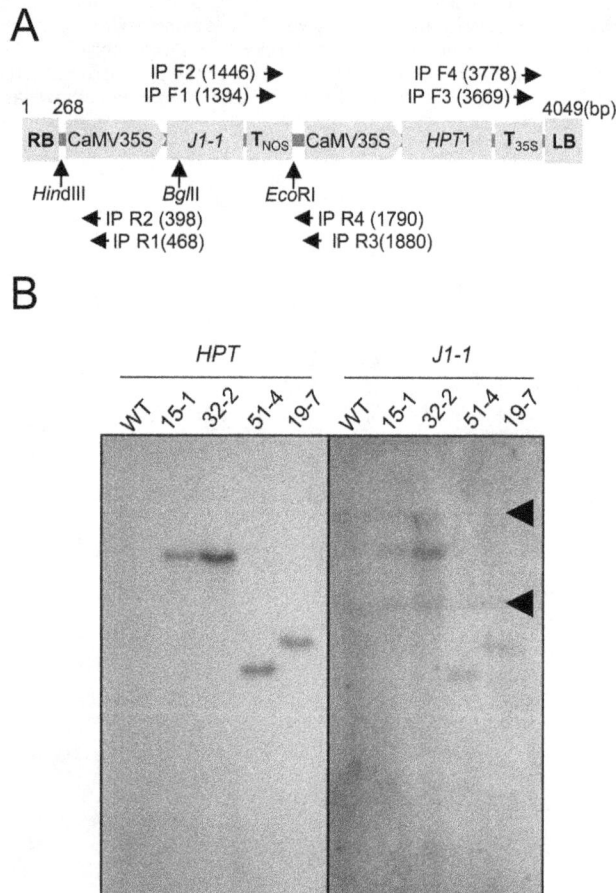

Figure 3. Southern blot analysis of transgenic pepper plants carrying *J1-1* gene. A, Schematic diagram of the T-DNA representing restriction enzyme sites and primer sites for i-PCR. LB, T-DNA left border repeat; RB, T-DNA right border repeat; *HPT1*, hygromycin phosphotransferase I; CaMV35S, CaMV 35S promoter; T$_{NOS}$, transcriptional terminator of nopaline synthase (*NOS*); T$_{35S}$, CaMV 35S transcriptional terminator. **B,** Southern blot analysis. gDNA was digested with *Hind*III, and hybridized with ^{32}P-labeled *HPT1* probe (left) or rehybridized with the *J1-1*gene (right). WT, non-transformed wild-type pepper plant. Arrowheads indicate endogenous J1-1 bands.

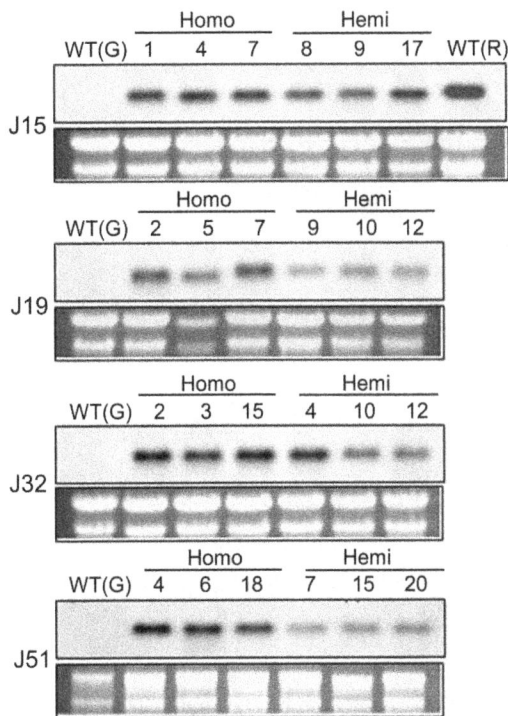

Figure 4. Northern blot analysis of unripe fruits from transgenic pepper lines. Lane 1, green fruit (G) from wild-type (WT) plant as a negative control; lanes 2–4, three T₁ transgenic plants representing homozygous progenies; lanes 5–7, three T₁ transgenic plants representing hemizygous progenies; lane 8, ripe fruit (R) from WT plant as a positive control. ^{32}P-labeled J1-1 was used as a probe, and total RNAs were shown as loading controls in lower panels.

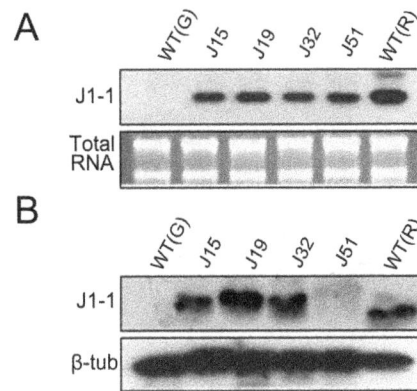

Figure 5. Expression of the J1-1 in the unripe pepper fruits of transgenic plants. A, Northern blot analysis of the *J1-1* transcript. Total RNA from each T₂ progeny was hybridized to a radiolabeled *J1-1* probe. Lane 1, unripe green fruits (G) of non-transformed wild-type (WT) pepper plant as a negative control; lanes 2–5, four T₂ transgenic lines representing homozygous progenies; lane 6, ripe fruits (R) of non-transformed pepper as a positive control. **B,** Immunoblot analysis of the J1-1 protein. Total soluble proteins from T₂ progenies were subjected to immunoblot analysis with polyclonal anti-J1-1 antibody. Total RNA and β-tubulin were shown as loading controls.

spite of the exclusion of the T-DNA junction fragments. In addition, gDNA sequence analysis of the T-DNA insertion sites of three transgenic lines did not reveal any similarity compared with known EST or gDNA sequences in the public database.

J1-1 Transcript and Protein Accumulation in Transgenic Pepper Plants

Since unripe green fruits are extremely vulnerable to infection by the hemibiotrophic anthracnose fungus, we generated transgenic plants overexpressing J1-1 constitutively to control the disease. Before investigating fungal resistance of the transgenic

pepper plants, the expression of J1-1 was initially examined in the unripe fruits of transgenic plants (T₂) at both mRNA and protein levels, as compared with that in wild-type ripe fruit as a positive control. The results showed that a considerable amount of the *J1-1* transcript was detected in the unripe green fruits of the transgenic lines, but not detected in those of the non-transformed control plant (Figure 5A). Marked accumulations of the J1-1 proteins were also observed in the green fruits of the transgenic lines except line J51, which revealed a slightly detectable amount of J1-1 protein (Figure 5B). From detailed observation with the J32 and J51 transgenic lines, a strong band of the *J1-1* transcript was detected in the unripe fruits of the both lines, indicating that the mRNA expression is normal in the J51 line (Figure S5A). Moreover, the expression of the *J1-1* transcript was increased in both lines at 24 hours after inoculation (HAI) with *C. gloeosporioides*. Unlike mRNA expression, J1-1 protein was detectable in J51 line only after fungal infection at 24 HAI, whereas J32 showed highly accumulated J1-1 protein in both healthy and infected fruits (Figure S5B). In the non-transformed control plants, a slight increase in *J1-1* mRNA was shown in the unripe fruit at 24 HAI,

Table 1. Sequence analysis of T-DNA/gDNA junctions in transgenic pepper lines by i-PCR.

Line	Border	Enzyme[a]	gDNA (bp)[b]	Deletion (bp)[c]	Primers
J15	RB	*Eco*RI	1961	52	IP F1/IP R1
	LB	*Bgl*II	951	344	IP F4/IP R3
J19	RB	*Eco*RI	98	63	IP F1/IP R1
	LB	*Hind*III	1581	289	IP F3/IP R2
J51	RB	*Eco*RI	ND	ND	IP F1/IP R1
	LB	*Eco*RI	1776	55	IP F3/IP R3

ND, not determined.
[a]Restriction enzyme used for gDNA rescue.
[b]Length of rescued gDNA flanking the T-DNA border.
[c]Deleted length at endpoint of the T-DNA.

but the fruit showed no induction of its protein. This result suggests that J1-1 protein might be unstable in healthy unripe fruits while the transcript was durable.

Expression of Jasmonic Acid (JA) Biosynthetic and Pathogenesis–Related (PR) Genes in J1-1 Transgenic Pepper Fruits

Previously, the expression of *J1-1* gene was shown to be up-regulated in the unripe and ripe pepper fruits by exogenous treatment of methyl jasmonate [28]. In addition, it has been reported that endogenous level of JA rises with the onset of ripening in several fruits [41]. Thus, it could be expected that JA signaling involves in J1-1 induced disease resistance in pepper fruits. In the present study, the expression of JA biosynthetic genes, such as lipoxygenase (*LOX*), allene oxide cyclase (*AOC*), and fatty acid hydroperoxide lyase (*HPL*) were examined in the unripe green fruits of transgenic lines as well as green and red fruits of non-transformed pepper as controls. Results showed that three JA-related genes were highly expressed in the green fruits of transgenic lines (Figure 6A). The *LOX* gene was expressed more abundantly in the green fruits of transgenic lines, while the expression levels of *AOC* and *HPL* genes were approximately similar to those in the ripe fruits of non-transformed pepper. Since it has been known that disease resistance is related to the expression of defense-related genes such as PR genes, we further investigated the expression of two PR genes: methyl jasmonate (MeJA)-treatment induced *CaPR*10 gene [42] and SA-induced *PepThi* gene [28]. The results showed that both genes were expressed in the unripe green fruits of transgenic lines, whereas no expression was observed in the green fruits of non-transformed pepper (Figure 6B). In the transgenic pepper fruits, the expression levels of MeJA-inducible *CaPR*10 were higher than those in the red fruits of non-transformed pepper. In contrast, the expression levels of SA-inducible *PepThi* were lower than those in the red fruits of non-transformed pepper. These results suggest that overexpression of J1-1 in the unripe fruits induced PR genes, which might be responsible for the disease resistance of the transgenic peppers. Moreover, the results also suggest that J1-1 proteins induced in the transgenic plants were affected by both JA and SA signalings, in which JA is more important than SA.

Fungal Resistance of the Transgenic Pepper Plants against *C. gloeosporioides*

To assess the efficacy of J1-1 protein against the anthracnose fungus, *in vivo* assay was conducted using the unripe fruits of four transgenic pepper lines. Spores were inoculated directly on the surface of detached green fruits and then observed for lesion development and spore formation. Within 24 hours, germinated conidium developed an appressorium and then penetrated into the cuticle layers in the unripe fruits of non-transformed wild-type pepper (Figure 7A). Prominent penetration marks were shown in the surroundings of the infection hypha on the outer surfaces and extensive fungal growth was observed in the lumen of fruit cells, which resulted in maceration and cell death at 5 days after infection (Figure 7B). On the contrary, the early infection process was compromised in the transgenic fruit, representing reduced cuticle penetration (Figure 7C). Moreover, the fungus was unable to colonize further in the transgenic pepper cells (Figure 7D).

Nine days after inoculation, non-transformed wild-type fruits showed typical sunken disease symptoms of which spreading lesions were covered with soaked spores (Figure 8A). In contrast, transgenic fruits revealed very low frequency of lesion formation compared to the non-transgenic fruits (Figure 8B). Interestingly,

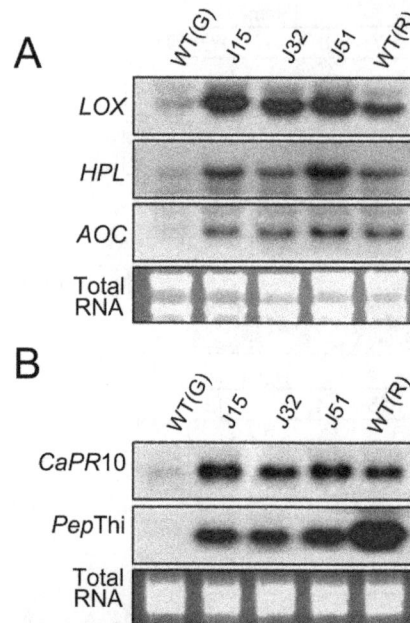

Figure 6. Expression of JA-biosynthesis related genes (A) and pathogenesis-related genes (B) in transgenic pepper fruits. Total RNAs were extracted from the unripe fruits of T$_2$ transgenic pepper lines (J15, J32, and J51). 10 µg of total RNA was separated in a formaldehyde/agarose gel, transferred onto nylon membrane, and hybridized to radiolabeled respective probes. WT (G), non-transgenic unripe fruits as a negative control; WT (R) non-transgenic ripe fruits as a positive control.

necrotic lesions were hardly observed on the unripe fruit of J15 and J32 transgenic lines. In the case of the J19 and J51 lines, inoculated fruits tend to develop intermediate sized lesions with arid surface, implying limited spore formation. Thus, spore production was measured in the lesion to verify whether symptom restriction in transgenic plants was caused by inhibited fungal colonization (Figure 8C). After 9 days of incubation, the number of spores in all transgenic pepper lines was drastically lower than that of control plants, in which the J15 and J32 transgenic plants showed lower spore formation than other lines. This observation is consistent with the size of lesion on the inoculated unripe fruits. The J51 transgenic plants showing less restricted lesion development showed reduction by half in spore formation compared with the wild-type plant. Consequently, a strong correlation was observed between J1-1 protein and fungal resistance in the transgenic plants. These results confirmed that the unripe fruits accumulating a high level of J1-1 protein showed elevated resistance, indicating that lesion and spore developments were retarded by the action of J1-1 protein. Taken together, the results suggest that overexpression of *J1-1* in pepper plants leads to the restriction of fungal colonization by inhibiting fungal growth and spore production, and demonstrate that the J1-1 protein has a protective activity that prevents the spread of anthracnose symptoms in unripe pepper fruits.

Discussion

This study finds new evidence that defensin is associated with a physiological process during phytopathogen interaction (Figure 1). Immunohistochemical study showed that massive J1-1 protein occurred in epidermal cells invaded by fungus. A noticeable amount of the peptide was found over the cell surface and

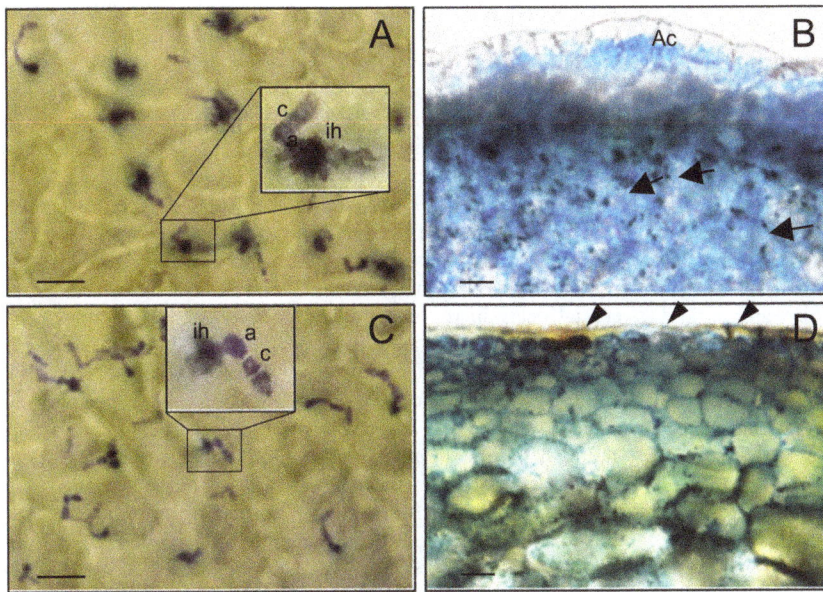

Figure 7. Inhibition of fungal growth in transgenic pepper fruits. A & C, Microscopic observation of fungal penetration at the infected area in the non-transformed (A) or J15 transgenic (C) unripe pepper fruit at 24 hr after inoculation with *C. gloeosporioides*. Fungus was stained with 0.1% toluidine blue. **B & D,** Cross sections of infection sites in the non-transformed (B) and J15 transgenic (D) fruits at 5 day after inoculation. Lactophenol-trypan blue was used for staining. a, appressorium; ih, infection hypha; c, conidium; Ac, acervuli. Arrowheads indicate spores and arrows indicate mycelia. Bar = 25 μm.

surrounding the invading fungal conidium. Serial observation revealed that the peptide was excreted to the outside of the fruit. In addition, the recombinant J1-1/GST fusion protein showed inhibitory activity on the growth and development of the anthracnose fungus (Figure 2). This implies that the pepper defensin, J1-1, retains its biological activity similar to other defensins that exhibit antifungal effects [43]. Consequently, the

initial contact of J1-1 with fungus can restrict fungal growth so that lesion formation might be effectively arrested during the early infection process.

Immunoblot analyses of J1-1 in various pepper tissues showed the expression of the protein in the ripe red fruit (Figure 1A). In addition, there was another defensin band with a higher size in flowers. This defensin is different from J1-1, because we could not

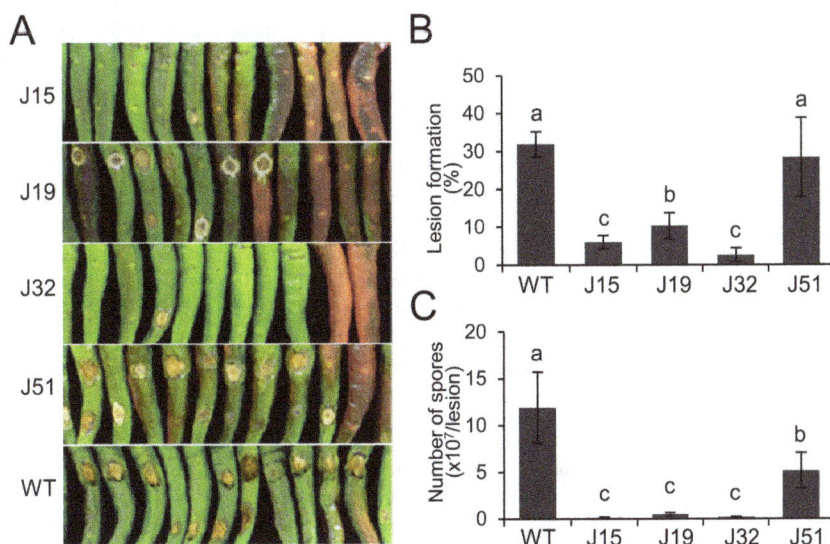

Figure 8. Fungal resistance of transgenic pepper fruits challenged with *C. gloeosporioides*. A, Representative photographs of unripe pepper fruits 9 days after infection with the anthracnose fungus. Green mature fruits from transgenic lines and wild-type control plants were inoculated with spores. J15, J19, J32 and J51, homozygous T_2 transgenic pepper lines; WT, non-transformed unripe fruits as a negative control. **B,** The rate of lesion development from inoculated spots on infected fruits. **C,** Number of spores in a lesion of the infected fruits. Fifty unripe mature fruits were infected at two spots. The number of spores was counted in the infected area at 9 days after infection. The data are presented as means ± SD from three independent estimations. Means with different letters in each column are significantly different at P<0.05.

detect the transcription of J1-1 in flowers (Figure S1). These results suggest the presence of two defensins in pepper, which is consistent with a previous report [27]. It has been shown that two defensin genes exist in the pepper genome, J1-1 (Gene accession no. X95363) and J1-2 (Gene accession no. X95730). Therefore, the defensin band in flowers is suspected as the other defensin gene (J1-2). Further studies will be necessary to elucidate the functional differences between flower-specific and fruit-specific defensins.

The present results showed that the production of J1-1 protein is regulated by ripening stages (Figure 1B and S2). To understand ripening related expression of J1-1, we retrieved promoter sequence of *J1-1* and analyzed for binding sites for transcription factors using PlantPAN (http://plantpan.mbc.nctu.edu.tw). The promoter contains the sequence motives related to known target sites for multiple consensus sequences for AtMYC2 and an ethylene responsive element (ERELEE4). These sequence motives are known to be involved in JA and ethylene signaling, respectively. Considering that pepper is known as non-climacteric fruit, this is consistent with our previous report that the expression of *J1-1* gene is up-regulated in the unripe pepper fruits by exogenous treatment of mehyl jasmonate [28].

During the T-DNA integration process, a short stretch of DNA can be deleted at the ends of the T-DNA or at the integration site of plant gDNA [44,45]. To investigate whether such events have a special feature in pepper transformation, we cloned the flanking regions and examined the nucleotide sequences in the transgenic pepper lines that carry a single copy of T-DNA. In agreement with previous reports, deletions at the RB and LB of the T-DNAs were observed in pepper transgenic plants (Table 1 and S3). In our case, the length of deleted DNA varied according to the transgenic event, and a more extensive deletion of border sequences was observed, especially in the LB of T-DNA. The results might be related to the transformation efficiency being extremely low in pepper. Pepper is known to be a very recalcitrant plant to transform, with transformation efficiency reported to be as low as 0.05%–0.6% [38,46]. According to the sequence of the T-DNA/gDNA junctions, improper maintenance of T-DNA significantly occurred in both the LB and RB in transgenic pepper plants. In the worst case, 28 bp was lost at the 3′ end of the *HPT* gene in the T-DNA integrated in J15. This may explain why the transformation of pepper is so difficult when using the *Agrobacterium*-mediated method. Severe loss of the border sequence of T-DNA may disturb the stable transformation process or the selection procedures during pepper transformation.

Once transgenic plants are established, the transgenes should be stably expressed in the plant over generations. To minimize genetic variation caused by the positional effect of T-DNA integration and complex transgene structure, it is necessary to screen genetically stable transformants to guarantee transgene inheritance in their progenies. However, the level and pattern of transgene expression may differ widely among transgenic pepper plants. In the present study, the mRNA level of the transgene was invariable among transgenic plants, but the accumulation of J1-1 protein was compromised in each transgenic line (Figure 5). According to the protein gel blot analysis, J1-1 protein was not maintained constantly in the unripe fruits during phytopathogen interaction (Figure S5). Consequently, a correlation between protein instability and the developmental state of fruits was observed for the J1-1 protein, indicating that the protein was seemingly not durable in the unripe fruit. Unlike other lines, the J51 line did not accumulate the J1-1 protein in healthy unripe fruit and showed transient accumulation of the transgene product after fungal infection (Figure 5B and S5B). This was unexpected because defensins are known to have inherent stability arising from

the characteristic structure known as the CSαβ motif [19]. Pepper J1-1 protein was also relatively stable under high-temperature heating at 90°C for 10 min, representing 75% of antifungal activity (Figure 2). This discrepancy in protein stability between *in vivo* and *in vitro* might arise from the post-translational modification that is involved in the processing of J1-1 protein in a developmentally regulated manner.

The present study demonstrates that the J1-1 protein provided effective resistance of pepper fruits against fruit-specific anthracnose fungus (Figure 8). Symptom development in J15 and J32 compared to that in non-transformed control fruit revealed a crucial delay in the onset of the disease. By contrast, J51, with lower J1-1 accumulation, displayed similar lesion formation to the non-transformed control fruit, but reduced sporulation (Figure 8B and 8C). This might be explained that the J1-1 protein was detected in the J51 transgenic pepper fruits after fungal infection, although the level was significantly reduced (Figure S5). Considering that the J1-1 protein was not detected in the non-transformed control fruits, the J1-1 protein level induced in the J51 transgenic pepper fruits might be effective to reduce sporulation but not enough to reduce lesion formation. Collectively, our results indicate the significance of the J1-1 protein during phytopathogen interaction.

During the symptom development on the green pepper fruits by infection with the anthracnose fungus, the peripheral regions of infection sites tended to turn red in transgenic fruits (Figure 8A). This might be explained by the increased expression of JA-biosynthetic genes in the transgenic fruits (Figure 6A). Previously, exogenous JA treatment was shown to accelerate chlorophyll degradation but β-carotene accumulation in tomato [47], and endogenous level of JA was reported to be risen coincidently with the onset of ripening in apple and tomato fruit [41]. Therefore, elevated JA synthesis from the induction of JA-biosynthetic genes by J1-1 overexpression might account for green-to-red color change in the fruits after infection.

The present results also suggest that up-regulation of JA-biosynthetic genes in the unripe transgenic fruits might induce the expression of a defense-related gene such as the *CaPR*10 (Figure 6B). In addition, *C. gloeosporioides* is a hemibiotroph that start out as biotroph, but switched to necrotroph. SA-dependent responses are typically associated with resistance to biotrophs, whereas JA and ethylene synergistically regulate defense against necrotrophs. Thus, it is likely that, in association with the expression of other defense-related proteins, the constitutive expression of J1-1 resulted in sustainable tolerance levels of the transgenic plants to the fungus. However, further studies will be necessary to determine how overexpression of J1-1 protein contributes to the up-regulation of JA-biosynthetic genes. Conclusively, a pepper defensin, J1-1, exhibiting antimicrobial activities are quite versatile for biotechnological purposes to provide biological protection to pepper fruits.

Supporting Information

Figure S1 The expresseion of J1-1 gene in various organs of *C. annuum*. The transcript levels were analyzed in leaf (Le), flower (Fl), unripe fruit (UF), and ripe fruit (RF) of *C. annuum* by RT-PCR. rRNA was shown as a loading control.

Figure S2 Developmentally regulated J1-1 production in pepper fruits during ripening at stages I through V. Stage I, green fruit; stage II, early breaker fruit; stage III, turning fruit: stage IV, purple fruit; stage V, red fruit. Total soluble proteins from the fruit were subjected to SDS-PAGE, blotted onto a PVDF membrane, and incubated with polyclonal J1-1 antibody. β-tubulin was shown as a loading control.

Figure S3 J1-1 recombinant protein affect the development of *C. gloeosporioides*, in vitro. A Appressorium formation. **B** Spore germination. Spore suspensions were amended with 10 μL of the GST/J1-1 recombinant protein or heated protein to final concentrations of 0.001, 0.01, 0.1, and 1 mg·mL^{-1}. The protein was heated by incubating at 90°C for 10 min. A minimum of 100 spores were counted per replicate. Each value represents the mean ± SD of three replicates. Means with different letters in each column are significantly different at P<0.05.

Figure S4 Development of transgenic plants from pepper explants. A Pepper seeds were germinated in the dark and incubated on a half strength MS medium for 6 days. **B** Hypocotyl and cotyledonary explants were pre-incubated on callus induction medium for two days. **C** After *Agrobacteria* infection, the explants were incubated in the callus induction medium containing 20 mg·L^{-1} hygromycin and 400 mg·L^{-1} cefotaxime. **D** Callus was incubated on the shoot induction media containing 10 mg·L^{-1} hygromycin. **E** The regenerated shoots were transferred onto a root inducing media. **F** Regenerated putative transgenic plant.

Figure S5 Expression of the *J1-1* in infected unripe pepper fruits. A Northern blot analysis. Unripe fruits from transgenic and wild-type plants at 0 and 24 hours after inoculation (HAI) with *C. gloeosporioides* were used in this analysis. **B**

Immunoblot analysis. Total soluble proteins from T$_2$ progenies were subjected to immunoblot analysis with polyclonal anti-J1-1 antibody. WT, infected unripe fruits of wild type; J32 and J51, infected unripe fruits of respective transgenic plants. rRNA and β-tubulin were shown as loading controls.

Table S1 Primers used in this study.

Table S2 Segregation ratios for hygromycin resistance in the progenies of transgenic peppers.

Table S3 Rescued sequences of T-DNA/gDNA junctions in the J15, J19 and J51 transgenic lines. RB, Right border; LB, Left border. Plant genomic DNA sequence is indicated in gray. The primer sequences used for i-PCR are underlined, and the restriction sites are in italic.

Acknowledgments

We thank the Kumho Life Science Laboratory for providing plant growth facilities. We also thank Yun-Jeong Han for valuable discussion and comments on the manuscript writing.

Author Contributions

Conceived and designed the experiments: HHS JIK YSK. Performed the experiments: HHS Sangkyu Park Soomin Park YSK. Analyzed the data: BJO KB OH JIK YSK. Contributed reagents/materials/analysis tools: KB OH JIK YSK. Wrote the paper: JIK YSK.

References

1. Coll NS, Epple P, Dangl JL (2011) Programmed cell death in the plant immune system. Cell Death Differ 18: 1247–1256.
2. Dixon RA (2001) Natural products and plant disease resistance. Nature 411: 843–847.
3. Jones JDG, Dangl JL (2006) The plant immune system. Nature 444: 323–329.
4. Glazebrook J (2005) Contrasting mechanisms of defense against biotrophic and necrotrophic pathogens. Annu Rev Phytopathol 43: 205–227.
5. Fritig B, Heitz T, Legrand M (1998) Antimicrobial proteins in induced plant defense. Curr Opin Immunol 10: 16–22.
6. Kitajima S, Sato F (1999) Plant pathogenesis-related proteins: molecular mechanisms of gene expression and protein function. J Biochem 125: 1–8.
7. Broekaert WF, Terras FRG, Cammue BPA, Osborn RW (1995) Plant defensins: novel antimicrobial peptides as components of the host defense system. Plant Physiol 108: 1353–1358.
8. Thomma BP, Cammue BP, Thevissen K (2002) Plant defensins. Planta 216: 193–202.
9. Terras FR, Eggermont K, Kovaleva V, Raikhel NV, Osborn RW, et al. (1995) Small cysteine-rich antifungal proteins from radish: their role in host defense. Plant Cell 7: 573–588.
10. Almeida MS, Cabral KM, Kurtenbach E, Almeida FC, Valente AP (2002) Solution structure of *Pisum sativum* defensin 1 by high resolution NMR: plant defensins, identical backbone with different mechanisms of action. J Mol Biol 315: 749–757.
11. Stotz HU, Thomson JG, Wang Y (2009) Plant defensins: defense, development and application. Plant Signal Behav 4: 1010–1012.
12. Lay FT, Anderson MA (2005) Defensins - components of the innate immune system in plants. Curr Protein Pept Sci 6: 85–101.
13. Chen JJ, Chen GH, Hsu HC, Li SS, Chen CS (2004) Cloning and functional expression of a mungbean defensin VrD1 in *Pichia pastoris*. J Agric Food Chem 52: 2256–2261.
14. Liu YJ, Cheng CS, Lai SM, Hsu MP, Chen CS, et al. (2006) Solution structure of the plant defensin VrD1 from mung bean and its possible role in insecticidal activity against bruchids. Proteins 63: 777–786.
15. Shahzad Z, Ranwez V, Fizames C, Marquès L, Le Martret B, et al. (2013) Plant Defensin type 1 (PDF1): protein promiscuity and expression variation within the *Arabidopsis* genus shed light on zinc tolerance acquisition in *Arabidopsis halleri*. New Phytol 200: 820–833.
16. Spelbrink RG, Dilmac N, Allen A, Smith TJ, Shah DM, et al. (2004) Differential antifungal and calcium channel-blocking activity among structurally related plant defensins. Plant Physiol 135: 2055–2067.
17. Sagaram US, Pandurangi R, Kaur J, Smith TJ, Shah DM (2011) Structure-activity determinants in antifungal plant defensins MsDef1 and MtDef4 with different modes of action against *Fusarium graminearum*. PLoS One 6: e18550.
18. Thevissen K, Ferket KK, Francois IE, Cammue BP (2003) Interactions of antifungal plant defensins with fungal membrane components. Peptides 24: 1705–1712.
19. Valente AP, de Paula VS, Almeida FC (2013) Revealing the properties of plant defensins through dynamics. Molecules 18: 11311–11326.
20. Hayes BM, Bleackley MR, Wiltshire JL, Anderson MA, Traven A, et al. (2013) Identification and mechanism of action of the plant defensin NaD1 as a new member of the antifungal drug arsenal against *Candida albicans*. Antimicrob Agents Chemother 57: 3667–3675.
21. Lay FT, Brugliera F, Anderson MA (2003) Isolation and properties of floral defensins from ornamental tobacco and petunia. Plant Physiol 131: 1283–1293.
22. Oomen RJ, Seveno-Carpentier E, Ricodeau N, Bournaud C, Conejero G, et al. (2011) Plant defensin AhPDF1.1 is not secreted in leaves but it accumulates in intracellular compartments. New Phytol 192: 140–150.
23. De Coninck B. Cammue BPA, Thevissen K (2013) Modes of antifungal action and *in planta* functions of plant defensins and defensin-like peptides. Fungal Biol Rev 26: 109–120.
24. Tesfaye M, Silverstein KA, Nallu S, Wang L, Botanga CJ, et al. (2013) Spatio-temporal expression patterns of *Arabidopsis thaliana* and *Medicago truncatula* defensin-like genes. PLoS One 8: e58992.
25. Fils-Lycaon BR, Wiersma PA, Eastwell KC, Sautiere P (1996) A cherry protein and its gene, abundantly expressed in ripening fruit, have been identified as thaumatin-like. Plant Physiol 111: 269–273.
26. Dracatos PM, van der Weerden NL, Carroll KT, Johnson ED, Plummer KM, et al. (2014) Inhibition of cereal rust fungi by both class I and II defensins derived from the flowers of *Nicotiana alata*. Mol Plant Pathol 15: 67–79.
27. Meyer B, Houlne G, Pozueta-Romero J, Schantz ML, Schantz R (1996) Fruit-specific expression of a defensin-type gene family in bell pepper. Upregulation during ripening and upon wounding. Plant Physiol 112: 615–622.
28. Oh BJ, Ko MK, Kostenyuk I, Shin B, Kim KS (1999) Coexpression of a defensin gene and a thionin-like via different signal transduction pathways in pepper and *Colletotrichum gloeosporioides* interactions. Plant Mol Biol 41: 313–319.

29. Manandhar JB, Hartman GL, Wang TC (1995) Conidial germination and appressorial formation of *Colletotrichum capsici* and *C. gloeosporioides* isolates from pepper. Plant Dis 79: 361–366.

30. Than PP, Prihastuti H, Phoulivong S, Taylor PW, Hyde KD (2008) Chilli anthracnose disease caused by *Colletotrichum* species. J Zhejiang Univ Sci B 9: 764–778.

31. Pakdeevaraporn P, Wasee S, Taylor PWJ, Mongkolporn O (2005) Inheritance of resistance to anthracnose caused by *Colletotrichum capsici* in *Capsicum*. Plant Breeding 124: 206–208.

32. Pelegrini PB, Franco OL (2005) Plant gamma-thionins: novel insights on the mechanism of action of a multi-functional class of defense proteins. Int J Biochem Cell Biol 37: 2239–2253.

33. Abdallah NA, Shah D, Abbas D, Madkour M (2010) Stable integration and expression of a plant defensin in tomato confers resistance to fusarium wilt. GM Crops 1: 344–350.

34. Aerts AM, Thevissen K, Bresseleers SM, Sels J, Wouters P, et al. (2007) *Arabidopsis thaliana* plants expressing human beta-defensin-2 are more resistant to fungal attack: functional homology between plant and human defensins. Plant Cell Rep 26: 1391–1398.

35. Gao AG, Hakimi SM, Mittanck CA, Wu Y, Woerner BM, et al. (2000) Fungal pathogen protection in potato by expression of a plant defensin peptide. Nat Biotechnol 18: 1307–1310.

36. Lee SB, Li B, Jin S, Daniell H (2011) Expression and characterization of antimicrobial peptides Retrocyclin-101 and Protegrin-1 in chloroplasts to control viral and bacterial infections. Plant Biotechnol J 9: 100–115.

37. Ntui VO, Thirukkumaran G, Azadi P, Khan RS, Nakamura I, et al. (2010) Stable integration and expression of wasabi defensin gene in "Egusi" melon (*Colocynthis citrullus* L.) confers resistance to Fusarium wilt and Alternaria leaf spot. Plant Cell Rep 29: 943–954.

38. Ko MK, Soh H, Kim KM, Kim YS, Im K (2007) Stable production of transgenic pepper plants mediated by *Agrobacterium tumefaciens*. HortScience 42: 1425–1430.

39. Kim KD, Oh BJ, Yang J (1999) Differential interactions of a *Colletotrichum gloeosporioides* isolate with green and red pepper fruits. Phytoparasitica 27: 97–106.

40. Oh BJ, Kim KD, Kim YS (1998) A microscopic characterization of the infection of green and red pepper fruits by an isolate of *Colletotrichum gloeosporioides*. J Phytopathology 146: 301–303.

41. Fan X, Mattheis JP, Fellman JK (1998) A role for jasmonates in climacteric fruit ripening. Planta 204: 444–449.

42. Park CJ, Kim KJ, Shin R, Park JM, Shin YC, Paek KH (2004) Pathogenesis-related protein 10 isolated from hot pepper functions as a ribonuclease in an antiviral pathway. Plant J 37: 186–198.

43. Aerts AM, Francois IE, Cammue BP, Thevissen K (2008) The mode of antifungal action of plant, insect and human defensins. Cell Mol Life Sci 65: 2069–2079.

44. Chilton MD, Que Q (2003) Targeted integration of T-DNA into the tobacco genome at double-stranded breaks: new insights on the mechanism of T-DNA integration. Plant Physiol 133: 956–965.

45. Kumar S, Fladung M (2002) Transgene integration in aspen: structures of integration sites and mechanism of T-DNA integration. Plant J 31: 543–551.

46. Lee YH, Kim HS, Kim JY, Jung M, Park YS, et al. (2004) A new selection method for pepper transformation: callus-mediated shoot formation. Plant Cell Rep 23: 50–58.

47. Perez AG, Sanz C, Richardson DG, Olias JM (1993) Methyl jasmonate vapor promotes β-carotene synthesis and chlorophyll degradation in 'Golden Delicious' apple peel. J Plant Growth Regulat 12: 163–167.

16

Variations of *CITED2* Are Associated with Congenital Heart Disease (CHD) in Chinese Population

Yan Liu[1,2,ϑ], Fengyu Wang[3,ϑ], Yuan Wu[4,ϑ], Sainan Tan[5], Qiaolian Wen[1,2], Jing Wang[6], Xiaomei Zhu[1,2], Xi Wang[1,2], Congmin Li[3]*, Xu Ma[1,2,7]*, Hong Pan[1,2]*

1 Graduate School, Peking Union Medical College, Beijing, China, 2 National Research Institute for Family Planning, Beijing, China, 3 Henan Research Institute of Population and Family Planning, Key Laboratory of Population Defects Intervention Technology of Henan Province, Zhengzhou, China, 4 Cardiac Surgery Department, Xiamen Heart Center, Organ Transplantation Institute of Xiamen University, Xiang'an District, Xiamen, China, 5 Key Laboratory of Genetics and Birth Health of Hunan Province, Family Planning Institute of Hunan Province, Chang sha, China, 6 Department of Medical Genetics, School of Basic Medical Sciences, Capital Medical University, Beijing, China, 7 World Health Organization Collaborating Centre for Research in Human Reproduction, Beijing, China

Abstract

CITED2 was identified as a cardiac transcription factor which is essential to the heart development. *Cited2*-deficient mice showed cardiac malformations, adrenal agenesis and neural crest defects. To explore the potential impact of mutations in *CITED2* on congenital heart disease (CHD) in humans, we screened the coding region of *CITED2* in a total of 700 Chinese people with congenital heart disease and 250 healthy individuals as controls. We found five potential disease-causing mutations, p.P140S, p.S183L, p.S196G, p.Ser161delAGC and p. Ser192_Gly193delAGCGGC. Two mammalian two-hybrid assays showed that the last four mutations significantly affected the interaction between *p300CH1* and *CITED2* or *HIF1A*. Further studies showed that four *CITED2* mutations recovered the promoter activity of *VEGF* by decreasing its competitiveness with *HIF1A* for binding to *p300CH1* and three mutations decreased the consocation of *TFAP2C* and *CITED2* in the transactivation of *PITX2C*. Both *VEGF* and *PITX2C* play very important roles in cardiac development. In conclusion, we demonstrated that *CITED2* has a potential causative impact on congenital heart disease.

Editor: Robert Dettman, Northwestern University, United States of America

Funding: This work was supported by the National Basic Research Program of China (2010CB529504), the National Natural Science Foundation of China (81300131) and the Applied Basic Research Program of Qinghai Province (QH2013-z-744). The funders had no role in study design, data collection and analysis, decision to publish, or preparation of the manuscript.

Competing Interests: The authors have declared that no competing interests exist.

* E-mail: 13838377996@163.com (CL); 13174483538@126.com (HP); nicgr@263.net (XM)

ϑ These authors contributed equally to this work.

Introduction

Congenital heart disease (CHD) is a most common defect caused by abnormal cardiac formation in fetuses and has become the leading reason of childhood mortality with an incidence around 1%[1–3]. In the past decades, a series of CHD-causing genes have been identified such as *NKX2-5*, *TBX5*, *GATA4* and *CITED2* [4–6]. It has been confirmed that their mutations can cause cardiac malformations through affecting the transcription activity of critical genes involved in heart development pathways.

CITED2 (Cbp/p300-interacting transactivator, with Glu/Asp-rich carboxy-terminal domain, 2) is one member of a new conserved family of transcriptional activators which includes four members: *CITED1* (*Msg1*), *CITED2* (*Mrg1/p35srj*), *CITED3* and *CITED4* (*Mrg2*) [7]. *CITED2* is a nuclear protein which binds closely to the CH1 region of *p300* and *CBP* by its CR2 region (including a conserved 32-amino acid sequence [8]). Meanwhile, many other transcription factors and transcription regulating factors such as *HIF1A*, *RXRα*, *NFk*, *Mdm2*, *Ets-1* and *Stat2* also bind to the CH1 region of *CBP/p300* [9,10]. Thus *CITED2* may act as a pivotal transcriptional modulator to regulate the expression of some specific genes. For example, *CITED2* decreased the expression of *HIF1A* (Hypoxia Inducible Factor 1) through its competitive binding to *CBP/p300CH1*[11,12], consequently interfering the transcription of genes induced by *HIF1A* such as *VEGF* (vascular endothelial growth factor) [13]. It has been confirmed that the overexpression of *vegf* is the main factor resulting in cardiac malformation in *cited2*[−/−] mice [14].

Besides being a transcriptional repressor of *HIF1A*, *CITED2* acts as a transcriptional coactivator of *TFAP2* (transcription factor AP2, also called *Tcfap2*) [15]. Mutations of *TFAP2A* and *TFAP2B* result in neural tube, cranial ganglia defects and cardiac malformations [16,17]. This suggested that the coactivation of *TFAP2* with *p300*, *CITED2* and *CREBBP* is essential for the normal development of those structures. As a critical transcription factor, *TFAP2* can affect the transcription of many genes, including *PITX2C* (Paired-Like Homeodomain 2 C)which is critical in Nodal-*PITX2C* pathways [18]. In addition, it has been detected that *TFAP2* isoforms and *CITED2* work together on the *PITX2C* promoter1 which controls the expression of *PITX2C* in the heart of embryonic mice. The mice experiments already indicated that knocking out *pitx2c* gene can lead to valve defects, body wall dysraphism, gastroschisis, ectopia cordis and other multiple organs polymorphous defects [19].

CITED2 gene mutation in human congenital heart disease was first reported by Sperling *et al* [20] in 2005. They identified 3 mutations which alter the amino acid sequence and studied their association with *HIF1A* and *TFAP2C*. Their study confirms that *CITED2* is an important transcription factor in heart development and provides new insights into the molecular mechanism of congenital heart defects. Later, Yang *et al* found 3 new mutations in Chinese patients with congenital heart disease (2010) [21]and Chen *et al* [22]demonstrated another 3 new mutations in European CHD patients. Recently, Xu *et al* found 3 *CITED2* gene mutations, their research showed that *CITED2* gene mutations and methylation may play an important role in CHD. In their study, most of these mutations were in SRJ region. The mutations in our study were identified for the first time and located in SRJ region as well. Our work aimed to determine whether the new mutations also affect *HIF1A* or *TFAP2C* and finally lead to an abnormal expression of *VEGF* or *PITX2C* which play an important role in heart development.

Materials and Methods

Ethics statement

The study protocol conformed to the ethical guidelines of the 1975 Declaration of Helsinki and was approved by the Ethics Committee of the National Research Institute for Family Planning. Written informed consent was obtained from patients' parents or guardians.

Subjects

The study population comprised 700 patients who were diagnosed with CHD based on anthropometric measurement, physical examination for malformation and dysmorphism, and radiological evaluation. The patients with a phenotype of VSD, TOF and ASD accounted for 43.71%, 8.42% and 12% respectively. 250 unrelated healthy children were used as controls. Peripheral blood was collected from each affected individual and their parents and controls were from 6 months to 12 years old and most of them volunteered to participate in the study.

We sequenced the whole *CITED2* ORF in 700 CHD patients (Table 1) and 250 healthy controls recruited from Lanzhou University, Beijing Children's Hospital, Zhengzhou Children's Hospital, Henan provincial Chest Hospital and Children's Hospital of Fudan University.

Mutational analysis and bioinformatics

Genomic DNA was extracted from peripheral blood leukocytes using standard methods. The human *CITED2* gene is located on 6q24.1 and is encoded by two exons. One of the exons and splice sites of *CITED2* were amplified by polymerase chain reaction (PCR) using two pairs of *CITED2* gene-specific primers (Table 2). PCR products were sequenced using the appropriate PCR primers and the Big Dye Terminator Cycle Sequencing kit (Applied Biosystems, Foster City, CA, USA) and run on an automated sequencer, ABI 3730XL (Applied Biosystems), to perform mutational analysis.

Site-directed mutagenesis and plasmid construction

Human *CITED2* and *HIF1A* cDNA were obtained from OriGene True-Clone, and *TFAP2C* cDNA was purchased from GeneCopoeia. *CITED2* mutations were constructed by using the Quick Change Lightning Site-Directed Mutagenesis kit (Strata gene, La Jolla, CA, USA). Then the introduced mutations were confirmed by DNA sequence.

The WT and mutant *CITED2* were amplified by PCR from cDNA and inserted into the pEGFP-N1 vector (BD Biosciences, Palo Alto, CA, USA). The ORF of *HIF1A* and *TFAP2C* were also amplified by PCR from cDNA and inserted respectively into the pcDNA3.1(+) vector (Invitrogen, Carlsbad, CA, USA) to create the expression plasmid pcDNA3.1-*HIF1A* and pcDNA3.1-*TFAP2C*.

A 1300-bp fragment of the p300-CH1, *PITX2C* promoter and an 870-bp segment of *VEGF* promoter amplified by PCR from Human genomic DNA were cloned respectively into the GAL4-pCMX vector and the luciferase reporter PLG3-basic vector. GAL4-*HIF1A* was constructed by cloning DNA fragments into GAL4-pCMX vector at the Ecorv and Nhel sites. All primers of the PCRS were list in Table 2.

The VP16-pCMX vector with the potent transactivating domain of HSV, the promoter pGL3-basic vector with 4×GAL4 DNA-binding sites and the GAL4-pCMX vector containing GAL4-DBD were provided by Dr. Ronald M. Evans (Salk Institute for Biological Studies, USA).

Cell culture and transient transfection

293T and Hela cells were maintained in Iscove's modified Dulbecco's medium supplemented with 10% fetal bovine serum, 100 mg/ml penicillin, and 100 mg/ml streptomycin in a humidified atmosphere containing 5% CO_2 at 37°C. Transfection was carried out using a standard calcium phosphate method or Lipofectamine 2000 (Invitrogen Corporation, Carlsbad, CA, USA).

Table 1. Patients with congenital heart disease included in the study.

Phenotype	Total(n = 700)
Ventricular septal defect(VSD)	306
Tetralogy of Fallot(TOF)	59
Atrial septal defect(ASD)	84
Patent ductus arteriosus(PDA)	21
Pulmonal atresia or stenosis(PS)	21
double outlet right ventricle(DORV)	11
Aortic coarctation(COA)	4
Pulmonary hypertension(PH)	2
Other complex cardiac malformations	192

Table 2. Primers used for PCR.

Name	Primer pair
Primers for *CITED2*	F CCGGCTGTGTTATGAGTGGTAG
	R AGTTGGGGGTTTGATTTCTTTC
Middle Primer for *CITED2*	TCGGAAGTGCTGGTTTGTC
Primers for P140S	F TGCCGGATTTGCACTCTGCTGCA GGCCAC
	R GTGGCCTGCAGCAGAGTGCAAAT CCGGCA
Primers for S183L	F GCTCTGGCAGCAGCTTGGGCGGCG
	R CGCCGCCCAAGCTGCTGCCAGAGC
Primers for S196G	F AACAGCGGCGGCGGCGGCGGCAGCG GCAACA
	R TGTTGCCGCTGCCGCCGCCGCCGCC GCTGTT
Primers for Ser161delAGCAGC	F TGCAACCCCAAGCACGGCGGCAGCA GCACC TGCAACCCCAAGCACGGCGGCAGCAGCACC
	R GGTGCTGCTGCCGCCGTGCTTGGGG TTGCA
Primers for Ser192_Gly193delAGCGGC	F CGCGGGCAGCAGCAACGGCGGCAGC GGCAGCGGCAACAT
	R ATGTTGCCGCTGCCGCTGCCGCCGTT GCTGCTGCCCGCG
pEGFP-*CIITED2*	F GGGGTACCATGGCAGACCATATGATG
	R CGGGATCCCGACAGCTCACTCTGCTGG
pCDNA3.1(+)-*CITED2*	F CGGGGTACCTATGGCAGACCATATGA TGGC
	R TGCTCTAGAGTCAACAGCTCACTCTGCTG
pCMX-GAL4-*CITED2*	F CGGATATCAATGGCAGACCATATGA TGGC
	R CTAGCTAGCTCAACAGCTCACTCTGCT
pCMX-GAL4-*HIF1A*	F CGGATATCAATGGAGGGCGCCGGCG
	R CTAGCTAGCTCAGTTAACTTGATCCAA AGCT
pCMX-VP16-*P300CH1*	F CGCGGATCCTATGGCCGAGAATGTGG TGGAAC
	R CTAGCTAGCCCAACGGGTGCTCCAGT CAAA
pCDNA3.1(+)-*HIF1A*	F CGGGGTACCTATGGAGGGCGCCGGC
	R TGCTCTAGATCAGTTAACTTGATCCAAAGC
pCDNA3.1(+)-*TFAP2C*	F CGGGGTACCACGCCGGACGCCATGTTG
	R TGCTCTAGACTCTCCTAACCTTTCTTC GTTCC
PGL3basic-*VEGF* promoter	F GGGGTACCTTTGGGTTTTGCCAGACT
	R CCGCTCGAGAGGGAGGGAGCAGGAATAG
PGL3basic-*PITX2C* promoter	F GGGGTACCGGGGACAAAAGGACTTTC
	R CCGCTCGAGCCCTGTTGGCCTAACATC

Subcellular localization

Hela cells were seeded in 12-well tissue culture plates 20 h prior to transfection at approximately 60% confluency. GFP-*CITED2* expression constructs containing wild-type and mutant *CITED2* were transfected using Lipofectamine 2000, according to the manufacturer's instructions. The empty vector pEGFP-N1 was transfected as a control. Forty hours after transfection, the cells were fixed and permeabilised in 4% paraformaldehyde for 15 min, 0.1% Triton X-100 for 20 min and the DNA was stained with 0.5 μg/ml DAPI for 3 min at room temperature. The cells were observed by fluorescence microscopy. All steps were operated in lucifugal conditions.

Mammalian two-hybrid assay and transcriptional assays

Mammalian two-hybrid assay plasmids including pCMX-VP16-*p300*, TK promoter reporter plasmid, the Renilla luciferase control plasmid pREP7-RLu and pCMX- GAL4-*CITED2* (wild-type or mutant) or pCMX-GAL4-*HIF1A* were contransfected into 293T cells. Thirty hours after transfection, cells were washed and lysed in passive lysis buffer (Promega, Madison, WI, USA) and the transfection efficiency was normalised to paired Renilla luciferase activity by using the Dual Luciferase Reporter Assay System (Promega, Madison, WI, USA) according to the manufacturer's instructions.

In addition, the Dual LuciferaseReporter Assay System was used to study the effect of *CITED2* on the transcription of *VEGF* and *PITX2C*. Plasmids consisting of the Renilla luciferase control plasmid pREP7-RLu, pcDNA3.1-*CITED2* (wild-type or mutant), PGL3-*VEGF*-pro and pcDNA3.1-*HIF1A* or PGL3-*PITX2C*-pro and pcDNA3.1-*TFAP2C* were contransfected into 293T cells. Thirty hours after transfection, cells were treated the same way as above.

Statistical analysis

The results represent the means of three independent experiments performed in triplicate, and the bars denote the S.D. The independent-samples t test was adopted to determine statistical significance of unpaired samples. All data were analyzed by Prism Demo 5 software.

Table 3. Position of variations

Coding position	Amino acid position	Phenotype of mutation carrier
c.C418T	p. P140S,Pro-Ser	F4
c.C548T	p. S183L,Ser-Leu	VSD
c.A586G	p. S196G,Ser-Gly	VSD
c.481–483delAGC	p.Ser161delAGC	ASD
c.574–579delAGCGGC	p.Ser192_Gly193delAGCGGC	VSD

Results

Genetic and bioinformatics analysis

From a total of 700 non-syndromic CHD patients, we identified five novel *CITED2* nucleotide alterations (two amino acid deletions and three amino acid substitutions, table3). Three mutations (c.C548T, c.A586G and c.574-59delAGCGGC) were found in one, one and four patients with Ventricular septal defect (VSD) respectively. One mutation (c.C418T) was detected in one patient with Tetralogy of Fallot (TOF) and another mutation (c.481–483delAGC) was detected in one patient with Artrial septal defect (ASD).

All potential pathogenic mutations have not been reported in the NCBI dbSNP and are not included in the 1000 Genome Project database (http://browser.1000genomes.org/).

The result of sequence alignment of *CITED2* proteins among several species showed that three acid substitutions were located at highly conserved regions among different species (human,

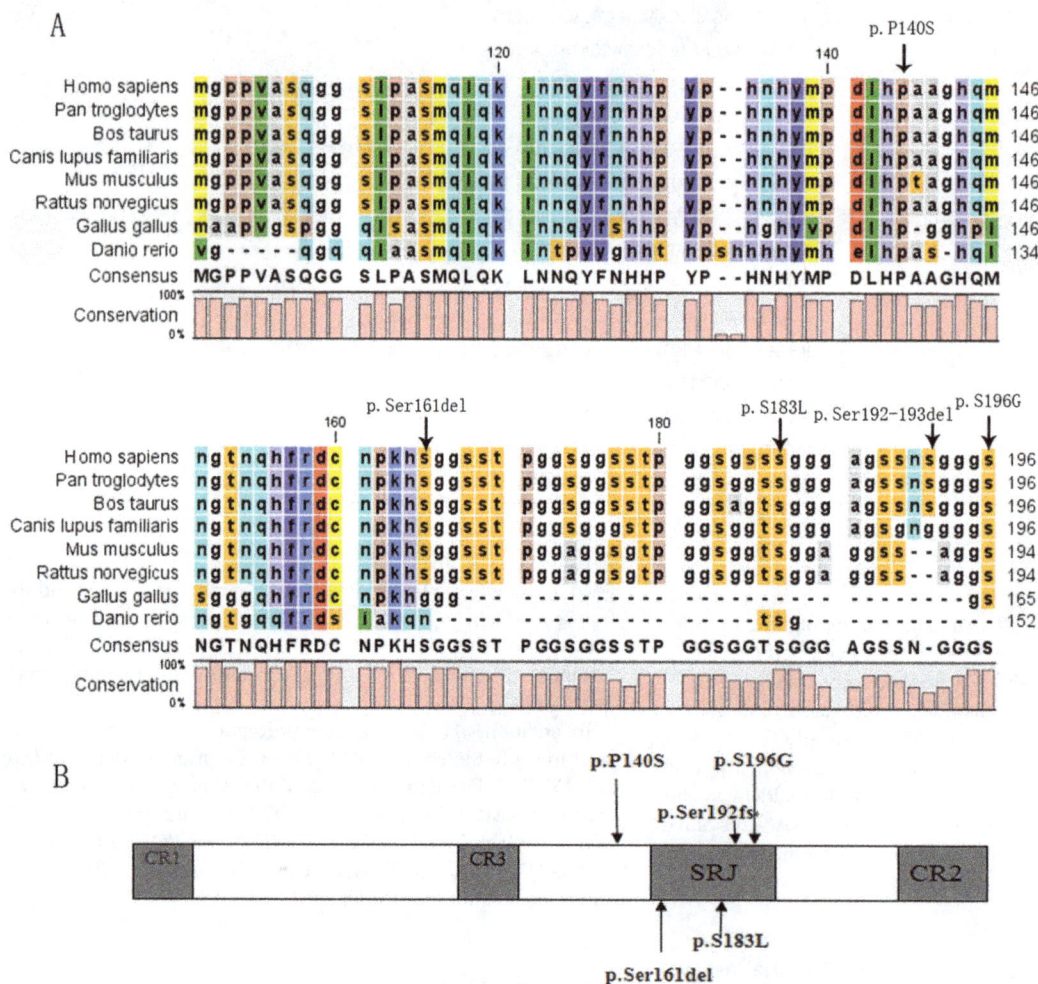

Figure 1. Structure of *CITED2*. A: Sequence alignment of *CITED2* proteins among several species. The figure showed that three acid substitutions were located at highly conserved regions among many species (human, chimpanzee, mice, dog, cattle, rat, chicken and zebrafish). **B**: Position of mutations in the *CITED2* protein identified in CHD patients. *CITED2* has three conserved regions CR1-3 and serine-glycine rich junction (SRJ). All other mutations were located in SRJ except p.P140S.

A

pCMX-VP16-P300CH1+pCMX-GAL4-CITED2

B

pCMX-VP16-P300CH1+pCMX-GAL4-HIF1a

C

D

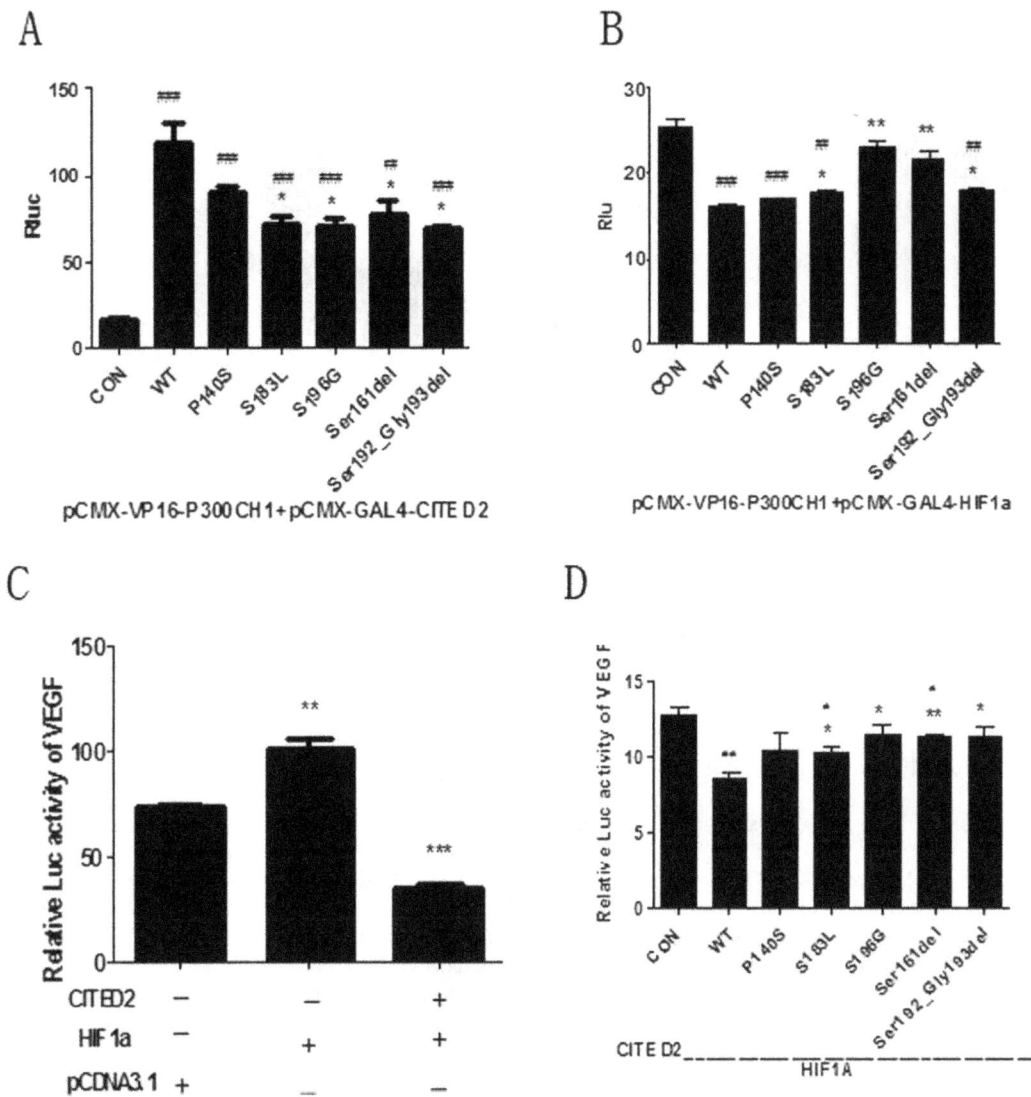

Figure 2. Effect of CITED2 mutations on the transcriptional activation of HIF1A to its target gene VEGF. A: Effect of mutations on CITED2-p300CH1 interactions. We cotransfected 293T cells with pCMX-VP16-p300CH1, TK promoter reporter plasmid, and the Renilla luciferase internal control plasmid, as well as empty vector pCMX-GAL4, GAL4-CITED2 wild-type, and the mutants. The significance of differences was calculated using the independent-samples t test. (*p<0.05, **p<0.01 versus. wt-type, #p<0.05, ##p<0.01 versus.empty vector pCMX-GAL4.) **B:** Effect of mutations on HIF1A-p300CH1 interactions. Cotransfection of pCMX-VP16-p300CH1, pCMX-GAL4-HIF1A, TK promoter reporter plasmid, and the Renilla luciferase internal control plasmid, as well as empty vector pcDNA3.1 (+)-CITED2 wild-type, and the mutant. (* p<0.05, ** p<0.01 versus wt-type, # p<0.05, ## p<0.01 versus. empty vector pcDNA3.1 (+)) **C:** Effect of wt-type on the transcriptional activation of VEGF. Transfected the VEGF reporter plasmid and the expression vector for HIF1A, CITED2 or pcDNA3.1 were transfected together in 293 T cells. The luciferase activity was normalized to Renilla activity.* p<0.05, **p<0.01 versus the untreated group (n = 3). **D:** Effect of CITED2 mutants on transcription activation of VEGF compared with CITED2-wt. The rest report plasmids were same as above. (*p<0.05, **p<0.01 versus wt-type, #p<0.05, ##p<0.01 versus empty vector pcDNA3.1(+)). The results represent the means of 3 independent experiments performed in triplicate and the significance of differences was calculated using independent-samples t test.(CITED2 = Cbp/p300-interacting transactivator, with Glu/Asp-rich carboxy-terminal domain, 2, HIF1A = Hypoxia Inducible Factor 1, VEGF = vascular endothelial growth factor)

chimpanzee, mice, dog, cattle, rat, chicken and zebrafish) and two amino acid deletions were not located at highly conserved regions among these species (Figure 1).

CITED2 mutations decrease HIF1A repression leading to up-regulation of VEGF expression

Two mammalian two-hybrid assays were used to evaluate whether the mutation affected the interaction between every two of CITED2, p300CH1 and HIF1A (Figure 2). Cotransfection of both VP16- P300 and wild-type GAL4-CITED2 with the TK

promoter reporter plasmid led to a nearly 10-fold increase in luciferase activity compared with VP16-P300 and empty vector of CMX-GAL4 (t test, p<0.01). The luciferase activity of p. P140S mutant was even the same as the wt-type, However, cotransfection of VP16-P300 and the four mutants (p.S183L, p.S196G, p.Ser161delAGC, p.Ser192_Gly193delAGCGGC) GAL4-CITED2 showed weakened luciferase activity (t test, p<0.05) (Figure 2A) compared with wt-type. These findings indicated that the four mutations diminished protein-protein interactions be-

A

B

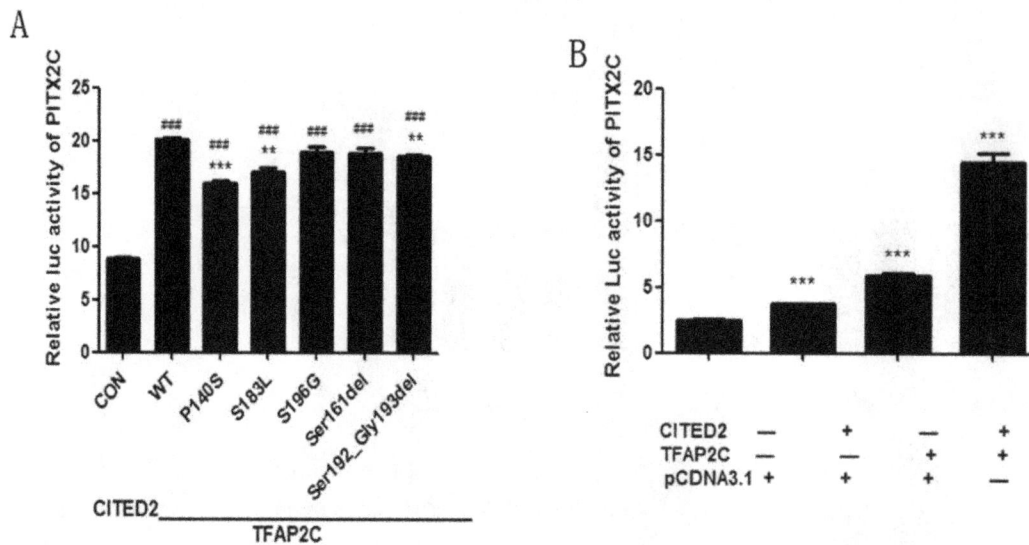

Figure 3. Effect of *CITED2* **variants on the cooperation between** *CITED2* **and** *TFAP2C* **in the transactivation of the** *PITX2C.* **A:** Effect of *CITED2* mutations on the transcription activation of *PITX2C*. (*$p<0.05$, **$p<0.01$ versus wt-type, #$p<0.05$, ##$p<0.01$ versus.empty vector pcDNA3.1(+)). **B:** *CITED2*-wt and *TFAP2C* working on the transcriptional activation of *PITX2C*. *PITX2C* reporter plasmid and the expression vector for *TFAP2C*, *CITED2*, or pcDNA3.1 alone were transfected respectively in 293 T cells. The luciferase activity was normalized to Renilla activity.(* $p<0.05$, **$p<0.01$ versus the untreated group (n = 3)).

tween p300 and *CITED2*, but the p.P140S mutant didn't alter the interactions.

Another mammalian two-hybrid assay was operated and analyzed to further evaluate whether the repression of *HIF1A* - p300 complex was influenced by *CITED2* mutation. The result showed that the luciferase activity of wt-type was only 60% of the control (t test, $p<0.01$) (Figure 2B). Compared with wild-type, the luciferase activity of mutants increased obviously except the p.P140S mutant. In conclusion, *CITED2* mutations weaken the *HIF1A* repression by diminishing the protein-protein interactions between *p300CH1* and *CITED2* on the one hand and by enhancing the interactions between *p300CH1* and *HIF1A* on the other hand.

As *HIF1A* can induce vascular endothelial growth factor (*VEGF*) potently, we supposed that *CITED2* mutations influenced the transcription of *VEGF* through their effect on *HIF1A*. This was confirmed by our dual luciferase assay (Figure 2C). Wild-type *CITED2* caused an approximately 32% decrease of activity compared with the control (t test, $p<0.01$). P140S showed no difference with wild-type in luciferase activity. As for the other four mutants, Ser161delAGCAGC showed an observable promotion of *VEGF*-promoter resulting in higher luciferase activity than wild-type (t test, $p<0.01$) and the rest mutants showed few differences compared with wt-type (t test, $p<0.05$) (Figure 2D).

CITED2 mutations impair *TFAP2C* coactivation resulting in abnormal transactivation of *PITX2C*

As a transcriptional coactivator of *TFAP2*, *CITED2* influenced cardiac left-right patterning by regulating the left-right patterning Nodal-*PITX2C* pathway. *PITX2C* is a critical gene of the Nodal-*PITX2C* pathway and controls the location of heart and intestines in embryo. Our study showed that *CITED2* mutations resulted in decreased luciferase activity of PITX2 by diminishing the coactivation of *CITED2* and *TFAP2C*. The luciferase activity of three mutants were decreased obviously compared with wt. (p.P140S vs. wt-type 80% (t test, $p<0.01$), p.S183L vs. wt-type 85% (t test, $p<0.01$), p.Ser192_Gly193delAGCGGC vs. wt-type

92% (t test, $p<0.01$)) (Figure 3A). The rest two mutants coactivated *TFAP2C* to the same level as wt-type.

In addition, we designed another test to prove the *TFAP2* coactivation with *CITED2*. The result showed that cotransfection of empty vector of pcDNA3.1 (+) with the luciferase reporter PGL3-*PITX2C*-pro was the lowest in all groups including pcDNA3.1-*TFAP2C* or wt-type pcDNA3.1-*CITED2* only and both of them (Figure 3B).

In conclusion, *CITED2* mutations contributed to the abnormal transactivation of *PITX2C*.

Impact of *CITED2* mutations on Subcellular Localization

To further study whether the functional changes are caused by changed subcellular localization of the protein, the transfections were performed using N-terminal GFP fusion constructs of wt and mutant *CITED2*, followed by fluorescence microscopy. The result indicated that the effects of *CITED2* mutations on *VEGF* and *PITX2C* were not caused by the incorrect localization of the protein. Whether in wt or mutant of *CITED2* the proteins were discovered mainly in nucleus and a lesser degree in the cytoplasm of Hela (Figure S1).

Discussion

Previous researches of *cited2-/-* mice confirmed that *cited2* plays a critical role in the development of heart and is essential for the normal creation of the left–right axis. *Cited2-/-* embryos showed a series of cardiac malformations such as VSD, ASD, outflow tract abnormalities and abnormal heart looping.

We screened the coding region and splice sites of the *CITED2* gene in 700 Chinese CHD patients. Two potential pathogenic amino acid deletions (p.Ser161delAGCAGC and p.Ser192_-Gly193delAGCGGC) and three potential pathogenic amino acid substitutions variants (p. P140S, p. S183L and p. S196G) were identified. These three regional highly conserved substitutions (conserved among Humans, chimpanzee, mice, dog, cattle, rat, chicken and zebrafish) were not identified in control group or the variant databases. Therefore, we supposed that these three mutations were possibly causative. Since, SRJ region is a research hot spot at present, the two potential pathogenic amino acid deletions in our study were found in SRJ region. As a result, the necessity of this study is highlight. Although the CHD phenotype was not seen in SRJ-deficient mice as observed in mutation carrying patients, we supposed that this could be due to species differences [23,24]in the function of *CITED2*, or some other unidentified factors[25] might interact with *CITED2* and modify its phenotype. Alternatively, it is also possible that CHD were present earlier in life but spontaneously closed at a later time in SRJ-deficient mice.

Mammalian two-hybrid analysis permits the semi-quantitative assessment of protein-protein interactions occurring within living cells. Cotransfection of wt or mutant *CITED2* and *p300CH1* in 293Tcells, the binding between *CITED2* and *p300CH1* activated the TK report gene expression in vivo. The functional study greatly supported the hypothesis that the mutations are causative and might affect the formation of heart. The last four mutated proteins (p. S183L, p. S196G, p.Ser161delAGCAGC and p.Ser192_Gly193delAGCGGC) showed significantly decreased reporter gene activation ability compared with wt-type. However, an opposite phenomenon occurred by transfecting *p300CH1*, *HIF1A* and wt or mutant *CITED2* together in cells. Taken together,the results indicated that the four mutated proteins decreased the interaction between *CITED2* and *p300CH1* compared with wt-type,causing a weakened competitive binding to p300 CH1 of *CITED2*. The increased interaction between *HIF1A* and *p300CH1* could up- regulate the promoter activity of *VEGF* according to our dual luciferase experiment.

Our study also showed that three mutations decreased the consocation of *TFAP2C* and *CITED2* in the transactivation of *pitx2c*, an essential gene of the left–right axis establishment confirmed in mice and chick embryo. The mice experiments already indicated that knocking out *pitx2c* gene can lead to valve

defects, body wall dysraphism, gastroschisis, ectopia cordis and other multiple organs polymorphous defects. In addition, there was no evidence that *CITED2* mutations were involved in the incorrect location of the protein in the subcellular localization experiment.

In conclusion, we identified five novel mutations among 700 CHD patients by screening the coding region and splice sites of the *CITED2* gene. To confirm our hypothesis that the mutations were pathogenic, we investigated the function and mechanism of them. Our study revealed that four mutations influenced the transcription regulatory properties of *VEGF* and three mutations reduced costimulation capacity to promote *PITX2C*. Further research showed that four *CITED2* mutations recovered the promoter activity of *VEGF* [26]caused by its decreased competitiveness with *HIF1A* to bind the *p300CH1*. Furthermore, three mutations also decreased the consocation of *TFAP2C* and *CITED2* in the transactivation of *PITX2C*. Our study confirmed that *CITED2* is a disease-causing gene of CHD and its mutations can result in the cardiac malformations.

Supporting Information

Figure S1 Subcellular localization of *CITED2*. Localization of wild-type and mutant *CITED2* GFP-fusion protein in transfected Hela cells were observed by fluorescent microscope. The empty vector pEGFP-N1 was transfected as a control. All figures were drawn by fluorescence microscopy and Adobe Photoshop CS5.

Author Contributions

Conceived and designed the experiments: XM HP YL. Performed the experiments: YL. Analyzed the data: YL XZ XW. Contributed reagents/materials/analysis tools: FW YW ST QW JW XZ CL. Wrote the paper: YL.

References

1. Hoffman JI, Kaplan S (2002) The incidence of congenital heart disease. J Am Coll Cardiol 39: 1890–1900.
2. Crider KS, Bailey LB (2011) Defying birth defects through diet? Genome Med 3: 9.
3. Blue GM, Kirk EP, Sholler GF, Harvey RP, Winlaw DS (2012) Congenital heart disease: current knowledge about causes and inheritance. Med J Aust 197: 155–159.
4. Xiong F, Li Q, Zhang C, Chen Y, Li P, et al. (2013) Analyses of GATA4, NKX2.5, and TFAP2B genes in subjects from southern China with sporadic congenital heart disease. Cardiovasc Pathol 22: 141–145.
5. Ching YH, Ghosh TK, Cross SJ, Packham EA, Honeyman L, et al. (2005) Mutation in myosin heavy chain 6 causes atrial septal defect. Nat Genet 37: 423–428.
6. Garg V, Kathiriya IS, Barnes R, Schluterman MK, King IN, et al. (2003) GATA4 mutations cause human congenital heart defects and reveal an interaction with TBX5. Nature 424: 443–447.
7. Andrews JE, O'Neill MJ, Binder M, Shioda T, Sinclair AH (2000) Isolation and expression of a novel member of the CITED family. Mech Dev 95: 305–308.
8. Li Q, Pan H, Guan L, Su D, Ma X (2012) CITED2 mutation links congenital heart defects to dysregulation of the cardiac gene VEGF and PITX2C expression. Biochem Biophys Res Commun 423: 895–899.
9. Yin Z, Haynie J, Yang X, Han B, Kiatchoosakun S, et al. (2002) The essential role of Cited2, a negative regulator for HIF-1alpha, in heart development and neurulation. Proc Natl Acad Sci U S A 99: 10488–10493.
10. Xu M, Wu X, Li Y, Yang X, Hu J, et al. (2014) CITED2 mutation and methylation in children with congenital heart disease. J Biomed Sci 21: 7.
11. Freedman SJ, Sun ZY, Kung AL, France DS, Wagner G, et al. (2003) Structural basis for negative regulation of hypoxia-inducible factor-1alpha by CITED2. Nat Struct Biol 10: 504–512.
12. Amati F, Diano L, Campagnolo L, Vecchione L, Cipollone D, et al. (2010) Hif1alpha down-regulation is associated with transposition of great arteries in mice treated with a retinoic acid antagonist. BMC Genomics 11: 497.
13. Macdonald ST, Bamforth SD, Braganca J, Chen CM, Broadbent C, et al. (2013) A cell-autonomous role of Cited2 in controlling myocardial and coronary vascular development. Eur Heart J 34: 2557–2565.
14. Xu B, Doughman Y, Turakhia M, Jiang W, Landsettle CE, et al. (2007) Partial rescue of defects in Cited2-deficient embryos by HIF-1alpha heterozygosity. Dev Biol 301: 130–140.
15. Bamforth SD, Braganca J, Eloranta JJ, Murdoch JN, Marques FI, et al. (2001) Cardiac malformations, adrenal agenesis, neural crest defects and exencephaly in mice lacking Cited2, a new Tfap2 co-activator. Nat Genet 29: 469–474.
16. Satoda M, Zhao F, Diaz GA, Burn J, Goodship J, et al. (2000) Mutations in TFAP2B cause Char syndrome, a familial form of patent ductus arteriosus. Nat Genet 25: 42–46.
17. Bhattacherjee V, Horn KH, Singh S, Webb CL, Pisano MM, et al. (2009) CBP/p300 and associated transcriptional co-activators exhibit distinct expression patterns during murine craniofacial and neural tube development. Int J Dev Biol 53: 1097–1104.
18. Campione M, Ros MA, Icardo JM, Piedra E, Christoffels VM, et al. (2001) Pitx2 expression defines a left cardiac lineage of cells: evidence for atrial and ventricular molecular isomerism in the iv/iv mice. Dev Biol 231: 252–264.
19. Bamforth SD, Braganca J, Farthing CR, Schneider JE, Broadbent C, et al. (2004) Cited2 controls left-right patterning and heart development through a Nodal-Pitx2c pathway. Nat Genet 36: 1189–1196.
20. Sperling S, Grimm CH, Dunkel I, Mebus S, Sperling HP, et al. (2005) Identification and functional analysis of CITED2 mutations in patients with congenital heart defects. Hum Mutat 26: 575–582.
21. Yang XF, Wu XY, Li M, Li YG, Dai JT, et al. (2010) [Mutation analysis of Cited2 in patients with congenital heart disease]. Zhonghua Er Ke Za Zhi 48: 293–296.
22. Chen CM, Bentham J, Cosgrove C, Braganca J, Cuenda A, et al. (2012) Functional significance of SRJ domain mutations in CITED2. PLoS One 7: e46256.
23. Li DY, Whitehead KJ (2010) Evaluating strategies for the treatment of cerebral cavernous malformations. Stroke 41: S92–94.
24. Ruiz-Perez VL, Blair HJ, Rodriguez-Andres ME, Blanco MJ, Wilson A, et al. (2007) Evc is a positive mediator of Ihh-regulated bone growth that localises at the base of chondrocyte cilia. Development 134: 2903–2912.
25. Bentham J, Michell AC, Lockstone H, Andrew D, Schneider JE, et al. (2010) Maternal high-fat diet interacts with embryonic Cited2 genotype to reduce

Pitx2c expression and enhance penetrance of left-right patterning defects. Hum Mol Genet 19: 3394–3401.

26. Agrawal A, Gajghate S, Smith H, Anderson DG, Albert TJ, et al. (2008) Cited2 modulates hypoxia-inducible factor-dependent expression of vascular endothe-lial growth factor in nucleus pulposus cells of the rat intervertebral disc. Arthritis Rheum 58: 3798–3808.

17

Identification of a Retroelement from the Resurrection Plant *Boea hygrometrica* That Confers Osmotic and Alkaline Tolerance in *Arabidopsis thaliana*

Yan Zhao[1,9], Tao Xu[1,9], Chun-Ying Shen[1], Guang-Hui Xu[1], Shi-Xuan Chen[1], Li-Zhen Song[1,2], Mei-Jing Li[1], Li-Li Wang[1], Yan Zhu[1], Wei-Tao Lv[1], Zhi-Zhong Gong[3], Chun-Ming Liu[2], Xin Deng[1]*

1 Key Laboratory of Plant Resources, Institute of Botany, Chinese Academy of Sciences, Beijing, China, 2 Key Laboratory of Plant Molecular Physiology, Institute of Botany, Chinese Academy of Sciences, Beijing, China, 3 State Key Laboratory of Plant Physiology and Biochemistry, College of Biological Sciences, China Agricultural University, Beijing, China

Abstract

Functional genomic elements, including transposable elements, small RNAs and non-coding RNAs, are involved in regulation of gene expression in response to plant stress. To identify genomic elements that regulate dehydration and alkaline tolerance in *Boea hygrometrica*, a resurrection plant that inhabits drought and alkaline Karst areas, a genomic DNA library from *B. hygrometrica* was constructed and subsequently transformed into *Arabidopsis* using binary bacterial artificial chromosome (BIBAC) vectors. Transgenic lines were screened under osmotic and alkaline conditions, leading to the identification of Clone L1-4 that conferred osmotic and alkaline tolerance. Sequence analyses revealed that L1-4 contained a 49-kb retroelement fragment from *B. hygrometrica*, of which only a truncated sequence was present in L1-4 transgenic *Arabidopsis* plants. Additional subcloning revealed that activity resided in a 2-kb sequence, designated *Osmotic and Alkaline Resistance 1 (OAR1)*. In addition, transgenic *Arabidopsis* lines carrying an *OAR1*-homologue also showed similar stress tolerance phenotypes. Physiological and molecular analyses demonstrated that *OAR1*-transgenic plants exhibited improved photochemical efficiency and membrane integrity and biomarker gene expression under both osmotic and alkaline stresses. Short transcripts that originated from *OAR1* were increased under stress conditions in both *B. hygrometrica* and *Arabidopsis* carrying *OAR1*. The relative copy number of *OAR1* was stable in transgenic *Arabidopsis* under stress but increased in *B. hygrometrica*. Taken together, our results indicated a potential role of *OAR1* element in plant tolerance to osmotic and alkaline stresses, and verified the feasibility of the BIBAC transformation technique to identify functional genomic elements from physiological model species.

Editor: Manoj Prasad, National Institute of Plant Genome Research, India

Funding: This work was supported by the National Natural Science Foundation of China (31270312 and 30970431), the National High Technology Research and Development Program of China (863 Program, 2007AA021403), the National Basic Research Program of China (973 Program, 2012CB114302), and the Ministry of Agriculture of China (2009ZX08009-060B). The funders had no role in study design, data collection and analysis, decision to publish, or preparation of the manuscript.

Competing Interests: The authors have declared that no competing interests exist.

* E-mail: deng@ibcas.ac.cn

⑨ These authors contributed equally to this work.

Introduction

Drought, alkaline and high calcium are major environmental factors in South-west China Karst landforms that limit plant growth and crop productivity [1,2]. Improving crop tolerance to environmental stresses is thus beneficial for both the agriculture and ecosystem dynamics in the region. Drought stress has been intensively studied in plants, resulting in the identification of a large number of genes that play a potential role in tolerance mechanisms [3]. In contrast, tolerance to alkaline and high calcium stress has not been intensively studied in plants.

Many plant species in the Gesneriaceae family such as Boea hygrometrica, Haberlea rhodopensis, Ramando myconi, Metapetrocosmea peltata, Chirita heterotricha, Oreocharis flavida, and Paraboea rufescens [4] are well adapted to the Karst region, and grow in shady limestone crevices where the soil is alkaline [5,6]. B. hygrometrica, H. rhodopensis and R. myconi, are also known as

resurrection plants, a category of plants that are able to tolerate full desiccation (leaf relative water content <10%) and are viable after rehydration within 48 h [7–9].

Previously, it was reported that the thylakoid pigment-protein complexes and pigment contents were highly stable during desiccation and rehydration in *B. hygrometrica*, but were irreversibly lost in the desiccated leaves of a non-resurrection Gesneriaceae species *Chirita heterotrichia* [7]. The stabilization of photosynthetic apparatus during water deficit had also been reported on resurrection plants in *Haberlea spp.* and *Ramando spp.* [8,9], indicating that these plants may have evolved distinct adaptive mechanisms to cope with desiccation. Plant responses to water deficit are complex, and these responses can be synergistically or antagonistically modified by the superimposition of other stresses [13]. It is unknown whether the concurrent environment factors such as high calcium and alkali evoke common adaptive mechanisms that dehydration triggers in these Gesneriaceae resurrection plants.

Genome-level regulation such as chromatin modification and assembly, transposable elements, small RNAs, and non-coding RNAs are involved in plant stress responses [14,15].With the exception of *CDT-1* from *Craterostigma plantagineum* [16–18], genomic elements that regulate stress tolerance in resurrection plants have not been identified, which is largely due to the lack of genome sequence data and genetic analysis tools.

Binary bacterial artificial chromosome (BIBAC) vectors were developed for transformation of large genomic DNA fragments into plants. This technology can be used overcome the technical limitations in species where genetic transformation and genome sequences are not available [19–21]. Therefore it became a useful tool for phenotype-based screening of genomic elements. For example screening of a BIBAC library from *Thellungiella halophila* led to the identification of a clone with a 120–130-kb insert that is associated with improved salt tolerance [22]. Screening of a BIBAC library from *Leavenworthia alabamica* also led to the identification of 84 20-kb genomic clones with phenotypic effects such as short fruit and aborted seeds [23]. In this study, a BIBAC library was constructed with *B. hygrometrica* genomic DNA and used for generation of transgenic populations. A BIBAC clone that conferred osmotic or alkaline tolerance was identified, and the resident functional element that might be responsible for the improved osmotic and alkaline tolerance was assigned.

Materials and Methods

BIBAC library construction

B. hygrometrica plants were collected from a self-bred population grown in green house conditions in our lab. Therefore no specific permission was required for these collections. This study did not involve endangered or protected species. Leaves of young *B. hygrometrica* plants were ground in liquid nitrogen. The isolation of the nuclear DNA was conducted as described by Zhang et al. [24]. Nuclear DNA in the plugs was partially digested by *Bam*HI and analyzed by pulsed-field gel electrophoresis (PFGE) using a WD-2010 apparatus (Beijing Liuyi Instrument Factory, China) on 1% agarose gels in 0.5×TBE buffer at a 5-s pulse time of 6 V/cm, at 15°C for 15 h. Restriction fragments in a range from 40 to 90-kb were collected and ligated with the vector pCLD04541 [25], which was completely digested and dephosphorylated. Ligated DNA was transformed into *E. coli* strain DH10B electrocompetent cells (Gibco-BRL, USA) by electroporation using a Cell Porator and Voltage Booster System (Gibco-BRL, USA) as described by Zhang et al [26]. About 4,600 clones were obtained from selection media, and were arrayed in 12×384-well microtiter plates and maintained in −80°C.

Analysis of BIBAC clones

Random clones from the BIBAC library were grown overnight at 37°C in LB medium containing 15 mg/L tetracycline (Amresco, USA). Plasmid DNA was isolated with the alkaline lysis method. Insert fragments were released from the pCLD04541 vector by digestion with *Not*I (TaKaRa, Japan) and subjected to PFGE performed as described above. The insert sizes of these clones were estimated using a lambda DNA ladder as the molecular-weight standard.

Estimation of nuclear genome size

The absolute amount of nuclear DNA (i.e. genome size) in *B. hygrometrica* was estimated using flow cytometry analysis. The experimental material consisted of leaves of *B. hygrometrica*, with

Arabidopsis thaliana serving as an internal reference standard. The histogram of relative DNA content was obtained after flow cytometric analysis of nuclei of *B. hygrometrica* and *Arabidopsis*, which were isolated, stained, and analyzed simultaneously. *B. hygrometrica* 2C DNA content = (*Boea* G1 peak mean)/(*Arabidopsis* G1 peak mean)×*Arabidopsis* 2C DNA content (*Arabidopsis* 1C = 125 Mb) [27].

Transformation of BIBAC clones into *Agrobacterium* and *Arabidopsis*

The BIBAC clone was sequenced to obtain terminal sequences that were used to design primers for clone-specific designation. Plasmid DNA of BIBAC clones was isolated with the alkaline lysis method and transformed into *A. tumefaciens* strain GV3101 via electroporation. The electroporated GV3101 clones were selected on YEB medium containing 50 mg/L rifampincin, 50 mg/L kanamycin, and 35 mg/L gentamycin. Random colonies were incubated in liquid YEB medium with antibiotics (as above) for 2 days at 28°C with shaking at 170 rpm and confirmed by PCR using primer pairs designed according to the appropriate clone-specific terminal sequences. *Agrobacterium* clones transformed with L1-4 were incubated in liquid YEB medium with antibiotics (as above) at 28°C, with shaking at 170 rpm. When the OD_{600} value of the culture increased to about 0.8, 500 µl of the culture was transferred to a 50 ml culture for continued growth. These cultures were subcultured 4 times, for approximately 10 hours for each passage. 1 µL of the first culture and 1 µL of the fifth culture were used to analyze the integrity of the BIBAC DNA in *Agrobacterium* using the multiple-marker PCR-based method [20].

Floral-dip transformation of *Arabidopsis* was conducted using the Columbia ecotype [28]. Transgenic plants were confirmed by PCR amplification using the BIBAC terminal specific primers. Homozygous lines were selected through two further rounds of selection on plates containing 50 mg/L kanamycin. Kanamycin-resistant T_1 plants were transferred to soil and seeds were collected. These T_2 seeds were sown on plates containing 50 mg/L kanamycin. The ratio of green to yellow seedlings of each line was analyzed with a Chi-square test. The lines with a 3:1 ratio of survival on kanamycin were selected and grown to maturity. T_3 seeds were collected and sown again on plates containing 50 mg/L kanamycin, and lines with 100% survival were considered as homozygous. The relative copy numbers and expression levels of the transgenes were checked by quantitative real-time PCR using genomic DNA and reversed transcribed DNA from RNA from the transgenic plants as templates, respectively. Only T_3 seeds of homozygous lines were used for further experiments.

Thermal asymmetric interlaced PCR (Tail-PCR) [29] was used to determine the insertion site of L1-4 in transgenic line L1-4-2. Genomic DNA was extracted using the CTAB method [30], and used as the template for Tail-PCR. Three primers, SP1, SP2, and SP3 were designed according to the adjacent sequences of the multiple cloning site of pCLD04541, using AD1 as the degenerate primer (Table S1 in File S1.).

Examination of stress tolerance in transgenic lines

T_3 seeds of transgenic lines were surface sterilized and placed on 1/2 Murashige & Skoog (MS) agar plates, and then cultivated vertically at 22°C for germination with a 16 h light/8 h dark cycle. 3 day-old seedlings were transferred to 1/2 MS agar plates that were saturated overnight with different concentrations of PEG 8000 (25% and 40% (w/v) PEG) for osmotic screening [31]; to 1/2 MS agar plates which were adjusted with potassium hydroxide to pH 5.6, 8.5, or 9.0 respectively, for alkaline screening; and to agar plates containing 60 mM or 80 mM $CaCl_2$ for high-calcium stress screening. The location of the seed lots were kept consistent in each treatment in one set of experiments but arranged randomly in different sets of repetitions. For each treatment, at least three independent experiments were conducted with at least three plates with 6 seedlings per line per plate was assayed. Photographs were taken, total root length and physiological parameters were determined after 2 weeks of growth. The total length of primary root and lateral roots were measured with ImageJ software. The empty vector control (pCLD04541) transformed plants showed no difference compared to the wild-type (Col-0) on PEG and alkali stress condition (Figure S1 in File S1.), therefore were not included in the subsequent phenotyping test. For soil dry treatment, T_3 seeds of transgenic lines were germinated on 1/2 MS agar plates and 5 day-old seedlings were transferred to soil in pots for dehydration for 7 days. Three replicates of 25 seedlings were tested in each treatment. In all cases, wild-type seed batches that were generated at the same time as the transgenic seed lines were germinated and transferred in parallel with the transgenic plants as controls.

Physiological parameter determination

Leaf relative water content (RWC) was estimated according the following formula: (RWC, %) = (fresh weight − dry weight)/(turgid weight − dry weight) × 100. Photochemical efficiency (Fv/Fm), and the extent of electrolyte leakage were measured as described previously [10]. Leaves were fixed with absolute ethyl alcohol for 2–3 min and used for stomata observation with light microscopy (B204LED, China). The experiments were performed twice, with three independent leaves for each treatment at each time point.

Shotgun sequencing

The BAC plasmid of L1-4 was purified using the QIAGEN Large-Construct Kit. Ultrasonically-broken and sheared BAC DNA (1.5–3-kb) was ligated into the pUC19 vector and transformed into *E. coli* strain Top10. The generated shotgun subclones were then sequenced from both ends using the dideoxy chain termination method using BigDye Terminator Cycle Sequencing V3.1 Ready Reaction (Applied Biosystems) on ABI3730xl Capillary Sequencing machines (Applied Biosystems). The subclones were sequenced to generate 8–10 fold coverage. The Phred-Phrap program (University of Washington, Seattle, WA, USA; http://www.phrap.org/phredphrapconsed.html) was used to assemble the shotgun sequences and gap-closing of each BAC [32,33].

Subclone library construction

The plasmid DNA of BIBAC L1-4 was partially digested by *Sau*3A. Fragments between the sizes of 5 and 8-kb were collected and ligated into the pCLD04541 plasmid which was completely digested by *Bam*HI and dephosphorylated. The recombinant DNA was transformed into *E. coli* EPI300 competent cells by electroporation. S3, S21, S32, and S35 were identified by PCR with different pairs of primers: 1 kF and 1 kR, 3 kF and 3 kR, 4 kF and 4 kR, 5 kF and 5 KR, 14 kF and 14 kR, 19 kF and 19 kR, 44 kF and 44 kR, 46 kF and 46 kR, and 47 kF and M13R-48, respectively (Table S1 in File S1.).

Sequence analysis

The open reading frames (ORFs) in the genomic sequences were predicted with the gene-finding tools FGENESH (http://linux1.softberry.com) using the *Arabidopsis* dataset with default settings. Repetitive elements were searched with the program RepeatMasker (http://www.repeatmasker.org/cgi-bin/WEBRepeatMasker). The retrotransposons were identified with LTR_ Finder (http://tlife.fudan.edu.cn/ltr_finder/). The gene identities were predicted with BLAST analysis (http://blast.ncbi.nlm.nih.gov).

RNA isolation and Real-time PCR

Total RNA was isolated using the TRIzol method with RNAiso Plus (Takara, D9108B). After digestion with DNase I, it was reverse transcribed into cDNA with Oligo(dT)18 primer using M-MLV reverse transcriptase, and used as template for PCR amplification. *ACTIN2* and *18S* were used as internal references for transgenic lines and *B. hygrometrica* gene expression determination, respectively. Real-time PCRs were performed in a Mastercycler ep realplex apparatus (Eppendorf, Hamburg, Germany) with SYBR Green Realtime PCR Master Mix (TOYOBO, Japan). Specific primers were listed in Table S1 in File S1.

Determination of *OAR1* relative copy number

Relative copy number of *OAR1* was assessed by quantitative real-time PCR using OAR1-1F and OAR1-1R primers and genomic DNA from *B. hygrometrica* and transgenic *Arabidopsis* lines S21-3 and S21-14. *NPTII* gene in the transgene cassette was used as an internal control for determination of the relative copy number of *OAR1* in transgenic *Arabidopsis* lines. *NPTII* gene in the plasmid DNA of the empty vector pCLD04541 was used as an external control for determination of the relative copy number of *OAR1* in the *B. hygrometrica* genome.

Accession numbers

The *OAR1* nucleotide sequence has been submitted to NCBI with accession number KF425673.

Results

Construction and characterization of a *B. hygrometrica* BIBAC library

A BIBAC genomic library was constructed for *B. hygrometrica* from nuclear DNA, containing about 4,600 clones in total. To estimate the quality of this library, the insert sizes of 50 randomly selected clones from the library were analyzed by digesting the plasmid DNA with *Not*I and subsequent separation by pulsed-field gel electrophoresis (PFGE). The results are presented in Figure 1 (A–D). The majority of these clones contained inserts with sizes ranging from 40 to 90-kb. Because 10% of the clones have inserts larger than 100-kb, the average insert size was 62-kb. Among these 50 clones, one did not contain an insert, accounting for 2% of the clones analyzed. Thus the frequency of clones lacking an insertion was lower than that of the BIBAC libraries made for tomato [34], petunia [35], rice [36], *Arabidopsis thaliana* ecotype Landsberg [37], or chickpea [38], which were 10%, 6.5%, 4.8%, 17.6%, and <5%, respectively. The low frequency of empty constructs and the fact that each BAC clone contained different insertion sizes implies that the *B. hygrometrica* BIBAC library was of good quality.

To determine the genome coverage of this library, the nuclear DNA content of *B. hygrometrica* was estimated by flow cytometry. Using *Arabidopsis* as an internal reference, the haploid genome size of *B. hygrometrica* was calculated to be 240 Mb (Figure 1E). Based on the average insert size of 62-kb the coverage of the library is approximately 1.18 haploid genome equivalents.

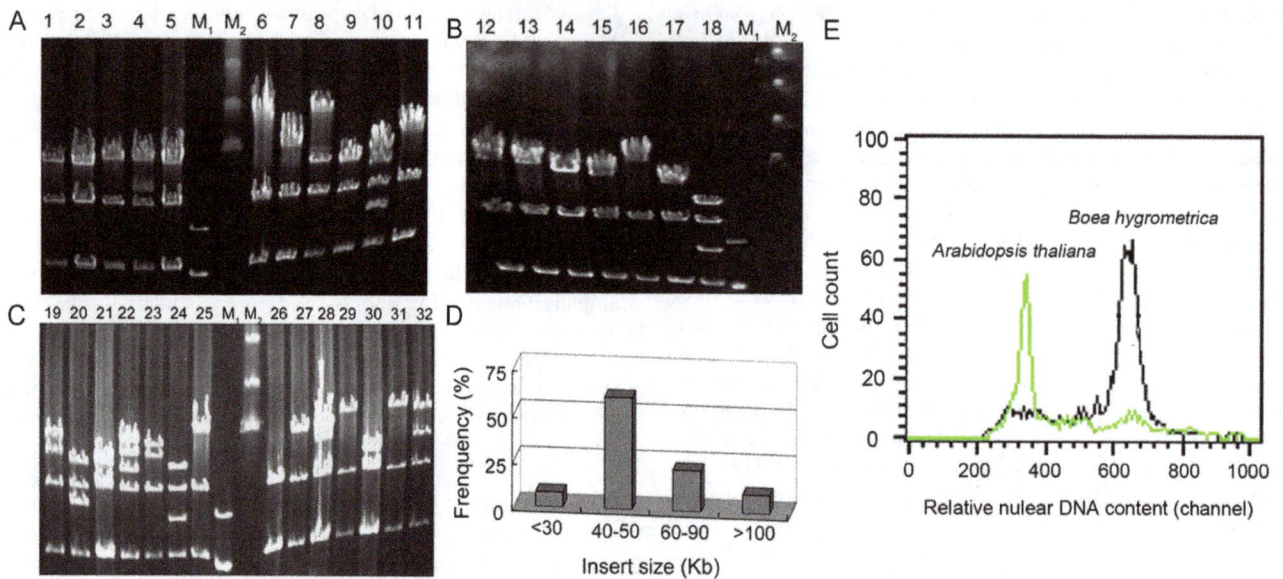

Figure 1. Analysis of insert size for the *Boea hygrometrica* BIBAC library and determination of genome size of *B. hygrometrica* by flow cytometry. (A–C) Pulsed-field gel electrophoresis (PFGE) patterns of 32 representatives of the 50 random BIBAC clones that were digested with *Not*I. PFGE gels were stained with ethidium bromide. M_1, marker with bands of size 3, 5 and 8-kb; M_2, Lamda ladder PFG marker with band sizes of 48.5, 97, 145 and 194-kb. (D) Insert size distributions of the 50 clones randomly selected from the *B. hygrometrica* BIBAC library. (E) Determination of genome size of *B. hygrometrica* by flow cytometry. *Arabidopsis* served as internal reference standard. *B. hygrometrica* 2C DNA content = (*Boea* G1 peak mean)/ (*Arabidopsis* G1 peak mean) ×*Arabidopsis* 2C DNA content, the ratio of G1 peak means (*B. hygrometrica*: *Arabidopsis*) was 1.86, hence the 2C DNA amount of *B. hygrometrica* was estimated as 480 Mb (*Arabidopsis* 1C = 125 Mb).

Construction of populations of transgenic *Arabidopsis* lines carrying *B. hygrometrica* BIBAC clones

288 BIBAC clones were subjected to BAC terminal sequencing to obtain sequence information for designing primers that could be used to generate markers of each clone. Among these, 172 unique BIBAC plasmids were selected for transformation into *Agrobacterium* and subsequently into *Arabidopsis*, as indicated by distinct PCR products using the clone-specific primers. In total, 43 BIBAC clones were successfully transformed into *Arabidposis* and 213 transgenic lines were generated. For 37 of these lines, >3 independent transformants were obtained and subsequently used for phenotypic analyses.

BIBAC clone L1-4 conferred osmotic and alkaline tolerance in transgenic *Arabidopsis*

To identify genomic DNA fragments of *B. hygrometrica* that are able to improve osmotic, alkali, and/or high calcium tolerance, T_3 generation transgenic plants were subjected to screening under osmotic, alkaline, or high calcium stress conditions. The results revealed that transgenic lines carrying the BIBAC clone L1-4 exhibited tolerance to both osmotic and alkali stresses. These transgenic plants grew well on 1/2 MS plates (pH 5.6), showing no obvious difference from the wild-type. However, transgenic plants exhibited better growth as shown by the bigger rosette, longer total roots (primary and more lateral roots together) on alkaline media (pH 8.5) or media with 25% PEG, as compared to the wild-type and "empty vector" control plants (Figure 2, Figure S1 in File S1.). The alkali resistant phenotype was further confirmed by the higher ratio of plant survival in transgenic lines on alkaline media (pH 9.0) than that in wild-type (Figure S2 in File S1.). On PEG-mediated osmotic stress conditions, the transgenic plants exhibited thicker and longer lateral roots compared to the wild type (Figure 2). The osmotic resistant phenotype was further supported

by the higher survival rates of the transgenic plants of L1-4 than the wild-type under soil drought conditions (Figure S3 in File S1). However, these transgenic lines did not show tolerance to high calcium (data not shown). These data suggested that a functional fragment of genomic DNA carried by BIBAC clone L1-4 was able to confer osmotic and alkaline tolerance in *Arabidopsis*.

To check if the observed phenotypes of transgenic plants resulted from an insertional mutagenesis event, the insertion site of L1-4 in transgenic line L1-4-2, which exhibited the strongest osmotic and alkaline tolerance among the transgenic lines, was examined. The result indicated that L1-4 DNA inserted in the interval of a non-coding region between two alpha tubulin (TUA) genes in *Arabidopsis* chromosome 5,897 bp up-stream of *TUA3* (reverse orientation) and 1703 bp up-stream of *TUA5* (forward orientation) on chromosome 5 (Figure S4A in File S1). Semi-quantitative RT-PCR revealed that *TUA3* and *TUA5* expression was reduced in L1-4 transgenic plants (Figure S4B in File S1).

L1-4 contained a cluster of nested LTR-retrotransposons

The insert size of BIBAC L1-4, as determined by a complete *Not*I digestion and PFGE, was approximately 50-kb (Figure S5 in File S1). To identify the specific genetic element that is responsible for the osmotic and alkaline pH-resistant phenotypes, shotgun sequencing was performed to obtain the full sequence information of the insert DNA in L1-4. Sequence assembly resulted in a continuous 49,387 bp contig with the GC content of 45.49%. Using the sequence to query the GenBank nucleotide database, we detected several discontinuous DNA fragments with similarities to the genomic sequences of *Mimulus guttatus*, grape, tomato, populous, and pineapple, with the highest similarity of 70% and query coverage of up to 17%. Besides, TBLASTX analysis did not detect any similarity to known proteins with the exception of gag-pol polyproteins [39], which are generally components of retro-transposons (retroelements).

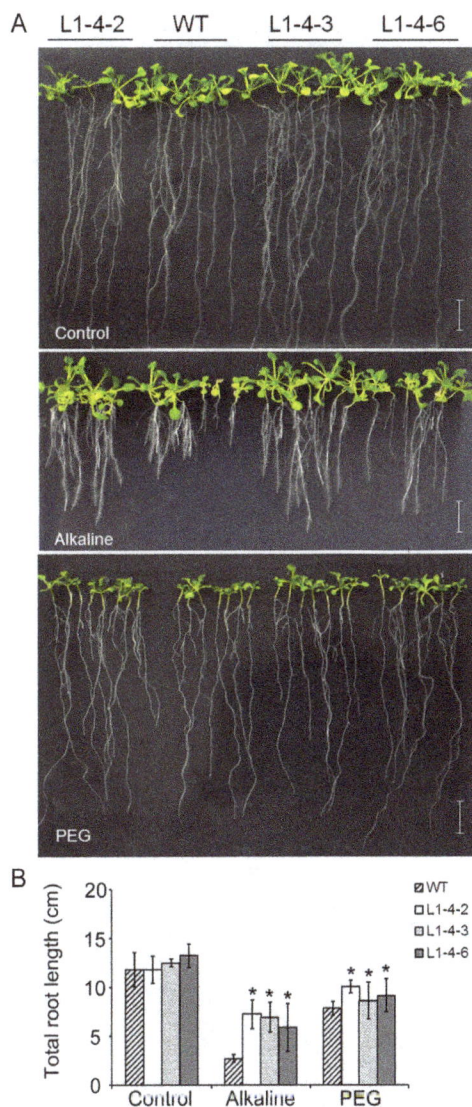

Figure 2. Phenotypes of L1-4 transgenic plants under alkaline and osmotic stress. (A) Seedlings grown on agar plates containing 1/2 MS with pH5.6 (control), pH8.5 and soaked with 25% PEG 8000. (B) Total root length of transgenic and wild-type plants grown on alkaline media (pH 8.5) and osmotic stress with 25% PEG 8000 for 14 d. 3 day-old seedlings were transferred to control media or media with alkaline pH or 25% PEG 8000 media. Wild-type seed batches that were generated at the same time as the transgenic seed lines were germinated and transferred in parallel with the transgenic plants as controls. n = 18, Bar = 1cm, data are shown as means ± SD.

To identify possible protein-coding sequences, the FGENESH algorithm was used for ORF prediction in the L1-4 insert. This resulted in the identification of 11 ORFs, 10 of which were complete (Figure 3A). Standard BLASTP analysis failed to find any homologues for ORF2, 3, or 8; but it revealed homologues to long terminal repeats (LTRs) for ORF1, 4, and 9, as well as gag-pol polyproteins of LTR-retrotransposons for ORF5, 6, 7, and 11, suggesting that L1-4 may indeed consist of LTR-retrotransposons, which is a major type of transposons (transposable elements, TEs) in higher plants [40]. In total, 3 gypsy (TE1, TE3 and TE4), and 1 copia (TE2) were predicted in the L1-4 contig according to the organization of the *gag* and *pol* genes. LTR finder and RepeatMasker were then used to predict the corresponding

LTR of the identified retrotransposons. Only one pair of LTRs was identified as clamped LTRs of TE1, while the others were solo-LTR or incomplete. TE1 and TE4 were highly homologous (>95%) in different orientations. TE2 and TE3 were located inside the LTR pair-clamped region downstream of the 3' terminus of TE1 (Figure 3A). Thus the whole contig of L1-4 was constituted of four nested and truncated retrotransposons. Such genomic structure has been reported previously in both plants and animals [41–43]. The middle sections of both TE1 and TE4 show a remarkable resemblance to *Arabidopsis* LTR-retrotransposons *AtGP1* and *AtGP2* [44]. Besides, 130 dimer- and 42 trimer-microsatellites were detected in the 49-kb sequence. These data showed that L1-4 contained a region of the *B. hygrometrica* genome with short tandem repeated transposons and nested transposons.

L1-4 became truncated in the transgenic *Agrobacterium* and *Arabidopsis*

Large DNA fragments are unstable after transformation into *Agrobacterium* [45]. The existence of multiple transposon-coding sequences in L1-4 raised the possibility of DNA deletion, re-insertion, and rearrangement. To check if the transgenic plants carried an intact fragment of the L1-4 BIBAC clone, primers were designed according to the shotgun sequence to amplify individual regions of the L1-4 sequence. As shown in Figure 3B, correct PCR products were obtained with all tested primer pairs when L1-4 plasmid DNA was used as template. However, when genomic DNA from three independent L1-4 transgenic lines were used as templates, correct PCR products were produced only with the primers within 0–4 and 47–49-kb of L1-4 insert. This suggested that only two terminal sequences of L1-4 had been integrated into the *Arabidopsis* genome in the transgenic lines. The loss of the middle part of the L1-4 sequence in transgenic plants was further confirmed by sequencing of the amplified products, which showed continuous sequence homology from 4,096 to 4,298 bp and from 47,317 to 48,145 bp in the L1-4 contig. Thus it was evident that the transgenic lines harbor only an internally truncated L1-4 sequence, which was 6,369 bp in length, as shown in Figure 3C. To distinguish it from the intact plasmid L1-4 sequence, the truncated 6,369 bp fragment was designated as L1-4*.

Because all three independent transgenic lines exhibited the same deletion, we speculated that the deletion had occurred in *Agrobacterium*. To address this possibility, the insert DNA in the L1-4 transformed *Agrobacterium* strain stored in −80°C for different periods was analyzed. The PCR amplification of different regions of L1-4 indicated that the inserted DNA was intact in *Agrobacterium* cultured directly from the stock in −80°C for both six months and three years (Figure S6A–C in File S1). However, when *Agrobacterium* were continuously subcultured in liquid medium at 28°C, some regions of L1-4 were no longer detectable in some of the randomly selected clones (Figure S6D–E in File S1). These results indicated that the large insert DNA in the plasmid of *Agrobacterium* was stable when stored in −80°C for three years, but not during sequential liquid subculturing at 28°C. This is probably the reason for the truncation of L1-4 in transgenic *Arabidopsis*.

Identification of the subclones containing partial sequences of L1-4*

To dissect the particular genomic element that was responsible for the improved osmotic and alkaline tolerance of transgenic *Arabidopsis*, a subclone library was constructed containing 5–8 kb fragments of the L1-4 large insert DNA produced by partial digestion using *Sau*3A. The library consisted of 200 clones and covered >20 times of the L1-4 insert sequence. Subclones

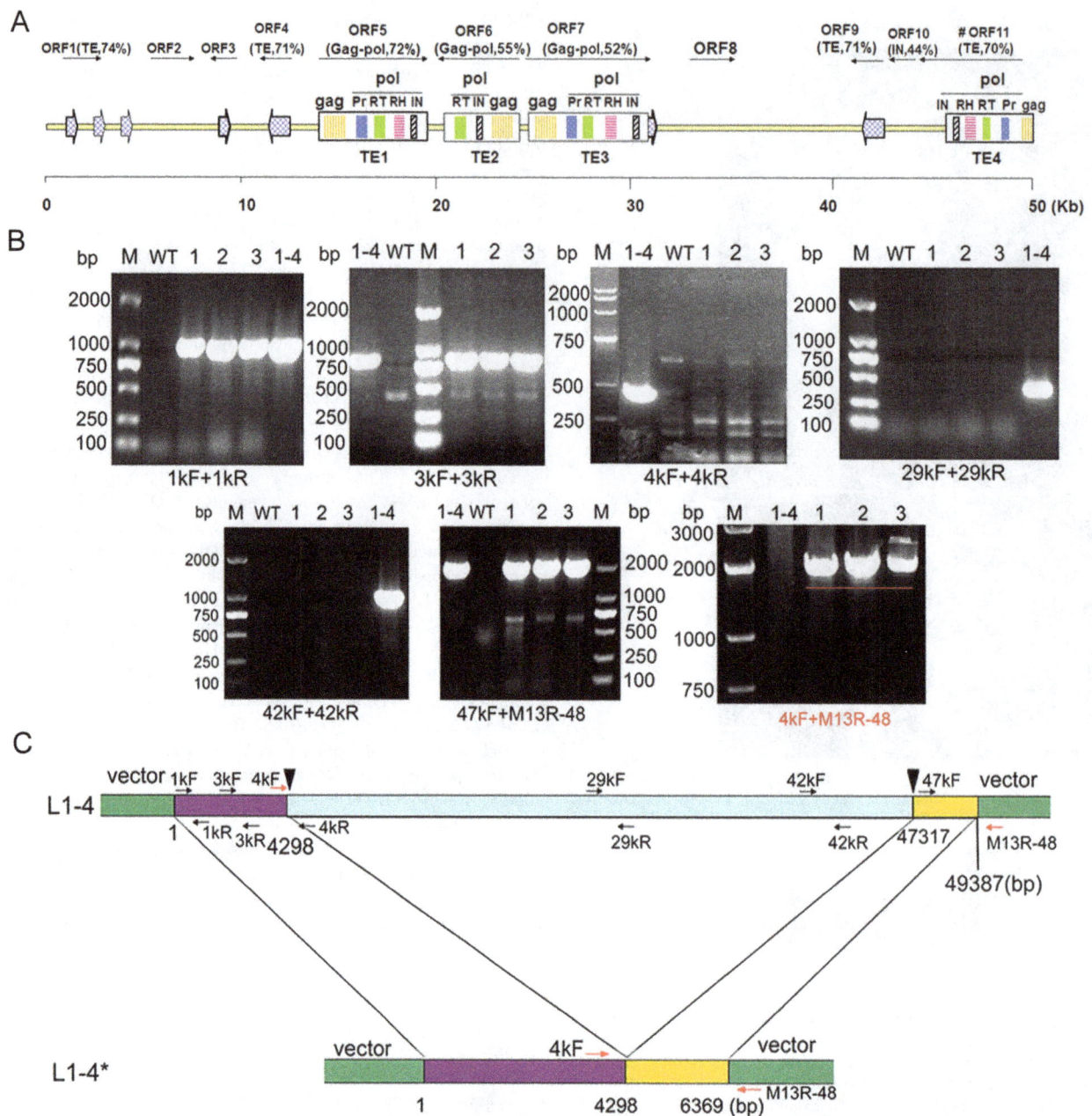

Figure 3. Annotation of insert sequence in BIBAC clone L1-4 and truncation of L1-4 in the transgenic plants. (A) ORFs predicted with the program FGENESH were annotated by TBLASTX, transcriptional orientation is indicated with an arrow. A schematic diagram was drawn to show the position of the predicted LTRs (grid arrows) and retrotransposons. The intact LTR-retrotransposon identified by LTR-finder is indicated in bold. gag, capsid-like protein; pol, polpolyprotein; IN, integrase; PR, pepsin-like aspartate proteases; RH, RNase H; RT, Reverse transcriptase; LTR, long terminal repeat; TE, transposable element. (B) Assay of PCR amplification using specific primers corresponding to different regions in L1-4. Primers were designed to amplify fragments located at 1, 3, 4, 14, 29, 42, and 47–49-kb of L1-4, as indicated on the bottom of each picture and listed in (C) and Table S1 in File S1. M, DNA marker with the size of 100, 250, 500, 750, 1000 and 2000 bp; 1–4, plasmid DNA of BIBAC clone L1-4; WT, genomic DNA of Arabidopsis wild-type Col-0; 1, 2, 3, genomic DNA of three independent transgenic lines of L1-4. The PCR fragments were separated on 1% agarose gels. (C) Schematic diagram of L1-4 and L1-4*. The positions and orientation of primers are indicated by arrows.

containing partial sequences of L1-4* were identified by PCR and confirmed by sequencing. The identified subclones S32 and S35 carried sequence corresponding to 1–4,298 bp of L1-4. Subclone S21 carried sequence corresponding to 46,317–49,387 bp of L1-4 (Figure 4). These constructs were transformed into Arabidopsis, and 20, 12, and 2 transgenic lines were obtained for the S21, S32, and S35, respectively.

Identification of the genetic loci responsible for osmotic and alkaline tolerance

Two transgenic lines from each construct of S21, S32, and S35 were assayed for osmotic and alkaline tolerance along with L1-4 transgenic and wild-type plants. Most of these transgenic lines grew similarly to the wild-type on 1/2 MS plates with or without PEG, or adjusted to pH 5.6 or 8.5. Only the plants harboring S21

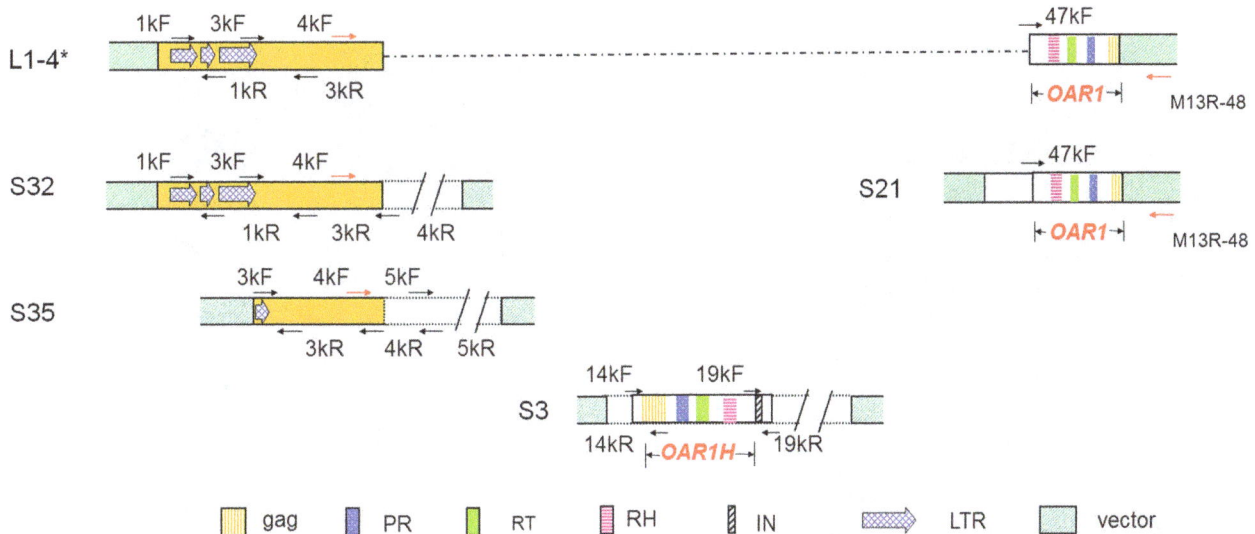

Figure 4. Schematic diagram of subclones containing partial sequences of L1-4*. Pairs of primers were designed according to sequence of L1-4, as indicated in Figure 3C. Subclone S21 contains 46–49-kb of L1-4 and 4–6-kb of L1-4*; subclones S32 and S35 contain 0–4-kb of L1-4 and L1-4*; subclone S3 contains 14–19-kb of L1-4. gag, capsid-like protein; pol, polpolyprotein; IN, integrase; PR, pepsin-like aspartate proteases; RH, RNase H; RT, Reverse transcriptase; LTR, long terminal repeat; TE, transposable element.

displayed tolerance to alkaline and osmotic stresses, similar to the L1-4 transgenic plants, as indicated by the vigorous growth of shoots and improved root growth, including longer primary roots and more lateral roots, on the alkaline and PEG plates (Figure 5). The transgenic plants grow normally in soil, showing no obvious difference from the wild-type under unstressed conditions. However, when the seedlings were grown under soil drought condition in parallel with line L1-4-2 and the wild-type plants, the survival rates of S21 and L1-4 transgenic plants were significantly higher than the wild-type (Figure S3 in File S1), despite that the drought-resistant phenotype was observed only when the plants were young.

These observations indicated that the genetic element that conferred plant alkaline and drought tolerance in L1-4* was also located in S21 subclone. Alignment of the L1-4* and S21 sequences revealed that an overlapping sequence of 2076 bp, representing the 47,317–49,387 bp region of L1-4, hereby designated Osmotic and Alkaline Resistance 1 (OAR1). OAR1 contains part of ORF11 of L1-4, which encodes a partial gag-pol transcript for RNase H, reverse transcriptase and protease (Figure 4).

Because OAR1 is highly homologous to TE1 that is located in the 14–19-kb region of L1-4, subclone S3 containing TE1 (containing an OAR1-homologous sequence in the middle, OAR1H) was identified and transformed into Arabidopsis (Figure 4). Similar to the S21 transgenic plants, transgenic plants carrying S3 also displayed the improved tolerance to alkaline and osmotic stress treatments (Figure 6). Furthermore, over-expression of OAR1 under the control of the 35S promoter revealed a similar osmotic and alkaline stress tolerance phenotype (Figure S7 in File S1). In total, 4 independent transgenic lines of L1-4, 4 transgenic lines of S21, 5 transgenic lines of S3 and 3 lines of 35S::OAR1 displayed similar phenotypes, which suggested that the phenotype was not simply the result of positional effects. Thus, our results demonstrate that OAR1 and OAR1H in L1-4 and its subclones are functional elements to confer plant tolerance to alkaline pH and drought stresses.

On the other hand, no difference was detected when the transgenic plants harboring S32 and S35 were compared to the wild-type under non-stressed or PEG/alkaline-stressed conditions (Figure S8 in File S1). As these plants were produced, propagated and phenotypic assayed in parallel with S21 transgenic plants, the failure of detecting visible PEG and alkaline resistant phenotypes in S32 and S35 transgenic plants not only enabled us to define the functional genetic element in L1-4* to OAR1 that was common with L1-4* and S21, but also provided good negative controls to help to eliminate the possible effects of the antibiotic resistance marker on the PEG and alkaline resistant phenotypes of the L1-4 and S21 transgenic plants.

Physiological characterization of the osmotic and alkaline tolerance in transgenic plants harboring OAR1

Transgenic plants harboring OAR1 were further analyzed by the electrolyte leakage and photochemical efficiency using two representative lines that displayed the strongest osmotic and alkaline tolerance, L1-4-2 and S21-3, in parallel with the wild-type. The results revealed no difference in electrolyte leakage or photochemical efficiency (Fv/Fm) between the transgenic and wild-type plants grown under unstressed conditions. However, under alkaline or osmotic stresses, the electrolyte leakage increased to 51–53% and the Fv/Fm declined to 0.4 and 0.6 in wild-type plants under osmotic and alkaline stresses, while that of transgenic lines remained around 29–35% and 0.7, respectively (Figure 7). The difference between the wild-type and transgenic plants harboring OAR1 was significant, indicating that the stress-triggered decline of cytomembrane integrity and photochemical efficiency was prevented in the transgenic plants harboring OAR1, which was consistent with the observed growth phenotype. Examination of the relative water content (RWC) and the stomata in the plants surviving on PEG plates showed no difference in these parameters between transgenic line S21-3 and the wild-type plants under unstressed and PEG-stressed conditions (Figure S9 in File S1). Thus, the survival of OAR1-containing transgenic plants on PEG plates might be due to the maintenance of membrane integrity and

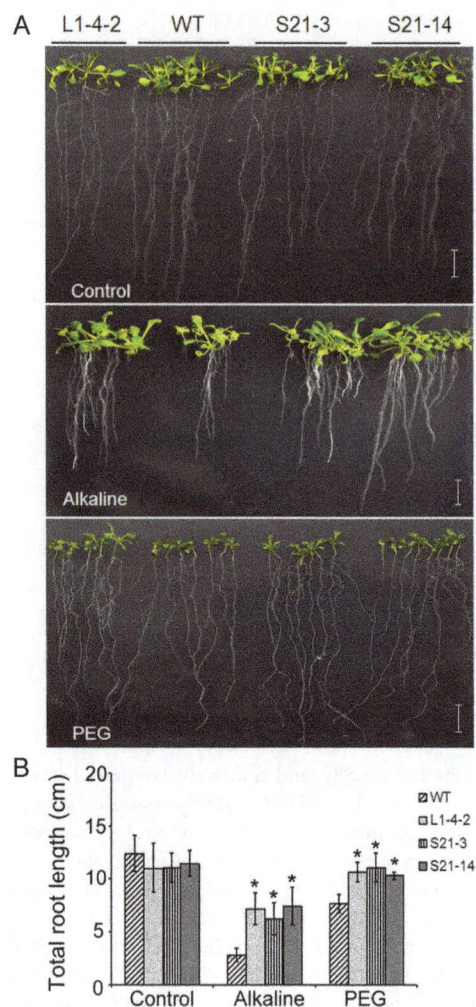

Figure 5. Phenotype comparison of the wild-type and transgenic plants harboring L1-4 and S21 under alkaline and osmotic stresses. (A) Seedlings grown on 1/2 MS agar plates adjusted to pH 5.6 and pH 8.5, and 1/2 MS agar plates saturated with 25% PEG 8000. (B) Total root length of plants on 1/2 MS agar plates adjusted to pH 5.6 and pH 8.5, and 25% PEG 8000 agar plates for 14 d.

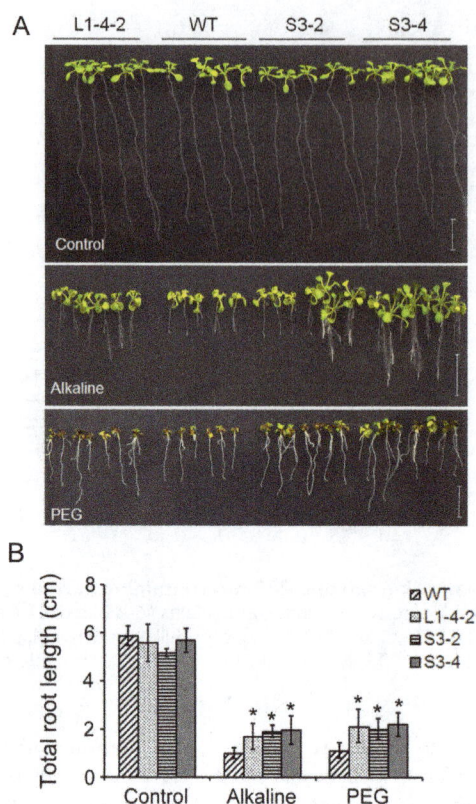

Figure 6. Phenotype comparison of the wild-type and transgenic plants harboring L1-4 and S3 under alkaline and osmotic stresses. (A) Seedlings grown on 1/2 MS agar plates adjusted to pH 5.6 (control) and pH 9.0, and 1/2 MS agar plates saturated with 40% PEG 8000. (B) Total root length of the wild-type and transgenic plants on 1/2 MS agar plates adjusted to pH 5.6 and pH 9.0, and on PEG 8000 agar plates for 10 d. n = 18, Bar = 1 cm. Data are shown as means ± SD.

photochemical efficiency and not due to differences in rates of water loss.

The expression and copy number of *OAR1* in transgenic *Arabidopsis*

To understand the molecular mechanisms that *OAR1* may function in osmotic and alkaline resistance, we first determined if *OAR1* element was transcribed in transgenic *Arabidopsis*. Despite that there was no any known promoter sequence adjacent to S21 insert in the BIBAC vector, quantitative RT-PCR had detected high levels of two short transcripts corresponding to 0.4–0.5 kb (designated as *OAR1-2*, by primer pair of OAR1-2F and OAR1-2R) and 1.7–1.9 kb (designated as *OAR1-1*, by primer pair of OAR1-1F and OAR1-1R) regions of *OAR1* sequence in both S21-3 and S21-14 under unstressed condition and osmotic and alkaline-stressed conditions (Fig. 8A, B). No product was amplified when primers OAR1-1F and OAR1-2R were used to amplify the long transcript corresponding to 0.4–1.9 kb regions of *OAR1* sequence in transgenic plants under unstressed or stressed

condition, indicating that *OAR1-1* and *OAR1-2* were either transcribed separately, or spliced from a long unstable transcript of *OAR1* in the transgenic plants. It is noticed that expression levels of *OAR1-1* and *OAR1-2* was higher in osmotic and alkaline stressed plants compared to that in unstressed plants (Fig. 8A, B).

Furthermore, relative gene copy number was examined by genomic quantitative PCR to check if the phenotype of the transgenic lines were dependent on copy number and if *OAR1* element was capable of transposition. Data indicated that the relative copy number was <1 in both S21-3 and S21-14 after normalized to NPTII gene which was co-transformed with *OAR1* within the T-DNA left and right borders in the same vector (Fig. 8C). Considering that the two transgenic lines exhibited a 3:1 ratio of survival on kanamycin in T2 generation and 100% survival on kanamycin in T3 generation, it was likely that only single copy of *OAR1* existed in the genome of S21-3 and S21-14 plants and no transposition of *OAR1* had occurred. Furthermore, the copy number of *OAR1* in S21-3 and S21-4 remained unaltered under osmotic stress but slightly decreased under alkaline stress. Because retrotransposons usually transpose in a "copy-paste" manner, the unchanged copy number indicated that *OAR1* element was not transposed in S21 transgenic plants.

Marker gene expression in *S21* transgenic *Arabidopsis* plants

To examine whether the insertion of *OAR1* element influenced stress responsive gene expression, several marker genes in osmotic

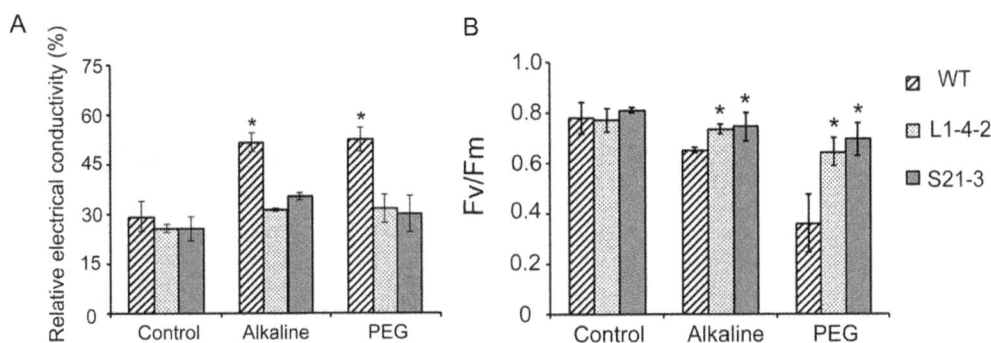

Figure 7. Physiological characterization of the L1-4 and S21 transgenic plants. (A) Electrical conductivity and chlorophyll fluorescence Fv/Fm characteristics (B) of seedlings grown on 1/2 MS (control) and alkaline (pH 9.0), or PEG plates (40% PEG 8000) following 2 weeks of growth. n = 18. Data are shown as means ± SD.

and alkaline stress pathways were checked for expression changes, along with 4 genes related to photosynthesis. The data have shown that all test genes, with an exception of *AHA2*, increased transcription in S21-3 plants under unstressed conditions, and remained higher than that in the wild type under both osmotic and alkaline stresses (Fig. 9D). This is consistent to the observed stress

tolerant phenotype and the stabilized photosynthetic apparatus in the transgenic plants.

OAR1 transcription and copy number in *B. hygrometrica*

To understand the molecular nature and the mechanisms that *OAR1* functions in its native host genome of *B. hygrometrica*, the

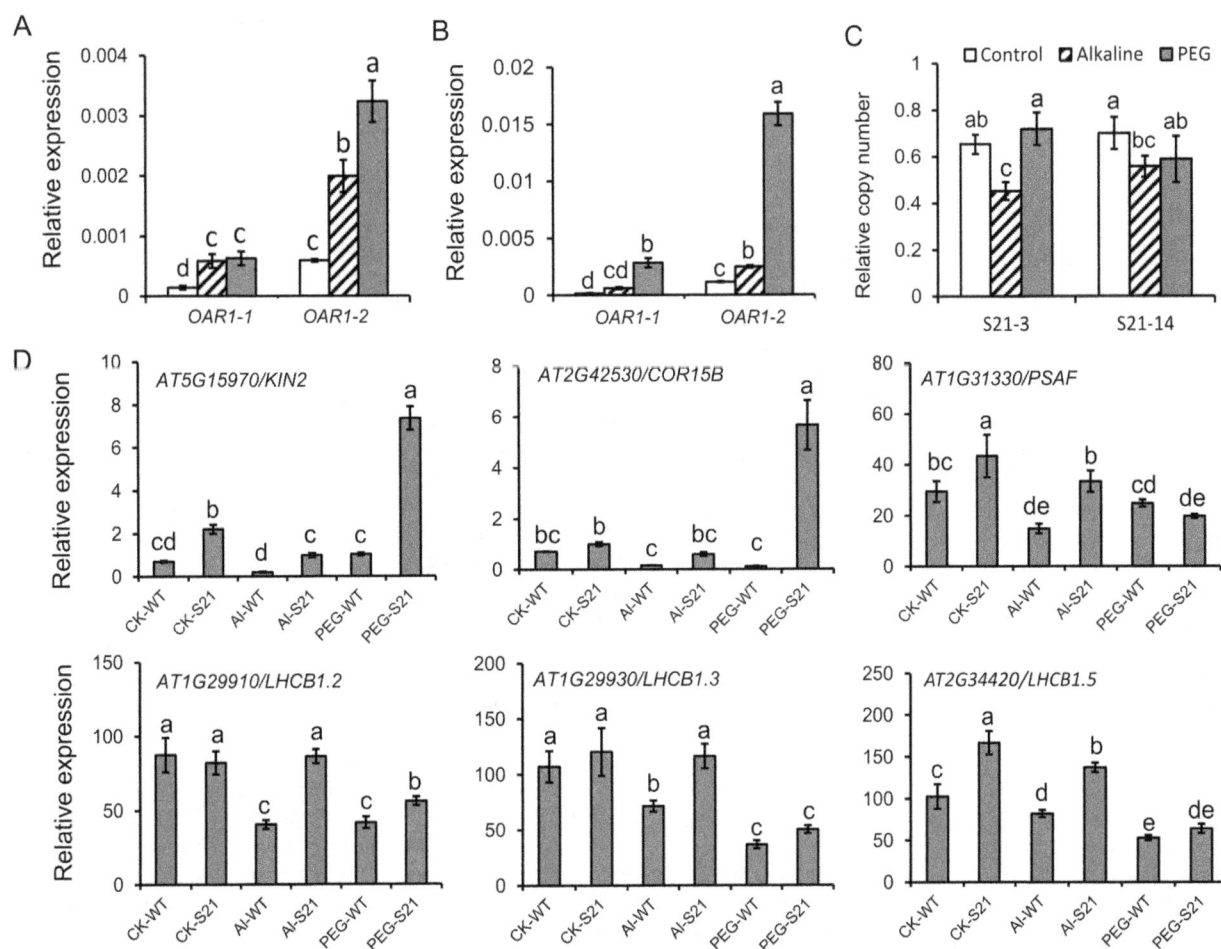

Figure 8. The expression and relative copy numbers of *OAR1* element, and marker gene expression in transgenic *Arabidopsis* under osmotic and alkaline stresses. (A, B) The expression of *OAR1* element in S21-3 (A) and S21-14 (B); (C)The relative copy numbers of *OAR1* element in S21-3 and S21-14; (D) The expression of stress marker gene and photosynthesis related gene in S21-3 and S21-14. Data are shown as means ± SD.

A

B

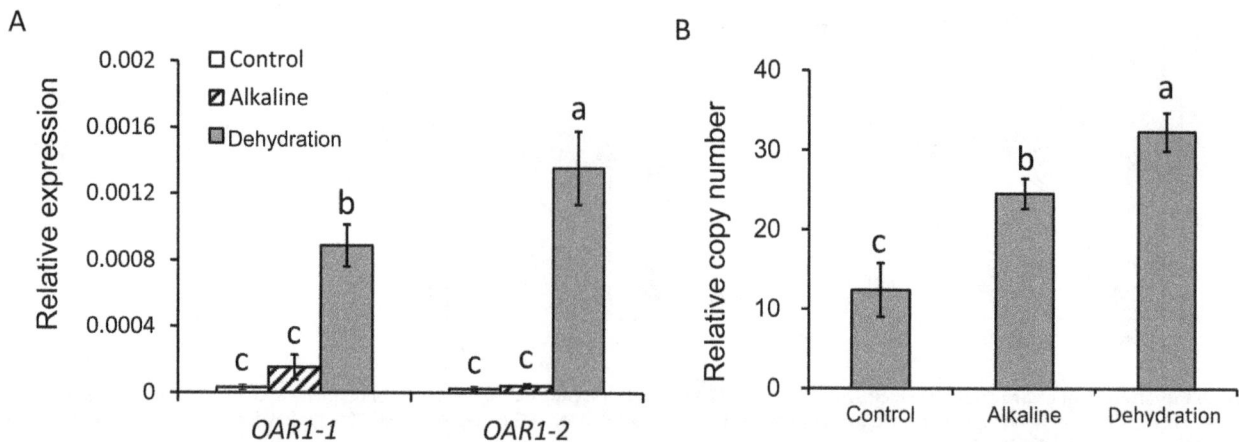

Figure 9. The expression and relative copy numbers of *OAR1* element in *B. hygrometrica* under osmotic and alkaline stresses. (A) The expression of *OAR1* element, *OAR1-1* and *OAR1-2* in *B. hygrometrica*; (B) relative copy number of *OAR1*. Data are shown as means ± SD.

expression levels of *OAR1* and the relative copy number have been determined. The results showed that *OAR1* transcription levels were increased slightly in response to alkaline and significantly under dehydration stresses in *B. hygrometrica* (Fig.9A). *B. hygrometrica* genome contained as high as >10 copy numbers of *OAR1*, and the relative copy number was increased to 2–3 fold under alkaline and PEG treatments (Fig. 9B). Considering that *OAR1* was located within a stretch of nested retroelement regions in the *B. hygrometrica* genome, as revealed by the sequence of BIBAC clone L1-4, our data indicated that *OAR1* might locate in an active retroelement cluster in its host genome.

Discussion

In species which survive in extreme natural environments, certain adapted traits are expected to be established during evolution, to allow the species to cope with harsh conditions, such as osmotic, alkaline, and salt stresses [46–48]. Recent studies have shown the involvement of chromatin modification and assembly, distantly located regulatory elements, complex loci, transposable elements, small RNAs, and non-coding RNAs in the regulation of various plant stress responses [49,50]. Unraveling stress-associated genomic regulatory mechanisms in these plants will enable future molecular manipulation of highly stress-tolerant crops. Limestone-inhabiting *B. hygrometrica* is one of the few resurrection species in the Gesneriaceae family that the vegetative tissues can survive desiccation and recover upon rehydration [7,10]. Thus it is a good resource for isolation of functional genes and genomic elements for crop breeding to improve tolerance to drought and alkaline stresses. In this paper, we identified *OAR1*, a functional element located in a 49-kb nested LTR-retrotransposon fragment that conferred osmotic and alkaline stress tolerance in transgenic *Arabidopsis*, taking the advantage of BIBAC library transformation system. This finding provides the first insight into the role of nested LTR-retrotransposon clusters in the genome of *B. hygrometrica* in environmental adaptation of this resurrection plant.

Transformation of large genomic DNA fragments into recipient plants has been used primarily to identify genes or quantitative trait loci (QTL) in species in which map-based cloning is not practical, allowing for the isolation of the *FILAMENTOUS FLOWER* gene [51,52]. This technique was soon applied for the discovery of novel genes and genomic elements from species lacking genome sequence information, genetic transformation methods, or mutation tools [21,22]. In this paper, this laborious

approach has been successfully applied to identified the 49-kb fragment that conferred stress tolerance, and herein the 2-kb *OAR1* element from *B. hygrometrica* by combination with shotgun sequencing, subcloning and phenotypic screening.

Photosynthesis has been recognized as the most sensitive process that was affected by water stress and other abiotic stresses such as salt and alkali [53]. The carbon balance, ROS homeostasis, energy generation, growth and survival of a plant under water stress depend heavily on the degree and velocity of photosynthesis decline during water depletion. Accumulating evidence demonstrated that in resurrection plants, complex mechanisms involving compatible solutes and desiccation-associated proteins such as LEAs and sHSPs, antioxidants, membrane protectants were triggered by water loss, which contribute to the stabilization of photosynthetic pigment-protein complexes and chlorophyll content, protection of photosynthetic apparatus and prevention of ROS accumulation from photosystems (PSI and PSII) [7,10,11,12,54,55]. For example, *BhLEA1* and *BhLEA2* are two dehydration-inducible *LEA* genes from *B. hygrometrica*. When overexpressed in tobacco, they conferred improved drought tolerance, higher photosystem II activity, lower membrane permeability and more stable ribulose-bisphosphate carboxylase (large subunit), light-harvesting complex II and photosystem II extrinsic proteins under drought stress [11]. Similarly, in this paper, the transgenic plants harboring *OAR1* from *B. hygrometrica* also displayed higher survival rates, better growth, high levels of membrane integrity and photochemical efficiency, and stable expression of photosynthesis related genes and drought-induced marker genes under PEG-mediated osmotic stress.

Drought tolerance and drought avoidance are two major mechanisms in drought resistance of higher plants [56]. No difference in rate of water loss was detected between transgenic and the wild-type plants, indicating that the transformation of the *B. hygrometrica* genome fragments containing *OAR1* into *Arabidopsis* conferred plant osmotic tolerance via stabilization of photosynthetic apparatus, instead of drought avoidance.

Transposable elements can modulate gene expression and regulatory patterns in various ways, and have been described as "distributed genomic control modules" [57] at the core of regulatory networks to specific stimuli [58]. The identification of an *OAR1* element that confers osmotic and alkaline stress tolerance, isolated from native desiccation and calcarenite tolerant *B. hygrometrica* has provided the first reverse genetics evidence for the possible function of this type of retroelements in plant tolerance to abiotic stresses. It is not clear so far by what mechanisms that *OAR1* function to maintain photochemical efficiency in stress tolerance in transgenic plants. What we have known is that *OAR1* is a part of an active LTR-retrotransposon cluster, and this type of retroelement had been identified from another resurrection plant *C. plantagineum* by T-DNA activation tagging, namely *CDT-1* (*desiccation-tolerant-1*) and its homologue *CDT-2* [16,17]. *CDT-1* could direct the synthesis of a double-stranded 21 bp short interfering RNA (siRNA), which triggered the regulatory pathway for desiccation tolerance through activation of stress-responsive genes [18]. Small-RNA coding as described for *CDT-1* present a possible analogous mechanism by which *OAR1* might regulate gene expression under stress conditions in transgenic plants.

Transcription of transposon genes without any known promoter-like sequence had been observed with retrotransposons such as *Sadhu* elements [59,60]. In this study, short transcripts generated from *OAR1* could be detected in both transgenic *Arabidopsis* and *B. hygrometrica*, suggesting this 2-kb element itself is capable of activating its transcription. Despite that both the transposition activity and transcription of *OAR1* were activated in *B. hygrometrica* in its dry and alkaline native habitat, the unchanged relative copy numbers in osmotic and alkaline-stressed plants indicated that the 2-kb *OAR1* element alone in S21 transgenic *Arabidopsis* plants was not transposable. Thus *OAR1* may function in a transposition-independent manner to confer plant osmotic and alkaline resistance in the transgenic plant. In other words, *OAR1* element may have an impact on the expression of certain category of genes, probably via encoding short transcripts. Further investigation on the function of these elements in plant stress tolerance, will aid further understanding of the possible mechanisms which gave rise to the evolution and development of desiccation tolerance in *B. hygrometrica* (and possibly also other Gesnericeae resurrection plants) in alkaline and dry habitats.

Acknowledgments

We are very grateful to Professor Hong-Bin Zhang (Department of Soil and Crop Sciences, Texas A and M University) for providing the BIBAC vectors and technical training. We thank Professor Dr. Kang Chong and Professor Dr. Yalong Guo (Institute of Botany, the Chinese Academy of Sciences) for constructive discussions, and Dr. Jonathan Phillips and Dr. John Moore for the aid in manuscript revision.

Author Contributions

Conceived and designed the experiments: ZZG CML XD. Performed the experiments: Y.Zhao TX CYS SXC GHX LZS MJL LLW Y.Zhu. Analyzed the data: Y.Zhao TX CYS SXC GHX WTL. Contributed reagents/materials/analysis tools: Y.Zhao TX CYS SXC GHX LZS MJL LLW Y.Zhu. Wrote the paper: Y.Zhao TX CML XD.

References

1. Liu C, Liu Y, Guo K, Fan D, Li G, et al. (2011) Effect of drought on pigments, osmotic adjustment and antioxidant enzymes in six woody plant species in karst habitats of southwestern China. Environ. Exp Bot 71: 174–183.

2. Wei Y (2012) Molecular diversity and distribution of arbuscular mycorrhizal fungi in karst ecosystem, Southwest China. Afr J Biotechnol 11(80): 14561–14568.

3. Shinozaki K, Yamaguchi-Shinozaki K (2007) Gene networks involved in drought stress response and tolerance. J Exp Bot 58: 221–227.

4. Xu WB, Liu Y, Gao HS (2009) *Chiritopsis jingxiensis*, a new species of *Gesneriaceae* from a Karst Cave in Guangxi, China. Novon 19: 559–561.

5. Petrova G, Dzhambazova T, Moyankova D, Georgieva D, Michova A, et al. (2014) Morphological variation, genetic diversity and genome size of critically endangered *Haberlea* (Gesneriaceae) populations in Bulgaria do not support the recognition of two different species. Plant Syst Evol 1: 29–41.

6. Rakic T, Quartacci MF, Cardelli R, Navari-Izzo F, Stevanovic B (2009) Soil properties and their effect on water and mineral status of resurrection *Ramonda serbica*. Plant Ecol 203: 13–21.

7. Deng X, Hu ZA, Wang HX, Wen XG, Kuang TY (2003) A comparison of photosynthetic apparatus of the detached leaves of the resurrection plant *Boea hygrometrica* with its non-tolerant relative *Chirita heterotrichia* in response to dehydration and rehydration. Plant Sci 165: 851–861.

8. Georgieva K, Szigeti Z, Sarvari E, Gaspar L, Maslenkova L, et al. (2007) Photosynthetic activity of homoiochlorophyllous desiccation tolerant plant Haberlea rhodopensis during dehydration and rehydration. Planta 225: 955–964.

9. Drazic G, Mihailovic N, Stevanovic B (1999) Chlorophyll metabolism in leaves of higher poikilohydric plants Ramonda serbica Panč. and Ramonda nathaliae Panč. et Petrov. during dehydration and rehydration. J Plant Physiol 154:379–384.

10. Jiang G, Wang Z, Shang H, Yang W, Hu Z, et al. (2007) Proteome analysis of leaves from the resurrection plant *Boea hygrometrica* in response to dehydration and rehydration. Planta 225: 1405–1420.

11. Liu X, Wang Z, Wang LL, Wu RH, Phillips J, et al. (2009) LEA 4 group genes from the resurrection plant *Boea hygrometrica* confer dehydration tolerance in transgenic tobacco. Plant Sci 176: 90–98.

12. Zhang ZN, Wang B, Sun S, Deng X (2013) Molecular cloning and differential expression of sHSP gene family members from the resurrection plant *Boea hygrometrica* in response to abiotic stresses. Biologia 68: 651–661.

13. Chaves MM, Pereira JS, Maroco J, Rodrigues ML, Ricardo CPP, et al. (2002) How plants cope with water stress in the field. Photosynthesis and growth. Ann. Bot 89: 907–916.

14. Shukla LI, Chinnusamy V, Sunkar R (2008) The role of microRNAs and other endogenous small RNAs in plant stress responses. Bba-Gene Regul Mech 1779: 743–748.

15. Ruiz-Ferrer V, Voinnet O (2009) Roles of plant small RNAs in biotic stress responses. Annu Rev Plant Biol 60: 485–510.

16. Furini A, Koncz C, Salamini F, Bartels D (1997) High level transcription of a member of a repeated gene family confers dehydration tolerance to callus tissue of *Craterostigma plantagineum*. Embo J 16: 3599–3608.

17. Smith-Espinoza CJ, Phillips JR, Salamini F, Bartels D (2005) Identification of further *Craterostigma plantagineum cdt* mutants affected in abscisic acid mediated desiccation tolerance. Mol Genet Genomics 274: 364–372.

18. Hilbricht T, Varotto S, Sgaramella V, Bartels D, Salamini F, et al. (2008) Retrotransposons and siRNA have a role in the evolution of desiccation tolerance leading to resurrection of the plant *Craterostigma plantagineum*. New Phytol. 179, 877–887.

19. Shibata D, Liu YG (2000) *Agrobacterium*-mediated plant transformation with large DNA fragments. Trends Plant Sci 5: 354–357.

20. Chang YL, Chuang HW, Meksem K, Wu FC, Chang CY, et al. (2011) Characterization of a plant-transformation-ready large-insert BIBAC library of *Arabidopsis* and bombardment transformation of a large-insert BIBAC of the library into tobacco. Genome 54: 437–447.

21. Ábrahám E, Salamó IP, Koncz C, Szabados L (2011) Identification of *Arabidopsis* and *Thellungiella* genes involved in salt tolerance by novel genetic system. Acta Biologica Szegediensis 55(1): 53–57.

22. Wang W, Wu Y, Li Y, Xie J, Zhang Z, et al. (2010) A large insert *Thellungiella halophila* BIBAC library for genomics and identification of stress tolerance genes. Plant Mol Biol 72: 91–99.

23. Correa R, Stanga J, Larget B, Roznowski A, Shu GP, et al. (2012) An assessment of transgenomics as a tool for identifying genes involved in the evolutionary differentiation of closely related plant species. New Phytol 193: 494–503.

24. Zhang HB, Zhao XP, Ding XL, Paterson AH, Wing RA (1995) Preparation of Megabase-Size DNA from Plant Nuclei. Plant J 7: 175–184.

25. Jones JD, Shlumukov L, Carland F, English J, Scofield SR, et al. (1992) Effective vectors for transformation, expression of heterologous genes, and assaying transposon excision in transgenic plants. Transgenic Res 1: 285–297.

26. Zhang HB, Scheuring CF, Zhang M, Zhang Y, Wu CC, et al. (2012) Construction of BIBAC and BAC libraries from a variety of organisms for advanced genomics research. Nat Protoc 7: 479–499.

27. Dolezel J, Bartos J (2005) Plant DNA flow cytometry and estimation of nuclear genome size. Ann Bot 95: 99–110.

28. Clough SJ, Bent AF (1998) Floral dip: a simplified method for *Agrobacterium*-mediated transformation of *Arabidopsis thaliana*. Plant J 16: 735–743.

29. Liu YG, Whittier RF (1995) Thermal asymmetric interlaced PCR - automatable amplification and sequencing of insert end fragments from P1 and YAC clones for chromosome walking. Genomics 25: 674–681.

30. Doyle JJ, Doyle JL (1990) Isolation of plant DNA from fresh tissue. Focus 12: 13–15.

31. Verslues PE, Agarwal M, Katiyar-Agarwal S, Zhu JH, Zhu JK (2006) Methods and concepts in quantifying resistance to drought, salt and freezing, abiotic stresses that affect plant water status. Plant J 45: 523–539.

32. Ewing B, Hillier L, Wendl M, Green P (1998) Basecalling of automated sequencer traces using phred. I. Accuracy assessment. Genome Res 8: 175–185.

33. Ewing B, Green P (1998) Basecalling of automated sequencer traces using phred. II. Error probabilities. Genome Res 8: 186–194.

34. Hamilton CM, Frary A, Xu YM, Tanksley SD, Zhang HB (1999) Construction of tomato genomic DNA libraries in a binary-BAC (BIBAC) vector. Plant J 18: 223–229.

35. McCubbin AG, Zuniga C, Kao TH (2000) Construction of a binary bacterial artificial chromosome library of *Petunia inflata* and the isolation of large genomic fragments linked to the self-incompatibility (S-) locus. Genome 43: 820–826.

36. Jones JDG, Shlumukov L, Carland F, English J, Scofield SR, et al. (1992) Effective vectors for transformation, expression of heterologous genes, and assaying transposon excision in transgenic plants. Transgenic Res. 1(6):285–97.

37. Chang YL, Henriquez X, Preuss D, Copenhaver GP, Zhang HB (2003) A plant-transformation-competent BIBAC library from the *Arabidopsis thaliana* Landsberg ecotype for functional and comparative genomics. Theor Appl Genet 106: 269–276.

38. Zhang X, Scheuring CF, Zhang M, Dong JJ, Zhang Y, et al. (2010) A BAC/BIBAC-based physical map of chickpea, *Cicer arietinum* L. BMC Genomics 11: 501.

39. Flavell AJ, Pearce SR, Heslop-Harrison JSP, Kumar A (1997) The evolution of Ty1-copia group retrotransposons in eukaryote genomes. Genetica 100: 185–195.

40. Havecker ER, Gao X, Voytas DF (2004) The diversity of LTR retrotransposons. Genome Biol 5.

41. Abe H, Sugasaki T, Terada T, Kanehara M, Ohbayashi F, et al. (2002) Nested retrotransposons on the W chromosome of the wild silkworm *Bombyx mandarina*. Insect Mol Biol 11: 307–314.

42. Leigh F, Kalendar R, Lea V, Lee D, Donini P, et al. (2003) Comparison of the utility of barley retrotransposon families for genetic analysis by molecular marker techniques. Mol Genet Genomics 269: 464–474.

43. Piegu JDG, Guyot R, Picault N, Roulin A, Saniyal A, et al. (2006) Doubling genome size without polyploidization: Dynamics of retrotransposition-driven genomic expansions in *Oryza australiensis*, a wild relative of rice. Genome Res 16: 1262–1269.

44. Lippman Z, May B, Yordan C, Singer T, Martienssen R (2003) Distinct mechanisms determine transposon inheritance and methylation via small interfering RNA and histone modification. Plos Biol 1: 420–428.

45. Song J, Bradeen JM, Naess SK, Helgeson JP, Jiang J (2003) BIBAC and TAC clones containing potato genomic DNA fragments larger than 100 kb are not stable in Agrobacterium. Theor Appl Genet 107: 958–964.

46. Bohnert HJ, Nelson DE, Jensen RG (1995) A daptations to environmental stresses. Plant Cell 7: 1099–1111.

47. Zheng J, Fu JJ, Gou MY, Huai JL, Liu YJ, et al. (2010) Genome-wide transcriptome analysis of two maize inbred lines under drought stress. Plant Mol Biol 72: 407–421.

48. Gao P, Bai X, Yang LA, Lv DK, Pan X, et al. (2011) osa-MIR393: a salinity- and alkaline stress-related microRNA gene. Mol Biol Rep 38: 237–242.

49. Boyko A, Kovalchuk I (2011) Genome instability and epigenetic modification - heritable responses to environmental stress? Curr Opin Plant Biol 14: 260–266.

50. Gutzat R, Scheid OM (2012) Epigenetic responses to stress: triple defense? Curr Opin Plant Biol 15: 568–573.

51. Liu YG, Shirano Y, Fukaki H, Yanai Y, Tasaka M, et al. (1999) Complementation of plant mutants with large genomic DNA fragments by a transformation-competent artificial chromosome vector accelerates positional cloning. Proc Natl Acad Sci USA 96: 6535–6540.

52. Sawa S, Watanabe K, Goto K, Kanaya E, Morita EH, et al. (1999) *FILAMENTOUS FLOWER*, a meristem and organ identity gene of *Arabidopsis*, encodes a protein with a zinc finger and HMG-related domains. Genes Dev 13: 1079–1088.

53. Wardlaw I, Porter H (1967) The redistribution of stem sugars in wheat during grain development. Aust J Biol Sci 20: 309–318.

54. Quartacci MF, Glisic O, Stevanovic B, Navari-Izzo F (2002) Plasma membrane lipids in the resurrection plant *Ramonda serbica* following dehydration and rehydration. J Exp Bot 53: 2159–2166.

55. Georgieva K, Sarvari E, Keresztes A (2010) Protection of thylakoids against combined light and drought by a lumenal substance in the resurrection plant *Haberlea rhodopensis*. Ann. Bot 105: 117–126.

56. Yue B, Xue WY, Xiong LZ, Yu XQ, Luo LJ, et al. (2006) Genetic basis of drought resistance at reproductive stage in rice: Separation of drought tolerance from drought avoidance. Genetics 172: 1213–1228.

57. Shapiro JA (2005) Retrotransposons and regulatory suites. Bioessays 27: 122–125.

58. Bui QT, Grandbastien MA (2012) LTR Retrotransposons as controlling elements of genome response to stress? Plant Transposable Elements, Top Curr Genet 24: 273–296.

59. Rangwala SH, Elumalai R, Vanier C, Ozkan H, Galbraith DW, et al. (2006) Meiotically stable natural epialleles of Sadhu, a novel *Arabidopsis* retroposon. Plos Genet 2: e36.

60. Suoniemi A, Narvanto A, Schulman AH (1996) The BARE-1 retrotransposon is transcribed in barley from an LTR promoter active in transient assays. Plant Mol. Bio. 31, 295–306.

18

Phytophthora sojae Effector PsCRN70 Suppresses Plant Defenses in *Nicotiana benthamiana*

Nasir Ahmed Rajput[9], **Meixiang Zhang**[9], **Yanyan Ru, Tingli Liu, Jing Xu, Li Liu, Joseph Juma Mafurah, Daolong Dou***

Department of Plant Pathology, Nanjing Agricultural University, Nanjing, China

Abstract

Phytophthora sojae, an oomycete pathogen, produces a large number of effector proteins that enter into host cells. The Crinklers (Crinkling and Necrosis, CRN) are cytoplasmic effectors that are conserved in oomycete pathogens and their encoding genes are highly expressed at the infective stages in *P. sojae*. However, their roles in pathogenesis are largely unknown. Here, we functionally characterized an effector *PsCRN70* by transiently and stably overexpressing it in *Nicotiana benthamiana*. We demonstrated that PsCRN70 was localized to the plant cell nucleus and suppressed cell death elicited by all the tested cell death-inducing proteins, including BAX, PsAvh241, PsCRN63, PsojNIP and R3a/Avr3a. Overexpression of the *PsCRN70* gene in *N. benthamiana* enhanced susceptibility to *P. parasitica*. The H_2O_2 accumulation in the *PsCRN70*-transgenic plants was reduced compared to the *GFP*-lines. The transcriptional levels of the defense-associated genes, including *PR1b*, *PR2b*, *ERF1* and *LOX*, were also down-regulated in the *PsCRN70*-transgenic lines. Our results suggest that PsCRN70 may function as a universal suppressor of the cell death induced by many elicitors, the host H_2O_2 accumulation and the expression of defense-associated genes, and therefore promotes pathogen infection.

Editor: Zhengyi Wang, Zhejiang University, China

Funding: NSF of Jiangsu Province (BK2012027); NSFC(31301613); Youth Science and Technology Innovation Fund of NJAU (KJ2013005). The funders had no role in study design, data collection and analysis, decision to publish, or preparation of the manuscript.

Competing Interests: The authors have declared that no competing interests exist.

* E-mail: ddou@njau.edu.cn

[9] These authors contributed equally to this work.

Introduction

Phytopathogens secrete a battery of effector proteins that are delivered inside host cells to promote infection [1,2,3]. Plants detect conserved microbial molecular signatures, termed pathogen-associated molecular patterns (PAMP), resulting in PAMP-trigger immunity (PTI). To counter PTI, pathogens evolve diverse effectors to suppress PTI and trigger susceptibility (effector-triggered susceptibility, ETS). When the effectors are recognized by the corresponding resistance (R) proteins in the host plants, effector-triggered immunity (ETI) in host cells is activated [4]. To overcome R protein-mediated immune responses, pathogens response by mutating or losing effectors, or by developing novel effectors that can suppress ETI [5]. For example, *P. infestans* secreted the effector SNE1 to suppress the R3a/Avr3a -mediated PCD [6]. In the absence of the R proteins, pathogen effectors exert virulence activity to interfere with plant immunity [1]. It has been shown that *in planta* expression of *Phytophthora* effectors, such as *P. infestans Avrblb2* [7], RxLR effector *PITG_03192* [8], *CRN8* [9], and *P. sojae Avh241*[10], enhanced susceptibility to pathogens.

Oomycete pathogens, such as *Phytophthora spp.*, cause a wide variety of devastating plant diseases globally [11]. For example, *P. infestans*, a pathogen of late blight of potato and tomato, was responsible for the Irish potato famine in the mid-nineteenth century; *P. sojae* causes soybean root and stem rot and leads to substantial yield losses annually [11]. *Phytophthora* are hemibiotrophs that initiate the infection cycle as biotrophs, during which the pathogen proliferates asymptomatically in the host, and at this stage pathogens must employ efficient mechanisms to evade and suppress plant immune responses. At the later stage, hemibiotrophs switch to a necrotrophic lifestyle by killing the host plants, and this process is presumably modulated by the coordinated secretion of factors such as lytic enzymes and cell-death inducers. However, the mechanisms underlying regulation of the switch from biotrophs to necrotrophs are still largely unknown. It was reported that *P. infestans* expresses effector *SNE1* at the early infective stage to suppress cell death induced by many other effectors, which may function to maintain the biotrophic phase. A high throughput functional assay for *P. sojae* effectors revealed that *Phytophthora* pathogens may produce effectors with contrasting activity to regulate the infection process [12].

The genomes of the *Phytophthora* species contain large repertoires of RxLRs and CRNs [13,14,15], which are two kinds of cytoplasmic effectors. RxLR effectors are defined by a conserved N-terminal motif, which enables delivery of effector proteins inside plant cells [16,17]. CRN effectors were first identified in *P. infestans* as proteins that resulted in a leaf-crinkling and cell-death phenotype in plants [18]. The CRN effector family showed extensive expansion in all sequenced *Phytophthora* species [13,14]. The average expression levels of CRNs were much higher than those of the RxLR genes, indicating that CRNs may play important roles in pathogenicity [19]. Evolutionary analyses uncovered that gene duplication and fragment recombination drive functional diversity of CRN family [19]. Analogous to RxLR effectors, the N-termini of CRN effectors harbor a conserved FLAK motif, which translocates effectors inside host cells [20]. The C-terminal region of CRN proteins is diverse and controls virulence [14]. CRN1 and CRN2 were identified from *P. infestans* following an *in planta* functional screen for candidate secreted proteins, and transient expression of these two CRNs induced cell death in plants [18]. *P. infestans* CRN8, which contains a kinase domain, targets plant nucleus to induce cell death [9]. CRN63 and CRN115 were identified from *P. sojae*, and they share high sequence similarity, however, they possess contrasting biological activities on host plants, in which PsCRN63 induces cell death and PsCRN115 suppresses cell death [21].

It was previously considered that the majority of CRN effectors caused cell death, however, it has been shown that few CRN effectors can trigger cell death when expressed *in planta* [19,22]. On the contrary, the majority of CRNs can suppress cell death triggered by PAMPs or other elicitors [19]. These observations suggested that CRN effectors may also act in the biotrophic phase, which promote infection of hemibiotrophic *Phytophthora*. Interestingly, most of the CRN effectors are localized in the plant cell nucleus when expressed *in planta* [20,22], indicating that CRN effectors may target and perturb host nuclear processes to achieve virulence. Alteration in subcellular localization of CRN effectors blocked their cell-death-inducing activity [20,21], which suggests that CRN effectors need target to plant cell nucleus to exert their biological functions. Recent progress showed that CRN effector family may target distinct subnuclear compartments and modify host cell signaling [23]. However, the roles of CRN effectors in virulence are still largely unknown.

Cytoplasmic effectors are translocated into host plant cells to interfere with plant immunity, and it has become an efficient strategy to study the virulence functions of effectors by expressing them *in planta* [7,8,9,10]. Here we identified a CRN effector *PsCRN70* from *P. sojae*, and expressed it in a model plant *Nicotiana benthamiana* by *Agrobacterium*-mediated transient expression and stable transformation. We showed that PsCRN70 can suppress cell death induced by many cell-death inducers, such as the mouse BAX, *P. sojae* RxLR effector Avh241, CRN effector PsCRN63, necrosis-inducing protein PsojNIP, and the R3a/Avr3a. Overexpression of *PsCRN70* gene enhanced susceptibility of *N. benthamiana* to *P. parasitica*, indicating that PsCRN70 positively contributes to virulence. Our results indicated that, in addition to causing cell death *in planta*, CRN effectors may also function as a suppressor of plant cell-death and defense responses.

Materials and Methods

Plant material, bacterial strain and growth condition

Nicotiana benthamiana seeds were surface sterilized by soaking in 75% ethanol for 30 s followed by in 1.0% sodium hypochlorite for 5 min. The seeds were rinsed 5 times with sterile water. They were subsequently spread onto petri dishes containing solid 1/2 MS medium. Plates were kept at 18 °C for 4 days in the dark and then at 22 °C for 10 days in 16/8 hour light/dark cycle. The seedlings about 2–3 cm long were transferred aseptically to the glass bottles to get 7–8 young leaves. *Escherichia coli* and *Agrobacterium tumefaciens* strains carrying the disarmed Ti plasmid were routinely grown on Luria-Bertani (LB) agar or broth at 37°C and 28°C, respectively. *P. sojae* isolate P6497 and *P. parasitica* isolate Pp016 was routinely cultured on V8 medium at 28°C.

Plasmid construction

PsCRN70 gene (submitted to Genbank; awaiting accession numbers) lacking the predicted signal peptide was amplified using cDNA from *P. sojae* through PCR with the forward primer: 5'-acgcgtcgacATGGTGACGATCGCGTGTG-3' and the reverse primer 5'-gctctagaTTAAGTACGACGGAGAATTC-3'. After digesting with the *Sal*I and *Xba*I restriction enzymes, the resulting PCR product was inserted into the expression vector pBinGFP2 [10]. For the PVX construct, the *PsCRN70* gene was amplified and inserted into the PVX vector pGR106 [24] using the *Sma*I and *Not*I restriction sites. The recombinant plasmids were confirmed by sequencing and introduced into *Agrobacterium* strains by electroporation.

Generation of the *PsCRN70*-transgenic *N. benthamiana*

The *PsCRN70* gene was introduced into *N. benthamiana* using the leaf-disc transformation approach as described previously [25]. Briefly, the *Agrobacterium* EHA105 carrying the *pBinGFP:PsCRN70* construct was incubated over night at 28°C with shaking at 220 rpm to an OD_{600} of 0.4–0.6. The healthy *N. benthamiana* leaf discs were co-cultured with the *Agrobacterium* suspension for 30 min in 20 mL of liquid MS medium, and then placed on a piece of sterile filter paper and cultured on non-selective callus induce medium (CIM) which contains 1 mg/L of 6-BA at 25°C in the dark for 3 days. After 3 days the infected explants were transferred to a fresh shoot induction medium (SIM) supplemented with 1 mg/L of 6-BA, 100 mg/L of kanamycin and 500 mg/L of carbenicillin at 25°C in the light for 25–30 days for shoot regeneration. Healthy shoots that reached a length of 1–2 cm tall were excised and transferred into a jar containing the selective rooting medium (RIM) supplemented with 100 mg/L of kanamycin, 500 mg/L of carbenicillin and 0.2 mg/L of IAA for root generation. Roots were obtained after 2–3 weeks in culture and transferred to soil under growth room conditions for seed set.

Agrobacterium-mediated transient expression

PVX constructs were transformed into *A. tumefaciens* strain GV3101 using the electroporation method [21]. For agroinfiltration assay, *Agrobacteria* containing the corresponding constructs were cultured in LB media containing 50 mg/mL of kanamycin at 28 °C with shaking at 220 rpm for 48 h. The culture was harvested and washed three times in 10 mM $MgCl_2$, and resuspended in 10 mM $MgCl_2$ to an OD_{600} of 0.3. Infiltration was performed on six-week-old *PsCRN70*-transgenic *N. benthamiana* and the *GFP*-control plants. Symptom development was monitored and photographs were taken 4–6 days post infiltration. The experiments were repeated at least three times.

Phytophthora parasitica inoculation assay

We used two approaches, transient and stable expression, to evaluate the role of PsCRN70 in suppression of plant immunity. For transient expression approach, we expressed the *PsCRN70* and *GFP* in *N. benthamiana* leaves using agroinfiltration method, and inoculated with *P. parasitica* zoospores 2 days post infiltration. Briefly, the *P. parasitica* zoospores were prepared as described previously [26] and *N. benthamiana* leaves were detached and placed in a plastic tray, then each leaf was inoculated with 20 μL of zoospore suspensions with a concentration of 100 zoospores per microliter on the abaxial surface of the leaf. Phenotype was monitored within 72 h, and photographs were taken 36 hours post-inoculation.

For the stable *PsCRN70*-transgenic *N. benthamiana*, we used the detached leaves and the whole seedlings of 5-week old to evaluate the role of PsCRN70 in plant defense. The resistant levels of the whole transgenic plants were assessed using the root-dip inoculation assay. Twenty plants for each T2 transgenic line (#1, #3, #4 and #12) were inoculated with the *P. parasitica* zoospore suspensions, and the *GFP*-transgenic lines were used as the control. The inoculated plants were kept in a moist chamber, and the disease progression was monitored within 10 days. At least 3 independent experiments were performed for this assay. Duncan's multiple range test (SPSS Statistical software version 16.0) was used for statistical analysis (P<0.01).

RNA extraction and quantitative RT-PCR

Total RNA was extracted from *N. benthamiana* leaves using the RNeasy Mini Kit (Qiagen) according to the manufacturer's instructions. The cDNA was generated using the PrimeScript RT reagent Kit (Takara). Real-time quantitative PCR was performed in 20- μL reactions including 20 ng of cDNA, 0.2 μM gene-specific primers, 0.4 ul ROX Reference Dye, 10 μL of SYBR Premix Ex Taq (Takara), and 6.8 μL of deionized water. PCR was performed on an ABI PRISM 7300 Fast Real-Time PCR System (Applied Biosystems) under the following conditions: 95°C for 30 s, 40 cycles of 95°C for 5 s and 60°C for 31 s to calculate cycle threshold values, followed by a dissociation program of 95°C for 15 s, 60°C for 1 min, and 95°C for 15 s to obtain the melt curves. The *N. benthamiana* EF1α gene was used as the internal reference gene for calculating relative transcript levels. An equal volume of cDNA was used for gene analysis expression of defense-related genes, using specific primers *PR1b* 5'- GTGGA-CACTATACTCAGGTG-3'/5'-TCCAACTTGGAAT-CAAAGGG-3', *PR2b* 5'-AGGTGTTTGCTATGGAATGC-3'/5'-TCTGTACCCACCATCTTGC-3', *ERF1* 5'-GCTCTTAACGTCGGATGGTC-3'/5'-AGCCAAACCC-TAGCTCCATT-3', *LOX* 5'-AAAACCTATGCCTCAAGAAC-3'/5'-ACTGCTGCATAGGCTTTGG-3' and *EF1α* 5'-AGAGGCCCTCAGACAAAC-3'/5'-TAGGTCCAAAGGTCA-CAA-3'. The induction ratio of treatment/control was then calculated by the $2^{-\Delta\Delta CT}$ approach. Student t-test was used for statistical analysis.

Confocal microscopy

To observe subcellular localization of PsCRN70, the transgenic *N. benthamiana* leaves were cut into small squares. The leaf squares or plant roots were then immersed into PBS buffer containing 5 μg/mL DAPI for staining of the nuclei for 5 min. Sildes carrying the samples were observed with a Zeiss LSM 710 confocal laser scanning microscope (CLSM). The excitation wavelength used for GFP was 488 nm and 405 nm for DAPI. The *GFP*-transgenic *N. benthamiana* leaves were used as the control. Images were progressed using the Zeiss 710 CLSM and Adobe Photoshop software packages.

Protein extraction and Western blot analyses

The *N. benthamiana* leaf tissues were ground in liquid nitrogen. The ground materials were mixed with protein extraction buffer [50 mM Hepes, 150 mM KCL, 1 mM EDTA, 0.1% Triton X-100, adjust pH to 7.5 with KOH] supplemented with 1 mM DTT, and 1× protease inhibitor mixture (Roche). Crude plant protein extracts were collected by centrifuging at 12,000×g at 4 °C for 15 min. Protein extracts were loaded on 12% SDS- polyacryl-amide gels and protein gel blot analyses were performed as reported previously [10]. Briefly, proteins were transferred from the gel to an Immobi-lon-P^{SQ} polyvinylidene difluoride membrane after electrophoresis. The membranes were washed in PBST (PBS with 0.1 Tween 20) for 2 min and then blocked in PBSTM (PBS with 0.1% Tween 20 and 5% non-fat dry milk) for 1 h. Mouse monoclonal antibody against GFP or HA was added into PBSTM and incubated for 90 min, followed by washing with PBST for three times. The membranes were then incubated in PBSTM with

a goat anti-mouse IRDye 800CW (Li-Cor) for 40 min. The membranes were washed for three times with PBST and then visualized using a LI-COR Odyssey scanner with excitation 700 and 800 nm.

DAB staining

H_2O_2 was visualized using the 3,3-diaminobenzi-dine (DAB) (Sigma) staining approach as described previously [27]. Leaves were inoculated with *P. parasitica* as described above; infected leaves 12 hours post inoculation were soaked in the DAB aqueous solution at 1 mg/ml and maintained at 25°C for 8 h. Leaf sections were cleared by boiling in 95% ethanol for 15 min, bleaching solution was replaced and leaves were incubated until the chlorophyll was completely bleached. DAB-staining experiments were independently repeated at least three times. Duncan's multiple range test were used for statistical analysis (P<0.01).

Results

Generation of the *PsCRN70*-transgenic *N. benthamiana*

Introduction of *PsCRN70* gene into *N. benthamiana* were achieved by *Agrobacterium*-mediated leaf disc transformation. Integration of the *PsCRN70* and *GFP* transgenes were confirmed by PCR analysis. Thirty independent transgenic lines (T0) including ten *GFP*-transgenic plants and twenty *PsCRN70*-transgenic plants were obtained, and fifteen of them were randomly selected and self-pollinated to produce T1 lines. RT-PCR analysis confirmed that the *GFP:PsCRN70* fusion genes were highly expressed in four (#1, #3, #4 and #12) independent T1 transgenic lines (Figure 1A). The seedlings of the transgenic progenies (T1 or T2 generations) were screened by checking fluorescent signal of GFP for further analyses. We confirmed the expression of GFP:PsCRN70 fusion protein using Western blot in transgenic lines, and the result showed that the fusion protein was correctly expressed (Figure 1B). Observation of the GFP fluorescence also validated the expression of the transgenes (Figure 1C). No visible differences in plant growth and other phenotypes were observed between the *PsCRN70*-transgenic plants and the *GFP*-transgenic lines, indicating that expression of *PsCRN70* did not affect the development of *N. benthamiana* under normal growth conditions.

It was reported that the majority of CRN proteins are localized in the plant cell nuclei [20,23]. To test the subcellular localization of the target proteins, we first analyzed whether PsCRN70 harbors an NLS using the cNLS Mapper software [28], and the result showed that PsCRN70 contained a typical NLS (YLARRKKR-KEE) in the C-terminal region, indicating that PsCRN70 may also target the plant cell nucleus. We then observed the distribution of the GFP fluorescent signal in the four T2 transgenic plants (#1, #3, #4 and #12) and the *GFP*-control lines under a confocal microscope using the leaf and root tissues, respectively. The GFP fluorescent signal in the *PsCRN70*-transgenic lines was dominantly distributed in the nucleus of the leaf epidermal cells (Figure 1C). The fluorescent signal in the root epidermal cells of the *PsCRN70*-transgenic *N. benthamiana* was also distributed mainly in the plant nucleus, which was further confirmed by co-localization with DAPI staining (Figure 1D). In contrast, the fluorescent signal in the *GFP*-transgenic lines was equally distributed in the cytoplasm and nucleus of the leaf and root epidermal cells (Figure 1D). Thus, we concluded that PsCRN70 is a plant nucleus-localized protein.

Expression of the *PsCRN70* in *N. benthamiana* enhance susceptibility to *P. parasitica*

To elucidate roles of PsCRN70 in *N. benthamiana* resistance, we constructed *PVX:PsCRN70* and *PVX:GFP*, and then transiently

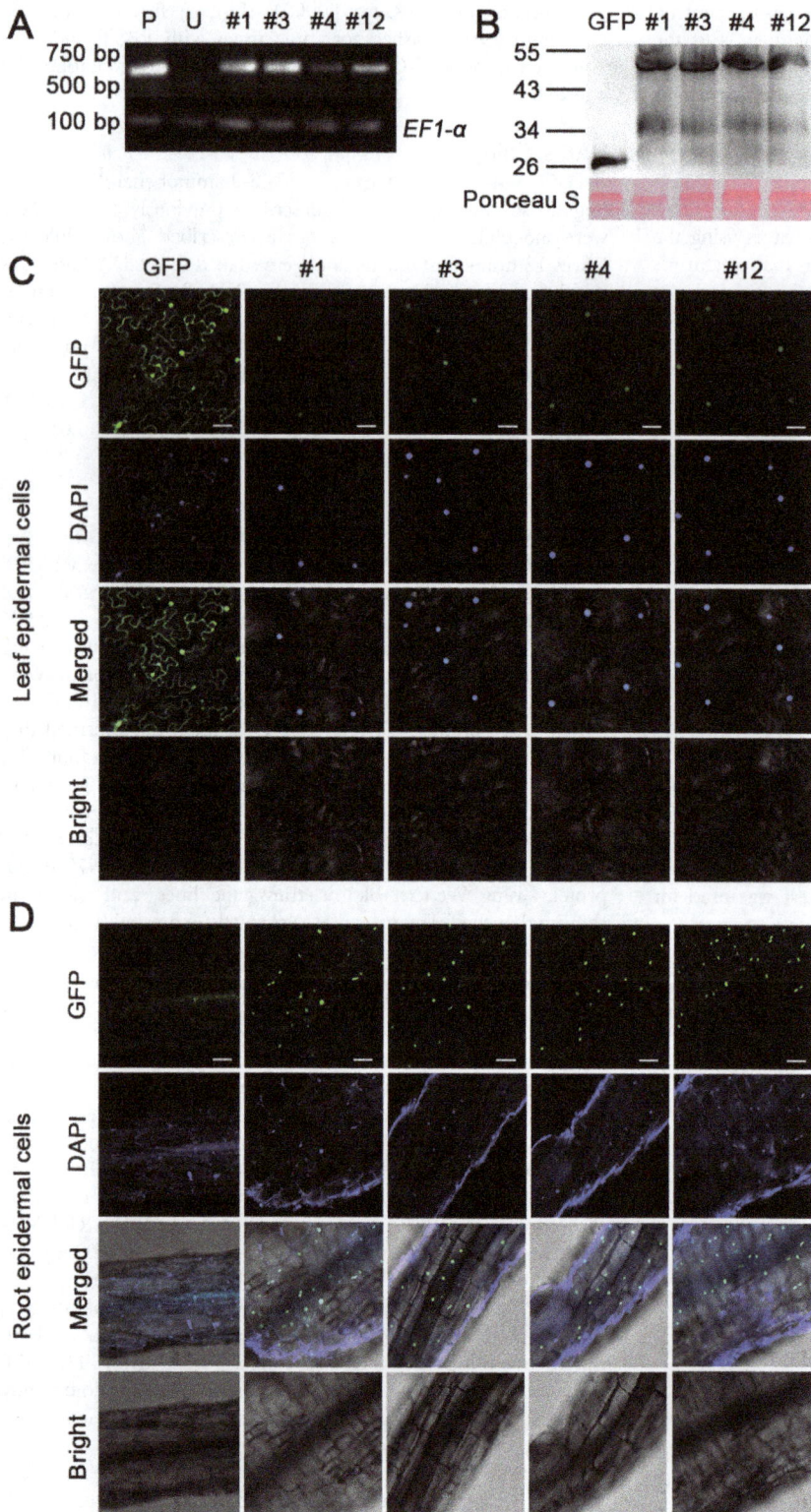

Figure 1. Characterizations of the *PsCRN70*-transgenic *N. benthamiana*. A. RT-PCR analysis of *PsCRN70* expression in independent transgenic lines. #1, #3, #4 and #12, four independent T1 transgenic lines; P, *pBinGFP:PsCRN70* plasmid as a positive control; U, untransformed plant as a negative control. The upper panel represents the 624 bp fragment of *PsCRN70* gene and the lower panel represents the 100 bp fragment of *EF1α* gene as the control. B. Western blot analysis of expression of GFP: PsCRN70 fusion protein in transgenic *N. benthamiana* using monoclonal antibody against GFP. Subcellular localization of PsCRN70 in the transgenic *N. benthamiana* leaf tissues (C) and roots (D). The pictures were taken using a confocal microscope. The scale bar indicates 50 μm.

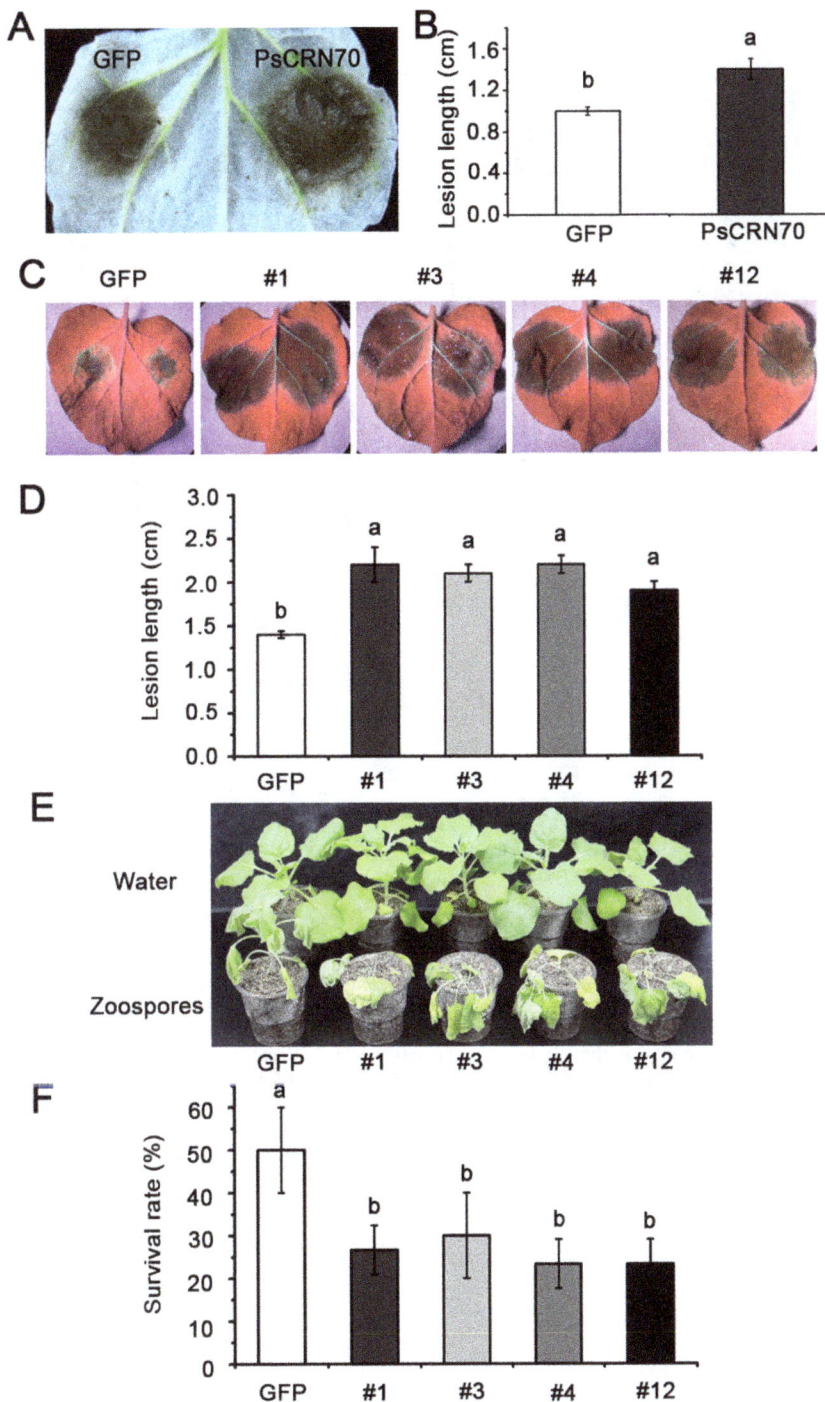

Figure 2. Suppression of the plant resistance by PsCRN70. A. Lesions on the *N. benthamiana* leaves expressing *PsCRN70* or *GFP*. A decolorized infected leaf was photographed 4 dpi. The experiments were repeated three times and shown with a representative image. B. Average lesion diameters on *N. benthamiana* leaves expressing the indicated genes inoculated with *P. parasitica*. Averages were calculated from four lesions per construct. Error bars represent standard errors. Different letters at the top of the columns indicate significant differences (P<0.01, Duncan's multiple range test). C. Phenotypes of the *PsCRN70*- and *GFP*-transgenic lines inoculated with *P. parasitica*. Detached leaves from five-weak old seedlings (T1 generation) of transgenic plants (#1, #3, #4, #12) were inoculated with the *P. parasitica* zoospores. Photographs were taken 36 hpi under a UV lamp. D. Lesion diameters on inoculated leaves measured 36 hpi. The experiments were repeated four times in all transgenic lines with similar results. Bars represent the standard deviation (SD). Different letters at the top of the columns indicate significant differences (P<0.01, Duncan's multiple range test). E. Phenotypes of the *PsCRN70*-transgenic plants inoculated with *P. parasitica* zoospores. The transgenic lines were inoculated with zoospores using the root dip method. Mock-inoculated seedlings were inoculated with sterile distilled water. Photographs were taken 4 days post inoculation. F. Survival rates of the transgenic plants inoculated with *P. parasitica* zoospores. The rates were measured at 4 dpi. The experiments were repeated four times in all transgenic lines with similar results. Twenty plants were used for each treatment in each experiment. Bars represent the standard deviation (SD). Different letters at the top of the columns indicate significant differences (P<0.01, Duncan's multiple range test).

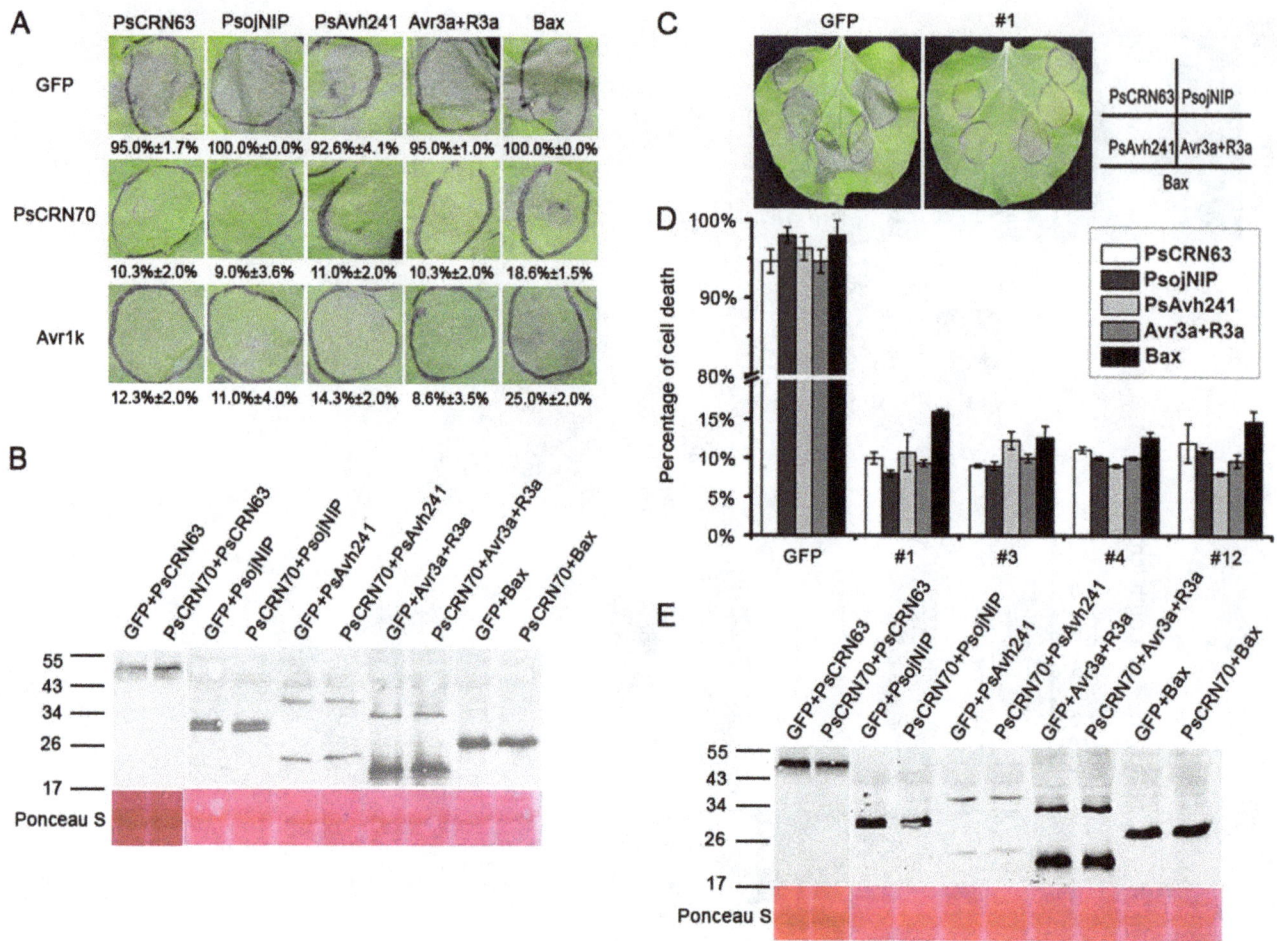

Figure 3. Suppression of cell death by PsCRN70. A. Transient expression assay of *PsCRN70* in *N. benthamiana* leaves. Agroinfiltration sites in each *N. benthamiana* leaf expressing *GFP* (upper panel, a negative control), *Avr1k* (lower panel, a positive control), and *PsCRN70* (middle panel) were challenged after 12 h with *A. tumefaciens* carrying the indicated cell death-inducers. Photographs were taken 5 d after cell death inducer infiltration. The data is the percentage of cell death sites from 8 infiltrated leaves based on three independent experiments. B. Western blot analysis of expression of the cell-death inducers in *N. benthamiana* leaves transiently co-expressed with *GFP* or *PsCRN70*. Antibody against the HA-epitope tag was used to detect the expression of *PsCRN63*, *PsojNIP*, *PsAvh241*, *Avr3a* and *Bax* when they were co-expressed with *GFP* or *PsCRN70*. C. Cell-death-suppression assay on leaves of the stable *PsCRN70*-transgenic *N. benthamiana*. D. The percentage of cell death sites. The percentages of cell death sites on the *PsCRN70*-transgenic lines were scored from 8 infiltrated leaves based on three independent experiments. E. Western blot analysis of expression of the cell-death inducers in leaves of *PsCRN70*-transgenic *N. benthamiana* two days post infiltration. The *PsCRN70*-transgenic lines (T2 generation of #1 was shown as an example) were infiltrated with *A. tumefaciens* containing the *PsCRN63*, *PsojNIP*, *PsAvh241*, *Avr3a/R3a* and *Bax*; the *GFP*-transgenic plant was used as the control. Photographs were taken 5 d after cell death inducer infiltration.

expressed them in *N. benthamiana* leaves. Two days post-infiltration, the infiltrated leaves were challenged with the *P. parasitica* zoospores. The detached leaf tissues expressing the *PsCRN70* developed larger lesion compared to those expressing the *GFP* control (Figure 2A). On leaf areas expressing the *GFP* gene, the average diameter of the lesions was ~1.0 cm 36 hours post-inoculation; however, on areas infiltrated with *PsCRN70*, the average lesion diameter was ~1.4 cm (Figure 2B). Thus, transient expression of the *PsCRN70* in *N. benthamiana* reduced its resistance to *P. parasitica*.

To validate the above observations, we also did inoculation assays of leaves and roots using stable *PsCRN70*-transgenic plants of T1 generation. The average lesion diameters were significantly larger in the *PsCRN70*-transgenic leaves than that in the *GFP*-transgenic leaves (Figure 2C, D). Root inoculations resulted in both *PsCRN70*- and the *GFP*-transgenic plants (T2 generation) displaying symptoms of wilting and stunting (Figure 2E). However, the survival rate of the *PsCRN70*-transgenic plants was significantly

lower than the *GFP*-transgenic plants (Figure 2F). Taken together, these results showed that PsCRN70 enhanced susceptibility of *N. benthamiana* to *P. parasitica* infection.

To determine whether expression of *PsCRN70* enhanced susceptibility of *N. benthamiana* to nonhost pathogen, we inoculated the transgenic *N. benthamiana* leaves (T2 generation) with *P. sojae* mycelial plugs. No obvious infection was found in both *GFP*- and *PsCRN70*-transgenic plants 5 days post inoculation (Figure S1), and this indicated that expression of *PsCRN70* in *N. benthamiana* did not affect its nonhost resistance to oomycete pathogen *P. sojae*.

PsCRN70 suppresses cell death in *N. benthamiana*

To characterize how PsCRN70 contributes to virulence, we tested its cell-death-manipulation activity using the *Agrobacterium*-mediated transient expression in *N. benthamiana*. This approach has been widely used for analyzing cell-death manipulation [14,21]. The cell-death inducing proteins included the mouse BAX [29],

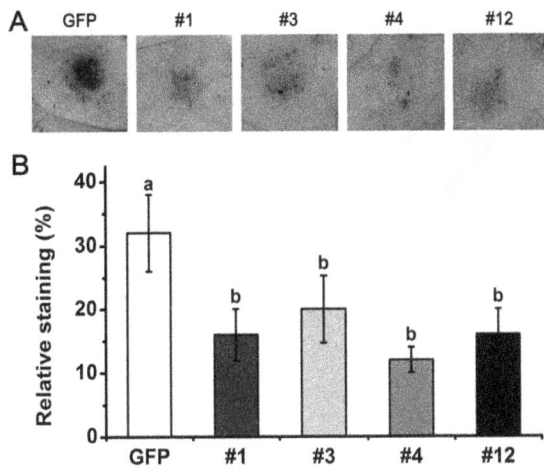

Figure 4. Suppression of the H₂O₂ accumulation in N. benthamiana by PsCRN70. A. DAB staining of the *P. parasitica*-inoculated *N. benthamiana* leaves. The H_2O_2 accumulation in the *PsCRN70*- and *GFP*- transgenic leaves were detected using DAB staining at 12 hpi. Photographs were taken after de-colorization of leaves with ethanol. B. The relative levels of DAB staining. The data were calculated by a combination of Photoshop and Quantity One for H_2O_2 accumulation in the indicated transgenic lines. The experiments were repeated three times in all transgenic lines with similar results. Four leaves were used for each treatment in each experiment. Bars represent the standard deviation (SD). Different letters at the top of the columns indicate significant differences ($P < 0.01$, Duncan's multiple range test).

the *P. sojae* necrosis-inducing protein PsojNIP [30], the RxLR effector PsAvh241 [10], the CRN effector PsCRN63 [21] and the R3a/Avr3a [31]. *P. sojae* effector Avr1k [32] can inhibit cell death induced by all the above elicitors and was used as a positive control. As expected, no cell death phenotypes were observed in the *Avr1k*-infiltrated leaves (Figure 3A). Expression of the *PsCRN70* gene also blocked cell death triggered by these elicitors 5 days post-infiltration. As a negative control, expression of the *GFP* gene did not suppress the cell death (Figure 3A). Western blot result showed that the cell-death inducers were expressed at similar levels when co-expressed with *GFP* control or *PsCRN70*, indicating that suppression of cell death was due to expression of *PsCRN70*. We further validated the results in the stable *PsCRN70*-transgenic *N. benthamiana*. In the transgenic line expressing *GFP*, cell death occurred at 5 days after infiltration with the above tested elicitors. However, cell death symptoms were only occasionally and weakly observed in the four *PsCRN70*-transgenic lines under the same conditions (Figure 3C, D). Western blot results also showed that the cell-death inducers were expressed at similar levels in *PsCRN70*-transgenic and *GFP*-lines (Figure 1E). Collectively, these results suggested that PsCRN70 may function as a broad cell death-suppressor to manipulate the plant immunity.

Expression of the *PsCRN70* impairs the H₂O₂ accumulation in *N. benthamiana*

H_2O_2, a kind of reactive oxygen species (ROS), plays an important role in plant defense responses [27]. To examine the role of PsCRN70 on the plant H_2O_2 accumulation, the H_2O_2 levels in the transgenic plants were detected at the early infective stages of *P. parasitica* using the DAB staining method. Weak staining was observed in the infected area of the *PsCRN70*-transgenic leaves (Figure 4A). The relative staining was signifi-cantly lower in the *PsCRN70*-transgenic *N. benthamiana* leaves

compared to that in the *GFP*-control plants (Figure 4B). This result suggested that PsCRN70 may promote *Phytophthora* infection by suppressing the H_2O_2 accumulation.

Expressions of the *PsCRN70* reduces the expressional levels of the plant defense-associated genes

To further assess the role of PsCRN70 in plant defense responses, we examined the transcriptional levels of the defense-associated genes in plants, including *PR1a* (*Pathogenesis-related protein*), *PR2b*, *ERF1* (*Ethylene response factor 1*) and *LOX* (*Lipoxygenase*) genes, among which, the *PR1a* and *PR2b* genes are markers in the salicylate-mediated signaling pathway [33]; the *ERF1* is a marker in the ethylene-mediated signaling pathway [34]; and the *LOX* is a marker in the jasmonate-mediated signaling pathway [35]. These genes are all involved in downstream of the defense signaling pathways. Expressional levels of all the four tested genes were significantly repressed in the *N. benthamiana* leaves transiently expressing *PsCRN70* compared to that in leaves expressing *GFP* (Figure 5A). The expression levels of the four genes also exhibited significant reduction in the stable *PsCRN70*-transgenic lines when challenged with *P. parasitica* zoospores (Figure 5B). These results suggest that PsCRN70 may repress the expression of the defense-associated genes in plants.

Discussion

Phytophthora pathogens encode a large number of RxLR and CRN effectors [13,14,15], however, virulence functions of CRN effectors are largely unknown. Overexpression of the *PsCRN70* in *N. benthamiana* enhanced susceptibility to *P. parasitica*, indicating that PsCRN70 contributes to pathogen virulence. DAB staining results showed that the H_2O_2 accumulation in the *PsCRN70*-transgenic plants were significantly lower than that in the control lines, indicating that PsCRN70 can promote *Phytophthora* infection by reducing H_2O_2 levels in plants. The role of H_2O_2 in plant defense responses has been extensively studied [36]. It has been adopted by many pathogens to promote infection by regulating H_2O_2 production in plants [1,37]. For example, *Ustilago maydis* secretes the effector Pep1 into the apoplast to suppress the H_2O_2 production, resulting in suppression of plant immunity [37]. Quantitative RT-PCR results showed that the marker genes from different hormone signaling pathways were significantly down-regulated in the *PsCRN70*-transgenic *N. benthamiana* compared to the control, which further confirmed that PsCRN70 significantly reduced plant defense responses. SA and JA signaling pathways usually act antagonistically in plant defense. We showed that PsCRN70 may suppress both pathways, indicating the effector protein exhibits broad suppression activities.

It was originally surmised that the expression of CRN effectors triggered cell death [18]. However, more recent studies suggest that only a few CRNs induce necrosis [19,22]. Our results showed that PsCRN70 can suppress cell death induced by many elicitors including the mouse BAX, *P. sojae* RxLR effector Avh241, CRN effector PsCRN63, necrosis-inducing protein PsojNIP, PCD triggered by the resistance protein R3a and the avirulence protein Avr3a. These results indicate that, similar to the SNE1 from *P. infestans* [6] and Avr1k [32] from *P. sojae*, PsCRN70 may function as a broad cell-death suppressor to promote *P. sojae* infection. These broad acting cell-death suppressor proteins will be useful tools in identifying the components of protein regulatory networks in immune signaling and cell death pathways.

It has been reported that the majority of CRN effectors are localized in the plant cell nuclei [20]. PsCRN70 is also located in the plant cell nucleus, indicating that function by members of the

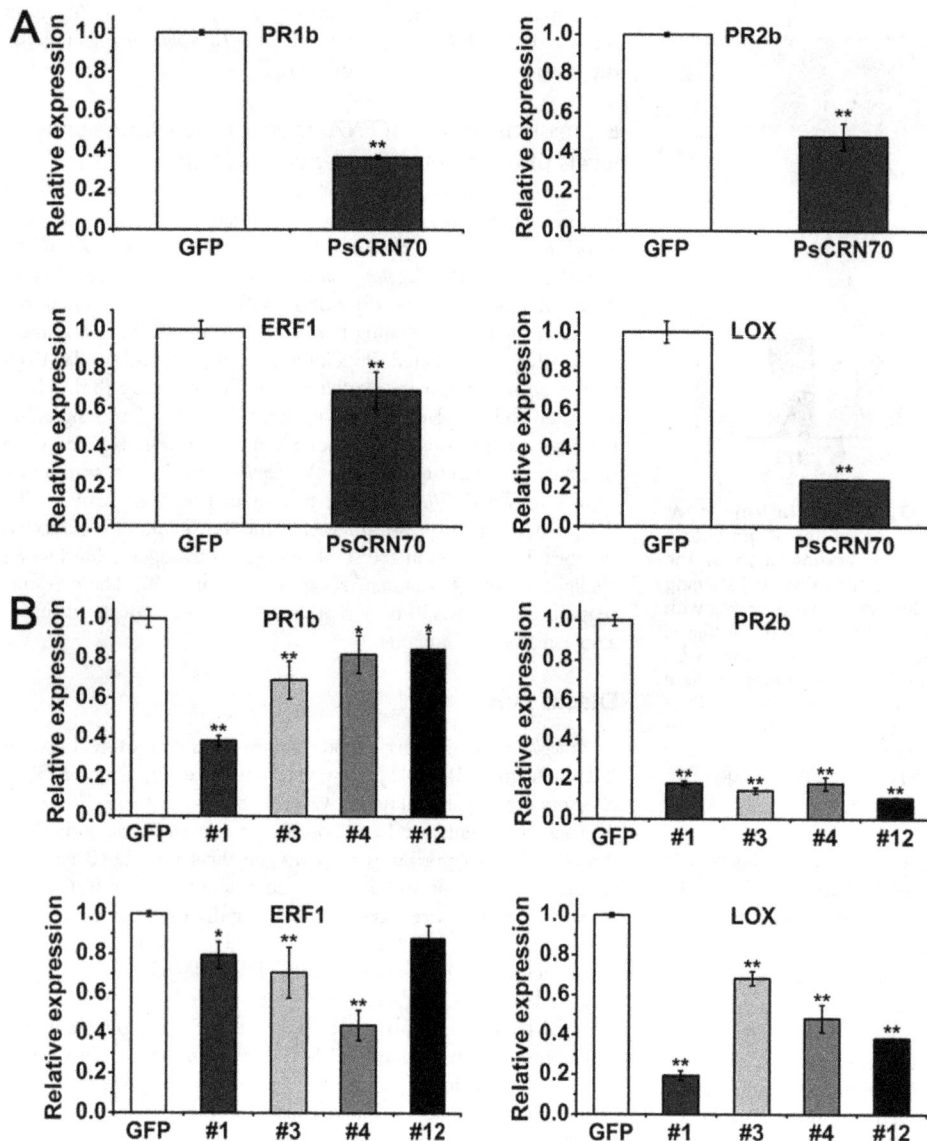

Figure 5. Down-regulation of the defense-associated genes in *N. benthamiana* by PsCRN70. A. Relative expression levels of the *PR1b*, *PR2b*, *ERF1* and *LOX* genes. The total RNA was extracted from the leaf tissues that transiently expressing *PsCRN70* and *GFP*, respectively, and the expression levels of the indicated genes were measured using qRT-PCR. B. Relative expression levels of the indicated genes in the stable transgenic *N. benthamiana*. The *N. benthamiana* leaves 36 hpi with *P. parasitica* zoospores were collected, and the gene expression levels were measured by qRT-PCR. The gene expression levels were normalized to the *EF1α* gene. Bars represent the standard deviation (SD), with significant difference (** for P< 0.01 and * for P<0.05, Student's *t*-test).

CRN effector family may function by manipulating the host nuclear processes to suppress the plant immune signaling. Subcellular localization of several cell-death-inducing CRN effectors demonstrated that they target distinct subnuclear compartments [23], indicating that CRNs may interfere with diverse nuclear targets. Alteration of nuclear targeting signals in several CRNs blocked their cell death-inducing activity [20,21]. However, it was unclear whether the nuclear subcellular localization of CRNs can account for the increase in effector virulence due to the expression of these proteins. In conclusion, we showed that PsCRN70 may function as a broad cell-death suppressor to promote *Phytophthora* infection, providing insight into the role of CRN effectors in pathogenicity.

Supporting Information

Figure S1 Phenotypes of the *PsCRN70*-transgenic plants inoculated with *P. sojae* mycelial plugs. Photographs were taken 5 days post inoculation.

Author Contributions

Conceived and designed the experiments: MZ DD. Performed the experiments: NAR MZ YR TL JX LL JJM. Analyzed the data: NAR MZ DD. Contributed reagents/materials/analysis tools: YR TL LL. Wrote the paper: NAR MZ DD.

References

1. Dou DL, Zhou JM (2012) Phytopathogen effectors subverting host immunity: different foes, similar battleground. Cell Host & Microbe 12: 484–495.
2. Hann DR, Gimenez-Ibanez S, Rathjen JP (2010) Bacterial virulence effectors and their activities. Curr Opin Plant Biol 13: 388–393.
3. Rafiqi M, Ellis JG, Ludowici VA, Hardham AR, Dodds PN (2012) Challenges and progress towards understanding the role of effectors in plant-fungal interactions. Curr Opin Plant Biol 15: 477–482.
4. Jones JDG, Dangl JL (2006) The plant immune system. Nature 444: 323–329.
5. Stergiopoulos I, de Wit PJ (2009) Fungal effector proteins. Annu Rev Phytopathol 47: 233–263.
6. Kelley BS, Lee SJ, Damasceno CMB, Chakravarthy S, Kim BD, et al. (2010) A secreted effector protein (SNE1) from Phytophthora infestans is a broadly acting suppressor of programmed cell death. Plant J 62: 357–366.
7. Bozkurt TO, Schornack S, Win J, Shindo T, Ilyas M, et al. (2011) Phytophthora infestans effector AVRblb2 prevents secretion of a plant immune protease at the haustorial interface. P Natl Acad Sci USA 108: 20832–20837.
8. McLellan H, Boevink PC, Armstrong MR, Pritchard L, Gomez S, et al. (2013) An RxLR effector from Phytophthora infestans prevents re-localisation of two plant NAC transcription factors from the endoplasmic reticulum to the nucleus. PLoS Pathog 9: e1003670.
9. van Damme M, Bozkurt TO, Cakir C, Schornack S, Sklenar J, et al. (2012) The Irish potato famine pathogen Phytophthora infestans translocates the CRN8 Kinase into host llant cells. PLoS Pathog 8.
10. Yu XL, Tang JL, Wang QQ, Ye WW, Tao K, et al. (2012) The RxLR effector Avh241 from Phytophthora sojae requires plasma membrane localization to induce plant cell death. New Phytol 196: 247–260.
11. Tyler BM (2001) Genetics and genomics of the oomycete-host interface. Trends Genet 17: 611–614.
12. Wang QQ, Han CZ, Ferreira AO, Yu XL, Ye WW, et al. (2011) Transcriptional programming and functional interactions within the Phytophthora sojae RXLR effector repertoire. Plant Cell 23: 2064–2086.
13. Tyler BM, Tripathy S, Zhang XM, Dehal P, Jiang RHY, et al. (2006) Phytophthora genome sequences uncover evolutionary origins and mechanisms of pathogenesis. Science 313: 1261–1266.
14. Haas BJ, Kamoun S, Zody MC, Jiang RHY, Handsaker RE, et al. (2009) Genome sequence and analysis of the Irish potato famine pathogen Phytophthora infestans. Nature 461: 393–398.
15. Lamour KH, Mudge J, Gobena D, Hurtado-Gonzales OP, Schmutz J, et al. (2012) Genome sequencing and mapping reveal loss of heterozygosity as a mechanism for rapid adaptation in the vegetable pathogen Phytophthora capsici. Mol Plant Microbe In 25: 1350–1360.
16. Dou DL, Kale SD, Wang X, Jiang RHY, Bruce NA, et al. (2008) RXLR-mediated entry of Phytophthora sojae effector Avr1b into soybean cells does not require pathogen-encoded machinery. Plant Cell 20: 1930–1947.
17. Whisson SC, Boevink PC, Moleleki L, Avrova AO, Morales JG, et al. (2007) A translocation signal for delivery of oomycete effector proteins into host plant cells. Nature 450: 115–118.
18. Torto TA, Li SA, Styer A, Huitema E, Testa A, et al. (2003) EST mining and functional expression assays identify extracellular effector proteins from the plant pathogen Phytophthora. Genome Res 13: 1675–1685.
19. Shen D, Liu T, Ye W, Liu L, Liu P, et al. (2013) Gene duplication and fragment recombination drive functional diversification of a superfamily of cytoplasmic effectors in Phytophthora sojae. PLoS ONE 8: e70036.
20. Schornack S, van Damme M, Bozkurt TO, Cano LM, Smoker M, et al. (2010) Ancient class of translocated oomycete effectors targets the host nucleus. P Natl Acad Sci USA 107: 17421–17426.
21. Liu TL, Ye WW, Ru YY, Yang XY, Gu BA, et al. (2011) Two host cytoplasmic effectors are required for pathogenesis of Phytophthora sojae by suppression of host defenses. Plant Physiol 155: 490–501.
22. Stam R, Jupe J, Howden AJM, Morris JA, Boevink PC, et al. (2013) Identification and characterisation CRN effectors in Phytophthora capsici shows modularity and functional diversity. PLoS ONE8.
23. Stam R, Howden AJ, Delgado-Cerezo M, TM MMA, Motion GB, et al. (2013) Characterization of cell death inducing Phytophthora capsici CRN effectors suggests diverse activities in the host nucleus. Front Plant Sci 4: 387.
24. Lu R, Malcuit I, Moffett P, Ruiz MT, Peart J, et al. (2003) High throughput virus-induced gene silencing implicates heat shock protein 90 in plant disease resistance. EMBO J 22: 5690–5699.
25. Gallois P, Marinho P (1995) Leaf disk transformation using Agrobacterium tumefaciens-expression of heterologous genes in tobacco. In: Jones H, editor. Plant Gene Transfer and Expression. Protocols: Springer New York pp. 39–48.
26. Zhang MX, Meng YL, Wang QH, Liu DD, Quan JL, et al. (2012) PnPMA1, an atypical plasma membrane H⁺-ATPase, is required for zoospore development in Phytophthora parasitica. Fungal Biol 116: 1013–1023.
27. ThordalChristensen H, Zhang ZG, Wei YD, Collinge DB (1997) Subcellular localization of H₂O₂ in plants. H₂O₂ accumulation in papillae and hypersensitive response during the barley-powdery mildew interaction. Plant J 11: 1187–1194.
28. Kosugi S, Hasebe M, Tomita M, Yanagawa H (2009) Systematic identification of cell cycle-dependent yeast nucleocytoplasmic shuttling proteins by prediction of composite motifs. P Natl Acad Sci USA 106: 13142–13142.
29. Lacomme C, Cruz SS (1999) Bax-induced cell death in tobacco is similar to the hypersensitive response. P Natl Acad Sci USA 96: 7956–7961.
30. Qutob D, Kamoun S, Gijzen M (2002) Expression of a Phytophthora sojae necrosis-inducing protein occurs during transition from biotrophy to necrotrophy. Plant J 32: 361–373.
31. Armstrong MR, Whisson SC, Pritchard L, Bos JIB, Venter E, et al. (2005) An ancestral oomycete locus contains late blight avirulence gene Avr3a, encoding a protein that is recognized in the host cytoplasm. P Natl Acad Sci USA 102: 7766–7771.
32. Song TQ, Kale SD, Arredondo FD, Shen DY, Su LM, et al. (2013) Two RxLR avirulence genes in Phytophthora sojae determine soybean Rps1k-mediated disease resistance. Mol Plant Microbe In 26: 711–720.
33. Lee S, Ishiga Y, Clermont K, Mysore KS (2013) Coronatine inhibits stomatal closure and delays hypersensitive response cell death induced by nonhost bacterial pathogens. PeerJ 1: e34.
34. Lorenzo O, Piqueras R, Sanchez-Serrano JJ, Solano R (2003) ETHYLENE RESPONSE FACTOR1 integrates signals from ethylene and jasmonate pathways in plant defense. Plant Cell 15: 165–178.
35. Wang CX, Zien CA, Afitlhile M, Welti R, Hildebrand DF, et al. (2000) Involvement of phospholipase D in wound-induced accumulation of jasmonic acid in Arabidopsis. Plant Cell 12: 2237–2246.
36. Petrov VD, Van Breusegem F (2012) Hydrogen peroxide-a central hub for information flow in plant cells. AoB Plants 2012: pls014.
37. Hemetsberger C, Herrberger C, Zechmann B, Hillmer M, Doehlemann G (2012) The Ustilago maydis effector Pep1 suppresses plant immunity by inhibition of host peroxidase activity. PLoS Pathog 8.

19

Overexpression of Phosphomimic Mutated OsWRKY53 Leads to Enhanced Blast Resistance in Rice

Tetsuya Chujo[1,◐,¤a], **Koji Miyamoto**[1,3,◐], **Satoshi Ogawa**[1], **Yuka Masuda**[1], **Takafumi Shimizu**[1,¤b], **Mitsuko Kishi-Kaboshi**[2], **Akira Takahashi**[2], **Yoko Nishizawa**[2], **Eiichi Minami**[2], **Hideaki Nojiri**[1], **Hisakazu Yamane**[1,3], **Kazunori Okada**[1]*

1 Biotechnology Research Center, The University of Tokyo, Bunkyo-ku, Tokyo, Japan, 2 Genetically Modified Organism Research Center, National Institute of Agrobiological Sciences, Tsukuba, Ibaraki, Japan, 3 Department of Biosciences, Teikyo University, Utsunomiya, Tochigi, Japan

Abstract

WRKY transcription factors and mitogen-activated protein kinase (MAPK) cascades have been shown to play pivotal roles in the regulation of plant defense responses. We previously reported that *OsWRKY53*-overexpressing rice plants showed enhanced resistance to the rice blast fungus. In this study, we identified OsWRKY53 as a substrate of OsMPK3/OsMPK6, components of a fungal PAMP-responsive MAPK cascade in rice, and analyzed the effect of OsWRKY53 phosphorylation on the regulation of basal defense responses to a virulence race of rice blast fungus *Magnaporthe oryzae* strain Ina86-137. An *in vitro* phosphorylation assay revealed that the OsMPK3/OsMPK6 activated by OsMKK4 phosphorylated OsWRKY53 recombinant protein at its multiple clustered serine-proline residues (SP cluster). When OsWRKY53 was coexpressed with a constitutively active mutant of OsMKK4 in a transient reporter gene assay, the enhanced transactivation activity of OsWRKY53 was found to be dependent on phosphorylation of the SP cluster. Transgenic rice plants overexpressing a phospho-mimic mutant of OsWRKY53 (OsWRKY53SD) showed further-enhanced disease resistance to the blast fungus compared to native *OsWRKY53*-overexpressing rice plants, and a substantial number of defense-related genes, including pathogenesis-related protein genes, were more upregulated in the *OsWRKY53SD*-overexpressing plants compared to the *OsWRKY53*-overexpressing plants. These results strongly suggest that the OsMKK4-OsMPK3/OsMPK6 cascade regulates transactivation activity of OsWRKY53, and overexpression of the phospho-mimic mutant of OsWRKY53 results in a major change to the rice transcriptome at steady state that leads to activation of a defense response against the blast fungus in rice plants.

Editor: Mingliang Xu, China Agricultural University, China

Funding: This work was supported by JSPS KAKENHI (grant no. 22380066) and by the Program for Promotion of Basic Research Activities for Innovative Biosciences (PROBRAIN). The funders had no role in study design, data collection and analysis, decision to publish, or preparation of the manuscript.

Competing Interests: The authors have declared that no competing interests exist.

* E-mail: ukokada@mail.ecc.u-tokyo.ac.jp

¤a Current address: Institute of Fundamental Sciences, Massey University, Palmerston North, New Zealand
¤b Current address: Center for Sustainable Resource Science, RIKEN, Yokohama, Kanagawa, Japan

◐ These authors contributed equally to this work.

Introduction

Plants protect themselves against pathogens through an innate immune system comprised of two layers [1–4]. The first layer is pathogen-associated molecular pattern (PAMP)-triggered immunity (PTI), activated by the recognition of PAMPs (e.g. bacterial flagellin and fungal chitin oligosaccharide) via plant pattern-recognition receptors. To prevent the colonization of plant tissue by pathogens, plants activate basal defense responses, such as synthesis of pathogenesis-related (PR) proteins and accumulation of phytoalexins, after perception of PAMPs by plant receptors. The second layer is effector-triggered immunity (ETI), which is a more accelerated defense response than PTI and is triggered by host-resistance (R) protein-mediated recognition of pathogen effectors. Activation of mitogen-activated protein kinases (MAPKs) and transcriptional regulation of defense-related gene expression are central to the induction of disease resistance in higher plants,

and a number of transcription factor families (e.g. WRKY) have been identified [5] as associated with plant defense responses.

MAPK cascades are signaling systems that are evolutionarily conserved among eukaryotes [6]. A basic MAPK cascade consists of 3 interconnected kinases: a MAPK, a MAPK kinase (MAPKK), and a MAPKK kinase (MAPKKK). In tobacco, 2 MAPKs, a wound-induced protein kinase (WIPK) and a salicylic acid-induced protein kinase (SIPK) are involved in both PAMP-triggered basal defense responses against fungal pathogens and ETI [7–12]. These 2 MAPKs share a common upstream MAPKK (NtMEK2) and function together in a single MAPK cascade. In *Arabidopsis*, it has been shown that AtMPK3 and AtMPK6, homologues of WIPK and SIPK respectively, are activated by recognition of PAMPs. They also share common upstream MAPKKs, namely AtMKK4 and AtMKK5 [13,14]. The AtMPK3/AtMPK6 cascade plays a role in activation of defense-related genes, generation of reactive oxygen species (ROS), hypersensitive response-like cell death, and production of an indole-derived phytoalexin, camalexin [14–18]. It has also been shown that rice WIPK and SIPK homologues, OsMPK3 and OsMPK6, are activated by a fungal chitin oligosaccharide elicitor via a MAPKK (OsMKK4) and have important roles in cell death, biosynthesis of diterpenoid phytoalexins and lignin accumulation [19].

WRKY proteins form a large family of plant-specific transcription factors, and there are 74 and 109 genes encoding WRKY proteins in the *Arabidopsis* and rice genomes respectively [20,21]. In numerous plant species, transcription of WRKY genes is strongly and rapidly upregulated in response to pathogen infection or treatment with either PAMPs or defense-related plant hormones [22]. WRKY proteins recognize the W-box elements ([T/C]TGAC[C/T]) in promoter regions of pathogen- or PAMPs-responsive genes like those encoding PR proteins, and modulate host defense against various phytopathogens as either positive or negative regulators [5,22–25]. WRKY proteins contain 1 or 2 almost invariant WRKY domains composed of the conserved WRKYGQK amino acid sequence at the N terminus followed by a zinc-finger motif ($CX_{4-7}CX_{22-23}HXH/C$). They are divided into 3 groups based on the number of WRKY domains present: 2 WRKY domains with a C_2H_2 zinc finger motif (group I); 1 WRKY domain with a C_2H_2 zinc finger motif (group II); and 1 WRKY domain with a C_2H/C zinc finger motif (group III) [26]. In the N-terminal region of several group I WRKY proteins, multiple clustered serine-proline residues (SP cluster), which can be putatively phosphorylated by MAPKs, are highly conserved (Fig. S1) [27–30].

Research to date has demonstrated that plant MAPK cascades regulate downstream gene expression through phosphorylation of group I WRKY proteins in defense-related signaling pathways. *Arabidopsis* AtWRKY33 exists in the nuclear complex with MAP KINASE 4 SUBSTRATE 1 (MKS1) and AtMPK4, and this ternary complex depends on MKS1. Complexes with MKS1 and AtWRKY33 are released after phosphorylation of MKS1 by the MAPK cascade AtMEKK1-AtMKK1/2-AtMPK4 in response to PAMP, and AtWRKY33 activates expression of *PAD3*, which is required for the synthesis of camalexin [31,32]. It has also been demonstrated that AtWRKY33 is phosphorylated within the SP cluster by AtMPK3/AtMPK6, resulting in a stimulation effect on camalexin production [33]. In tobacco, SIPK phosphorylates NtWRKY1, resulting in enhanced DNA-binding activity of NtWRKY1 to the W-box, and co-expression of SIPK and NtWRKY1 enhanced SIPK-induced cell death [34]. In *Nicotiana benthamiana*, NbWRKY8 was identified as a substrate of SIPK, WIPK, and NTF4. These MAPKs phosphorylated NbWRKY8 within the SP cluster, and enhanced both DNA-binding and transactivation activities of NbWRKY8. Ectopic expression of the phospho-mimicking mutant of NbWRKY8 induced expression of defense-related genes, such as 3-hydroxy-3-methylglutaryl CoA reductase 2, that are involved in the production of isoprenoid phytoalexins in solanaceous plants [29]. In rice, OsWRKY33 is phosphorylated by OsBWMK1 which results in enhanced DNA-binding activity to the W-box elements [35].

Previously, we demonstrated that expression of *OsWRKY53*, one of the rice group I WRKY protein genes, was rapidly induced by either a fungal chitin oligosaccharide elicitor treatment or by infection with the blast fungus *Magnaporthe oryzae*, and 3 tandem W-box elements in the promoter of this gene were essential for the elicitor response. We also found that overexpression of *OsWRKY53* upregulated several defense-related genes in rice cells and resulted in enhanced resistance to a virulence race of *M. oryzae* in rice plants [36,37]. OsWRKY53 has a conserved SP cluster in the N terminal region of the protein as found for other reported group I WRKY

proteins (Fig. S1). Given that OsWRKY53 is the closest homologue of NbWRKY8 in rice [29], we hypothesized that OsWRKY53 is regulated by OsMPK3 and OsMPK6 at the posttranslational level, as part of a basal defense-signaling pathway in rice.

Here, we report that posttranslational regulation of OsWRKY53 plays an important role in regulating the basal defense response of rice plants against rice blast fungus through activation of the expression of defense-related genes. The OsMKK4-OsMPK3/OsMPK6, components of a fungal PAMP-responsive MAPK cascade in rice, phosphorylates the SP cluster of OsWRKY53 *in vitro*. OsMPK6-mediated phosphorylation of OsWRKY53 did not alter the DNA-binding activity to W-box elements, but co-expression of OsWRKY53 with a constitutively active OsMKK4 increased transactivation activity in an SP cluster-dependent manner. Transgenic rice plants overexpressing a phospho-mimic mutant of *OsWRKY53* leads to enhancement of disease resistance to a virulence race of rice blast fungus compared to native *OsWRKY53*-overexpressing rice plants. In addition, transcriptome analysis revealed that substantial numbers of defense-related genes were upregulated in the phospho-mimic mutant of *OsWRKY53*-overexpressing rice plants without blast fungus infection.

Materials and Methods

Plants, chemical treatment, pathogen, and rice transformation

Suspension-cultured rice cells (*Oryza sativa* L. cv. 'Nipponbare') were maintained as described previously [36]. Six days after subculturing, a small portion of the rice cells was harvested for particle bombardment. Rice plants (*Oryza sativa* L. cv. 'Nipponbare') were used in this study. Rice plants were grown in a chamber following previously described protocols [38]. The blast fungus *M. oryzae* strain Ina86–137 (MAFF 101511, race, 007.0, virulent to Nipponbare) was used for infection, and water was adopted as a mock treatment in this study. Rice transformation was performed as described previously [39].

Plasmid construction

Plasmids containing OsWRKY53 SP cluster DNA fragments with alanine or aspartic acid substitutions at 6 serine residues in the cluster were generated by Takara Bio Inc. (Takara Bio Inc., Japan), resulting in pW53SA and pW53SD respectively.

To construct mutated *OsWRKY53* genes in which all 6 serine residues in SP cluster were substituted for alanine (*OsWRKY53SA*) or aspartic acid (*OsWRKY53SD*), the mutated SP cluster region was amplified from pW53SA or pW53SD by PCR using the primers W53-Ala F and W53-Ala R, or W53-Asp F and W53-Asp R, respectively. The amplified mutated SP cluster DNA fragments were used as reverse primers for PCR with forward primer W53-N Fw or 53 GAL4 F, respectively, to amplify *OsWRKY53* N-terminal regions containing the mutated SP cluster from pET-W53. After the second PCR, the amplified DNA fragments were used as forward primers for PCR with reverse primer W53-C Rv or 53 GAL4 R, respectively, to amplify whole *OsWRKY53SA* ORF or *OsWRKY53SD* ORF from pET-W53. The amplified DNA fragments were directly cloned into the pZErO2 vector (Invitrogen, CA, USA) and sequenced, resulting in pZE-W53SA or pZE-W53SD, respectively.

To construct the thioredoxin-6× histidine tag (Trx-His)-fused *OsWRKY53SA* gene, the *OsWRKY53SA* ORF was excised from pZE-W53SA by *EcoRV* and *HindIII* digestion and was inserted between corresponding sites of pET-32b(+) (Novagen, Germany) to generate pET-W53SA. To construct the Trx-His-fused *OsWRKY53SD* gene, the *OsWRKY53SD* ORF was amplified from pZE-W53SD by PCR using the primers W53-N and W53-R. The amplified DNA fragment was directly cloned into the pZErO2 vector (Invitrogen) to generate pZE-W53SD2. After performing a sequence check, the DNA fragment was excised from pZE-W53SD2 by *Eco*RV and *Hin*dIII digestion and was inserted between corresponding sites of pET-32b(+) to generate pET-W53SD.

To construct the plasmids containing *OsWRKY53* promoter region with W-box or mutated W-box elements, the corresponding promoter regions were amplified by PCR using the primers W53 Wbox Fw and W53 Wbox Rv from pZE-W53P2.0 or pZE-W53PmG, respectively. The amplified DNA fragments were directly cloned into a pT7Blue T-vector (Novagen) and sequenced, resulting in pT7-W53PW and pT7-W53mPW.

To construct the DNA-binding domain of the yeast transcription factor GAL4 (GAL4DB)-fused *OsWRKY53SA* gene, the *OsWRKY53SA* ORF was amplified from pZE-W53SA by PCR using the primers 53 GAL4 F and 53 GAL4 R. The amplified DNA fragment was directly cloned into the pZErO2 vector (Invitrogen) to generate pZE-W53SA2. After performing a sequence check, the DNA fragment was excised from pZE-W53SA2 by *Sma*I and *Sal*I digestion and was inserted between corresponding sites of 430T1.2 [40] to generate 35S-GAL4DB-W53SA. To construct GAL4DB-fused *OsWRKY53SD* gene, the *OsWRKY53SD* ORF was excised from pZE-W53SD by *Sma*I and *Sal*I digestion and inserted between corresponding sites of 430T1.2 [40] to generate 35S-GAL4DB-W53SD.

To construct Gateway destination vectors containing *Zea mays* polyubiquitin promoter and *Agrobacterium tumefaciens* nopaline synthase terminator, pUCAP/Ubi-NT [41] was digested with *Bam*HI and *Sac*I and blunt-ended. Then, reading frame cassette A (RfA) (Invitrogen) containing *att*R recombination sites flanking a *ccd*B gene and a chloramphenicol-resistance gene was cloned into blunt-ended pUCAP/Ubi-NT, resulting in pUbi_RfA_Tnos.

To construct effector plasmids in which GUS and a constitutively active form of OsMKK4, OsMKK4^{DD}, were under the control of the *Z. mays polyubiquitin* promoter, GUS and OsMKK4^{DD} were cloned into pUbi_RfA_Tnos from pENTR-GUS (Invitrogen) and pENTR-MKK4^{DD} using LR clonase II Enzyme mix (Invitrogen). The resultant plasmids were designated as pUbi_GUS_Tnos and pUbi_MKK4^{DD}_Tnos, respectively.

To construct a Gateway entry clone containing *OsWRKY53SD* ORF, *OsWRKY53SD* ORF was amplified by PCR using the primers 53 ORF Gateway F and OsWRKY53 pENTR R from 35S-GAL4DB-W53SD. The amplified DNA fragment was cloned into pENTR/D-TOPO (Invitrogen) according to manufacturer's protocol and sequenced, resulting in pENTR-W53SD.

To construct a binary vector in which *OsWRKY53* ORF and *OsWRKY53SD* ORF were under the control of the *Z. mays polyubiquitin* promoter for rice transformation, *OsWRKY53* ORF and *OsWRKY53SD* ORF were cloned into p2KG [42] from pENTR-W53 and pENTR-W53SD using LR clonase II Enzyme mix (Invitrogen). The resultant plasmids were designated as p2KG-W53 and p2KG-W53SD. A summary of the plasmids used in this study is provided in Table S1, and sequences of PCR primers used for plasmid construction are provided in Table S2.

Recombinant proteins, *in vitro* kinase assays, and immunoblot analysis

OsMKK4^{DD} and OsMPK3/6 were expressed as fusion proteins with an N-terminus poly-histidine tag using a bacterial expression system (Invitrogen) following the manufacturer's instructions [19].

OsWRKY53 variants were also expressed as fusion proteins with an N-terminus poly-histidine tag as described previously [36]. The proteins were purified by immobilized metal ion affinity chromatography. To detect the phosphorylation of OsWRKY53 variants by recombinant OsMPKs, 800 ng of each recombinant OsMPK was pre-incubated for 1 h at 25°C with 120 ng of OsMKK4^{DD} in phosphorylation buffer containing 20 mM HEPES-KOH (pH 7.5), 10 mM MgCl$_2$, 1 mM DTT, and 50 μM ATP, and then was incubated with 1.2 μg of each OsWRKY53 variant and 25 μM ATP for 1 h at 25°C. The reaction was stopped by the addition of Laemmli's SDS sample buffer and boiling. The above samples were subjected to SDS-PAGE on 10% (w/v) polyacrylamide gels, transferred to Immobilon-P Transfer Membrane (PVDF, 0.45 μm) (Millipore, MA, USA). Phosphorylated OsWRKY53 were detected by using Phos-tag Biotin BTL-104 (Wako, Japan) according to the manufacturer's instruction. OsWRKY53 variants, OsMPK3 and OsMPK6, were detected with anti-His antibody [dilution 1:3000 (v/v)] (GE Healthcare, UK), anti-OsMPK3 serum [dilution 1:3000 (v/v)] and anti-OsMPK6 serum [dilution 1:5000 (v/v)], respectively, as the primary antibody, and ECL anti-mouse IgG horseradish peroxidase-linked species-specific whole antibody (dilution 1:25,000 [v/v]) (GE Healthcare) and ECL anti-rabbit IgG horseradish peroxidase-linked species-specific whole antibody (dilution 1:25,000 [v/v]) (GE Healthcare) as the secondary antibody. Chemiluminescent detection was carried out with the Immobilon Western Chemiluminescent HRP Substrate (Millipore) according to the manufacturer's instruction.

Gel mobility shift assays

Double-stranded W-box or mutated W-box probes were amplified by PCR from pT7-W53PW or pT7-W53mPW using the primers W53 Wbox Fw and W53 Wbox Rv and end-labeled with ^{32}P by T4 polynucleotide kinase (Takara Bio). The probe was purified using Illustra Microspin G-25 columns (GE Healthcare). Phosphorylation of OsWRKY53 recombinant protein was described above in the "Recombinant proteins, in vitro kinase assays, and immunoblot analysis" section. The gel mobility shift assay (GMSA) reaction mixture comprised 12 mM HEPES-KOH (pH 8.0), 60 mM KCl, 4 mM MgCl$_2$, 1 mM EDTA, 12% glycerol, 1 mM DTT, 2.5 mM PMSF, 0.1 μg of recombinant OsWRKY53, and 2 μl of the probe in a final volume of 20 μl. To form DNA-protein complexes, the above mixed samples were incubated for 20 min on ice. Finally, samples were separated on a 6% polyacrylamide gel in 1× TBE at room temperature, and bands were visualized by autoradiography.

Transactivation assay

pUbi_GUS_Tnos and pUbi_MKK4^{DD}_Tnos, in which GUS and OsMKK4^{DD}, respectively, are under the control of the *Z. mays polyubiquitin* promoter, were used as the OsMKK4-variants effector plasmids. 430T1.2, 35S-GAL4DB-W53, 35S-GAL4DB-W53SA, and 35S-GAL4DB-W53SD, in which GAL4DB and GAL4DB-OsWRKY53 variants are under the control of the CaMV *35S* promoter, were used as the GAL4DB-OsWRKY53-variants effector plasmids. GAL4-TATA-LUC-NOS, which contains a firefly luciferase (LUC) gene, was used as a reporter plasmid. The plasmid pRL, which contains the *Renilla* LUC gene under the control of the CaMV *35S* promoter, was used as an internal control. Particle bombardment was carried out with the PDS-1000 He Biolistic Particle Delivery System (Bio-Rad, CA, USA). Suspension-cultured rice cells were used for the bombardment. In co-transfection assays, 1.2 μg of the GAL4DB-OsWRKY53-variants effector construct, 1.6 μg of the reporter plasmid, and

0.4 μg of pRL with and without 1.2 μg of the OsMKK4-variants effector construct were used for each bombardment. Luciferase assays were performed with the Dual-Luciferase Reporter Assay System (Promega, WI, USA) and a Centro LB960 plate reader (Berthold Japan, Japan) following the manufacturer's instructions. The ratio of LUC activity (firefly LUC/$Renilla$ LUC) was calculated to normalize values after each assay.

Quantitative RT-PCR

Total RNA was extracted from rice leaves of non-transformed control (NT), $OsWRKY53$-overexpressing ($W53$-OX), $OsWRKY53SD$-overexpressing ($W53SD$-OX) rice plants, and NT rice calli using an RNeasy Plant Mini Kit (Qiagen, Germany) and subjected to cDNA synthesis using a PrimeScript RT reagent Kit with gDNA Eraser (Takara Bio Inc.). Quantitative RT-PCR (qRT-PCR) was performed using TaqMan probe with THUNDERBIRD Probe qPCR Mix (TOYOBO, Japan) for $OsKSL4$ and a Power SYBR Green PCR Master Mix (Applied Biosystems, CA, USA) for the other target genes on an ABI PRISM 7300 Real-Time PCR System (Applied Biosystems). To calculate the transcript levels of characterized genes, the copy number of their mRNAs was determined by generating standard curves using a series of known concentrations of the target sequence. Ubiquitin domain-containing protein (UBQ, Os10g0542200) was used as an internal control to normalize the amount of mRNA. For each sample, the mean value from triplicate amplifications was used to calculate the transcript abundance. Sequences of PCR primers and TaqMan probe used for qRT-PCR analysis are provided in Table S3.

Pathogen inoculation and disease-resistance test

To examine fungal lesions and biomass, fungal inoculation of rice leaf blades was carried out as described previously [38] with some modifications. Fourteen-day-old rice plants were placed on moistened filter paper in plastic trays. Washed conidia of the blast fungus were suspended at a concentration of 1×10^5 cells mL^{-1} in distilled water for the disease-resistance test, and were sprayed on rice plants incubated at 25°C in the dark for 24 h, followed by 14 h-light/10 h-dark cycles for 5 days. Lesions on the fourth leaves were classified as described previously [43]: necrotic spots (resistant dark brown specks), intermediate lesions (yellow and brown lesions without a gray center), and susceptible lesions (areas with a gray center and emerging aerial hyphae and conidia). Blast disease development was quantified by measurement of $M. oryzae$ genomic DNA (encoding 28S rRNA) relative to rice genomic DNA (encoding the eEF-1α gene) using quantitative genomic PCR analysis [44]. Quantitative genomic PCR was performed using a SYBR Premix Ex Taq II (Takara Bio Inc.) on an MX3000P (Stratagene, CA, USA). Sequences of PCR primers used for qPCR analysis are provided in Table S4. Data are presented relative to the value in leaves of NT rice plants.

Microarray data acquisition and cluster analysis

Fourth leaves of NT, $W53$-OX, and $W53SD$-OX rice plants were harvested from 4 plants at the 4-leaf stage. Total RNA was isolated from the rice leaves using an RNeasy Plant Mini Kit (Qiagen) and subjected to fluorescence labeling according to the manufacturer's instructions. The RNA was labeled with a Cyanine 3 dye (Cy3). Aliquots of Cy3-labeled cRNAs (1650 ng each) were used for hybridization in a 60-mer rice oligo microarray with 44k features (Agilent Technologies, CA, USA). Four biological replicate sample sets were analyzed. The glass slides were scanned using a microarray scanner (G2565, Agilent), and resulting output files were imported into Feature Extraction software (ver. 11; Agilent). Data normalization and statistical analyses were performed using

Partek Genomics Suite software (ver. 6.5; Partek software, MO, USA). Data from the selected spots were imported into Multi-Experiment Viewer (MeV v4.8, http://www.tm4.org/mev/) for cluster analysis. A hierarchical clustering analysis based upon the average linkage and cosine correlation was then used to cluster genes on the y-axis using MeV.

Phytoalexin measurements

Rice leaves (15–50 mg FW) were soaked in 2 mL of phytoalexin extraction solvent (ethanol/water/acetonitrile/acetic acid, 79:13.99:7:0.01, v/v) and heated for 15 min twice in a glass tube with a screw cap. The extract was centrifuged (4°C, 15 min, $16,000 \times g$). The supernatant was collected and subjected to phytoalexin measurements by LC-ESI-MS/MS as described previously [45].

Results

In vitro phosphorylation of OsWRKY53 by OsMPK3 and OsMPK6

First, we focused on potential MAPK phosphorylation sites of OsWRKY53, especially Ser or Thr residues followed by Pro, which is a minimal consensus motif for MAPK phosphorylation [27,28]. The deduced amino acid sequence of OsWRKY53 possessed 6 SP and 2 TP motifs, 6 of which were concentrated in the N-terminal region as the SP cluster conserved among several group I WRKY proteins (Fig. 1A and Fig. S1). To investigate whether OsWRKY53 was phosphorylated by rice MAPKs, OsMPK3, OsMPK6 and a Thioredoxin-His-tagged OsWRKY53 recombinant proteins were expressed and purified from $E. coli$, and subjected to *in vitro* MAPK phosphorylation assay. As shown in Fig. 1B OsWRKY53 was phosphorylated by recombinant OsMPK3 and OsMPK6 activated by a constitutively active form of OsMKK4, OsMKK4^{DD}. We also confirmed this phosphorylation did not occur without OsMPK3 and OsMPK6.

As the SP cluster of OsWRKY53 was hypothesized to be essential for the phosphorylation by OsMPK3 and OsMPK6, we prepared a Thioredoxin-His-tagged OsWRKY53SA recombinant protein in which all 6 Ser residues in the SP cluster were substituted for Ala (Fig. 1A), and performed an *in vitro* MAPK phosphorylation assay. Compared to the results of the original experiment using native OsWRKY53 recombinant protein, activated recombinant OsMPK3 and OsMPK6 could no longer phosphorylate OsWRKY53SA (Fig. 1B). In addition, we detected single shifted band of native OsWRKY53 by using anti-His antibody in an OsMPK3/OsMPK6 dependent manner (Fig. 1B). Given that this band shift is correlated with the phosphorylation of native OsWRKY53 by OsMPK3/OsMPK6, it is suggested that nearly all native OsWRKY53 protein molecules are phosphorylated in our experimental condition. Taken together, these results indicate that OsMPK3 and OsMPK6 can phosphorylate OsWRKY53 efficiently, and that phosphorylation site of OsWRKY53 by OsMPK3 and OsMPK6 resides in the SP cluster in the N-terminal region.

Phosphorylation of OsWRKY53 by OsMPK6 does not alter its W-box binding activity

It has been reported that phosphorylation of the group I WRKY proteins NtWRKY1 and NbWRKY8 by MAP kinase enhanced their W-box specific DNA-binding activity [29,34]. To investigate whether phosphorylation of OsWRKY53 by OsMPK3/OsMPK6 enhanced W-box binding activity, we performed a gel mobility shift assay (GMSA). First, we analyzed the W-box-specific DNA binding activity of OsWRKY53 by using

A

B

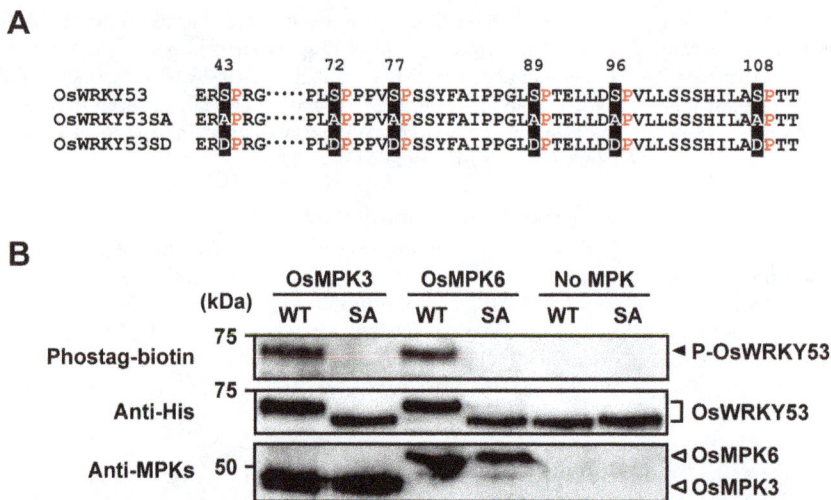

Figure 1. *In vitro* phosphorylation of OsWRKY53 by chitin-responsive OsMPKs. A, Putative MAPK phosphorylation sites in the SP cluster region of OsWRKY53, the loss-of-phosphorylation OsWRKY53 mutant with all 6 Ser substituted to Ala (OsWRKY53SA), and the phospho-mimicking OsWRKY53 mutant with all 6 Ser substituted to Asp (OsWRKY53SD). **B**, *In vitro* phosphorylation of OsWRKY53 by OsMPK3 and OsMPK6. Recombinant His-OsWRKY53 and His-OsWRKY53SA proteins were used as the substrate for rice mitogen-activated protein kinases (OsMPKs) activated by a constitutively active form of the rice MAP kinase OsMKK4^{DD}. Proteins separated by SDS-PAGE were blotted on membrane and probed with a Phostag-biotin antibody (top panel). The arrowhead indicates the position of phosphorylated OsWRKY53 (P-OsWRKY53). The membranes were reprobed followed by probing with an anti-His antibody to detect added substrate OsWRKY53 and OsWRKY53SA proteins (middle panel). OsMPKs in the reaction mixtures were also detected by immunoblot analysis with anti-OsMPK3 and anti-OsMPK6 antiserum (bottom panel). WT, native His-OsWRKY53; SA, His-OsWRKY53SA.

a W-box probe derived from the native promoter region of this gene. When incubated with the OsWRKY53 recombinant protein, a retarded band was observed, and this retardation of the labeled probe was abolished by competition using an unlabeled probe. In contrast, an unlabeled W-box mutated probe could not compete with the labeled probe (Fig. 2A).

The effect of the phosphorylation of OsWRKY53 by the rice MAP kinases on the W-box-specific DNA binding activity of OsWRKY53 was also tested by GMSA with phosphorylated OsWRKY53 protein. Because OsMPK3 and OsMPK6 showed approximately the same phosphorylation pattern for OsWRKY53, we used OsMPK6 to phosphorylate OsWRKY53 in this experiment. Recombinant OsWRKY53 protein was incubated with OsMPK6 alone, OsMKK4^{DD} alone, or OsMPK6 activated by OsMKK4^{DD}, and subjected to GMSA. As a result, there is no distinct difference in the W-box binding activity between the phosphorylated OsWRKY53 and native OsWRKY53 (Fig. 2B) We performed further GMSA analysis using native OsWRKY53, OsWRKY53SA and a phospho-mimic mutant of OsWRKY53 in which all 6 Ser residues in the SP cluster were substituted for Asp (OsWRKY53SD) (Fig. 1A). As shown in Fig. 2C, these three WRKY53 variants had similar W-box binding activity. Taken together, these results suggest that activated OsMPK6-mediated phosphorylation of OsWRKY53 did not affect its DNA binding activity.

OsMKK4^{DD} enhances transactivation activity of OsWRKY53

It has also been reported that *N. benthamiana* MEK2^{DD} enhanced transactivation activity of NbWRKY8 [29]. To investigate whether the rice OsMKK4^{DD} correspondingly enhances transactivation activity of OsWRKY53, we performed a transient reporter gene assay. We constructed effector plasmids that contained a *GUS* or *OsMKK4^{DD}* gene under the control of the maize *ubiquitin* promoter (Fig. 3A). We also constructed effector

plasmids that contained the CaMV *35S* promoter, driving a gene that encodes a fusion protein of the DNA-binding domain of the yeast transcriptional activator GAL4 and the full-length OsWRKY53 or OsWRKY53SA (GAL4DB-OsWRKY53 or GAL4DB-OsWRKY53SA), and a control plasmid encoding only GAL4DB (Fig. 3A). Each of the GAL4DB-OsWRKY53-variant effector plasmids or the control plasmid was delivered into rice cells along with a reporter plasmid *GAL4-TATA-LUC-NOS*, which contained 5 tandem repeats of a GAL4 binding site fused to the firefly *LUC* (*FLUC*), and either *GUS* or *OsMKK4^{DD}* effector plasmid by particle bombardment. As described in our previous report [36], coexpression of GAL4DB-OsWRKY53 with GUS showed more than 20-fold greater LUC activity compared to coexpression of GAL4DB with GUS as the negative control. On the other hand, LUC activity of rice cells coexpressing GAL4DB-OsWRKY53SA with GUS was reduced by 67% compared to those co-expressing GAL4DB-OsWRKY53 with GUS (Fig. 3B). Interestingly, coexpression of GAL4DB-OsWRKY53 with OsMKK4^{DD} significantly increased (more than doubled) the LUC activity (Fig. 3B). In contrast, coexpression of GAL4DB-OsWRKY53SA with OsMKK4^{DD} did not increase LUC activity (Fig. 3B). These results indicate that OsMKK4^{DD} can increase transactivation activity of OsWRKY53 in an SP cluster-dependent manner.

Given that OsWRKY53 is phosphorylated within the SP cluster by OsMPK3 and OsMPK6 activated by OsMKK4^{DD} *in vitro* (Fig. 1B), we expected that a phospho-mimic mutant of OsWRKY53 would show enhanced transactivation activity. To verify this hypothesis, we constructed an effector plasmid that encodes a fusion protein of the GAL4DB and OsWRKY53SD (GAL4DB-OsWRKY53SD) (Fig. 3A). Rice cells expressing GAL4DB-OsWRKY53SD showed more than 2-fold greater LUC activity compared to those coexpressing GAL4DB-OsWRKY53 (Fig. 3C). We also confirmed almost equal expression of *GAL4DB* and *GAL4DB-OSWRKY53* variants in these experiments by quantitative RT-PCR (qRT-PCR) analysis (Fig. 3D).

Overexpression of Phosphomimic Mutated OsWRKY53 Leads to Enhanced Blast Resistance in Rice

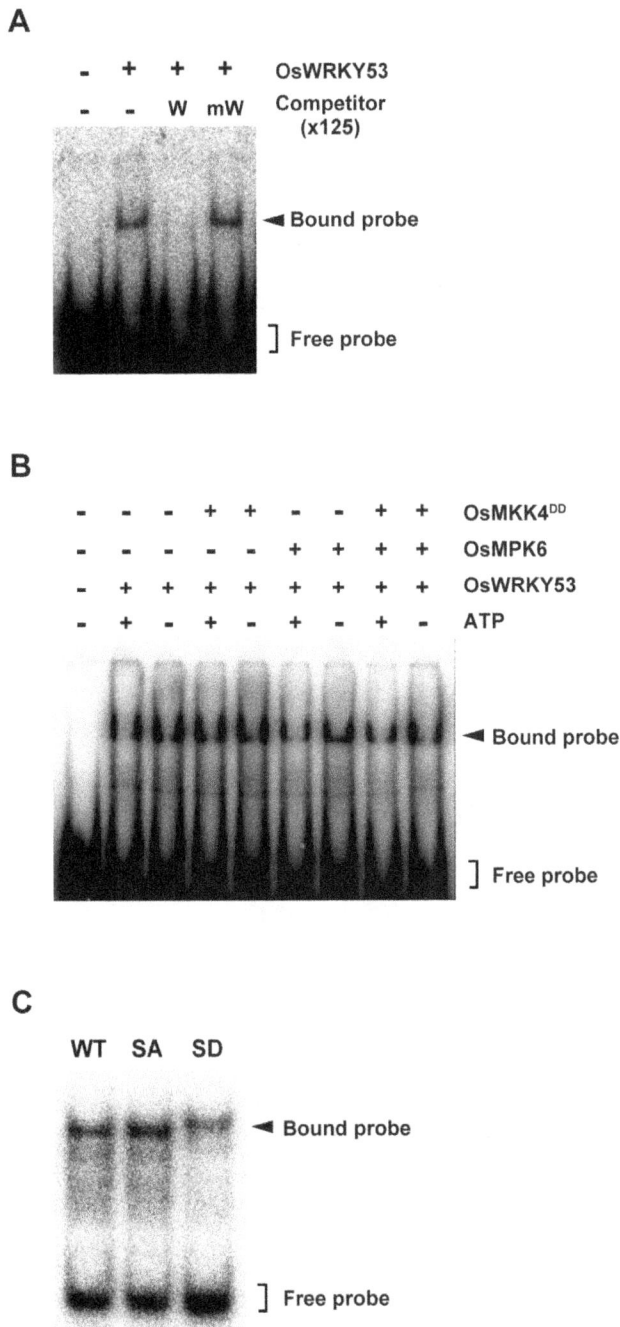

Figure 2. Phosphorylation of OsWRKY53 does not alter its W-box binding ability. A, W-box-specific DNA-binding activity of OsWRKY53. GMSA assay was performed using purified recombinant OsWRKY53 protein and ^{32}P-labeled W-box probe containing the W-box cis-elements in the *OsWRKY53* promoter. The specificity of the W-box binding activity was demonstrated by competition assay using 125-fold excess amount of unlabeled W-box probe (W) and mutated W-box probe (mW). **B**, Phosphorylation of OsWRKY53 does not enhance its W-box binding activity. Purified recombinant OsWRKY53 protein was phosphorylated using the OsMPK6 activated by OsMKK4^{DD}. GMSA was performed as in **A**. **C**, W-box binding activity of OsWRKY53 variant proteins. Purified recombinant OsWRKY53, OsWRKY53SA and OsWR-KY53SD proteins were subjected to GMSA. GMSA was performed as in **A**. WT, native His-OsWRKY53; SA, His-OsWRKY53SA; SD, His-OsWR-KY53SD.

Thus, these results strongly suggest that activation of OsMPK3 and OsMPK6 by OsMKK4 is an important mechanism for regulating the transactivation activity of OsWRKY53 *in vivo* by phosphorylation of OsWRKY53 within the SP cluster.

Overexpression of a phospho-mimic mutant of OsWRKY53 further enhances the basal defense against rice blast fungus

We previously showed that *OsWRKY53*-overexpressing transgenic rice plants exhibited enhanced resistance to a virulence race of *M. oryzae* [36]. Given the result here that a phospho-mimic mutant of OsWRKY53 (OsWRKY53SD) shows enhanced transactivation activity compared to the native OsWRKY53, we hypothesized that overexpression of *OsWRKY53SD* would confer further enhanced disease resistance against the rice blast fungus in rice plants. To test this hypothesis, we generated transgenic rice plants that expressed either native *OsWRKY53* or *OsWRKY53SD* under the control of the constitutive maize *ubiquitin* promoter and tested the transformants for resistance to a virulence race of rice blast fungus, *M. oryzae* strain Ina86–137. The constructs for *OsWRKY53*- or *OsWRKY53SD*-overexpression were introduced into the rice cultivar Nipponbare by *Agrobacterium*-mediated transformation, and the overexpression of *OsWRKY53* or *OsWR-KY53SD* in transgenic rice plants (*W53-OX* and *W53SD-OX*, respectively) was confirmed by qRT-PCR analysis (Fig. S2). Overexpression of *OsWRKY53* or *OsWRKY53SD* did not affect the growth and morphology of the transgenic rice plants (data not shown).

Next, we tested the transformants for resistance to the blast fungus. The transgenic rice plants were grown in a growth chamber, inoculated with conidia of the blast fungus, and disease symptoms were characterized 5 days later. We confirmed as previously reported that *W53-OX* rice plants showed enhanced resistance against the blast fungus compared to non-transformed control (NT) rice plants (Fig. S3). Thus, we compared resistance to the blast fungus between *W53-OX* and *W53SD-OX* rice plants directly. We did not see statistically-significant differences in average numbers of lesions in the total infected area between 2 independent *W53-OX* and 2 independent *W53SD-OX* samples (Fig. S4). Therefore, we categorized the disease lesions into 3 classes (Fig. 4A) and compared the ratio of lesions between *W53-OX* and *W53SD-OX* rice plants. Interestingly, the ratio of necrotic spot lesions in leaves of the *W53SD-OX* plants was significantly increased relative to that of *W53-OX* plants (Fig. 4B). Moreover, the ratio of susceptible lesions in *W53SD-OX* plants was significantly decreased relative to *W53-OX* plants (Fig. 4B). We also quantified fungal biomass in leaves of NT, *W53-OX* and *W53SD-OX* rice plants. Quantitative genomic PCR analysis demonstrated that significantly reduced levels of fungal DNA were detected from *W53SD-OX* plants compared with *W53-OX* plants. It was also indicated that fungal biomass levels tend to be decreased in *W53-OX* plants compared with NT plants (Fig. 4C). These results strongly suggest that overexpression of a phospho-mimic mutant of *OsWRKY53* further enhances the basal defense response against the virulence rice blast fungus compared to native *OsWRKY53*-overexpressing rice plants.

Genome-wide profiling of gene expression in *W53-OX* and *W53SD-OX* rice plants

To examine the transcriptional changes that accompany the enhanced disease resistance in *W53SD-OX* rice plants, we performed a genome-wide DNA microarray analysis using NT, *W53-OX*, and *W53SD-OX* rice plants, and gene expression was

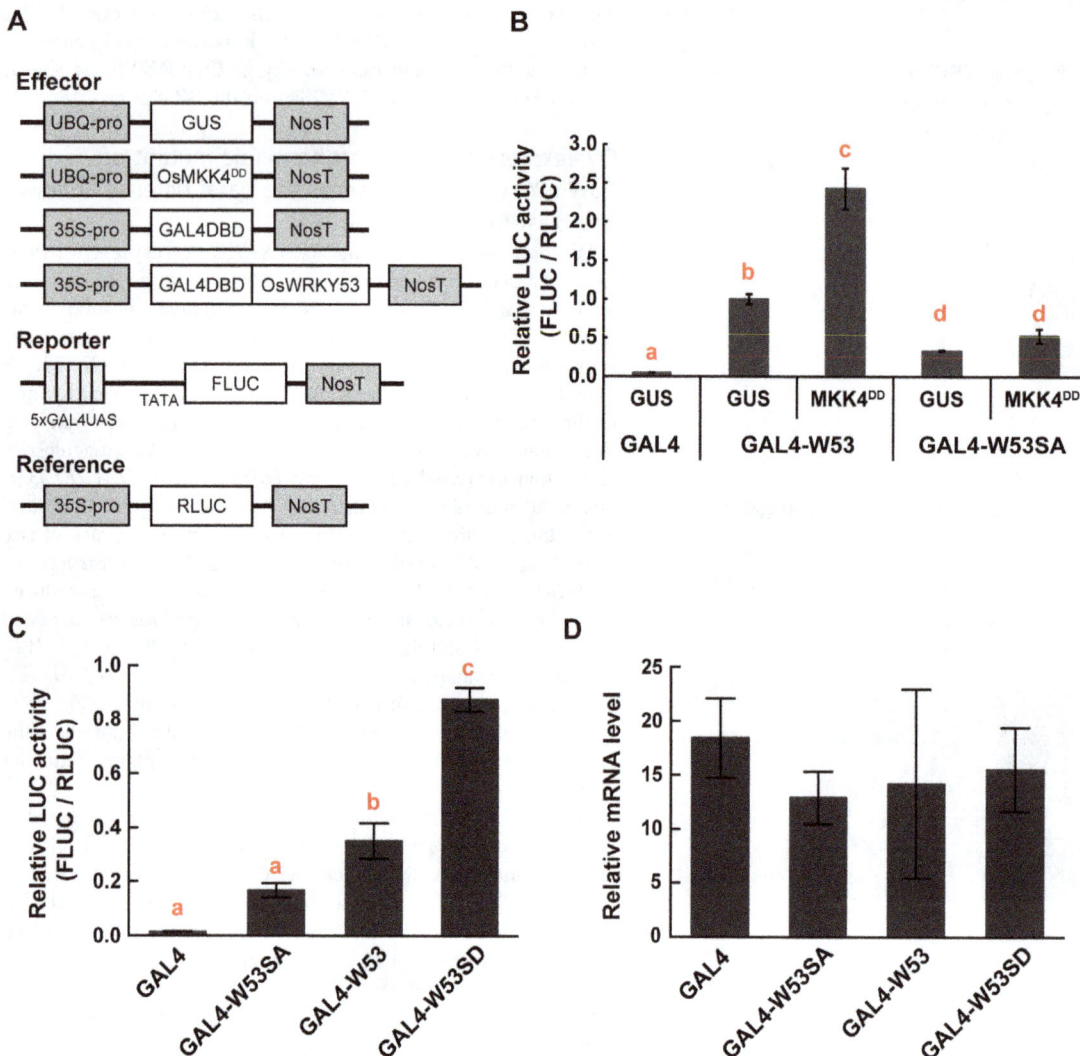

Figure 3. Post-translational regulation of OsWRKY53 transactivation activity by OsMKK4^DD. A, Diagrams of effector, reporter, and reference plasmids used in transient reporter gene assays. **B**, Regulation of transactivation activity of OsWRKY53 by OsMKK4^DD in cultured rice cells. A GUS construct was used as a negative control. Firefly LUC activity was normalized against that of *Renilla* LUC. Values of LUC activity are shown relative to those of GAL4-W53 + GUS (n = 4); *bars* indicate the standard error of the mean. Three independent experiments were performed, and a representative result is shown. Statistically different data groups are indicated using different letters ($p<0.01$ by One-way ANOVA with Tukey post hoc test). **C**, Enhanced transactivation of a phosphorylation-mimic mutant of OsWRKY53 in cultured rice cells. Transient reporter gene assay was performed as in **B**. Values of LUC activity are shown relative to those of GAL4-W53 (n = 4); *bars* indicate the standard error of the mean. Three independent experiments were performed, and a representative result is shown. Statistically different data groups are indicated using different letters ($p<0.05$ by One-way ANOVA with Tukey post hoc test). **D**, Expression analysis of *GAL4DB* and *GAL4DB-OsWRKY53* variants in cultured rice cells. qRT-PCR analysis was performed using total RNA isolated from rice cells after particle bombardment with effector, reporter and reference plasmids. Values indicate relative mRNA levels normalized to the expression of the *RLUC* gene (n = 4); *bars* indicate the standard error of the mean. W53, W53SA and W53SD indicate the native OsWRKY53, an OsWRKY53 mutants whose Ser residues in the SP cluster were substituted to Ala, and an OsWRKY53 mutant that mimics the phosphorylated form, respectively.

compared between *W53-OX* and *W53SD-OX*. These data have been deposited in the Gene Expression Omnibus in NCBI (http://www.ncbi.nlm.nih.gov/geo/; ID:GSE48500). We performed 4 biological replicates with 2 independent *W53-OX* (#8 and #10) or *W53SD-OX* (#38 and #40) rice-plant lines. The statistical analysis was performed on normalized data using the ANOVA-false discovery rate (ANOVA-FDR, q value ≤0.05) as calculated by Partek Genomics Suite (http://www.partek.com/), and we selected genes with changes in expression based on the criterion of a twofold increase or decrease in the average levels of fold change in *W53SD-OX* relative to expression levels in *W53-OX*. Based on this

criterion, 280 genes were upregulated and 135 genes were downregulated in *W53SD-OX* rice plants (Table S5).

Given that OsWRKY53 is a transactivator whose expression is upregulated upon blast infection [36], it is likely that its target genes involved in the blast resistance are also upregulated in NT rice plants upon blast fungus infection. Thus, we first focused on the 280 upregulated genes in *W53SD-OX* rice plants compared to *W53-OX* rice plants, and compared those with the results of our previous transcriptome analysis using NT rice plants with and without blast infection (http://www.ncbi.nlm.nih.gov/geo/; ID: GSE39635). This analysis showed that 151 out of the 280 upregulated genes in *W53SD-OX* rice plants were also upregulated

A

B

C

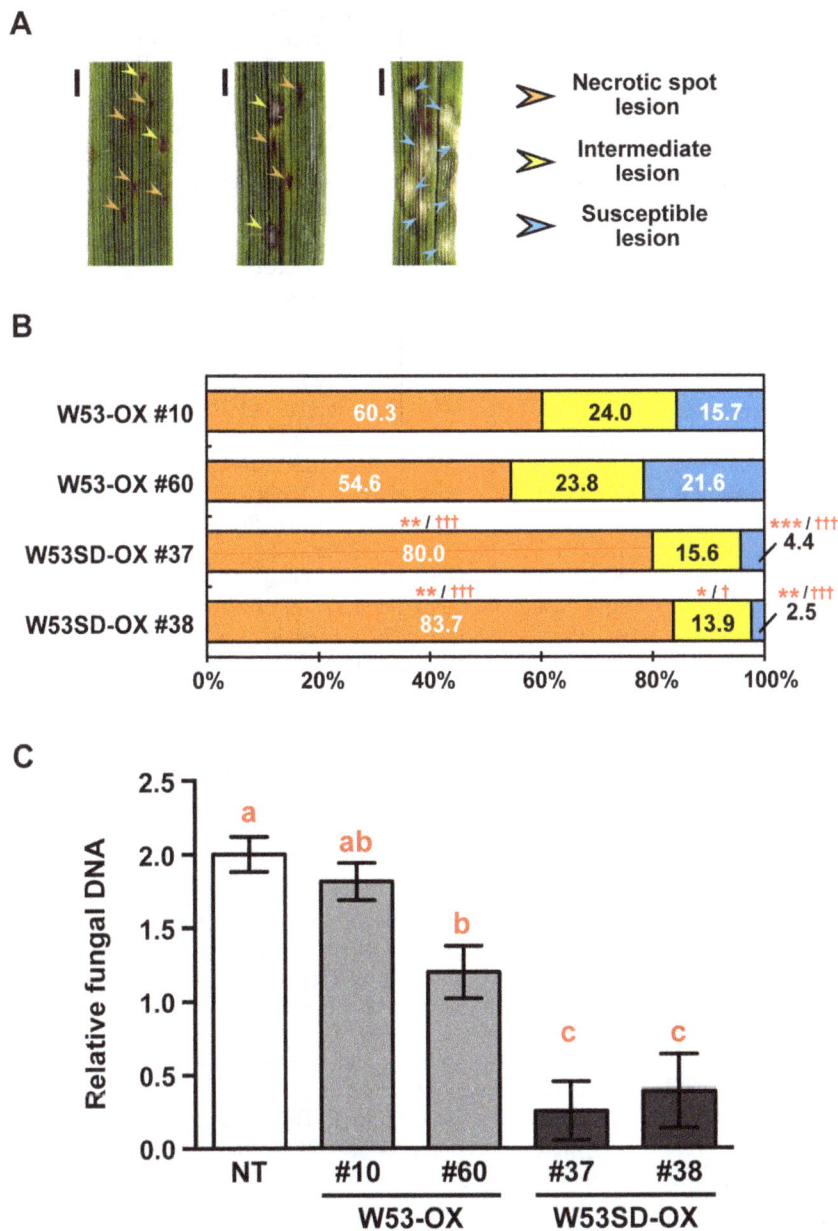

Figure 4. Overexpression of a phosphorylation-mimic mutant of *OsWRKY53* enhances the rice blast resistance of rice plants. A, Representative lesions of rice blast disease observed at 5 days post inoculation (dpi). Bars indicate 1 mm. **B**, Ratio of the classes of lesions in transgenic rice leaves infected with rice blast fungus *Magnaporthe oryzae* Ina86–137. Washed conidia of the blast fungus were suspended in 1 mM MES-NaOH (pH 5.7) and then were inoculated on leaves of *W53-OX* and *W53SD-OX* transgenic rice plants. The lesions were counted according to the classifications shown in **A**. Bars represent the ratio of lesions of each class to the total number of counted lesions in 3 or 4 individual leaf blades. Three independent experiments were performed, and a representative result is shown. *Asterisks* and *daggers* denote significant differences compared to *W53-OX* plants (*$p < 0.05$; **$p < 0.01$; ***$p < 0.001$ vs. *W53-OX* #10; $^{†}p < 0.05$; $^{†††}p < 0.001$ vs. *W53-OX* #60, by One-way ANOVA with Tukey post hoc test) **C**, Development of blast disease in leaf blades evaluated by quantitating *M. oryzae* genomic DNA. The amount of *M. oryzae* 28S rDNA relative to rice genomic *eEF1α* DNA was determined by quantitative PCR analysis. Values are represented as mean values ±SE for 6 leaf blades. Statistically different data groups are indicated using different letters ($p < 0.05$ by One-way ANOVA with Tukey post hoc test on log-transformed data). NT, non-transformant control rice plants; *W53-OX*, native *OsWRKY53*-overexpressing rice plants; *W53SD-OX*, phosphorylation-mimic mutant of *OsWRKY53*-overexpressing rice plants.

in NT rice plants infected with the blast fungus. Next, the 151 genes that showed increased expression upon blast infection were subjected to hierarchical clustering using the Pearson correlation and average linkage methods together with the microarray data of NT rice plants. Based on their expression pattern in the rice plants, we classified the 151 upregulated genes into the following 2 major groups: group I, genes whose expression increased in *W53SD-OX* plants compared to NT, and increased a little or not in *W53-OX* plants; group II, genes whose expression decreased in both *W53-OX* and *W53SD-OX* plants compared to NT. This analysis showed that most of the genes (94%) were included in the group I, and more than 70% of the group I genes showed increased expression pattern in a stepwise fashion in *W53-OX* and *W53SD-OX* plants compared to NT (Fig. S5, Table S5).

Overexpression of Os*WRKY53SD* enhances the activation of defense-related genes

Based on the results described above, it appears that a subset of defense-related genes is further activated in *W53SD-OX* rice plants compared to *W53-OX* plants. Thus, the 151 genes upregulated in *W53SD-OX* were organized into molecular function Gene Ontology (GO) categories based on their primary functions. Most of the genes grouped into the following categories: binding (GO:0005488), catalytic activity (GO:0003824), nucleic acid binding transcription factor activity (GO:0001071), and transporter activity (GO:0005215). But some of these genes grouped into the following 2 non-GO categories: defense-related genes and molecular function unknown (Table S5). Besides, almost all of the defense-related genes (22 out of 23 genes) among the 151 upregulated genes fell into the group I, including several PR protein genes such as β-1,3-glucanase, chitinase, and *PR-5* genes (Table 1). Given that *W53SD-OX* rice plants showed further-enhanced resistance to the virulence rice blast fungus, we finally focused on the group I genes, especially those relevant to defense response. To further validate the results of the microarray analysis, we performed qRT-PCR analysis using NT and each of 2 independent *W53-OX* and *W53SD-OX* rice plants. As shown in Fig. 5, the expression of most of these genes was significantly upregulated in *W53SD-OX* plants compared to NT plants, and some of these genes' transcripts were also accumulated in *W53-OX* plants compared to NT. We also noticed that *OsCPS4* and *CYP99A2*, which encoded biosynthetic enzymes of rice diterpenoid phytoalexins (momilactones), were included in the above defense-related genes (Table 1). Momilactones are synthesized from geranylgeranyl diphosphate (GGDP) through 2 cyclization and multiple oxidation steps [46,47]. We therefore examined the expression levels of these above 2 genes (*OsCPS4* and *CYP99A2*) and 3 additional genes (*OsKSL4*, *OsMAS* and *CYP99A3*) for momilactone biosynthesis by qRT-PCR analysis. We found that transcript levels of these momilactone-biosynthetic genes except *OsKSL4* were increased only in the *W53SD-OX* rice plants, a result consistent with the basal accumulation of momilactones in these plants (Fig. 6). Taken together, these results suggest that the potentiated induction of these defense-related genes at steady state provides further enhanced disease resistance of *W53SD-OX* rice plants against rice blast fungus at the time of infection compared to *W53-OX* rice plants.

Discussion

Regulation of OsWRKY53 function by the rice OsMKK4-OsMPK3/OsMPK6 cascade

We have shown here that OsWRKY53 is phosphorylated *in vitro* by the rice OsMPK3/OsMPK6 activated by OsMKK4, and that the SP cluster in the N-terminal region of OsWRKY53 is essential for the phosphorylation (Fig. 1). The SP cluster is shown to be highly conserved among several group I WRKY proteins in higher plants (Fig. S1), and also to be essential for phosphorylation of NbWRKY8 in *N. benthamiana* and AtWRKY33 in *A. thaliana* by the MEK2-SIPK/NTF4/WIPK and the AtMKK4/AtMKK5-AtMPK3/AtMPK6 cascades respectively [29,33]. It has also been shown that the interaction of NbWRKY8 with the MAPKs depends on its D domain [29,30]. Given that OsWRKY53 and AtWRKY33 possess the D domain as well, and the above 3 MAPKKs-MAPKs cascades from different plant species are functionally similar to one another, our findings strongly suggest

that the phosphorylation mechanism of the group I WRKY proteins is well conserved in both monocots and dicots.

Coexpression of OsWRKY53 with OsMKK4^{DD}, a constitutively active form of OsMKK4, increased transactivation activity of OsWRKY53 in an SP cluster-dependent manner (Fig. 3). We have also shown that OsWRKY53SD, a phospho-mimic mutant of OsWRKY53, had enhanced transactivation activity (Fig. 3), strongly suggesting that phosphorylation of OsWRKY53 within the SP cluster increases its transactivation activity. Similarly, a phospho-mimic mutant of NbWRKY8 also showed enhanced transactivation activity [29]. Given that OsWRKY53 was phosphorylated by OsMPK3/OsMPK6 activated by OsMKK4^{DD} *in vitro* (Fig. 1) and that OsMKK4^{DD} efficiently activated OsMPK3 and OsMPK6 *in vivo* [19], it is likely that OsMPK3/OsMPK6 activated by OsMKK4 phosphorylates OsWRKY53 within the SP cluster *in vivo*.

Whereas the enhanced transactivation activity of OsWRKY53SD, phosphorylation of OsWRKY53 by OsMPK6, did not alter its W-box binding activity (Fig. 2) as is the case with AtWRKY33 [33]. In contrast, phosphorylation of NbWRKY8 by MAPKs increases both W-box binding and transactivation activity [29]. In addition, phosphorylation of OsWRKY33 by OsBWMK1 enhances W-box binding activity, and coexpression of OsWRKY33 with OsBWMK1 shows enhanced transactivation activity [35]. These findings suggest that there are at least 2 different mechanisms that regulate the W-box binding activity of group I WRKY proteins even in the same plant species. The regulatory mechanisms for NbWRKY8 and OsWRKY33 DNA-binding activity via phosphorylation are still unclear; however, it has been reported that 2 VQ motif-containing proteins (SIB1 and SIB2) interact with AtWRKY33, resulting in stimulation of the W-box binding activity of AtWRKY33, and overexpression of SIB1 enhanced disease resistance to the necrotrophic pathogen *Botrytis cinerea* in an AtWRKY33-dependent manner [48]. In *Arabidopsis*, there are 34 VQ motif-containing protein genes, and expression of the majority of these is responsive to pathogen infection [49]. In the rice genome, a number of genes were found to encode VQ motif-containing proteins and our previous transcriptome data showed that some of these genes were upregulated in response to rice blast fungus infection [50]. Thus, one or more of these rice VQ motif-containing proteins may interact with OsWRKY53 to modulate W-box binding activity and contribute to disease resistance to rice blast fungus in *OsWRKY53*-overexpressing rice plants as reported previously [36]. The regulatory mechanisms for the enhanced transactivation activity of phosphorylated/phospho-mimic group I WRKY proteins remain to be elucidated. In mammals, it has been reported that phosphorylation of transcription factors by MAPKs modulates other intrinsic transcription factor activities, such as their affinities for coactivators [51,52]. Therefore, it will be important to examine whether the enhanced transactivation activity of the phospho-mimic mutant of OsWRKY53 is regulated by interaction with coactivators.

Overexpression of an OsWRKY53 phospho-mimic mutant alters the rice transcriptome and further enhances basal defense responses to the rice blast fungus

Overexpression of a phospho-mimic mutant of *OsWRKY53*, *OsWRKY53SD*, resulted in further-enhanced disease resistance to a virulence rice blast fungus *M. oryzae* strain Ina86−137 compared to native *W53-OX* rice plants (Fig. 4). This result implies that the observed difference in degree of disease resistance to blast fungus is correlated with the difference in transactivation activity between OsWRKY53 and OsWRKY53SD.

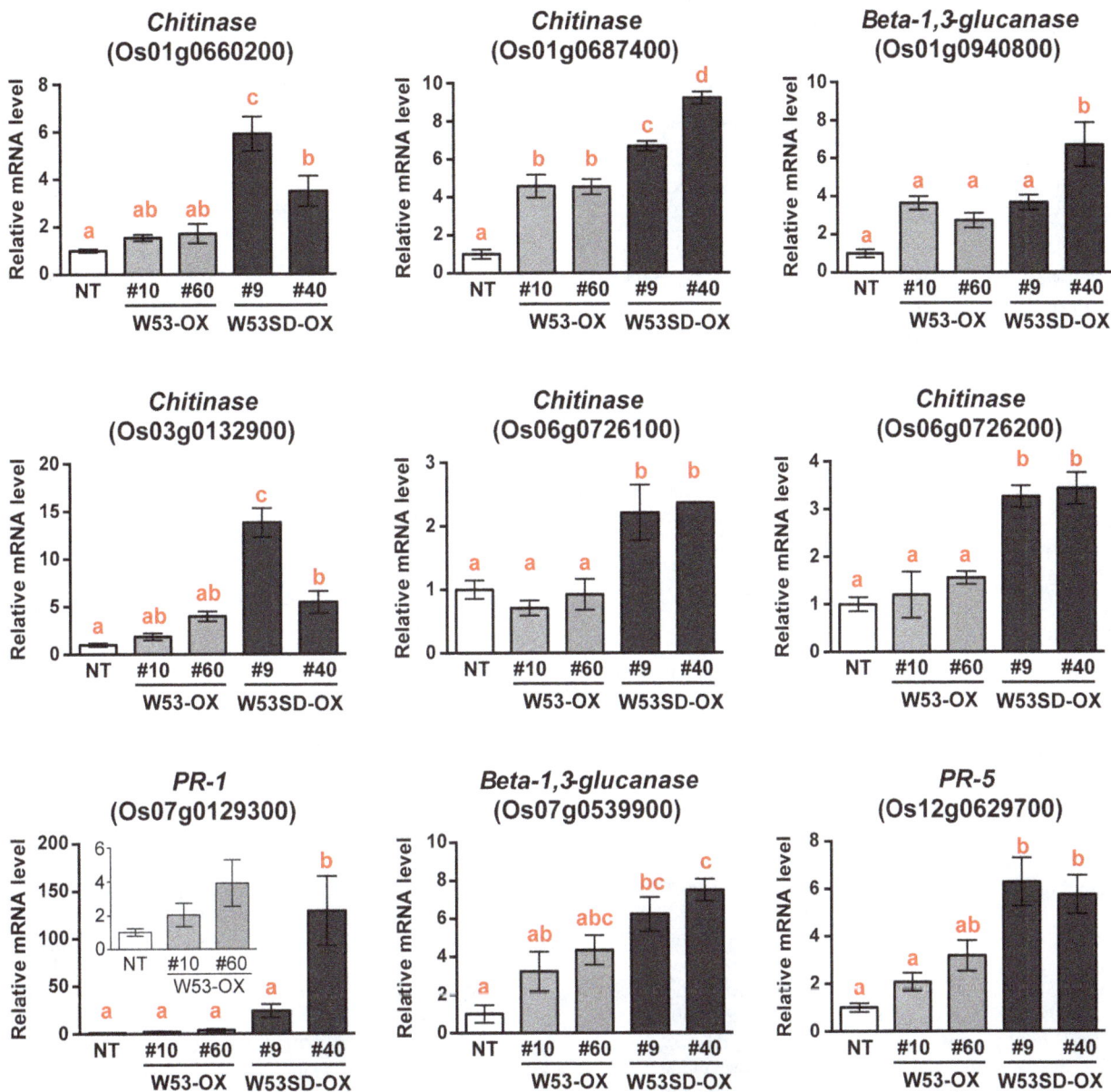

Figure 5. Expression analysis of defense-related genes belonging to group I upregulated genes in *W53SD-OX* rice plants. qRT-PCR analysis was performed using total RNA isolated from uninfected rice leaves. Values indicate relative mRNA levels normalized to the expression of the *UBQ* gene (n = 3); *bars* indicate the standard error of the mean. Three independent experiments were performed, and a representative result is shown. Statistically different data groups are indicated using different letters ($p < 0.05$ by One-way ANOVA with Tukey post hoc test). NT, non-transformed control; OX, *OsWRKY53*-overexpressing rice plants; SD-OX; *OsWRKY53SD*-overexpressing rice plants.

We performed a transcriptome analysis using NT, *W53-OX*, and *W53SD-OX* rice plants, and found that 151 out of the 280 upregulated genes in *W53SD-OX* rice plants compared to *W53-OX* plants were also upregulated in rice plants infected with the blast fungus in comparison with our previous transcriptome data. The 151 upregulated genes could be classified into 2 groups based on their expression patterns (Fig. S5). Given that OsWRKY53 is a transcriptional activator, genes in group I may be good candidates as direct targets of OsWRKY53. In fact, we found several W-box elements in promoters of most of the group I genes (Table S5).

Furthermore GO annotation analysis revealed that all the defense-related genes in the 151 upregulated genes, including beta-1, 3-glucanase, chitinase, and *PR-5*, belonged to group I

(Fig. 5 and Table S5), and several of these defense-related genes were also upregulated in uninfected *W53-OX* rice plants compared to NT (Fig. S5 and Fig. 5). Thus, it is plausible that upregulation of these defense-related genes in *W53-OX* rice plants partly contributes to enhanced rice blast resistance. Most notably, these defense-related genes were found to be more upregulated in *W53SD-OX* rice plants compared to *W53-OX* plants, as were the other group I genes (Fig. 5). Therefore, it is demonstrated that overexpression of *OsWRKY53SD* causes further induction of these defense-related genes in rice plants even without blast fungus infection, resulting in the further-elevated disease resistance of *W53SD-OX* rice plants.

Table 1. List of defense related genes belonging to Group I.

Locus ID	Accession Number	Description	Fold change (*W53SD-OX* vs *W53-OX*)	*q*-value
Os01g0660200	AK100973	Chitinase	2.69706	0.00050024
Os01g0687400	AK106178	Chitinase	2.14256	0.000193826
Os01g0713200	AK060113	Beta-1,3-glucanase	3.09387	0.000710591
Os01g0860500	AK103976	Chitinase	2.13807	0.0317721
Os01g0940700	AK070677	Beta-1,3-glucanase	2.6235	0.00416658
Os01g0940800	AK105972	Beta-1,3-glucanase	2.31721	0.0229786
Os01g0963000	AK102172	Peroxidase	2.19554	0.0167012
Os03g0132900	AK099355	Chitinase	2.88386	0.0029401
Os04g0178300	AY530101	OsCPS4	3.93669	0.0179749
Os04g0180400	AK071546	CYP99A2	3.09676	0.00288931
Os04g0688200	AK103558	Peroxidase	3.3661	0.037161
Os05g0384300	AK107183	Peptidase aspartic family protein	2.37293	0.00196873
Os05g0492600	AF456247	NBS-LRR resistance-like protein	2.53676	0.0306967
Os06g0726100	AK061280	Chitinase	3.44204	0.0106932
Os06g0726200	AK061042	Chitinase	2.14672	0.0353149
Os07g0129300	AF306651	PR-1	4.53826	0.0459025
Os07g0539900	AK071889	Beta-1,3-glucanase	4.34453	1.95454E-05
Os08g0124000	AK063293	Disease resistance protein family protein	2.11057	0.00277816
Os08g0202400	AK070769	Disease resistance protein family protein	2.02296	0.0364594
Os11g0686500	AK066559	Disease resistance protein family protein	2.30764	2.04996E-05
Os12g0628600	X68197	PR-5	2.91142	0.038663
Os12g0629700	AK099946	PR-5	4.29651	0.00171269

We also found that the group I gene, *PR-5* (Os12g0629700) was upregulated in *W53SD-OX* rice plants (Table 1). Recently, we reported that overexpression of *OsWRKY28* resulted in enhanced susceptibility to Ina86–137 and decreased accumulation of the identical *PR-5* transcripts in response to blast fungus infection [50]. Given that OsWRKY28 is a transcriptional repressor [50], these contrasting results may indicate that OsWRKY53 and OsWRKY28 act competitively to modulate the upregulated transcript levels of some defense-related genes for fine-tuning of the basal defense-response level against the rice blast fungus.

Momilactone biosynthetic genes were also upregulated in uninfected *W53SD-OX* rice plants, resulting in constitutive accumulation of momilactones only in *W53SD-OX* plants (Table 1; Fig. 6). Recently, we have shown that the *oscps4-tos17* mutant was more susceptible to rice blast fungus than was the non-transformed control, possibly due to lower levels of momilactones [53]. Therefore, it is assumed that momilactone accumulation may also be involved in the further-enhanced disease resistance of *W53SD-OX* plants to the fungus. We have also reported previously that these momilactone biosynthetic genes constitute a functional gene cluster and are regulated in a coordinate manner by a bZIP transcription factor, OsTGAP1 [46,54]. It has also been shown that conditional expression of OsMKK4[DD] induced biosynthesis of diterpenoid phytoalexins including momilactones. However, expression of *OsTGAP1* was not induced by OsMKK4[DD], and OsTGAP1 was not phosphorylated by the OsMKK4-OsMPK3/OsMPK6 cascade *in vitro* [19,55]. Given that OsWRKY53 is phosphorylated by the OsMKK4-OsMPK3/OsMPK6 cascade *in vitro*, it is likely that OsWRKY53 and OsTGAP1 are in different signaling pathways and regulate the gene cluster for momilactone

biosynthesis either cooperatively or in a different manner. Interestingly, it has also been demonstrated that both AtWRKY33 and NbWRKY8 were involved in regulation of phytoalexin biosynthetic genes [29,32,33], also suggesting conserved roles for group I WRKY proteins in basal defense responses in higher plants.

It has been shown that *OsWRKY53* is induced by a fungal chitin elicitor [36]. Given that the chitin elicitor activated the OsMKK4-OsMPK3/OsMPK6 cascade and OsMKK4[DD] induced the accumulation of *OsWRKY53* transcripts [19], it is likely that the OsMKK4-OsMPK3/OsMPK6 cascade activates OsWRKY53 at both the transcriptional and posttranslational levels, in response to fungal PAMPs. Our results suggest that phosphorylation of OsWRKY53 alters the rice transcriptome resulting in a further enhancement of disease resistance to the rice blast fungus, and the OsMKK4-OsMPK3/OsMPK6 cascade may play a crucial role in driving high-level activation of basal defense responses by the above dual-level regulation of OsWRKY53. Some of the defense-related genes in the group I, such as chitinases (Os01g0687400, Os06g0726100 and Os06g0726200) and *PR-5* (Os12g0629700), were also found to be upregulated by OsMKK4[DD] in suspension-cultured rice cells [19], supporting this hypothesis. There are several group I *WRKY* genes in the rice genome that possess the SP cluster (Fig. S1). Therefore, it will be important to analyze other group I *WRKY* genes and to identify MAPKs that phosphorylate those WRKY proteins so as to provide further insights into the biological roles of group I WRKY proteins in basal defense responses to the rice blast fungus.

A

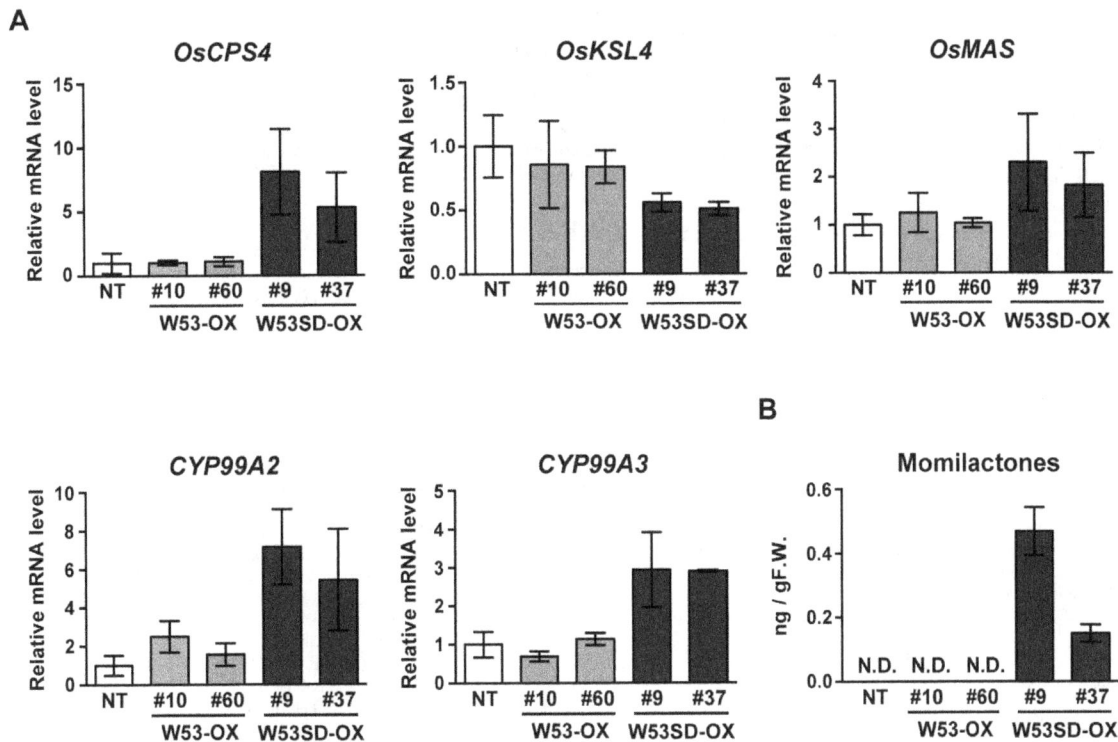

Figure 6. Phytoalexin accumulation in *W53SD-OX* rice plants. A, Expression analysis of momilactone biosynthetic genes. qRT-PCR analysis was performed using total RNA isolated from uninfected rice leaves. Values indicate relative mRNA levels normalized to the expression of the *UBQ* gene (n = 3); *bars* indicate the standard error of the mean. Three independent experiments were performed, and a representative result is shown. **B**, Accumulation of momilactones in uninfected *W53SD-OX* rice plants. Momilactone levels in the rice leaves were determined by LC-MS/MS. The results are the average of at least 3 independent experiments. *Bars* indicate the standard error of the mean. NT, non-transformed control; OX, *OsWRKY53*-overexpressing rice plants; SD-OX; *OsWRKY53SD*-overexpressing rice plants. N.D.: not detected.

Supporting Information

Figure S1 Alignment of the deduced amino acid sequences of group I WRKYs in rice and their homologues. The deduced amino acid sequences of *OsWRKY53/24/78/70/33/35/30*, *AtWRKY25/33*, *N. benthamiana NbWRKY8*, and *N. tabacum NtWRKY1* were aligned using the CLUSTAL W program. Highly conserved residues are shaded in black, and similar residues are shaded in gray. D domain, MAP kinase-docking domain; SP cluster, clustered serines or threonines followed by proline (SP or TP); WRKY domain, WRKY DNA-binding domain; NLS, putative nuclear localization signal.

Figure S2 Overexpression of the native *OsWRKY53* and a phosphorylation-mimic mutant of *OsWRKY53* in transgenic rice plants. qRT-PCR analysis was performed using total RNA extracted from transgenic rice plants. Values indicate relative mRNA levels normalized to the expression of the *UBQ* gene. The data are represented as mean values for at least 3 independent experiments; bars indicate the standard deviation of the mean. NT, non-transformant control rice plants; *W53-OX*, native *OsWRKY53*-overexpressing rice plants; *W53SD-OX*, phosphorylation-mimic mutant of *OsWRKY53*-overexpressing rice plants.

Figure S3 Overexpression of *OsWRKY53* enhances the disease resistance of rice plants to rice blast fungus. Ratio of the classes of lesions in transgenic rice leaves infected with the

rice blast fungus *Magnaporthe oryzae* Ina86–137 is shown. Washed conidia of the blast fungus were suspended in 1 mM MES-NaOH (pH 5.7) and then were inoculated on leaves of NT and *W53-OX* transgenic rice plants. The lesions were counted according to the classifications shown in Fig. 4A. Bars represent the ratio of lesions of each class to the total number of counted lesions in 3 or 4 individual leaf blades. NT, non-transformant control rice plants; *W53-OX*, native *OsWRKY53*-overexpressing rice plants.

Figure S4 Lesion numbers on leaves of *W53-OX* and *W53SD-OX* transgenic rice plants infected with the rice blast. Washed conidia of the blast fungus were suspended in 1 mM MES-NaOH (pH 5.7) and then were inoculated on leaves of *W53-OX* and *W53SD-OX* transgenic rice plants, and average numbers of lesions in the total infected area on 3 or 4 individual leaf blades of 2 independent *W53-OX* and 2 independent *W53SD-OX* transgenic rice plants infected with the rice blast are shown; *bars* indicate the standard error of the mean. Three independent experiments were performed, and a representative result is shown. There is no significant difference in the average numbers of lesions among samples by One-way ANOVA with Tukey host hoc test. *W53-OX*, native *OsWRKY53*-overexpressing rice plants; *W53SD-OX*, phosphorylation-mimic mutant of *OsWRKY53*-overexpressing rice plants.

Figure S5 Expression profiles of the genes whose expressions were upregulated in the phosphorylation-mimicking mutant of *OsWRKY53*-overexpressing rice

plants. NT, OX, and SD represent non-transformed control, native *OsWRKY53*-overexpressing, and phosphorylation-mimic mutant of *OsWRKY53* (*OsWRKY53SD*)-overexpressing rice plants, respectively. Each column of NT represents the mean of 4 biological replicates, and each column of OX and SD represents the mean of 8 biological replicates. Colors represent induction (red) and repression (blue), as indicated by the color bar. The values of heat maps are relative to those in uninfected NT samples. The columns are sorted by hierarchical clustering using the Pearson correlation and average linkage methods. Mock, uninfected control; 137, *M. oryzae* Ina86–137 infection.

Acknowledgments

We thank Dr. M. Takagi and Dr. M. Shikata for the 430T1.2, GAL4-TATA-LUC-NOS, and pRL plasmids; Dr. H. Takatsuji for the pUCAP/Ubi-NT plasmid; Dr. K. Shimamoto for the p2KG plasmid; and Dr. Y. Nagamura and Ms. R. Motoyama of the Rice Genome Resource Center for technical support with the microarray analysis. We also thank Prof. N. Shibuya and Prof. B. Scott for critical reading of the manuscript.

Author Contributions

Conceived and designed the experiments: TC HN HY KO. Performed the experiments: TC KM SO YM TS YN EM KO. Analyzed the data: TC KM SO TS. Contributed reagents/materials/analysis tools: MK AT. Wrote the paper: TC KM KO.

References

1. Ausubel FM (2005) Are innate immune signaling pathways in plants and animals conserved? Nat Immunol 6: 973–979.
2. Chisholm ST, Coaker G, Day B, Staskawicz BJ (2006) Host-microbe interactions: shaping the evolution of the plant immune response. Cell 124: 803–814.
3. Jones JD, Dangl JL (2006) The plant immune system. Nature 444: 323–329.
4. Zipfel C, Robatzek S (2010) Pathogen-associated molecular pattern-triggered immunity: veni, vidi? Plant Physiol 154: 551–554.
5. Rushton PJ, Somssich IE, Ringler P, Shen QJ (2010) WRKY transcription factors. Trends Plant Sci 15: 247–258.
6. Ichimura K, Shinozaki K, Tena G, Sheen J, Henry Y, et al. (2002) Mitogen-activated protein kinase cascades in plants: a new nomenclature. Trends Plant Sci 7: 301–308.
7. Seo S, Okamoto M, Seto H, Ishizuka K, Sano H, et al. (1995) Tobacco MAP kinase: a possible mediator in wound signal transduction pathways. Science 270: 1988–1992.
8. Zhang S, Klessig DF (1997) Salicylic acid activates a 48-kD MAP kinase in tobacco. Plant Cell 9: 809–824.
9. Zhang S, Klessig DF (1998) The tobacco wounding-activated mitogen-activated protein kinase is encoded by SIPK. Proc Natl Acad Sci U S A 95: 7225–7230.
10. Romeis T, Piedras P, Zhang S, Klessig DF, Hirt H, et al. (1999) Rapid Avr9- and Cf-9 -dependent activation of MAP kinases in tobacco cell cultures and leaves: convergence of resistance gene, elicitor, wound, and salicylate responses. Plant Cell 11: 273–287.
11. Jin H, Liu Y, Yang KY, Kim CY, Baker B, et al. (2003) Function of a mitogen-activated protein kinase pathway in N gene-mediated resistance in tobacco. Plant J 33: 719–731.
12. Tanaka S, Ishihama N, Yoshioka H, Huser A, O'Connell R, et al. (2009) The Colletotrichum orbiculare SSD1 mutant enhances Nicotiana benthamiana basal resistance by activating a mitogen-activated protein kinase pathway. Plant Cell 21: 2517–2526.
13. Asai T, Tena G, Plotnikova J, Willmann MR, Chiu WL, et al. (2002) MAP kinase signalling cascade in Arabidopsis innate immunity. Nature 415: 977–983.
14. Ren D, Yang H, Zhang S (2002) Cell death mediated by MAPK is associated with hydrogen peroxide production in Arabidopsis. J Biol Chem 277: 559–565.
15. Kroj T, Rudd JJ, Nurnberger T, Gabler Y, Lee J, et al. (2003) Mitogen-activated protein kinases play an essential role in oxidative burst-independent expression of pathogenesis-related genes in parsley. J Biol Chem 278: 2256–2264.
16. Kim CY, Zhang S (2004) Activation of a mitogen-activated protein kinase cascade induces WRKY family of transcription factors and defense genes in tobacco. Plant J 38: 142–151.
17. Liu Y, Ren D, Pike S, Pallardy S, Gassmann W, et al. (2007) Chloroplast-generated reactive oxygen species are involved in hypersensitive response-like cell death mediated by a mitogen-activated protein kinase cascade. Plant J 51: 941–954.
18. Ren D, Liu Y, Yang KY, Han L, Mao G, et al. (2008) A fungal-responsive MAPK cascade regulates phytoalexin biosynthesis in Arabidopsis. Proc Natl Acad Sci USA 105: 5638–5643.
19. Kishi-Kaboshi M, Okada K, Kurimoto L, Murakami S, Umezawa T, et al. (2010) A rice fungal MAMP-responsive MAPK cascade regulates metabolic flow to antimicrobial metabolite synthesis. Plant J 63: 599–612.
20. Eulgem T, Somssich IE (2007) Networks of WRKY transcription factors in defense signaling. Curr Opin Plant Biol 10: 366–371.

21. Ross CA, Liu Y, Shen QJ (2007) The WRKY gene family in rice (*Oryza sativa*). J Integr Plant Biol 49: 827–842.
22. Pandey SP, Somssich IE (2009) The role of WRKY transcription factors in plant immunity. Plant Physiol 150: 1648–1655.
23. Rushton PJ, Torres JT, Parniske M, Wernert P, Hahlbrock K, et al. (1996) Interaction of elicitor-induced DNA-binding proteins with elicitor response elements in the promoters of parsley PR1 genes. EMBO J 15: 5690–5670.
24. Eulgem T, Rushton PJ, Schmelzer E, Hahlbrock K, Somssich IE (1999) Early nuclear events in plant defence signalling: rapid gene activation by WRKY transcription factors. EMBO J 18: 4689–4699.
25. Yang PZ, Chen CH, Wang ZP, Fan BF, Chen ZX (1999) A pathogen- and salicylic acid-induced WRKY DNA-binding activity recognizes the elicitor response element of the tobacco class I chitinase gene promoter. Plant J 18: 141–149.
26. Eulgem T, Rushton PJ, Robatzek S, Somssich IE (2000) The WRKY superfamily of plant transcription factors. Trends Plant Sci 5: 199–206.
27. Cohen P (1997) The search for physiological substrates of MAP and SAP kinases in mammalian cells. Trends Cell Biol 7: 353–361.
28. Sharrocks AD, Yang SH, Galanis A (2000) Docking domains and substrate-specificity determination for MAP kinases. Trends Biochem Sci 25: 448–453.
29. Ishihama N, Yamada R, Yoshioka M, Katou S, Yoshioka H (2011) Phosphorylation of the *Nicotiana benthamiana* WRKY8 transcription factor by MAPK functions in the defense response. Plant Cell 23: 1153–1170.
30. Ishihama N, Yoshioka H (2012) Post-translational regulation of WRKY transcription facrors in plant immunity. Curr Opin Plant Biol 15: 431–437.
31. Andreasson E, Jenkins T, Brodersen P, Thorgrimsen S, Petersen NH, et al. (2005) The MAP kinase substrate MKS1 is a regulator of plant defense responses. EMBO J 24: 2579–2589.
32. Qiu JL, Fiil BK, Petersen K, Nielsen HB, Botanga CJ, et al. (2008) Arabidopsis MAP kinase 4 regulates gene expression through transcription factor release in the nucleus. EMBO J 27: 2214–2221.
33. Mao GH, Meng XZ, Liu YD, Zheng ZY, Chen ZX, et al. (2011) Phosphorylation of a WRKY Transcription Factor by Two Pathogen-Responsive MAPKs Drives Phytoalexin Biosynthesis in Arabidopsis. Plant Cell 23: 1639–1653.
34. Menke FL, Kang HG, Chen Z, Park JM, Kumar D, et al. (2005) Tobacco transcription factor WRKY1 is phosphorylated by the MAP kinase SIPK and mediates HR-like cell death in tobacco. Mol Plant Microbe Interact 18: 1027–1034.
35. Koo SC, Moon BC, Kim JK, Kim CY, Sung SJ, et al. (2009) OsBWMK1 mediates SA-dependent defense responses by activating the transcription factor OsWRKY33. Biochem Biophys Res Commun 387: 365–370.
36. Chujo T, Takai R, Akimoto-Tomiyama C, Ando S, Minami E, et al. (2007) Involvement of the elicitor-induced gene OsWRKY53 in the expression of defense-related genes in rice. Biochim Biophys Acta 1769: 497–505.
37. Chujo T, Sugioka N, Masuda Y, Shibuya N, Takemura T, et al. (2009) Promoter analysis of the elicitor-induced WRKY gene OsWRKY53, which is involved in defense responses in rice. Biosci Biotechnol Biochem 73: 1901–1904.
38. Ando S, Sato Y, Shigemori H, Shimizu T, Okada K, et al. (2011) Identification and characterization of 2'-deoxyuridine from the supernatant of conidial suspensions of rice blast fungus as an infection-promoting factor in rice plants. Mol Plant Microbe Interact 24: 519–532.

39. Toki S, Hara N, Ono K, Onodera H, Tagiri A, et al. (2006) Early infection of scutellum tissue with Agrobacterium allows high-speed transformation of rice. Plant J 47: 969–976.

40. Hiratsu K, Mitsuda N, Matsui K, Ohme-Takagi M (2004) Identification of the minimal repression domain of SUPERMAN shows that the DLELRL hexapeptide is both necessary and sufficient for repression of transcription in Arabidopsis. Biochem Biophys Res Commun 321: 172–178.

41. Shimono M, Sugano S, Nakayama A, Jiang CJ, Ono K, et al. (2007) Rice WRKY45 plays a crucial role in benzothiadiazole-inducible blast resistance. Plant Cell 19: 2064–2076.

42. Kitagawa K, Kurinami S, Oki K, Abe Y, Ando T, et al. (2010) A Novel Kinesin 13 Protein Regulating Rice Seed Length. Plant Cell Physiol 51: 1315–1329.

43. Ando S, Tanabe S, Akimoto-Tomiyama C, Nishizawa Y, Minami E (2009) The Supernatant of a Conidial Suspension of Magnaporthe oryzae Contains a Factor that Promotes the Infection of Rice Plants. J Phytopathol 157: 420–426.

44. Yokotani N, Sato Y, Tanabe S, Chujo T, Shimizu T, et al. (2013) OsWRKY76 is a rice transcriptional repressor playing opposite roles in blast disease resistance and cold stress tolerance. J Exp Bot 64: 5085–5097.

45. Shimizu T, Jikumaru Y, Okada A, Okada K, Koga J, et al. (2008) Effects of a bile acid elicitor, cholic acid, on the biosynthesis of diterpenoid phytoalexins in suspension-cultured rice cells. Phytochemistry 69: 973–981.

46. Shimura K, Okada A, Okada K, Jikumaru Y, Ko KW, et al. (2007) Identification of a biosynthetic gene cluster in rice for momilactones. J Biol Chem 282: 34013–34018.

47. Wang Q, Hillwig ML, Peters RJ (2011) CYP99A3: functional identification of a diterpene oxidase from the momilactone biosynthetic gene cluster in rice. Plant J 65: 87–95.

48. Lai ZB, Li Y, Wang F, Cheng Y, Fan BF, et al. (2011) Arabidopsis Sigma Factor Binding Proteins Are Activators of the WRKY33 Transcription Factor in Plant Defense. Plant Cell 23: 3824–3841.

49. Cheng Y, Zhou Y, Yang Y, Chi YJ, Zhou J, et al. (2012) Structural and Functional Analysis of VQ Motif-Containing Proteins in Arabidopsis as Interacting Proteins of WRKY Transcription Factors. Plant Physiol 159: 810–825.

50. Chujo T, Miyamoto K, Shimogawa T, Shimizu T, Otake Y, et al. (2013) OsWRKY28, a PAMP-responsive transrepressor, negatively regulates innate immune responses in rice against rice blast fungus. Plant Mol Biol 82: 23–37.

51. Mayr B, Montminy M (2001) Transcriptional regulation by the phosphorylation-dependent factor CREB. Nat Rev Mol Cell Bio 2: 599–609.

52. Yang SH, Sharrocks AD, Whitmarsh AJ (2003) Transcriptional regulation by the MAP kinase signaling cascades. Gene 320: 3–21.

53. Toyomasu T, Usui M, Sugawara C, Otomo K, Hirose Y, et al. (2014) Reverse-genetic approach to verify physiological roles of rice phytoalexins: characterization of a knockdown mutant of OsCPS4 phytoalexin biosynthetic gene in rice. Physiol Plant 150: 55–62.

54. Okada A, Okada K, Miyamoto K, Koga J, Shibuya N, et al. (2009) OsTGAP1, a bZIP Transcription Factor, Coordinately Regulates the Inductive Production of Diterpenoid Phytoalexins in Rice. J Biol Chem 284: 26510–26518.

55. Kishi-Kaboshi M, Takahashi A, Hirochika H (2010) MAMP-responsive MAPK cascades regulate phytoalexin biosynthesis. Plant Signal Behav 5: 1653–1656.

20

Undesired Small RNAs Originate from an Artificial microRNA Precursor in Transgenic Petunia (*Petunia hybrida*)

Yulong Guo[1], Yao Han[1], Jing Ma[1], Huiping Wang[1], Xianchun Sang[2], Mingyang Li[1]*

1 Chongqing Engineering Research Center for Floriculture, Key Laboratory of Horticulture Science for Southern Mountainous Regions, Ministry of Education, College of Horticulture and Landscape Architecture, Southwest University, Chongqing, China, **2** College of Agronomy and Biotechnology, Southwest University, Chongqing, China

Abstract

Although artificial microRNA (amiRNA) technology has been used frequently in gene silencing in plants, little research has been devoted to investigating the accuracy of amiRNA precursor processing. In this work, amiRNAchs1 (amiRchs1), based on the *Arabidopsis* miR319a precursor, was expressed in order to suppress the expression of *CHS* genes in petunia. The transgenic plants showed the *CHS* gene-silencing phenotype. A modified 5′ RACE technique was used to map small-RNA-directed cleavage sites and to detect processing intermediates of the amiRchs1 precursor. The results showed that the target *CHS* mRNAs were cut at the expected sites and that the amiRchs1 precursor was processed from loop to base. The accumulation of small RNAs in amiRchs1 transgenic petunia petals was analyzed using the deep-sequencing technique. The results showed that, alongside the accumulation of the desired artificial microRNAs, additional small RNAs that originated from other regions of the amiRNA precursor were also accumulated at high frequency. Some of these had previously been found to be accumulated at low frequency in the products of ath-miR319a precursor processing and some of them were accompanied by 3′-tailing variant. Potential targets of the undesired small RNAs were discovered in petunia and other Solanaceae plants. The findings draw attention to the potential occurrence of undesired target silencing induced by such additional small RNAs when amiRNA technology is used. No appreciable production of secondary small RNAs occurred, despite the fact that amiRchs1 was designed to have perfect complementarity to its *CHS-J* target. This confirmed that perfect pairing between an amiRNA and its targets is not the trigger for secondary small RNA production. In conjunction with the observation that amiRNAs with perfect complementarity to their target genes show high efficiency and specificity in gene silencing, this finding has an important bearing on future applications of amiRNAs in gene silencing in plants.

Editor: Christophe Antoniewski, CNRS UMR7622 & University Paris 6 Pierre-et-Marie-Curie, France

Funding: This work was supported by National Natural Science Foundation of China (31272199, http://www.nsfc.gov.cn/) and Fundamental Research Funds for the Central Universities (2362014xk10). The funders had no role in study design, data collection and analysis, decision to publish, or preparation of the manuscript.

Competing Interests: The authors have declared that no competing interests exist.

* E-mail: limy@swu.edu.cn

Introduction

MicroRNAs (miRNAs) are a class of small RNAs of around 21 nt (nucleotides) in length, which are generated from imperfect fold-back regions of long endogenous primary transcripts (pri-miRNAs). In plants, miRNAs repress gene expression at the transcriptional, post-transcriptional and translational levels [1]. They play pivotal roles in plant development [2,3], and are also involved in a range of other biological functions including hormonal regulation [4,5], nutrient homeostasis [6,7] and responses to various biotic and abiotic stresses [8–10].

Most plant miRNAs are transcribed by RNA polymerase II [11]. After transcription, the newly formed pri-miRNA transcripts must be capped and polyadenylated to promote stabilization, and sometimes they must also be spliced to promote the formation of stem-loop structural features [12]. The pri-miRNAs are then processed by DICER-LIKE (DCL) RNAaseIII endonucleases into short miRNA/miRNA* duplexes with 2nt 3′-overhangs. In *Arabidopsis*, pri-miRNA processing is mainly orchestrated by DCL1, with the assistance of the dsRNA-binding protein, HYPONASTIC LEAVES1 (HYL1), and the C2H2-zinc finger

protein, SERRATE, to improve the efficiency and precision of cleavage. DCL1 and HYL1, together with other associated factors, are co-localized in sub-nuclear regions termed Dicing-bodies (D-bodies), where miRNAs are processed [13,14]. The biogenesis of most animal and plant miRNAs usually begins with a cut at the base of their stem-loop structures, thereby yielding precursors (pre-miRNAs), which are then further cut to produce 21–22 nt miRNA duplexes [15]. RNA secondary structure affects both the accuracy and the productivity of plant miRNA processing. For the processing of *Arabidopsis* pri-miR172a, the 14- to 15-base-pair stem region below the miRNA/miRNA* duplex is essential, although small unpaired bulges that do not damage its linear structure are tolerated; and a loop is required, although mutations in the terminal loop are mostly neutral [16,17]. The structural features of most conserved plant pri-miRNAs are similar to those of pri-miR172a and thus it is assumed that they have a similar processing mechanism [16,18]. However, for the 'long fold-back' pri-miRNAs, pri-miR159 and pri-miR319, the processing mechanism is different from that undergone by animal miRNAs and most plant conserved miRNAs; it begins with a cleavage next to the terminal loop, and then DCL1 cuts three more times at 20–

22 nt intervals until the miRNA/miRNA* duplex is released [19]. In contrast to the processing of pri-miR172a, the precursor sequences below the miRNA/miRNA* duplex are dispensable for pri-miR319 processing, but the conserved upper stem is critical [19].

Following their release from D-bodies, the miRNA/miRNA* duplexes are stabilized by the addition of a methyl group at their 3'-end, catalyzed by the methyltransferase protein, HUA ENHANCER1 [20]. Subsequently, the miRNA strands are loaded onto ARGONAUTE-containing, RNA-induced silencing complexes (RISCs), and the miRNA* strands are generally degraded. Plant miRNA-loaded RISCs recognize their target genes by highly complementary pairing between the miRNA and its target mRNA [1]; thus, a family of plant miRNAs can only repress the expression of a few target genes, usually duplicated genes. After a plant RISC has recognized its target mRNA, cleavage of the target at the central region of the predicted hybrid can usually be observed. Both the mechanism of action of cleavage and the sites at which it can occur have been validated in a range of studies using technologies such as 5' RNA ligase-mediated rapid amplification of 5' cDNA ends (5' RLM-RACE) [21] and degradome sequencing [22].

In addition to resulting in mRNA degradation, the miRNA-mediated cleavage of target mRNAs can in some cases trigger the biogenesis of phased, secondary small interfering RNAs (phasiRNAs) [23], which can in turn silence additional genes, leading to a cascade of gene silencing.

A few classes of plant miRNA precursor have been successfully engineered to silence genes of interest, by replacing natural miRNAs with specifically modified or "designed" miRNA molecules, termed artificial miRNAs (amiRNAs) [24–28]. AmiRNAs engineered by modifications of *Arabidopsis MIR319a* precursor, the precursor backbone most commonly used, have been successfully used to induce gene silencing in *Arabidopsis* [29], tobacco [25], eggplant [30], soybean [31], *Medicago* [32,33], and *Physcomitrella patens* [34]. Recently, this technology has also been used to silence genes in the vegetative cells of pollen in *Petunia inflata* [35]. When amiRNA gene silencing technology is used in plants, it is usually assumed that the amiRNA precursor is specifically processed to produce a single mature amiRNA [24]. Data relating to the processing of amiRNA precursors are scant, however, so the accuracy of amiRNA precursor processing requires further investigation.

In the study reported here, deep sequencing and modified 5' RLM-RACE technology have been used to investigate the accuracy of amiRNA precursor processing in petunia. It is demonstrated that, in addition to the production of amiRNAs possessing the intended sequences, the processing of amiRNA precursors can lead to the abundant accumulation of additional small RNAs that may have potentially detrimental effects on unintended gene targets.

Materials and Methods

Plant materials

Inbred V26 *Petunia hybrid* was used as the recipient of amiRchs1 transgene construct. Inbred V26 was a generous gift from Prof. Manzhu Bao (Huazhong Agricultural University, China). Transgenic and wild type plants were grown side by side under a 16/8-h photoperiod in a greenhouse equipped with high-pressure sodium lights, at an intensity of 100–200 $\mu mol \cdot m^{-2} \cdot s^{-1}$.

Production of transgenic plants expressing the CHS-amiRNA construct

At the start of this work, when it was decided to use amiRNA to suppress the expression of petunia genes, petunia had not been

included in Web MicroRNA Designer (WMD) [24]. Thus, artificial microRNAchs (amiRchs) molecules were designed following the rules for amiRNA design introduced by Schwab et al. [24]: i.e. 21 nt long; a uridine at position 1 and an adenine at position 10; positions 2 and 12 to have no mismatch to the target; and the amiRNA to display 5' instability relative to its corresponding miRNA*. The amiRNA candidate sequences designed in this way were then submitted to the RNAcofold WebServer (http://rna.tbi.univie.ac.at/cgi-bin/RNAcofold.cgi), in order to predict the free energy of formation of the amiRNA/target hybrid. One of the candidate amiRNA sequences, amiRchs1, was then selected for the production of the amiRNA silencing construct used for the investigation.

The amiRchs1 precursor was synthesized as described by Schwab et al. [24]. Routine molecular cloning procedures were used for plasmid construction, and the primers used are listed in Table 1. The *Arabidopsis* miR319a precursor (pRS300, a present from Prof. Detlef Weigel, Max Planck Institute for Developmental Biology, Tübingen, Germany) was used as the backbone for amiRchs1 construction and expression. An overlap PCR method was used to substitute the designed amiRchs1 sequence for the natural miRNA sequence and, at the same time, to modify the miRNA* region. The PCR procedure was carried out using a Mastercycler 5331 (Eppendorf, Hamburg, Germany), using pRS300 plasmid DNA as the template, together with the oligonucleotide sequences I, II, III, IV and the general primers A and B. The PCR products were then cloned into pMD19-T (Takara, Dalian, China) and the resulting clones were sequenced, using T3 primer, in order to select clones that were free of errors introduced by PCR.

The amiRchs1 precursor fragment was released by the use of *Pst*I and *Bam*HI and then cloned into an intermediate vector under the control of 2×35S promoters and a 35S poly-A sequence, which were amplified from vector pSAT3 [36]. The expression box containing the amiRchs1 precursor sequence was then released using *Sal*1 and *Bam*H1 and inserted into the multiple cloning site (MCS) of pCAMBIA2301 to produce the pCMF-amiRchs1 vector. The resultant vector was then introduced into *Agrobacterium tumefaciens* strain GV3101 by electroporation.

Plant genetic transformation

Plant genetic transformation and the regeneration of transformants were performed as described by Jorgensen et al. [37]. Briefly, V26 leaf-discs were immersed in *Agrobacterium* suspension for about 5 min, blotted dry, plated on co-cultivation medium, and then incubated in darkness at 22°C for 48 h. The explants were then transferred to selection medium. Four weeks later, callus islets were excised from the mother leaf-segments, and sub-cultured separately. When adventitious shoots appeared, the shoots regenerated from each callus islet were considered to constitute an independent transformation line. The shoots were excised and sub-cultured on rooting medium. All media used were as reported by Jorgensen et al. [37].

Total RNA extraction and real-time RT-PCR analysis

Total RNA was extracted from the petals of opening flower buds of V26 and transgenic plants (Figure 1A). Tissues were frozen with liquid nitrogen and extracted with TRIzol Reagent (Invitrogen, Carlsbad, CA, USA), used according to the manufacturer's directions. The concentration and quality of total RNA were analyzed using a NanoDrop 2000 spectrophotometer (Thermo Scientific, Waltham, MA, USA) and by gel electrophoresis. Total RNA was treated with RNase-free DNase I (Roche, Penzberg, Germany) and then reverse-transcribed to cDNA using PrimeScript

Table 1. PCR primers used in this study.

Name	Sequence(5'→3')	Comments
amiRchs1-I	gatgttggtacatcatgagtcgctctctcttttgtattcc	amiRchs1 precursor synthesis
amiRchs1-II	gagcgactcatgatgtaccaacatcaaagagaatcaatga	amiRchs1 precursor synthesis
amiRchs1-III	gagcaactcatgatgaaccaacttcacaggtcgtgatatg	amiRchs1 precursor synthesis
amiRchs1-IV	gaagttggttcatcatgagttgctctacatatatattcct	amiRchs1 precursor synthesis
Primer A	ctgcaaggcgattaagttgggtaac	General primer for amiRNA precursor synthesis
Primer B	gcggataacaatttcacacaggaaacag	General primer for amiRNA precursor synthesis
qUBQ-F	tggaggatggaaggactttgg	qRT-PCR
qUBQ-R	caggacgacaacaagcaacag	qRT-PCR
qCHSA-F	ggcgcgatcattataggttc	qRT-PCR
qCHSA-R	tttgagatcagcccaggaac	qRT-PCR
qCHSJ-F	aaagtttagtggaggcattcc	qRT-PCR
qCHSJ-R	tccatactcactcaagacatg	qRT-PCR
CHSA5R-1	gtagttcctaaaccttctttggctgag	5' RACE (round 1) to map cleavage sites
CHSA5R-2	tgagcaatccagaatagagagttccaa	5' RACE (round 2) to map cleavage sites
CHSJ5R-1	agagacactatggagcacaacagtt	5' RACE (round 1) to map cleavage sites
CHSJ5R-2	tagagttccagtcagaaatgcccaat	5' RACE (round 2) to map cleavage sites
RACE5-1	cgactggagcacgaggacactga	General GeneRacer 5' Primer (round 1)
RACE5-2	ggacactgacatggactgaaggagta	General GeneRacer 5' Nested Primer (round 2)
amiRchs5R-1	tgagcgaaaccctataagaaccctaa	5' RACE (round1) to map processing intermediates
amiRchs5R-2	acgaaggcagcatatatgtcacttag	5' RACE (round2) to map processing intermediates

RT Master Mix (Takara). Specific primers (qCHSA-F and qCHSJ-R, qCHSJ-F and qCHSJ-R, Table 1) for the amplification of *CHS-A* or *CHS-J* were synthesized as reported by Koseki et al [38]. Primers for the amplification of *UBQ* (qUBQ-F and qUBQ-R, Table 1) were synthesized according to Mallona et al. [39]. Each qRT- PCR reaction was performed in a 10 μL volume containing 0.5 μL of cDNA, 0.5 μL of each primer (10 μmol/L) and 5 μL of 2×SsoFast EvaGreen Supermix (Bio-Rad, Hercules, CA, USA), and was carried out on a CFX96 Real-time PCR Detection System (Bio-Rad) under the following conditions: 95°C for 30 s, followed by 39 cycles (each of 95°C for 5 s, then 60°C for 5 s), followed by melt curve analysis. The data were normalized using ubiquitin (*UBQ*) as endogenous control and analyzed to calculate relative expression values (ΔΔCt mode) as described by Schmittgenl and Livak [40]. Three technical and three biological replicates were performed for each sample and the standard deviation was calculated.

Cleavage site mapping

To map the cleavage sites of the *CHS-A* and *CHS-J* target mRNAs, a modified procedure for 5' RLM-RACE was carried out using the GeneRacer kit (Invitrogen), as described previously [21]. Five micrograms of total RNA from wild-type and amiRchs1 transgenic petals, respectively (without any prior treatment with calf intestine alkaline phosphatase and tobacco acid pyrophosphatase), were ligated to the adapter for 5' RLM-RACE. Amplification was carried out as described in the manufacturer's instruction manual. The initial PCR (round 1) was performed using the 5' RACE outer primer (RACE5-1, Table 1) from the manufacturer and a gene-specific outer primer (CHSA5R-1 or CHSJ5R-1, Table 1). Nested PCR (round 2) was performed using 1/50 of the initial PCR products as the template, together with GeneRacer 5' Nested Primer (RACE5-2, Table 1) and a gene-specific primer

(CHSA5R-2 or CHSJ5R-2, Table 1). The PCR products were cleaned using an AxyPrep PCR Clean-up kit (Axygen, Union City, CA, USA) and cloned using a pMD19-T cloning kit (Takara). DNA sequencing was undertaken by BGI (Genomics Institute of Science and Technology Co., Ltd, Shenzhen, China).

To detect the processing intermediates of the amiRchs1 precursor, the same procedure was used as described above for the mapping of the cleavage sites of the *CHS* target mRNAs. The gene-specific PCR primers were amiRchs5R-1 and amiRchs5R-2 (Table 1). For the detection of PCR-amplified fragments by gel electrophoresis, the PCR products were resolved on 10% (w/v) polyacrylamide gels and detected by silver staining.

Deep-sequence analysis of small RNAs

One microgram of high-quality total RNA from petals of the transgenic line 603-8 was ligated to 5'- and 3'-adaptors, reverse transcribed, and then amplified by PCR (12 cycles), using a Truseq Small RNA sample preparation kit (Illumina, Santiago, CA, USA) according to the manufacturer's protocol. The library of small RNAs was purified by electrophoresis on a 6% Novex TBE PAGE gel (Invitrogen). Following quantification using TBS380 (Turner BioSystems, Sunnyvale, CA, USA), the nucleotide sequence of the amplified cDNA was analyzed using Illumina Hiseq 2000 (2×100 bp read length). The nucleotide sequence data have been deposited in the NCBI Sequence Read Archive under the accession number SRP036869. Only left side reads were used for analysis in this work.

The raw reads were trimmed by removing low-quality reads (Q value<20), adapter sequences, reads with ambiguous bases 'N', and fragments of less than 18 nt in length, using a Fastx-Toolkit (http://hannonlab.cshl.edu/fastx_toolkit/). The filtered small RNAs of 18–32 nt in length were then mapped onto the nucleotide sequences of the amiRchs1 precursor (allowing only

Figure 1. Phenotype of amiRchs1 transgenic flowers. (A) Opening flower bud of amiRchs1 transgenic plants (total RNA was extracted at this stage). (B) V26 (wild-type) flower. (C) amiRchs1 transgenic flower. (D–E) qRT-PCR detection of mRNA levels of the *CHS-A* (D) and *CHS-J* (E) gene in V26 and amiRchs1 transgenic petals. Data were normalized against petunia ubiquitin gene and were means of three biological pools (each with three technical replicates); the error bars indicate SD.

perfect matches), and onto the *CHS-A* (GenBanK database accession X14591) and *CHS-J* (X14597) gene regions, respectively (allowing 4 mismatches).

Two libraries of Illumina SBS sequencing data [41] for small RNAs from i) the petals of line V26 (GSM433598) and from ii) the petals of transgenic V26 constitutively expressing the *CHS-A* coding sequence and displaying white (silenced) flowers (GSM346607) were downloaded from the Gene Expression Omnibus (GEO) database. They were analyzed in parallel. When necessary, small RNA data [42] obtained from the 454 Sequencing (http://www.petunia_smrna.leeds.ac.uk/) of petunia flower buds were searched.

Small RNA target prediction

TargetSearch integrated into WMD3 Web Server (http://wmd3.weigelworld.org/) was used to search for potential targets of the undesired small RNAs identified in this study. TargetSearch is based on GenomeMapper (http://www.1001genomes.org), a sequence alignment tool. The search was carried out using default parameters (i.e., Mismatches: 5, Apply microRNA filter: yes, Perfect-match-dG cutoff: 70%, Hybridization temperature: 23°C, Folding program: RNAcofold, Direction: reverse, Allow gaps: no). Small RNAs of more than 19 nt in length were used for the search. In searching for targets in the *Petunia* genus, whole genomes were chosen, which included EST or transcript releases of *P. hybrida*, *P. axillaris* and *P. inflata* (*P. axillaris* and *P. inflata* are putative parents of *P. hybrida*). In searching for targets in tobacco, tomato and potato, EST releases of *Nicotiana tabacum* EST NtGI-7.0, *Solanum lycopersicum* EST LGI-13.0 and *Solanum tuberosum* EST StGI-13.0 were chosen, respectively.

Results

CHS gene silencing by an amiRNA in *Petunia hybrida*

In order to suppress the expression of genes in petunia using an amiRNA, the amiRchs1 molecule was designed and its precursor was synthesized as described by Schwab et al. [24]. The engineered amiRchs1 was found to pair with *CHS-A* (X14591) from nt3037 to nt3057, with a mismatch and a G-U wobble at the 3′ end of amiRchs1, and a ΔG for heterodimer binding of −31.38 kcal/mol (Figure 2A). It paired perfectly with *CHS-J* (X14597) from nt2661 to nt2681, and the ΔG for heterodimer binding was −37.02 kcal/mol (Figure 2B). The secondary

structure of the amiRchs1 precursor predicted by Mfold [43] was very similar to that found for the *Arabidopsis* miR319a precursor (Figure 3A and B). A difference between them was that an extra-base pair occurred below the miRNA/miRNA* duplex in the amiRchs1 precursor (Figure 3A and B). This extra-base pair was introduced by replacing a 20 bp sequence in *MIR319a* with a 21bp designed sequence when amiRchs1 was engineered according to the procedure of Schwab et al [24,44].

The amiRchs1 construct was introduced into V26 (Figure 1B and C) and two other genotypes (Figure S1). Transgenic lines showing altered flower color were produced in all three cases. Amongst the ten transgenic lines regenerated from V26, six lines produced flowers with conspicuous color alterations. Their petals contained randomly located white or pale sectors, or were nearly white. Because these transgenic plants displaying phenotypic changes were self-sterile, they were pollinated with V26 pollen. A line (603–8) that produced offspring in a ratio of approximately one wild-type plant to one mutant plant was used in this study (Figure 1C).

Two CHS genes (*CHS-A* and *CHS-J*) are active in petunia floral tissues, and the level of *CHS-A* gene expression is higher than that of *CHS-J* [45,46]. The mRNA levels for both genes were analyzed in the petals of opening flower buds. For both *CHS-A* and *CHS-J*, mRNA accumulation was clearly reduced in the amiRchs1 transgenic line (Figure 1D and E), confirming the occurrence of *CHS-A* and *CHS-J* RNA degradation.

AmiRchs1-directed cleavage of CHS-A and CHS-J mRNAs

Using *CHS-A* or *CHS-J* gene-specific primers respectively, 5′ RLM-RACE produced only one clear band (Figure 2C). In the case of *CHS-A*, sequencing results showed that of 23 clones, 21 mapped onto *CHS-A* (Figure 2A), and the other two were non-specific amplification products. Moreover, most of the *CHS-A* transcript cleavages occurred between the sites complementary to the 10th and 11th positions of the engineered (also identified by deep sequencing, see below) amiRchs1, although one cleavage occurred upstream of the most frequent cleavage site, and four occurred downstream; these may have represented aberrant mRNAs. In the case of *CHS-J*, of the 29 clones sequenced, 25 mapped to *CHS-J*. Cleavage occurred only between the sites complementary to the 10th and 11th positions of amiRchs1 (Figure 2B). These results indicated that the artificially synthesized

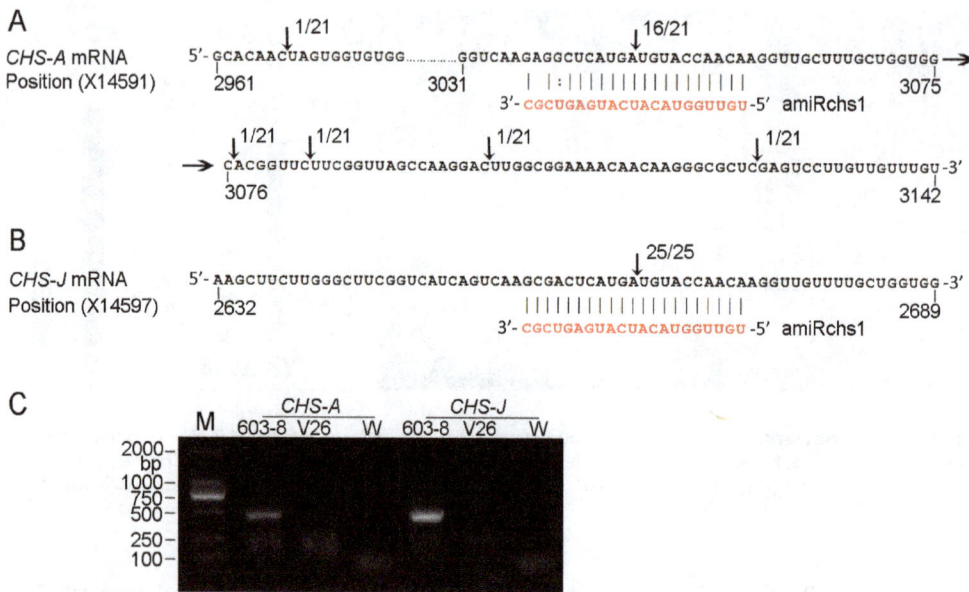

Figure 2. Mapping of target cleavage products by 5′ RLM-RACE. The cleavage sites and the number of sequenced clones corresponding to each site are indicated by arrows. For most of the sequenced clones, the 5′ end was at the expected position, opposite to nucleotides 10–11 of amiRchs1. (A) Cleavage of *CHS-A* mRNA. (B) Cleavage of *CHS-J* mRNA. (C) Agarose gel showing products after 5′ RLM-RACE PCR amplification. M, Marker; W, water.

amiRchs1 were effective in guiding the RISCs to their targets and that they resulted in cleavage at the predicted sites.

Detecting processing intermediates of the amiRchs1 precursor by RACE

RACE PCR to detect processing intermediates of the amiRchs1 precursor generated more than four products, of different sizes (Figure 3C). Two of these products (bands 3 and 4) were clearly accumulated to a higher level than the others. Sequencing of the RACE products revealed that they corresponded to distinct cleavage sites along the predicted fold-back region of the amiRchs1 precursor (Figure 3C), and most of these sites corresponded to the ends of small RNAs identified by deep sequencing. The two most frequent sites corresponded to the two high-intensity bands (bands 3 and 4). The distance between the cleavage sites was mainly 20-23 nt, which is consistent with the rule that DCL1 cleaves pri-miRNA at 21 nt intervals [15]. These results showed that the processing of the amiRchs1 precursor in petunia is consistent with the processing mechanism of the ath-miR319a precursor in *Arabidopsis*; thus, processing begins with a cleavage below the terminal loop, which is followed by three more DCL1 cleavages towards the base of the stem-loop structure of the precursor [19]. However, the accuracy of cleavage appears to be reduced and extra intermediates are produced.

Mapping small RNAs to the amiRchs1 precursor

Small RNAs in amiRchs1-transgenic petunia petals were analyzed using the deep-sequencing technique. Sequencing of the small RNA library produced a total of 11,076,276 Illumina reads from 603–8 petals. Following quality control and the removal of non-miRNA sequences representing other RNA species such as rRNA, tRNA and snRNA, 9,387,095 high-quality reads were obtained, ranging from 18 nt to 32 nt in length. The sequences ranged mainly from 19 nt to 25 nt in size, with two peaks at 21 nt and 24 nt (Figure S2). After collapsing identical sequences, 2,661,170 unique clones were extracted. Unless

otherwise indicated, only those clones with more than five sequencing reads were used in the analysis below, in order to reduce the potential for the introduction of sequencing errors.

Amongst the total of 14,668 reads that matched the predicted amiRchs1 precursor in transgenic petals, two prominent peaks were observed, of 21 nt (especially) and 22 nt, respectively (Figure 4). This was consistent with the fact that plant microRNAs are processed by DCL1 or DCL4 [47]. A unique small RNA sequence was isolated, based on 3,885 reads of the predicted amiRchs1 sequence. Of all the sequences that matched the amiRchs1 precursor, this was the most frequently occurring, indicating that the synthesized microRNA precursor functioned effectively in transgenic petunia.

The position and abundance of the small RNAs that mapped to the amiRchs1 precursor are shown in Figure 3B and Figure 5. For comparison, small RNA deep-sequencing data for *Arabidopsis MIR319a* were retrieved from the miRBase database (v20, http://miRbase.org/index.shtml), and incorporated into the ath-miR319a precursor. According to our results and based also on previous studies [19,48,49], we defined seven blocks of sequence along the miR319a precursor (Figure 3A and B). The most abundant sequence reads within each block were taken as the representative sequence for the block as a whole. All unique sequences having more than a 15 nt overlap with the representative sequence, and extending for no more than 6 nt beyond the representative sequence at either end, were considered to belong to the same block. Thus, Block1 (B1) corresponded to miR319a.1 (guide strand) region, and B1* to miR319a.1* (passenger strand); B2 corresponded to miR319a.2, and B2* to miR319 a.2*; B3 corresponded to the region between B1 and B2, and B3* to the region between B1* to B2*; and finally B3+ corresponded to a region longer than B3, in which the 3′ ends of B3 sequences stretched into the middle of B1.

More than 86% of the small RNA sequences originating from the ath-miR319a precursor matched the B1 region, and those that matched the B1*, B2, B2* and B3* regions occurred at low

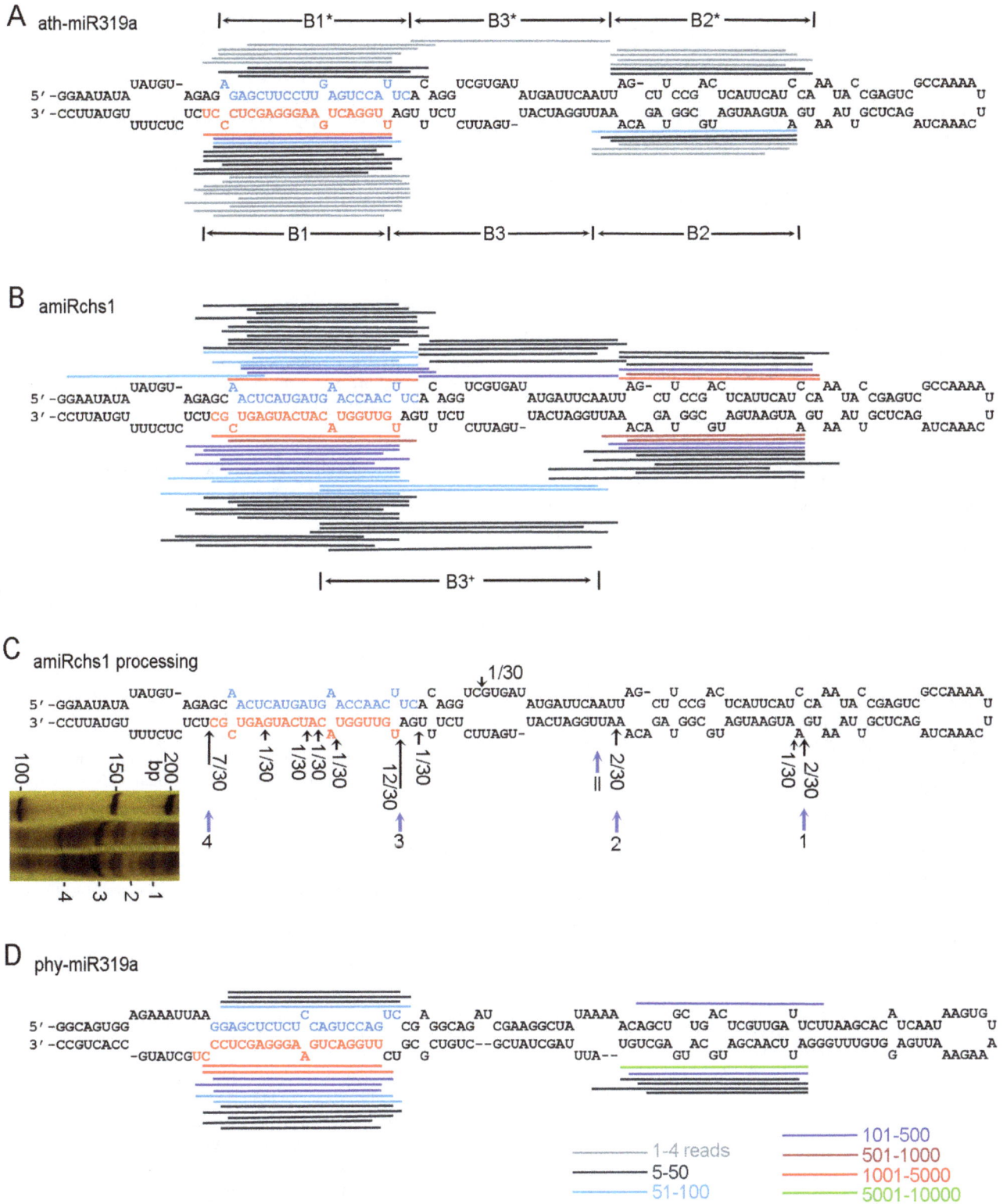

Figure 3. Processing of amiRchs1 and miR319 precursors. (A) Small RNA sequences from the miRBase database (v20 http://miRbase.org/index. shtml) were incorporated into the predicted stem-loop structure of the *Arabidopsis* miR319a precursor. Small RNAs cloned fewer than 5 times are also indicated. (B) Small RNA sequences from amiRchs1 transgenic petals incorporated into the scheme for the amiRchs1 precursor. Only small RNAs cloned more than 5 times are indicated. (C) Processing intermediates detected using 5′ RLM-RACE PCR amplification. The positions of cleavage sites, as revealed by 5′ RACE, and the number of sequenced clones corresponding to each site, are indicated by black arrows. The four sites marking the origins of B1 and B2 small RNAs, and corresponding to the four marked fragments in the polyacrylamide gel, are indicated by blue arrows (Sites 1, 2, 3 and 4). Site II corresponds to the cleavage site that marks the origin of the most frequent B1 small RNAs, but the processing intermediates had not been sampled by random sequencing of RACE PCR products. Left inset: Polyacrylamide gel showing fragments after 5′ RACE. (D) Small RNA sequences from amiRchs1 transgenic petals incorporated into the scheme for the petunia miR319a precursor. Only small RNAs cloned more than 5

times are indicated. B1, B1*, B2, and B2* correspond to the regions producing the sequences previously designated as miR319a.1, miR319a.1*, miR319a.2 and miR319a.2* [48,49], respectively. B3 corresponds to the region between B1 and B2, B3* corresponds to the region between B1* to B2*, and B3+ corresponds to a region longer than B3, in which the 3' ends of B3 sequences stretched into the middle of B1.

frequency (Figure 3A and Figure 5). By contrast, microRNA sequences arising from the B1 region of the amiRchs1 precursor accounted for only 43% of the total, with a concomitant increase in the relative proportions of small RNA sequences from other regions (Figure 3B and Figure 5). Although no sequences belonging to the B3 region were retrieved when deep-sequencing data for ath-*MIR319a* from miRBase were used, Bologna et al. [19] had previously identified B3-region small RNAs in another publicly available database. Therefore, a search of the deep-sequencing data was undertaken for small RNA clones that belonged to the B3 region and were represented by fewer than five reads. Seven unique sequences, with a total of 11 reads, were identified that belonged to the B3 region, indicating that B3-region small RNAs accumulated at a low level. A few B3+ small RNA sequences that overlapped with the B3-region sequences were also identified; these had 3' ends extending into the central regions of B1 small RNAs, so that they were longer than the B3-region small RNAs (Figure 3B and Figure 5). These deep-sequencing results indicated that processing of the amiRchs1 precursor in petunia produced small RNAs that arose from the same regions as the small RNAs produced from the processing of the ath-miR319a precursor in *Arabidopsis*, but that the relative proportions of the small RNAs derived from the different regions were altered, and that abundant unwanted small RNAs were produced.

To investigate whether this production of additional small RNAs from the amiRchs1 precursor could in fact have arisen through the processing of petunia miR319 homologue precursors, deep-sequencing data for: i) wild-type petunia (GSM433598 and http://www.petunia_smrna.leeds.ac.uk/), and ii) line V26 with co-suppressed *CHS* genes (GSM346607), were searched using the amiRchs1 precursor sequence. No sequences were identified that showed extensive complementarity to the additional small RNAs. Highly accumulated sequences (≥5 reads) that were complementary to amiRchs1 (which is complementary to petunia *CHS*) were, however, found in GSM346607. Secondly, the GenBank databases were searched using the ath-miR319a (identical to ath-miR319b) sequence to determine whether petunia miR319 precursors could be found. From the petunia EST database, FN011712 was found, containing a fragment showing perfect complementarity to miR319a. Analysis of the secondary structure of FN011712 mRNA using Mfold [43] showed that it encoded a stem-loop region similar to that of the miR319a precursor (Figure 3D), indicating that FN011712 represented a homologue of *MIR319*. Since it encoded a mature microRNA identical to ath-

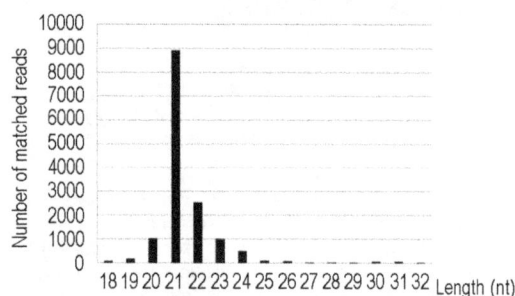

Figure 5. Relative proportions of small RNAs arising from different regions along the miR319 precursors. The regions are the same as indicated in Figure 3.

miR319a and ath-miR319b, but the secondary structure of its precursor was more similar to that of the ath-*MIR319a* precursor than to that of the ath-*MIR319b* precursor (Figure S3), it was therefore designated *phy-MIR319a*. We next searched our deep-sequencing data for small RNA sequences that originated from the phy-miR319a precursor and incorporated them into its stem-loop structure (Figure 3D). This revealed that these small RNAs were concentrated within the B1, B1*, B2 and B2* regions, and that the relative proportions of small RNAs that originated from these regions were different from the corresponding relative proportions observed in the case of the ath-miR319a precursor (Figure 5). Because the sequence of the phy-miR319a precursor differed from that of the amiRchs1 precursor, it was unlikely that the processing of petunia miR319 precursors would have contributed to the observed accumulation of additional small RNAs showing sequence homology to the amiRchs1 precursor. In addition, using all the undesired small RNA sequences identified in this study, a search was undertaken for microRNA hairpin precursor sequences in all the organisms deposited in miRBase (v20). Sequences showing 100% identity were only found from *Arabidopsis thaliana* and *Arabidopsis lyrata*; therefore, the possibility was minimal that these undesired small RNA sequences were coincidentally produced from other petunia miR319 precursors that have not been identified.

Perfect pairing between amiRchs1 and its *CHS-J* target did not induce secondary small RNAs

A previous study had suggested that perfect pairing between a microRNA and its target would prime the biogenesis of secondary small RNAs [50], and it had been suggested that three mismatches to their target genes should be deliberately introduced into the 3' regions of amiRNAs so as to reduce the likelihood that amiRNAs would trigger the production of secondary RNAi [24]. Amplification of an initial amiRNA signal by secondary small RNAs has also been observed in *Physcomitrella patens* [34]. The sequence of amiRchs1 matched that of *CHS-J* (X14597) perfectly between nt2661 and nt2681, and we therefore searched our deep sequencing data for small RNA sequences that showed matches to *CHS-J* and *CHS-A* transcripts. With the exception of the amiRchs1 sequence, no highly accumulated small RNA sequence (≥5 reads) showing extensive complementarity to *CHS-A* or *CHS-J* transcripts was found, indicating that the amiRchs1 guided target

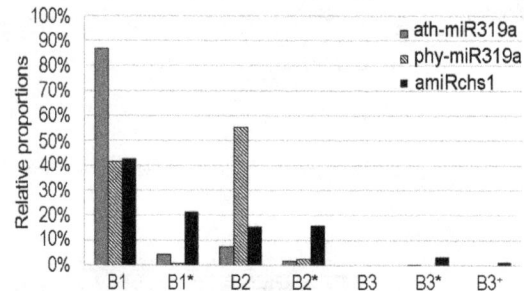

Figure 4. Size distribution of small RNAs mapped to the amiRchs1 precursor.

cleavage did not induce the biogenesis of phasiRNAs. This result was also confirmed by the 5'-RLM-RACE results: only one clear band was detected for *CHS-A* and *CHS-J*, respectively (Figure 2C). Interestingly, from our deep sequencing data and from GSM433598, we identified a few unique sequences that matched the 5' promoter sequence or the intron region of *CHS-A* (Table S1).

Small RNAs displaying tailing modification

From the deep-sequencing data of 603-8 transgenic petals, we also identified a class of small RNAs that could each be divided into two parts: a 5' genome-matched component (5GMC) [51], which matched perfectly to the amiRchs1 precursor, and a 3' "tail" component. The 5GMC sequences were between 17nt and 31 nt long, and their tails were mostly 1 nt, with the longest being 3 nt; the sequences arising from B1, B1*, B2 and B3+ comprised 116, 142, 311 and 12 reads, respectively (Table 2). The most frequently occurring "tail" nucleotide was uridine (45%), followed by adenine (28%), cytosine (16%) and guanine (11%). Because the 5GMC sequences of these small RNAs corresponded to the regions from which previously-identified small RNAs arising and they accumulated at relatively high levels, we concluded that they were tailed variants of the small RNAs arising from the amiRchs1 precursor. We also identified tailed small RNA variants arising from the phy-miR319a precursor; however, these originated only from B1 (501 reads) and B1* (23 reads). Their tails were 1–2 nt long, consisting predominantly of uridine (84%), with smaller proportions of cytosine (14%) and guanine (2%) (Table 3).

Potential targets found by target search

A target search using default parameters found 72 potential target sequences in the *Petunia* genus (Table S2). Since different sequence IDs in the PlantGDB (http://www.plantgdb.org/) and GeneIndex (http://compbio.dfci.harvard.edu/tgi/) databases may represent the same transcript sequence, in reality the total number should be lower than 72. Nevertheless, this analysis indicated the existence of potential targets in *Petunia* transcripts. Alignments between a small number of selected 21 nt small RNAs and some of their potential mRNA targets in *Petunia hybrida* are shown in Figure 6. When pairing with CV299538, amiRchs1-B2*-5 (Figure 6A) met the required criteria for amiRNA design [24]. The parameters of the other three pairs presented in Figure 6B, C and D met the (less stringent) criteria for plant microRNA target selection proposed by Schwab et al. [52]. These criteria include the empirical parameters for target recognition: i.e. no mismatch at claevage site (positions 10 and 11); no more than one mismatch at positions 2-12; no more than 4 mismatches downstream of position 13 and no more than two in a row; low overall free energy of targets paired with their corresponding miRNAs (at least 70% compared with a perfect match) [24,52]. Many other transcripts from the *Petunia* genus shown in Table S2 were also potential targets and met these criteria for target selection.

Since genomic information for petunia is very limited, we searched the genomes of tobacco, tomato and potato to determine whether potential targets would be found if the same collection of undesired small RNAs were to be produced in these plant species. There were 61, 33 and 39 potential targets identified from tobacco (GeneIndex NtGI7.0), tomato (LGI-13.0) and potato (StGI-13.0), respectively (Table S2). In a few cases, transcripts were potential targets for overlapping small RNAs. For example, all of the tobacco ESTs TC123041, TC126640, TC164324 and TC124235 were potential targets for all of the overlapping small RNAs amiRchs1-B2-4 (24nt), B2-5 (23nt), B2-6 (22nt), B2-7 (21nt) and B2-8 (20nt) (Table S2). Alignments between the most abundant

small RNA from each block (B2, B2*, B3* and B1*) and some of their mRNA targets are shown in Figure 6E–H. These results indicated that if additional small RNAs were to be produced in other Solanaceae plants besides petunia, they might similarly have potential detrimental functions.

Discussion

An artificial microRNA was designed for the silencing of petunia *CHS* genes, following the rules set by Schwab et al. [24]. When the synthesized amiRchs1 precursor was constitutively expressed in petunia, the resulting transgenic flowers showed the *CHS* gene-silencing phenotype, the predicted artificial microRNA was accumulated preferentially, the *CHS-A* and *CHS-J* genes were each cleaved exactly at the predicted site, and the accumulation of *CHS-A* and *CH-J* full-length transcripts was reduced. These results indicated that the amiRchs1 functioned as expected and demonstrated that artificial microRNA technology based on *Arabidopsis* microRNA precursors could be used successfully in *Petunia hybrida*.

On the other hand, high levels of unwanted, additional miRNA-like RNAs accumulated. Three factors may have contributed to this unexpected phenomenon. First, the replacement of the natural miR319a by the engineered amiRchs1 may have led to changes in the structural features of the miRNA precursors, with implications for the accuracy of cleavage and for the modification and the stability of the cleavage products. An extra-base pair was introduced below the miRNA/miRNA* duplex (Figure 3A and B) when the amiRchs1 was synthesized using the procedure of Schwab et al [44]. This extra-base pair may influence the processing of amiRNA precursors. However, since miR319 precursors are processed from loop to base, and the precursor sequences below the miRNA/miRNA* duplex are dispensable [19], the role of this extra-base pair would be expected to be limited. Secondly, there may conceivably exist subtle differences between species in the machinery of miRNA processing. It was observed that the B2-region small RNAs arising from the phy-miR319a precursor accumulated at higher frequency than those originating from the ath-miR319a precursor (Figure 5). This difference could, however, have resulted from structural differences between the miRNA precursors. It is therefore not possible at present to make a distinction between the possible effects of structural differences and species differences. Thirdly, overexpression of the amiRchs1 precursor may have overwhelmed the endogenous microRNA processing machinery, thereby compromising the accuracy of processing. At present, our knowledge of the processing of artificial microRNA precursors is limited. Even with the use of carefully selected amiRNAs designed with the use of WMD, the success of amiRNA-based gene silencing only approached 75% [44]. No method has yet been developed for predicting whether additional small RNAs will be produced in large quantities. Therefore, at least under some conditions, when artificial microRNA technology is used to suppress gene expression, high levels of additional small RNAs will be produced.

Derivatives of artificial microRNAs and extra miRNA-like RNAs with modified "tails" were identified from small RNAs of 603-8 transgenic petals. Since 'the 3' truncation and tailing take place while miRNAs are in association with ARGONAUTE1 (AGO1), either during or after RISC assembly' [53], these tailed small RNAs found in transgenic petals indicated that the additional small RNAs could be in association with AGO proteins. As miRNAs are loaded onto AGO proteins to silence target genes, these small RNAs may be turned into components of active RISCs, and play a role in regulating gene expression. Possible

Table 2. Small RNAS arising from the amiRchs1 precursor with 3′ tail.

Unique ID	Sequence (3′ tail nucleotides are capitalized)	Reads	Position	Total reads
4529255	tgttggtacatcatgagtcg**T**	28	B1	116
127628	tgttggtacatcatgagtcgc**A**	23	B1	
5944084	tgttggtacatcatgagtcgc**C**	18	B1	
4124961	tgttggtacatcatgagtcgctct**T**	10	B1	
4831892	tgttggtacatcatgagtcgct**T**	10	B1	
1311807	tgttggtacatcatgagtc**T**	9	B1	
2744001	tgttggtacatcatgagtcgctct**A**	8	B1	
4085913	tgttggtacatcatgagtcg**TTC**	5	B1	
5323907	tgttggtacatcatgagtcgc**CT**	5	B1	
5665375	aactcatgatgaaccaacttc**T**	41	B1*	142
1874545	—ctcatgatgaaccaacttcac**T**	27	B1*	
927297	---tcatgatgaaccaacttcac**T**	46	B1*	
4528246	----catgatgaaccaacttc**T**	28	B1*	
4270474	aatgaatgatgcggtagacaaa**A**	103	B2	311
4389442	aatgaatgatgcggtagacaaat**C**	48	B2	
5577805	aatgaatgatgcggtagacaaat**G**	47	B2	
1049979	aatgaatgatgcggtagacaaa**C**	32	B2	
5744117	aatgaatgatgcggtagacaaat**A**	22	B2	
3006666	aatgaatgatgcggtagacaa**T**	12	B2	
5142252	aatgaatgatgcggtagac**T**	10	B2	
5674106	aatgaatgatgcggtagacaa**TT**	9	B2	
6175468	aatgaatgatgcggtagacaaa**G**	9	B2	
4479594	aatgaatgatgcggtagaca**T**	7	B2	
1992832	aatgaatgatgcggtagacaa**G**	6	B2	
3318766	aatgaatgatgcggtagacaaa**AC**	6	B2	
3095025	ttggatcattgattctctttgatgttggtac**T**	7	B3+	12
1404229	ttggatcattgattctctttgatgttggtac**C**	5	B3+	

target genes of miR319b.2 small RNAs have been identified in the *Arabidopsis* genome [49]. On account of the high complexity of most plant genomes (and given that the genomes of most plants are more complex than that of *Arabidopsis*), it is possible that target genes for these additional small RNAs may exist in a number of plant species. Using TargetSearch integrated in WMD3 to explore possible targets of small RNAs, potential targets of B2, B1*, B2* and B3* small RNAs were discovered in petunia and other

Table 3. Small RNAs arising from the phy-miR319a precursor with 3′ tail.

Unique ID	Sequence (3′ tail nucleotides are capitalized)	Reads	Position	Total reads
1326087	-ttggactgaagggagctcc**TT**	112	B1	501
2403353	cttggactgaagggagctcc**T**	95	B1	
6094764	cttggactgaagggagctcc**TT**	80	B1	
2661774	-ttggactgaagggagctccct**T**	73	B1	
4673852	cttggactgaagggagctccc**C**	54	B1	
3325799	-ttggactgaagggagctccc**C**	40	B1	
1418357	-ttggactgaagggagctcc**T**	36	B1	
4245887	cttggactgaagggagctccct**T**	11	B1	
1975016	attcaacgatgcatgagctg**G**	12	B1*	23
4917471	attcaacgatgcatgagctg**C**	6	B1*	
2934896	attcaacgatgcatgagctgt**T**	5	B1*	

A
```
UUAUGGAAGAGUCGGAAGCUU   CV299538 (P. hybrida)
|||:| |||||||||||||
CCUACUUACUCAGCCUUCGAU   amiRchs1-B2*-5 (21nt)
```
ΔG=-32.2 kcal/mol (70.2%) Reads=110

B
```
UGCAAUCUCAUCACGACCUGG   TC6334 (P. hybrida)
|  |||| |||||||||||
AACUUAGUAUAGUGCUGGACA   amiRchs1-B3*-5(21nt)
```
ΔG=-31.7 kcal/mol (76.4%) Reads=17

C
```
GUGUGGUUGGUUCAUCAUGCA   TC13041 (P. hybrida)
|||  :|||||||||||||
CACUUCAACCAAGUAGUACTC   amiRchs1-B1*-2 (21nt)
```
ΔG=-33.2 kcal/mol (77.6%) Reads=344

D
```
GGAAGGUUGUUCGUCAUGAGG   CV298960 (P. hybrida)
||||  |  |||:|||||||
ACUUCAACCAAGUAGUACUCA   amiRchs1-B1*-4(21nt)
```
ΔG=-31.2 kcal/mol (74.5%) Reads=325

E
```
GGUUGUGAACUGCAUCAUUCAUU   TC124235 (N. tabacum)
: |||| ||:||||||||||||
UAAACAGAUGGCGUAGUAAGUAA   amiRchs1-B2-5 (23nt)
```
ΔG=-31.6 kcal/mol (72.2%) Reads=990

F
```
AGGAUGAUAGAGUCGGGAGCUA   TC124001 (N. tabacum)
|||||| |||||||:|||||
ACCUACUUACUCAGCCUUCGAU   amiRchs1-B2*-3 (22nt)
```
ΔG=-41.4 kcal/mol (86.5%) Reads=1447

G
```
UGUGAAUUAUAUCACGAACUGC   BE922479 (S. tuberosum)
|||||:||||||||| |||
UAACUUAGUAUAGUGCUGGACA   amiRchs1-B3*-4 (22nt)
```
ΔG=-30.9 kcal/mol (71.8%) Reads=411

H
```
GAGGUGGGACCAUCAUGAGUU   TC201862 (S. tuberosum)
||:|| || ||||||||||
CUUCAACCAAGUAGUACUCAA   amiRchs1-B1*-9 (21nt)
```
ΔG=-28.8 kcal/mol (70.3%) Reads=2015

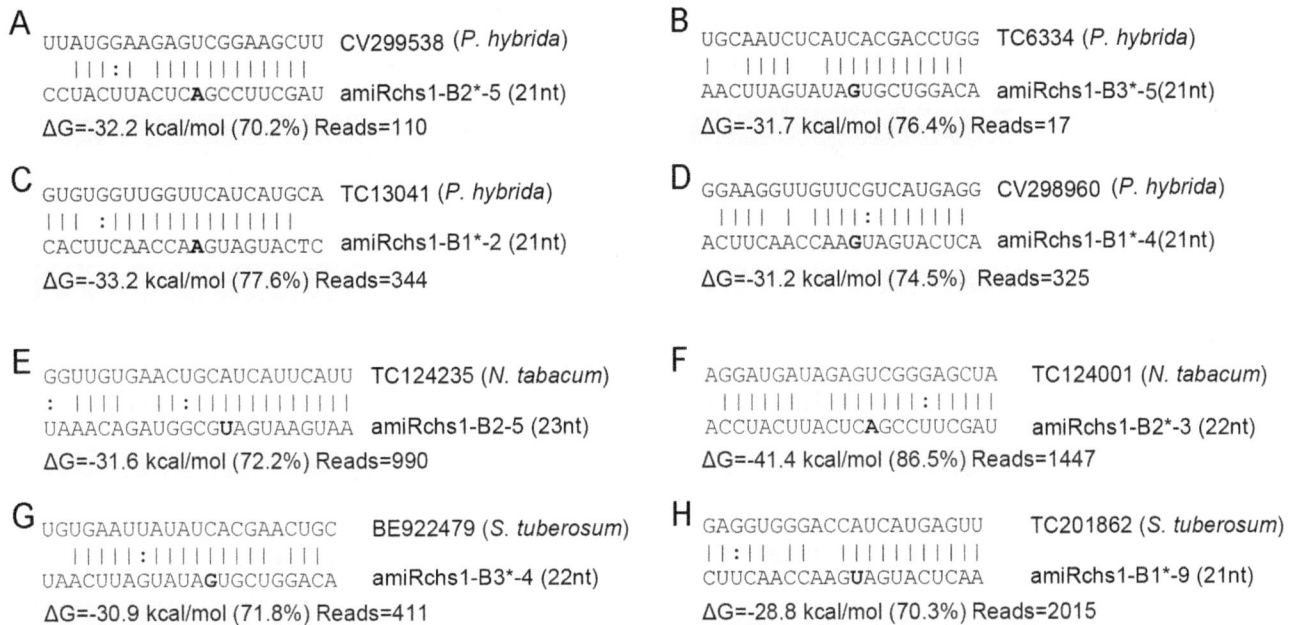

Figure 6. Alignments of selected small RNAs and some of their potential mRNA targets. Included are small RNA length, hybridization energy, percentage of free energy compared to a perfectly complementary target and small RNA reads identified in this study.

Solanaceae plants (Figure 6, Table S2). For many transcripts, the parameters of pairing with these undesired small RNAs were found to meet the criteria for plant microRNA target selection (Figure 6, Table S2) [52]. If undesired small RNAs are produced only at low frequency, their influence may be negligible; if, on the other hand, they are likely to be accumulated to high levels, it is desirable to take into account their potential functions and effects. In using artificial microRNA gene-silencing technology in plants, it must be recognized that the biogenesis of artificial microRNAs may generate additional small RNAs with the potential to affect unintended targets. One consequence of this is that the phenotypic consequences of the expression of amiRNAs require very careful interpretation.

The initial design rules for the generation of artificial microRNAs stipulated a mismatch within the amiRNA/target duplex at the 3′ end of the amiRNA in order to avoid the production of secondary small RNAs [24], yet the results presented here indicated that perfect pairing between amiRchs1 and *CHS-J* transcript did not result in the biogenesis of phasiRNAs. Recent studies have shown that the structural determinant for plant phasiRNA production is not in fact perfect pairing between a microRNA and its target, but the presence of asymmetrically positioned, "bulged" bases in the miRNA/miRNA* duplex [54]. The results of the present study are consistent with this model. Park et al. [55] reported that the use of an artificial miRNA having perfect complementarity to its target gene achieved highly efficient gene silencing; similarly, amiRNAs designed to have perfect complementarity to viral gene targets have shown high specificity [56]. Thus, perfect complementarity to

the target can in future be incorporated into the design of amiRNAs, in order to increase the specificity and efficiency of amiRNA-induced gene silencing.

Supporting Information

Figure S1 Phenotype of amiRchs1 transgenic flowers from 'GL8' and '10V1' genotypes.

Figure S2 Size distribution of small RNA clones from amiRchs1 transgenic petals.

Figure S3 Alignment of mature miR319 sequences and the fold-back structures of miR319 precursors.

Table S1 Small RNAs mapped to the promoter and intron regions of *CHS-A* (X14591).

Table S2 Potential mRNA targets of the undesired small RNAs in petunia, tobacco, tomato and potato.

Author Contributions

Conceived and designed the experiments: YG ML. Performed the experiments: YH HW JM XS YG. Analyzed the data: YG. Contributed reagents/materials/analysis tools: YG ML. Wrote the paper: YG.

References

1. Rogers K, Chen X (2013) Biogenesis, turnover, and mode of action of plant microRNAs. Plant Cell 25: 2383–2399.
2. Jones-Rhoades MW, Bartel DP, Bartel B (2006) MicroRNAs and their regulatory roles in plants. Annu Rev Plant Biol 57: 19–53.
3. Zhang B, Pan X, Cobb GP, Anderson TA (2006) Plant microRNA: a small regulatory molecule with big impact. Dev Biol 289: 3–16.
4. Guo HS, Xie Q, Fei JF, Chua NH (2005) MicroRNA directs mRNA cleavage of the transcription factor NAC1 to downregulate auxin signals for *Arabidopsis lateral* root development. Plant Cell 17: 1376–1386.
5. Schommer C, Palatnik JF, Aggarwal P, Chételat A, Cubas P, et al. (2008) Control of jasmonate biosynthesis and senescence by miR319 targets. PLoS biology 6: e230.

6. Liang G, Yang F, Yu D (2010) MicroRNA395 mediates regulation of sulfate accumulation and allocation in *Arabidopsis thaliana*. Plant J 62: 1046–1057.

7. Pant BD, Buhtz A, Kehr J, Scheible WR (2008) MicroRNA399 is a long-distance signal for the regulation of plant phosphate homeostasis. Plant J 53: 731–738.

8. Ehya F, Monavarfeshani A, Mohseni Fard E, Karimi Farsad L, Khayam Nekouei M, et al. (2013) Phytoplasma-responsive microRNAs modulate hormonal, nutritional, and stress signalling pathways in Mexican lime trees. PLoS One 8: e66372.

9. Sun G, Stewart CN, Jr., Xiao P, Zhang B (2012) MicroRNA expression analysis in the cellulosic biofuel crop switchgrass (*Panicum virgatum*) under abiotic stress. PLoS One 7: e32017.

10. Eldem V, Celikkol Akcay U, Ozhuner E, Bakir Y, Uranbey S, et al. (2012) Genome-wide identification of miRNAs responsive to drought in peach (*Prunus persica*) by high-throughput deep sequencing. PLoS One 7: e50298.

11. Xie Z, Allen E, Fahlgren N, Calamar A, Givan SA, et al. (2005) Expression of Arabidopsis *MIRNA* genes. Plant Physiol 138: 2145–2154.

12. Szarzynska B, Sobkowiak L, Pant BD, Balazadeh S, Scheible WR, et al. (2009) Gene structures and processing of *Arabidopsis thaliana* HYL1-dependent pri-miRNAs. Nucleic Acids Res 37: 3083–3093.

13. Song L, Han MH, Lesicka J, Fedoroff N (2007) *Arabidopsis* primary microRNA processing proteins HYL1 and DCL1 define a nuclear body distinct from the Cajal body. Proc Natl Acad Sci U S A 104: 5437–5442.

14. Fang Y, Spector DL (2007) Identification of nuclear dicing bodies containing proteins for microRNA biogenesis in living *Arabidopsis* plants. Curr Biol 17: 818–823.

15. Liu C, Axtell MJ, Fedoroff NV (2012) The helicase and RNaseIIIa domains of Arabidopsis Dicer-Like1 modulate catalytic parameters during microRNA biogenesis. Plant Physiol 159: 748–758.

16. Mateos JL, Bologna NG, Chorostecki U, Palatnik JF (2010) Identification of microRNA processing determinants by random mutagenesis of *Arabidopsis MIR172a* precursor. Curr Biol 20: 49–54.

17. Werner S, Wollmann H, Schneeberger K, Weigel D (2010) Structure determinants for accurate processing of miR172a in *Arabidopsis thaliana*. Curr Biol 20: 42–48.

18. Song L, Axtell MJ, Fedoroff NV (2010) RNA secondary structural determinants of miRNA precursor processing in *Arabidopsis*. Curr Biol 20: 37–41.

19. Bologna NG, Mateos JL, Bresso EG, Palatnik JF (2009) A loop-to-base processing mechanism underlies the biogenesis of plant microRNAs miR319 and miR159. EMBO J 28: 3646–3656.

20. Yu B, Yang Z, Li J, Minakhina S, Yang M, et al. (2005) Methylation as a crucial step in plant microRNA biogenesis. Science 307: 932–935.

21. Llave C, Xie Z, Kasschau KD, Carrington JC (2002) Cleavage of *Scarecrow-like* mRNA targets directed by a class of *Arabidopsis* miRNA. Science 297: 2053–2056.

22. Addo-Quaye C, Eshoo TW, Bartel DP, Axtell MJ (2008) Endogenous siRNA and miRNA targets identified by sequencing of the *Arabidopsis* degradome. Curr Biol 18: 758–762.

23. Fei Q, Xia R, Meyers BC (2013) Phased, secondary, small interfering RNAs in posttranscriptional regulatory networks. Plant Cell 25: 2400–2415.

24. Schwab R, Ossowski S, Riester M, Warthmann N, Weigel D (2006) Highly specific gene silencing by artificial microRNAs in *Arabidopsis*. Plant Cell 18: 1121–1133.

25. Alvarez JP, Pekker I, Goldshmidt A, Blum E, Amsellem Z, et al. (2006) Endogenous and synthetic microRNAs stimulate simultaneous, efficient, and localized regulation of multiple targets in diverse species. Plant Cell 18: 1134–1151.

26. Warthmann N, Chen H, Ossowski S, Weigel D, Hervé P (2008) Highly specific gene silencing by artificial miRNAs in rice. PLoS One 3: e1829.

27. Molnar A, Bassett A, Thuenemann E, Schwach F, Karkare S, et al. (2009) Highly specific gene silencing by artificial microRNAs in the unicellular alga *Chlamydomonas reinhardtii*. Plant J 58: 165–174.

28. Zhao T, Wang W, Bai X, Qi Y (2009) Gene silencing by artificial microRNAs in *Chlamydomonas*. Plant J 58: 157–164.

29. Hauser F, Chen W, Deinlein U, Chang K, Ossowski S, et al. (2013) A genomic-scale artificial microRNA library as a tool to investigate the functionally redundant gene space in *Arabidopsis*. Plant Cell 25: 2848–2863.

30. Toppino L, Kooiker M, Lindner M, Dreni L, Rotino GL, et al. (2011) Reversible male sterility in eggplant (*Solanum melongena* L.) by artificial microRNA-mediated silencing of general transcription factor genes. Plant Biotechnol J 9: 684–692.

31. Melito S, Heuberger AL, Cook D, Diers BW, MacGuidwin AE, et al. (2010) A nematode demographics assay in transgenic roots reveals no significant impacts of the Rhg1 locus LRR-Kinase on soybean cyst nematode resistance. BMC Plant Biol 10: 104.

32. Verdonk JC, Sullivan ML (2013) Artificial microRNA (amiRNA) induced gene silencing in alfalfa (*Medicago sativa*). Botany-Botanique 91: 117–122.

33. Haney CH, Long SR (2010) Plant flotillins are required for infection by nitrogen-fixing bacteria. Proc Natl Acad Sci U S A 107: 478–483.

34. Khraiwesh B, Ossowski S, Weigel D, Reski R, Frank W (2008) Specific gene silencing by artificial microRNAs in *Physcomitrella patens*: an alternative to targeted gene knockouts. Plant Physiol 148: 684–693.

35. Sun P, Kao TH (2013) Self-incompatibility in *Petunia inflata*: the relationship between a self-incompatibility locus F-box protein and its non-self S-RNases. Plant Cell 25: 470–485.

36. Tzfira T, Tian G-W, Vyas S, Li J, Leitner-Dagan Y, et al. (2005) pSAT vectors: a modular series of plasmids for autofluorescent protein tagging and expression of multiple genes in plants. Plant Mol Biol 57: 503–516.

37. Jorgensen RA, Cluster PD, English J, Que Q, Napoli CA (1996) Chalcone synthase cosuppression phenotypes in petunia flowers: comparison of sense vs. antisense constructs and single-copy vs. complex T-DNA sequences. Plant Mol Biol 31: 957–973.

38. Koseki M, Goto K, Masuta C, Kanazawa A (2005) The star-type color pattern in *Petunia hybrida* 'Red Star' flowers is induced by sequence-specific degradation of chalcone synthase RNA. Plant Cell Physiol 46: 1879–1883.

39. Mallona I, Lischewski S, Weiss J, Hause B, Egea-Cortines M (2010) Validation of reference genes for quantitative real-time PCR during leaf and flower development in *Petunia hybrida*. BMC Plant Biol 10: 4.

40. Schmittgen TD, Livak KJ (2008) Analyzing real-time PCR data by the comparative CT method. Nature Protocols 3: 1101–1108.

41. De Paoli E, Dorantes-Acosta A, Zhai J, Accerbi M, Jeong DH, et al. (2009) Distinct extremely abundant siRNAs associated with cosuppression in petunia. RNA 15: 1965–1970.

42. Tedder P, Zubko E, Westhead DR, Meyer P (2009) Small RNA analysis in *Petunia hybrida* identifies unusual tissue-specific expression patterns of conserved miRNAs and of a 24mer RNA. RNA 15: 1012–1020.

43. Zuker M (2003) Mfold web server for nucleic acid folding and hybridization prediction. Nucleic Acids Research 31: 3406–3415.

44. Ossowski S, Schwab R, Weigel D (2008) Gene silencing in plants using artificial microRNAs and other small RNAs. Plant J 53: 674–690.

45. Koes RE, Van Blokland R, Quattrocchio F, Van Tunen AJ, Mol J (1990) Chalcone synthase promoters in petunia are active in pigmented and unpigmented cell types. Plant Cell 2: 379–392.

46. Koes RE, Spelt CE, Mol JN (1989) The chalcone synthase multigene family of *Petunia hybrida* (V30): differential, light-regulated expression during flower development and UV light induction. Plant Mol Biol 12: 213–225.

47. Rajagopalan R, Vaucheret H, Trejo J, Bartel DP (2006) A diverse and evolutionarily fluid set of microRNAs in *Arabidopsis thaliana*. Genes Dev 20: 3407–3425.

48. Zhang W, Gao S, Zhou X, Xia J, Chellappan P, et al. (2010) Multiple distinct small RNAs originate from the same microRNA precursors. Genome Biol 11: R81.

49. Sobkowiak L, Karlowski W, Jarmolowski A, Szweykowska-Kulinska Z (2012) Non-canonical processing of *Arabidopsis* pri-miR319a/b/c generates additional microRNAs to target one RAP2.12 mRNA isoform. Front Plant Sci 3: 46.

50. Parizotto EA, Dunoyer P, Rahm N, Himber C, Voinnet O (2004) In vivo investigation of the transcription, processing, endonucleolytic activity, and functional relevance of the spatial distribution of a plant miRNA. Genes Dev 18: 2237–2242.

51. Zhao Y, Yu Y, Zhai J, Ramachandran V, Dinh TT, et al. (2012) The *Arabidopsis* nucleotidyl transferase HESO1 uridylates unmethylated small RNAs to trigger their degradation. Curr Biol 22: 689–694.

52. Schwab R, Palatnik JF, Riester M, Schommer C, Schmid M, et al. (2005) Specific effects of microRNAs on the plant transcriptome. Dev Cell 8: 517–527.

53. Zhai J, Zhao Y, Simon SA, Huang S, Petsch K, et al. (2013) Plant microRNAs display differential 3' truncation and tailing modifications that are ARGO-NAUTE1 dependent and conserved across species. Plant Cell 25: 2417–2428.

54. Manavella PA, Koenig D, Weigel D (2012) Plant secondary siRNA production determined by microRNA-duplex structure. Proc Natl Acad Sci U S A 109: 2461–2466.

55. Park W, Zhai J, Lee JY (2009) Highly efficient gene silencing using perfect complementary artificial miRNA targeting AP1 or heteromeric artificial miRNA targeting AP1 and CAL genes. Plant Cell Rep 28: 469–480.

56. Niu QW, Lin SS, Reyes JL, Chen KC, Wu HW, et al. (2006) Expression of artificial microRNAs in transgenic *Arabidopsis thaliana* confers virus resistance. Nat Biotechnol 24: 1420–1428.

21

Detection of the Virulent Form of AVR3a from *Phytophthora infestans* following Artificial Evolution of Potato Resistance Gene *R3a*

Sean Chapman[1,9], Laura J. Stevens[1,2,3,9], Petra C. Boevink[1,3], Stefan Engelhardt[2,3], Colin J. Alexander[4], Brian Harrower[1], Nicolas Champouret[5], Kara McGeachy[1], Pauline S. M. Van Weymers[1,2,3], Xinwei Chen[1,3], Paul R. J. Birch[1,2,3], Ingo Hein[1,3]*

1 Cell and Molecular Sciences, James Hutton Institute, Invergowrie-Dundee, United Kingdom, 2 Division of Plant Sciences, University of Dundee at James Hutton Institute, Invergowrie-Dundee, United Kingdom, 3 Dundee Effector Consortium, Invergowrie-Dundee, United Kingdom, 4 Biomathematics and Statistics Scotland, Invergowrie-Dundee, United Kingdom, 5 J.R. Simplot Company, Simplot Plant Sciences, Boise, Idaho, United States of America

Abstract

Engineering resistance genes to gain effector recognition is emerging as an important step in attaining broad, durable resistance. We engineered potato resistance gene *R3a* to gain recognition of the virulent AVR3a^{EM} effector form of *Phytophthora infestans*. Random mutagenesis, gene shuffling and site-directed mutagenesis of *R3a* were conducted to produce R3a* variants with gain of recognition towards AVR3a^{EM}. Programmed cell death following gain of recognition was enhanced in iterative rounds of artificial evolution and neared levels observed for recognition of AVR3a^{KI} by R3a. We demonstrated that R3a*-mediated recognition responses, like for R3a, are dependent on SGT1 and HSP90. In addition, this gain of response is associated with re-localisation of R3a* variants from the cytoplasm to late endosomes when co-expressed with either AVR3a^{KI} or AVR3a^{EM} a mechanism that was previously only seen for R3a upon co-infiltration with AVR3a^{KI}. Similarly, AVR3a^{EM} specifically re-localised to the same vesicles upon recognition by R3a* variants, but not with R3a. R3a and R3a* provide resistance to *P. infestans* isolates expressing AVR3a^{KI} but not those homozygous for AVR3a^{EM}.

Editor: Frederik Börnke, Leibniz-Institute for Vegetable and Ornamental Crops, Germany

Funding: LS is supported by The James Hutton Institute and the University of Dundee through a joint PhD student bursary. PSMVW is supported by the U.S. Department of Agriculture through National Institute of Food and Agriculture (NIFA) project 2011-68004-30154 (http://www.csrees.usda.gov/). NC is employed by J.R. Simplot Company, Simplot Plant Sciences, 5369 West Irving Street, Boise, ID 83706, USA. This work was funded by the Rural & Environment Science & Analytical Services (RESAS) Division of the Scottish Government (http://www.scotland.gov.uk/Topics/Research/About/EBAR/research-providers) and the Biotechnology and Biological Sciences Research Council (BBSRC) (http://www.bbsrc.ac.uk/home/home.aspx) through the joint projects CRF/2009/SCRI/SOP & BB/H018441/1 (IH), BB/K018299/1 and RESAS funded work package 6.4. The funders had no role in study design, data collection and analysis, decision to publish, or preparation of the manuscript.

Competing Interests: Author Nicolas Champouret is employed by J.R. Simplot Company.

* Email: Ingo.Hein@hutton.ac.uk

9 These authors contributed equally to this work.

Introduction

In a process known as effector triggered immunity (ETI), plant disease resistance (*R*) genes can facilitate immunity to phylogenetically diverse and unrelated pathogens that express cognate effector molecules [1]. Effectors that are recognised by *R* genes, either directly or indirectly, and provoke successful plant defences are genetically defined as avirulence (*Avr*) genes. The best described group of plant *R* gene products contains a nucleotide-binding (NB) domain and leucine-rich repeats (LRRs), collectively known as NB-LRRs [2]. NB-LRRs are strictly regulated by plants as, upon their activation, many elicit programmed cell death (PCD) as part of the hypersensitive response (HR) which prevents further spread of disease in plant tissues [3]. Together with effectors, *R* genes are at the forefront of host/pathogen co-evolution [4]. NB-LRRs are one of the largest gene families in plants and more than 750 members have recently been described in potato [5–6]. The organisation of many NB-LRRs into physically-linked clusters is providing insight into their evolution, which can involve duplication followed by diversification.

In agriculture, successful deployment of *R* genes to control important diseases in crop plants has so far been hampered by the ability of pathogens often to rapidly circumvent detection by the host plant's innate immune system. Advances in studying pathogen effector diversity coupled with the ability to engineer *R* genes offers the opportunity to develop more durable resistances that specifically target essential effectors and known variants [7–8].

The *Phytophthora infestans* effector AVR3a is an essential effector for this pathogen to cause late blight on potato. Stable silencing of *Avr3a* in the *P. infestans* isolate 88069 significantly reduces infection in susceptible *Solanum tuberosum* (potato) cv. Bintje and in the model solanaceous plant species *Nicotiana benthamiana* [9–10]. Two forms of AVR3a are prevalent in current *P. infestans* isolates and differ in only two amino acids within the mature protein; AVR3aE^{80}M^{103} (AVR3a^{EM}) and AVR3aK^{80}I^{103} (AVR3a^{KI}) [11–12]. AVR3a^{KI} elicits ETI upon recognition by the potato resistance protein R3a, a member of the coiled-coil (CC) NB-LRR gene family [13]. This response is evaded by AVR3a^{EM} which consequently is free to promote virulence [9,11,13]. However, a weak R3a-dependent response to AVR3a^{EM} can be observed under UV light [14]. The mechanism of R3a-mediated recognition of AVR3a has been investigated recently [15]. Upon activation by AVR3a^{KI}, both the effector protein and R3a rapidly re-localize from the host cytoplasm to late endosomes, components of the endocytic pathway, which is thought to be a prerequisite for subsequent HR development. The un-recognised AVR3a^{EM} form of the effector does not cause this re-localisation. There is no evidence of direct interaction between AVR3a^{KI} and R3a, but bimolecular fluorescence complementation (BiFC) assays reveal that the two proteins are in close proximity at late endosomes [15].

Artificial evolution has previously been used to alter the recognition specificity of the potato CC-NB-LRR resistance protein Rx to gain recognition of different strains of *Potato virus X* (PVX) and a distantly related virus, *Poplar mosaic virus* (PopMV) [16–17]. Random mutagenesis, screening for beneficial mutations and designed amalgamation of these mutations has been used to generate transgenic plants of the model species *N. benthamiana* that are resistant to previously virulent strains. DNA shuffling, also known as directed evolution, was first developed in the early 1990 s and has since been used to generate a wide variety of novel genes and proteins [18]. DNA shuffling has previously been used in the functional analysis of the resistance gene *Pto* [19]. In this study, the LRR of *R3a* has been subjected to error-prone PCR and iterative rounds of DNA shuffling to identify gain-of-recognition variants (R3a*) through functional screening in *N. benthamiana*. *R3a** gene products with varying degrees of AVR3a^{EM} recognition were generated in three rounds of DNA shuffling and a subsequent round of site-directed mutagenesis. The best-performing clones from each round were taken forward for further analysis and compared to wild-type R3a. R3a* variants demonstrated significantly improved gain-of-recognition towards AVR3a^{EM} that manifested itself as a gain of re-localisation to late endosomes, but not yet as a gain of resistance.

Materials and Methods

Construction of plasmid vectors

The plasmid pBinPlus.R3a [13] was amplified with the primer pairs R3a-5-Asc/R3a-1564-Bam-M and R3a-1564-Bam-P/R3a-3-Not (Table S1) and the *Asc*I/*Bam*HI and *Bam*HI/*Not*I digested amplification products cloned in a three way ligation in to *Asc*I/*Not*I digested binary vector pGRAB [20] to produce pGRAB.35S::R3a. A derivative of the former plasmid, pGRAB.-R3a::R3a, was produced in which the *Cauliflower mosaic virus* 35S (35S) promoter sequence was replaced with the *R3a* promoter sequence. This derivative was produced by digestion of pBin-Plus.R3a with *Pme*I and *Xho*I, and insertion of the released fragment, containing 2358 bases upstream of the translational start site, in to pGRAB.35S::R3a that had been treated with *Bsp*EI, T4 DNA polymerase and *Xho*I in order.

Mutagenesis and DNA shuffling

Shuffling of 2283 bp of the R3a LRR region was performed as described by Stemmer [18]. First round PCRs were carried out with the primer pair R3a-1564-Bam-P/R3a-3-Not in the presence of 0.5 mM MnCl$_2$ as described by Leung *et al.* [21] with *Taq* DNA polymerase (Roche, Mannheim, Germany). Following DNaseI treatment, fragments of circa 500 bp were reassembled through forty rounds of primer-less thermo-cycling. Gel-purified products of circa 2.3 kb were amplified through thirty cycles of PCR and, following gel-purification, digested with *Bam*HI and *Not*I prior to cloning in to pGRAB.35S::R3a digested with the same enzymes. The ligated population was amplified in *E. coli* resulting in a population with a complexity of circa 125 K, a vector background of circa 14% and a base mutation rate of 0.43%. Aliquots of plasmid, gel-purified on account of plasmid instability, were transformed in to *A. tumefaciens* cells. In the second and third rounds *PfuUltra* II Fusion HS DNA polymerase (Stratagene, La Jolla, CA, USA) was used in thermo-cycling reactions to produce shuffled populations with lower mutation rates, less than 0.05%. The second and third round populations had complexities of circa 200 K and 150 K, respectively, and both had a vector background of circa 10%.

Site-directed mutagenesis

A mixture of two templates, the wild-type gene and clone Rd1-2 from the first round of shuffling containing the Q931R codon substitution, was used in PCR amplifications with the primer pairs R3a-1564-Bam-P/R3a-1740W-M, R3a-1740W-P/R3a-1841W-M, R3a-1841W-P/R3a-2743W-M, R3a-2743W-P/R3a-3028-M. The products of the primary PCRs were used in an overlap PCR reaction with the flanking primer pair R3a-1564-Bam-P/R3a-3028-M. The product of the secondary PCR reaction was digested with *Bam*HI and *Aat*II and cloned between the same unique sites of pGRAB.R3a::R3a. A population of circa 50K clones, with a vector background of 1% and random base mutation rate of 0.05% was produced.

Plant growth conditions

N. benthamiana plants were grown in a glasshouse with a 16 h day period at 22°C and an 8 h night period at 18°C. Supplementary lighting was provided below 200 W m^{-2} and screening above 450 W m^{-2}.

Screening of mutated R3a clones

DNA populations prepared from *E. coli* were transformed in to *Agrobacterium tumefaciens* strain AGL1 [22] carrying the helper plasmids pSoup and pBBR1MCS1.VirG$_{N54D}$ [23] for screening. *Agrobacterium* cultures grown from single transformed colonies were co-infiltrated with cultures of *Agrobacterium* transformed with pGRAB.35S::AVR3a^{EM} according to the method of Engelhardt *et al.* [15] in to *N. benthamiana* leaves with each of the components at the same final OD$_{600}$ of 0.1–0.5. Reference mixtures were infiltrated in to opposing half-leaves and between two and seven days post infiltration leaves were inspected to assess visible symptoms and plant auto-fluorescence under day-light and 365 nm illumination from a Blak-Ray lamp (UVP, Upland, CA, USA), respectively.

Symptom scoring

Circa five week old plants were used for symptom scoring with two adjacent, expanded leaves, both circa 90 mm in length, being used for infiltrations. Symptoms were scored on an arbitrary scale for nine days after infiltration. Symptom scores were plotted; the areas under the curve determined and mean scores calculated by dividing by the duration. A linear mixed modelling approach was adopted using GenStat for Windows, 16th edition (VSN International Ltd., Hemel Hempstead, UK). The data were analysed in two stages: first, the stability of the phenotypes over repeated experiments was examined by fitting a model with experiment as a fixed effect. Infiltration mixture, leaf age, position of infiltration site on leaf, experiment and their interactions were set as fixed effects with plant and leaf within an individual plant as the random effects. This allowed for terms in the model which specifically tested for significant interactions with experiment. Lack of any significant interaction with the infiltration mixtures would provide evidence that the relative responses of the phenotypes were consistent. Having determined that the infiltration mixtures behaved consistently over the experiments, the second stage fitted a model with experiment as a random effect with plant and leaf within plant nested below this. Multiple comparison tests then examined the differences in response amongst the infiltration mixtures.

Virus-induced gene silencing (VIGS) of SGT1 and HSP90

Tobacco rattle virus (TRV)-induced gene silencing in N. benthamiana was performed as described previously [14]. Agrobacterium cultures transformed with the binary TRV RNA1 construct, pBINTRA6, or the TRV RNA2 vector constructs PTV00, PTV:eGFP, PTV:HSP90 or PTV:SGT1 were re-suspended to $OD_{600} = 0.5$ for the RNA1 construct and $OD_{600} = 1.0$ for the RNA2 constructs. Re-suspended RNA1 and RNA2 cultures were mixed in a 1:1 ratio and infiltrated into non-cotyledonous leaves of N. benthamiana plants at the 5-leaf stage. For each of the biological replicates, six plants per treatment were used and six plants were used as non-TRV controls. Three weeks after treatment with the VIGS constructs, plants were infiltrated with culture mixtures ($OD_{600} = 0.5$) designed to express R3a, Rd2-1, Rd3-1 or Rd4-1 and AVR3a^{KI} or AVR3a^{EM}. HRs were scored at 6dpi.

Confocal laser scanning microscopy

Imaging was performed on a Leica TCS-SP2 AOBS microscope (Leica Microsystems) using HCX APO L, 40x/0.8, and 63x/0.9 water dipping lenses or a Zeiss 710 using a Plan APO 40x/1.0 water dipping lens. Images were collected using line by line sequential scanning. The optimal pinhole diameter and the same gain levels were used within experiments. YFP and CFP were imaged using 514 nm and 405 nm excitation, respectively, and emissions were collected between 520–563 nm and 455–490 nm, respectively. Photoshop CS5.1 software (Adobe Systems) was used for post-acquisition image processing.

Agrobacterium tumefaciens transient assays (ATTAs)

Functional Agrobacterium tumefaciens transient assays (ATTAs) were carried out in N. benthamiana. Cultures carrying pGRAB.R3a::R3a, pGRAB.R3a::Rd2-1, pGRAB.R3a::Rd3-1, pGRAB.R3a::Rd4-1 or pGRAB empty vector were re-suspended as described before to OD600 = 0.1 for each construct. Each of the five resuspensions was infiltrated into separate areas of leaves. Four leaves on each of sixteen plants were infiltrated in each replicate. Two days post infiltration, leaves were detached and infiltration sites inoculated with AVR3a^{KI} homozygous P. infestans isolate 7804.b or AVR3a^{EM} homozygous isolate 88069. Leaves were incubated in transparent sealed boxes at 100% humidity in a cool room and covered for the first 12 hours. Lesion sizes were measured up to 15dpi.

Production of transgenic potato plants

R3a wild-type gene and the three modified versions Rd2-1, Rd3-1 and Rd4-1 were cloned under R3a native regulatory elements in a pCambia-based binary vector with kanamycin resistance as a selectable marker, using standard restriction enzyme methods to create pSIM2093, pSIM3027, pSIM3028 and pSIM3029 respectively. Rpi-vnt1 under its native regulatory elements was cloned in to pSIM401 to generate pSIM1620. The binary vectors were then transformed into Agrobacterium strains AGL1 and LBA4404 for plant transformation. Ranger Russet was transformed as described in Duan et al. [24]. For each construct twenty to thirty lines were regenerated with kanamycin selection. These were tested for late blight resistance and assessed at 7dpi.

Results

Identification of R3a mutants with enhanced AVR3a^{EM} recognition

A binary vector, pGRAB.35S::R3a, containing the R3a open reading frame (Accession number AY849382.1) under the control of Cauliflower mosaic virus 35S promoter and terminator sequences was produced to allow specific mutation and shuffling of the R3a LRR domain. In this plasmid a unique BamHI site was introduced silently at nucleotide position 1567 in the ORF, ninety nucleotides upstream of the sequence encoding the LRR domain as denoted by Huang et al. [13]. Following mutagenesis and DNA shuffling of the LRR region, regenerated recombinant clones were screened for enhanced AVR3a^{EM} recognition through Agrobacterium-mediated transient expression in co-infiltrations with binaries expressing the virulent elicitor, AVR3a^{EM}. Clones were screened for enhanced recognition through comparison of visible symptoms of PCD and induced auto-fluorescence with reference to the responses produced by the wild-type R3a gene and AVR3a^{EM}. Clones with putatively improved recognition isolated from primary screens were screened again to confirm the phenotypic improvement and to eliminate auto-activators. In the first cycle, screening of approximately three thousand clones identified eleven R3a* clones with improved phenotypes. The eleven clones from the first round are referred to as Rd1-1 to Rd1-11 and contained, in addition to synonymous changes, base changes that resulted in between one and four amino acid substitutions (Fig 1a; Table S2). Three of the clones contained single amino acid substitutions (Rd1-1 [K920E], Rd1-2 [Q931R], Rd1-3 [R618Q]) identifying these changes as being responsible for enhanced recognition. Interestingly, the single amino acid substitution K920E in R3a was recently also identified in a complementary study by Segretin et al. [25] as a substitution that enhances recognition towards AVR3a^{EM}. The largest numbers of amino acid substitutions were found in LRRs #3 and #15 and included those found in the three clones with single substitutions (Fig. 1a).

A second round of shuffling was carried out to combine beneficial changes and remove deleterious substitutions using the eleven isolated clones, Rd1-1 to Rd1-11, and the wild-type gene as starting material. Screening of approximately six hundred recombinant clones identified four which gave responses comparable to or greater than the best performing clone from the first round, Rd1-1 [K920E]. All of these R3a* clones from the second round, Rd2-1 to Rd2-4, contained the amino acid substitution E620D in LRR #3 in addition to at least one other coding change (Fig. 1a; Table S2).

As recognition of AVR3a^{EM} improved and gave stronger responses it became more difficult to discriminate the differences in responses between modified clones with AVR3a^{EM} and also in comparison to the response of the wild-type R3a gene with AVR3a^{KI}. Therefore, a third round of shuffling, using the clones from the first and second rounds of shuffling and the wild-type

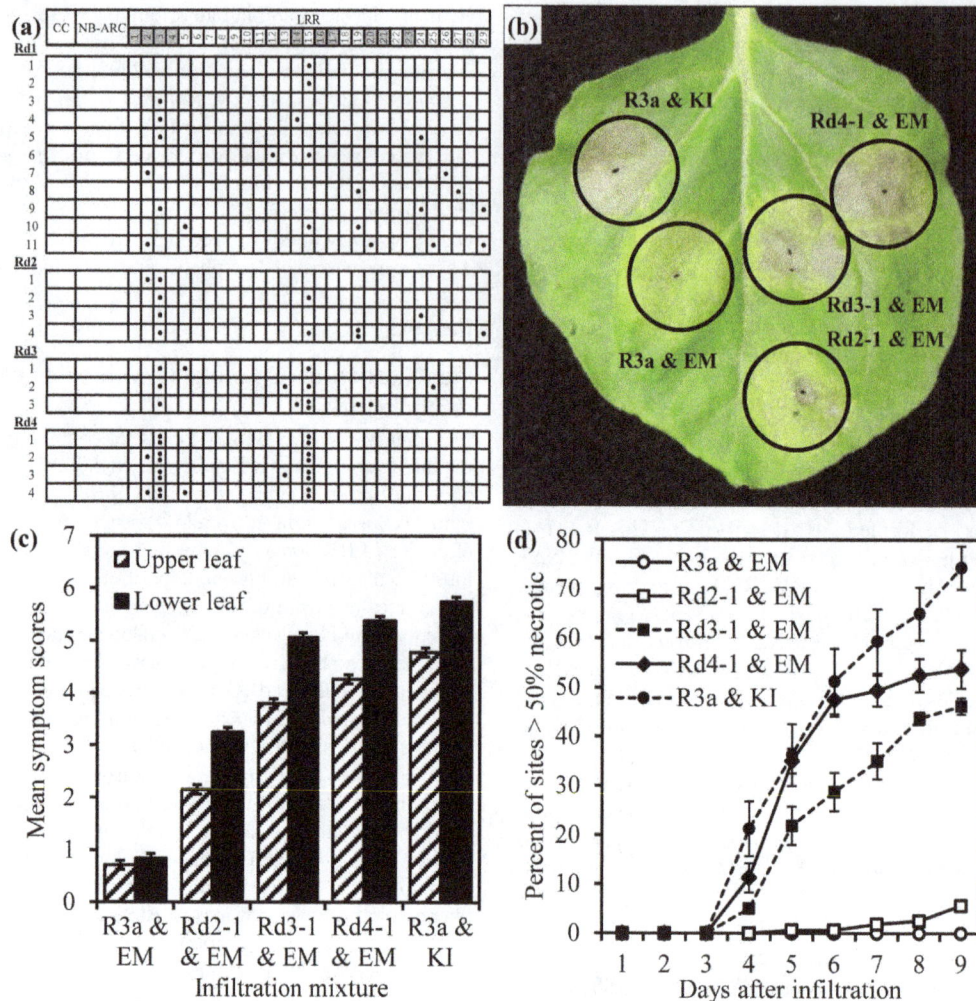

Figure 1. Four rounds of mutagenesis and shuffling identified R3a mutants with enhanced recognition of AVR3a^{EM} and disease responses (R3a*). (a) Schematic showing locations of non-synonymous mutations found in the LRRs of R3a* clones isolated from the four rounds of mutagenesis and shuffling (Rd1 to Rd4). LRRs containing amino acids under diversifying selection are shaded above. (b) Representative *N. benthamiana* leaf showing responses of best performing clones from second, third and fourth rounds (Rd2-1, Rd3-1 & Rd4-1) to AVR3a^{EM} (EM), compared to responses of wild-type R3a to AVR3a^{KI} (KI) and AVR3a^{EM} five days after co-infiltration with resistance genes under transcriptional control of *R3a* promoter. (c) Mean disease scores from the four experiments, each of nine days duration, for different infiltration mixtures in upper (hatched) and lower (solid) paired leaves. Error bars show +/− standard error. (d) Time-course of percentage of sites showing necrosis development, greater than 50% necrosis of individual infiltrated sites, for the five infiltrated mixtures. Mean percentages of the four experiments. Each experiment includes data for 40 infiltration sites (upper and lower leaves combined) and error bars show +/− standard errors.

gene as starting material, was performed. However, the shuffled LRR domains were cloned in to a binary vector, pGRAB.-R3a::R3a, containing the *R3a* gene or *R3a** variants under the wild-type *R3a* promoter, rather than the strong 35S promoter. The purpose of this was to protract the timing of the cell death responses and facilitate the discrimination of differences (Fig. 1b). Screening of approximately 300 clones from this third round population identified three R3a* clones, Rd3-1 to Rd3-3, that gave responses greater than the best performing clone from the second round of shuffling, Rd2-1 [T585A, E620D]. All of these clones contained the E620D change found in the second round clones and a number of other amino acid changes including at least one change in LRR #15. A previously conducted comparison of functional R3a with three paralogous, non-functional resistance gene analogues revealed 13 positions under diversification [13]. These positions, with the exception of one in the CC-domain, are located in LRRs 1 to 4 and 14 to 23.

Intriguingly, the majority of mutations present in the clones from the second and third rounds are within these LRRs (Fig. 1a).

One limitation of shuffling is in recombining beneficial mutations that are in close proximity due to the limited frequency of crossing-over. Therefore, as multiple mutations in LRRs #3 and #15 had been found to enhance AVR3a^{EM} recognition, an alternate approach was adopted. A population of site-directed mutants was produced using degenerate oligonucleotides that encoded different pairs of amino acids at positions 585 (T/A), 618 (R/Q), 620 (E/D), 918 (R/G), 920 (K/E), 923 (D/G) and 931 (Q/R) with the potential for 128 different permutations. Screening of approximately 400 clones from this population with the best performing clone from the third round of shuffling (Rd3-1 [E620D, L668P, Q931R]) as a reference, identified eight clones that produced responses with AVR3a^{EM} comparable to or greater than the reference. Five of these clones were found to have identical sequences. Representative unique clones, Rd4-1 to Rd4-4, contain the amino acid substitutions R618Q, E620D, K920E and Q931R, but lacked either of the designed substitutions R918G or D923G (Table S2).

Iterative rounds of shuffling have progressively improved AVR3a^{EM} recognition by R3a* variants, producing faster and stronger PCD responses upon co-infiltration

The enhanced recognition of AVR3a^{EM} was assessed more accurately in single leaf comparisons. The R3a* constructs Rd2-1, Rd3-1 and Rd4-1 were transiently expressed under the control of the native *R3a* promoter with AVR3a^{EM} and the responses compared with those produced by the wild-type gene with AVR3a^{EM} and AVR3a^{KI}. The best performing clone from the first round was not included in this analysis on account of the relatively poor response produced when it was under the control of the *R3a* promoter.

Symptom development on *N. benthamiana* was monitored for 9 days after infiltration in four independent experiments. In each experiment two adjacent, expanded leaves on each of twenty plants were infiltrated with the five infiltration mixtures in a circularly permuted arrangement to account for possible intra-leaf position effects. Symptoms were scored on an arbitrary scale ranging from 0 (no symptoms) to 10 (complete necrosis of the infiltrated area). A progressive increase in the recognition of AVR3a^{EM} was observed for the three R3a* clones with the necrotic response produced by Rd4-1 being close to that produced by the wild-type gene with AVR3a^{KI} (Fig. 1b).

A mixed model with experiment as a fixed effect was fitted to test for consistent responses of the infiltration mixtures over repeated experiments. The experiments showed significant differences in mean response (p = 0.001), but there was no significant interaction (p = 0.306) between experiment and infiltration mixture indicating that the relative responses of the phenotypes were stable over repeated experiments. In addition, there were no significant interactions between experiment and the other fixed effects.

Since the phenotypes were determined to be stable, a second analysis of the data with experiment as a random effect was now fitted. This showed highly significant effects (p<0.001) from infiltration mixture and leaf age (upper vs lower leaf). Further, there was a highly significant interaction (p<0.001) between infiltration mixture and leaf age because the combination of R3a and AVR3a^{EM} did not show a difference in mean scores between younger and older leaves, in contrast to the other combinations which showed significant differences (Fig. 1c). Position of infiltration site on leaf was non-significant and all other interactions of fixed effect were also non-significant. Multiple comparisons using Bonferroni correction with an experiment-wise significance level of 5% showed significant differences between the mean symptom scores for all infiltration mixtures within either younger or older leaves with a comparison-wise significance level of 0.0011 (Table S3). Ordering of the responses was the same for both younger and older leaves with R3a and AVR3a^{KI}> Rd4-1 & AVR3a^{EM}> Rd3-1 & AVR3a^{EM}> Rd2-1 & AVR3a^{EM}> R3a and AVR3a^{EM}. While the clone Rd2-1 from the second round gave symptoms when co-expressed with AVR3a^{EM}, it rarely produced an HR phenotype as shown in figure 1d, which shows the proportion of sites with more than 50% of the infiltrated area necrotic. Despite improved recognition of AVR3a^{EM} by the R3a* variants relative to wild-type R3a, their recognition of AVR3a^{KI} was not impaired (data not shown), indicating that this specificity had not been significantly attenuated. Further, when expressed from the strong 35S promoter in the absence of AVR3a^{EM} or AVR3a^{KI} none of the clones produced necrosis, indicating that they maintained

appropriate control mechanisms and were not auto-activators (Fig. S1).

R3a* recognition of Avr3a^{EM} is dependent on HSP90 and SGT1

Previous studies by Bos *et al.* [14] demonstrated that R3a-dependent recognition of AVR3a^{KI} involves both SGT1 (suppressor of the G2 allele of *skp1*) and HSP90 (heat shock protein 90) that are required for the activation of other R proteins [26–27]. Their involvement in the AVR3a^{EM}-dependent responses was tested through *Tobacco rattle virus* (TRV)-based gene silencing of *SGT1* and *HSP90* in *N. benthamiana* with TRV-based expression of truncated GFP (eGFP) as a control. Three biological replicates for R3a, Rd2-1, Rd3-1 and one for Rd4-1, with infiltrations in two leaves of each of six plants per TRV-based silencing construct, revealed that both SGT1 and HSP90 are required to mediate an HR upon R3a*-based recognition of AVR3a^{KI} and AVR3a^{EM} (Fig. 2, Fig. S2). HRs were abolished for all infiltrations on TRV:*SGT1* inoculated plants and there were almost no HRs recorded on TRV:*HSP90* inoculated plants (Fig. 2). The HRs were not affected on plants inoculated with TRV:*eGFP*.

Compared to TRV:*eGFP* inoculated plants, *SGT1* and *HSP90* silenced plants were morphologically stunted, a phenotype that had been reported previously [14]. Nevertheless, upon infection with the bacterial pathogen *Erwinia amylovora* that produces a SGT1 and HSP90 independent non-host response in *N. benthamiana* [28], all plants were able to mount the expected HR response (Fig. 2, Fig. S2).

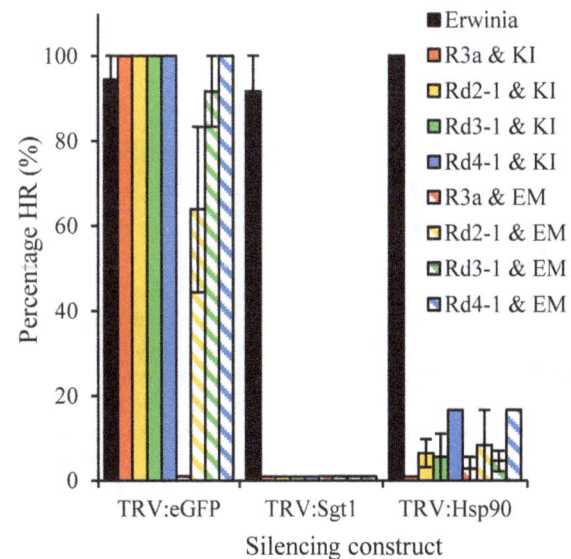

Figure 2. HR responses resulting from R3a* recognition of AVR3a^{EM} and AVR3a^{KI}, like those caused by wild-type R3a recognition of AVR3a^{KI}, are dependent on SGT1 and HSP90. SGT1- and HSP90-silenced plants were produced using TRV-based vectors. These plants and control plants inoculated with TRV:*eGFP* were infiltrated with different combinations of *Agrobacterium* cultures designed to express R3a, R3a* variants, AVR3a^{KI} (KI) or AVR3a^{EM} (EM). The percentage of sites (N = 12) showing HR responses six days after infiltration was recorded. The graph shows the mean percentages from three independent experiments with the exception that the dependence on HSP90 of Rd4-1 responses was only tested in a single experiment. The non-host bacterial pathogen *Erwinia amylovora* was used as a control for an SGT1- and HSP90-independnet HR response. Error bars show +/- standard error. Zero values have been transformed to 1% to facilitate their observation.

In a gain of mechanism, R3a* variants re-localize to late endosomes upon co-infiltration with Avr3a^{KI} or Avr3a^{EM}

In a previous study, Engelhardt *et al.* [15] demonstrated that, upon recognition of AVR3a^{KI} but not AVR3a^{EM}, wild-type R3a re-localizes from the host cytoplasm to specific late endosomes that can be labelled with the cyan fluorescent protein marker PS1-CFP [29]. This re-localization was found to be a pre-requisite for subsequent HR development for untagged R3a co-expressed with AVR3a^{KI} [15]. To study if R3a* variants with enhanced recognition of AVR3a^{EM} had gained the capacity to re-localize upon detection of AVR3a^{EM} and continued to exhibit this phenotype following detection of AVR3a^{KI}, N-terminal fusions of R3a* variants Rd2-1, Rd3-1 and Rd4-1 with yellow fluorescent protein (YFP) were generated as described previously with expression of these constructs driven by a 35S promoter. Western-blot analysis of protein extracts from inoculated tissue demonstrated the integrity of the fusion proteins (Fig. S3). As demonstrated for YFP-R3a wild-type fusions by Engelhardt *et al.* [15], YFP-R3a* fusions did not elicit HRs alone or in the presence of AVR3a^{KI} or AVR3a^{EM}, probably due to steric hindrance of the signalling domains of R3a (data not shown).

As anticipated, following transient expression in *N. benthamiana*, all YFP-R3a/R3a* fusions when expressed by themselves displayed cytoplasmic localizations (Fig. S4). In accord with the observations described by Engelhardt *et al.* [15], the localisation of YFP-R3a remained cytoplasmic upon co-infiltration with AVR3a^{EM} (Fig. 3), but changed to fast moving, PS1-CFP labelled vesicles, following recognition of AVR3a^{KI}. The YFP-R3a* fusions of Rd2-1, Rd3-1 and Rd4-1 proteins maintained this mechanistically characteristic re-localisation following co-expression with AVR3a^{KI} (Fig. 3). However, in contrast to YFP-R3a, all selected YFP-R3a* variants also displayed highly reproducible re-localization to PS1-CFP labelled vesicles after the perception of AVR3a^{EM} (Fig. 3; Fig S5).

AVR3a^{KI} and AVR3a^{EM} re-localize to endosomes upon co-infiltration with R3a* variants but not, in the case of AVR3a^{EM}, with wild-type R3a

As shown by Engelhardt *et al.* [15] AVR3a^{KI}, but not AVR3a^{EM}, also re-localizes from the cytoplasm to endosomes upon co-expression with R3a. This was demonstrated by N-terminal fusions of AVR3a^{KI} and AVR3a^{EM} to green fluorescent protein as well as by BiFC, also known as split-YFP, assays [9,15,30]. The latter revealed that wild-type R3a and AVR3a^{KI} are found in close proximity at PS1-CFP labelled vesicles [15].

To investigate if the vesicular co-association of R3a and AVR3a^{KI} was extended to the R3a* variants, BiFC was used to analyse and localize protein–protein interactions *in planta*. As described previously for wild-type R3a, the N-terminal portion of YFP, YN, was fused to the N-terminal end of the R3a* variants Rd2-1, Rd3-1 and Rd4-1. The constructs used to express the C-terminal portion of YFP, YC, fused to AVR3a^{KI} and AVR3a^{EM} were as described previously [15] with all constructs being transiently expressed in *N. benthamiana* from the 35S promoter.

In accord with previous findings [15], co-expression of YN-R3a with YC-AVR3a^{KI} gave strong YFP fluorescence, whereas co-expression with YC-AVR3a^{EM} did not give detectable YFP fluorescence (Fig. 4). Like the YN fusion to wild-type R3a, all the YN-R3a* fusions when co-expressed with AVR3a^{KI} gave strong, punctate, YFP signals (Fig. 4), but unlike the wild-type fusion also gave YFP fluorescence signals at PS1-CFP labelled vesicles when co-expressed with AVR3a^{EM} (Fig. 4; Fig. S6). This indicates that AVR3a^{EM} is also within close proximity of the re-localized R3a*

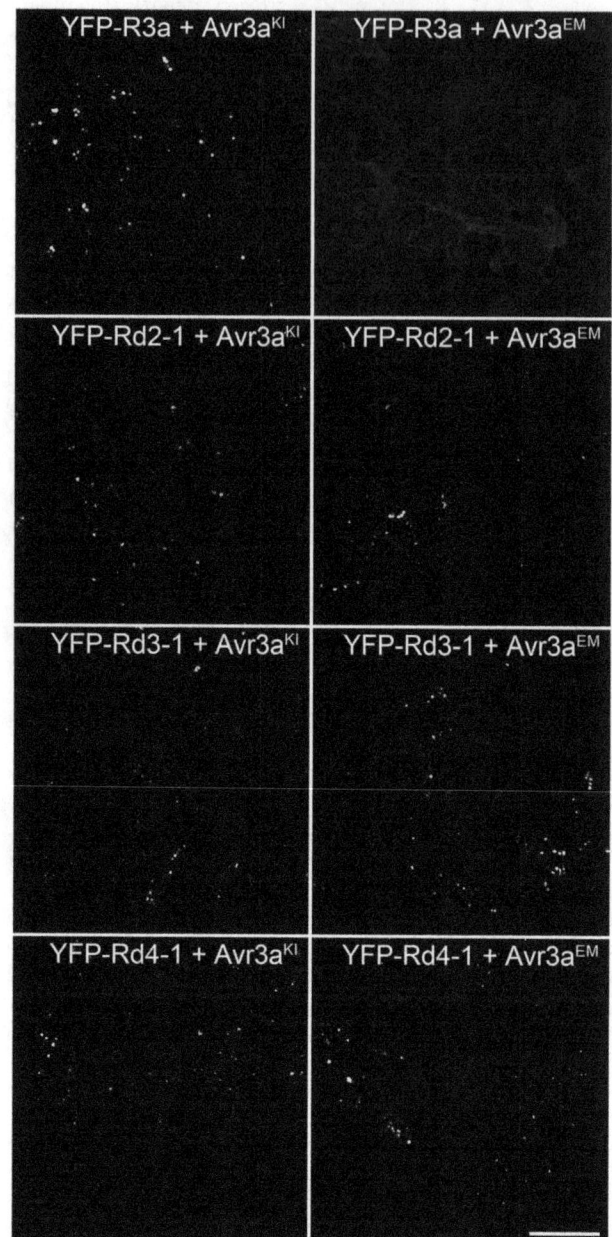

Figure 3. YFP fusions to R3a* variants re-localize to vesicles after the perception of both of AVR3a^{KI} and AVR3a^{EM}, whereas YFP-R3a remains cytoplasmic in the presence of AVR3a^{EM}. Two days after infiltration of mixtures of *Agrobacterium* cultures designed to express AVR3a^{KI}, AVR3a^{EM}, YFP-R3a or YFP fusions to the R3a* variants, infiltrated *N. benthamiana* leaf tissue was examined under a confocal laser scanning microscope. Scale bar = 50 μm.

gene products. Thus, in line with the gain of recognition of AVR3a^{EM} by the R3a* variants and subsequent necrosis responses, the R3a* variants and AVR3a^{EM} show the same mechanistic re-localization as observed for R3a and AVR3a^{KI}.

R3a* variants maintain resistance towards AVR3a^{KI}-expressing *P. infestans* isolates but have not gained resistance towards AVR3a^{EM} homozygous isolates

To evaluate if R3a* variants with gain of AVR3a^{EM} recognition and re-localisation mechanism yield effective disease resistance,

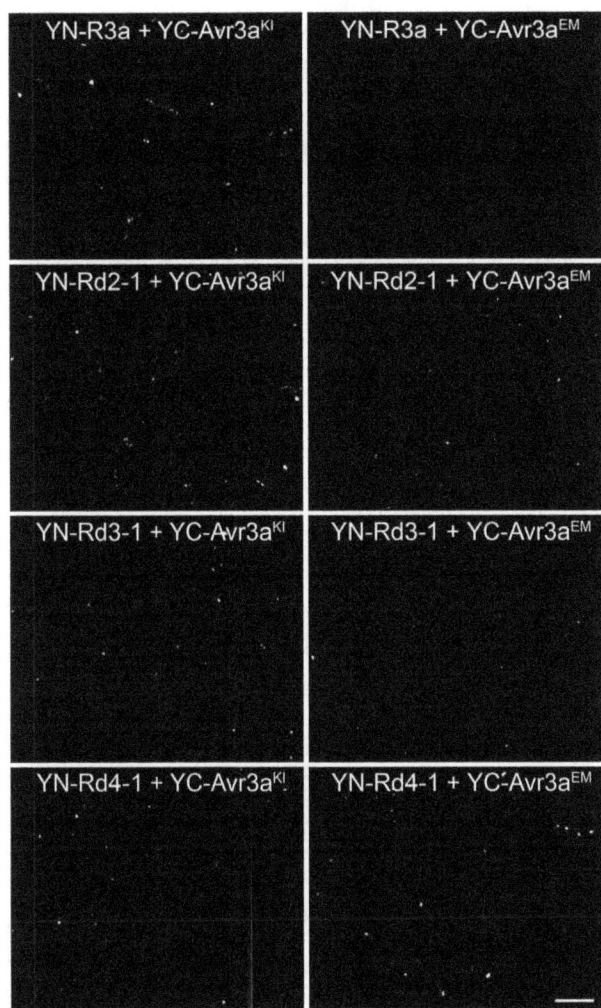

Figure 4. Both YC-AVR3a^{KI} and YC-AVR3a^{EM} when co-expressed with YN-R3a* fusions give vesicle associated YFP fluorescence like YC-AVR3a^{KI} and YN-R3a, whereas YC-AVR3a^{EM} and YN-R3a do not. Two days after infiltration of mixtures of *Agrobacterium* cultures designed to express YC-AVR3a^{KI}, YC-AVR3^{EM}, YN-R3a or YN fusions to the R3a* variants, infiltrated *N. benthamiana* leaf tissue was examined under a confocal laser scanning microscope. Representative images from two experiments. Scale bar = 50 µm.

transient and stable expression systems were utilised. *Agrobacterium tumefaciens* transient assays (ATTAs) in *N. benthamiana* have successfully been used to demonstrate function for late blight resistance gene products such as R2, Rpi_STO1 [31] and R3b [32]. Selected R3a* clones Rd2-1, Rd3-1 and Rd4-1 were transiently expressed in *N. benthamiana* using the *R3a* promoter in ATTAs alongside wild-type R3a and an empty vector control. Infiltrated leaf areas were challenged two days after infiltration with AVR3a^{KI} or AVR3a^{EM} homozygous *P. infestans* isolates via drop inoculation. Disease progression was monitored by measuring visible lesion diameters in multiple independent experiments and analysis of variance was carried out using GenStat on the data from individual experiments to test for significant differences. Multiple comparisons were performed using Bonferroni correction with a significance level of 5% and a comparison-wise error rate of 0.005.

In three experiments ATTA sites were inoculated with the AVR3a^{KI} homozygous *P. infestans* isolate 7804.b. In all three

experiments the wild-type R3a and the R3a* variants significantly reduced spread of *P. infestans* relative to the empty vector control and there were no significant differences between the different R3a forms (Fig. 5a). This result indicates that the selected mutations in the LRR do not impair the resistance induced by AVR3a^{KI}. Likewise ATTA sites were inoculated with the AVR3a^{EM} homozygous isolate 88069 [9] in five experiments. In four of the five experiments there were no significant differences in *P. infestans* spread between any of the R3a forms and the empty vector control (Fig. 5b). In the fifth experiment there was significantly increased spread with the empty vector control, but the R3a* variants showed no significant differences from the wild-type gene which does not provide resistance against AVR3a^{EM} homozygous isolates. Co-infiltrations of R3a, R3a*, AVR3a^{KI} and AVR3a^{EM} constructs were carried out contemporaneously in all experiments to confirm that the conditions were conducive to HR development (data not shown).

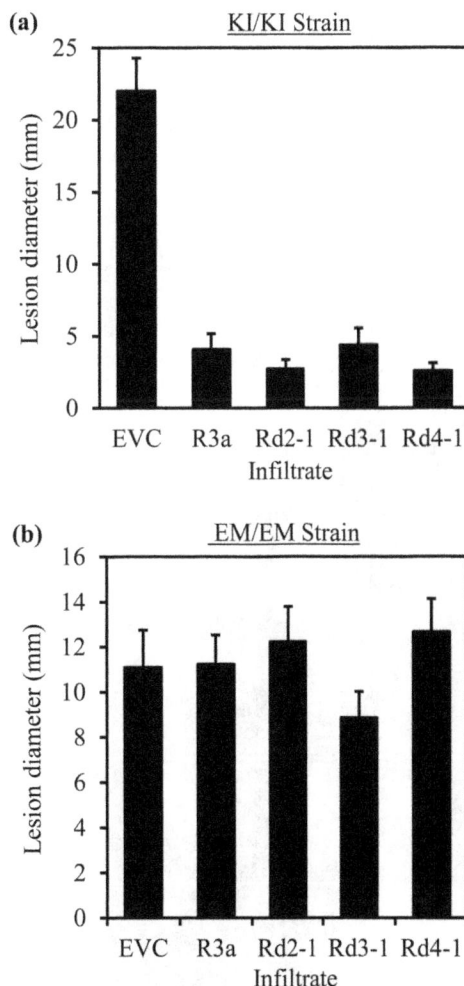

Figure 5. R3a and R3a* variants expressed from *Agrobacterium* reduce the spread of a *P. infestans* strain expressing AVR3a^{KI}, but not the spread of a strain expressing only AVR3a^{EM}. (a) Means of lesion diameters measured 12 days after drop inoculation of agro-infiltrated areas with strain 7804.b (KI/KI). (b) Means of lesion diameters measured 8 days after drop inoculation of agro-infiltrated areas with strain 88069 (EM/EM). (a) and (b) show representative experiments from sets of three and five repeated experiments, respectively. For both (a) and (b), error bars show +/− standard errors, N = 30. EVC indicates empty vector control.

To confirm these results and to rule out potential adverse effects and limitations of the ATTA system in *N. benthamiana*, transgenic potato plants were generated. The wild-type *R3a* gene and the *R3a** genes for Rd2-1, Rd3-1 and Rd4-1 were transformed into the potato cultivar Ranger Russet using the *R3a* promoter and terminator to regulate gene expression and stability. Transgenic Ranger Russet lines expressing R3a and the three R3a* variants were compared to Ranger Russet lines containing the *Rpi_vnt1* gene [33] and non-transgenic Ranger Russet plants as positive and negative controls, respectively. Transgenic lines were challenged with the Mexican isolate P6752, which is heterozygous for AVR3a^{KI} and AVR3a^{EM}, and the US isolate US-8 BF-6, which is homozygous for AVR3a^{EM}. The transgenic potato plants expressing R3a, Rpi_vnt1 and the R3a* variants, but not the non-transgenic Ranger Russet, demonstrated high levels of resistance towards the heterozygous isolate P6752 (Fig. 6). Thus, the transient ATTA data and the transgenic plants corroborate the conclusion that the selected mutations in the LRR do not negatively impact on resistance towards *P. infestans* isolates expressing Avr3a^{KI}. The transgenic Ranger Russet lines expressing Rpi_vnt1 provided resistance towards the AVR3a^{EM} homozygous isolate US-8 BF-6. However, none of the transgenic lines expressing the *R3a** or *R3a* genes, which had been shown to be resistant to isolate P6752, were able to control disease development of isolate US-8 BF-6 (Fig. 6).

Figure 6. R3a and R3a* variants expressed via the *R3a* promoter in transgenic plants protect the susceptible cultivar Ranger Russet from *P. infestans* strain P6752, which is heterozygous for AVR3a^{KI} and AVR3a^{EM}, but not from strain US-8 BF-6, which is homozygous for AVR3a^{EM}. Non-transgenic plants were used as a control for susceptibility. Transgenic plants expressing R3a or Rpi-vnt1 were used as positive controls for resistance to P6752 or US-8 BF-6, respectively. Representative plants were photographed at 11dpi.

Discussion

The relatively narrow genetic basis of clonal potato cultivars in agriculture provides pathogens such as *P. infestans* with sufficient opportunity to adapt and overcome inducible host resistant responses and to thus cause disease on a global scale. Resistance responses rely on the direct or indirect recognition of modified-self or pathogen-derived molecules [1]. For example, in the first layer of inducible resistance, also referred to as PAMP triggered immunity (PTI), conserved pathogen-associated molecular patterns (PAMPs) and/or damage-associated molecular patterns (DAMPs) are recognised [1,34,35]. Successful pathogens circumvent this recognition with the help of effectors that perturb host resistance responses and promote effector-triggered susceptibility (ETS). However, by being in close proximity to the host, pathogen effectors provide the innate plant immune system with another opportunity for detection that is dependent on the presence of cognate plant R proteins that subsequently yield ETI [1,36]. In nature, this closely entwined co-evolution between hosts and pathogens is evident in the diversification observed for effectors and *R* genes [4].

Indeed, *P. infestans* is known as a pathogen with 'high evolutionary potential' [37] and more than 560 RXLR-type candidate effectors have been described within the late blight pathogen genome [38]. The genomic organisation of RXLR effector genes, which are often in gene-poor and repeat-rich regions, is thought to facilitate their enhanced diversification by enabling non-homologous recombination. Oomycete RXLR-type effectors have been shown to evade detection by *R* gene products via transcriptional regulation [39], utilising functionally equivalent effectors that allow loss of recognised effectors [40], suppressor activity of unrelated effectors [41] and/or sequence diversity [11,42]. Sequence diversity underpins virulence or avirulence behaviour for the essential *P. infestans* effector AVR3a. AVR3a^{KI} determines avirulence on plants carrying *R3a* whereas AVR3a^{EM} promotes virulence. It is thought that only AVR3a^{KI} was present in the *P. infestans* strain responsible for the outbreak of late blight disease leading to the Irish Potato Famine in the 1840 s [43]. The AVR3a^{EM} allele may have come to dominate in *P. infestans* populations once the resistance gene *R3a* was deployed in potato crops, quickly usurping the AVR3a^{KI} allele. By using clonal potato varieties in current agriculture, the diversification of *R* genes is undermined and novel, naturally occurring resistances can only slowly be introgressed via breeding.

Functional *R* genes are often found in clusters that show evidence of duplication followed by diversification [44]. The phylogenetic NB-LRR gene grouping that contains homologs of the functional *R3a* gene in the potato genome has previously been described as CNL-8 [5]. This group is, after the *R2* cluster (CNL-5), the second largest *R* gene cluster in potato where more than 750 NB-LRR-like genes have been identified [5–6]. Functional *R3a*, a homolog of the tomato *I2* resistance gene that controls races of the fungus *Fusarium oxysporum* [45], was cloned alongside three paralogous sequences that provided insight into amino acid positions under diversification [13]. With one exception, positions under diversification reside in the LRR and cluster around two regions spanning LRRs 1 to 4 and 14 to 23 that were also identified as being important in this study.

Here we used, in addition to random mutagenesis and screening, DNA shuffling and targeted mutagenesis to enhance AVR3a^{EM} recognition by R3a*. DNA shuffling, which emulates the natural evolutionary processes of mutation, recombination and selection, has proven a highly effective method for evolving new specificities/properties for a wide range of proteins that cannot be rationally designed and is of particular use in identifying mutations that are beneficial in combination [18]. Artificial evolution of a resistance gene to broaden its specificity has previously been performed on the gene *Rx* that protects potato plants from strains of PVX. In an initial *Rx* study, mutagenesis of the LRR region was performed on the basis that this region is the primary determinant of recognition specificity. Screening identified four mutations in the LRR region that affected elicitor recognition and activation functions [16]. Introduction of the mutated *Rx* genes in to the model host *N. benthamiana* as transgenes extended resistance to a normally resistance-breaking strain of PVX, HB, and a distantly related carlavirus, PopMV [16].

Our primary screen identified eleven mutants with enhanced AVR3a^{EM} recognition containing in total 23 amino acid substitutions. However, only three of these (R618Q, K920E, Q931R), found in the clones with single amino acid substitutions, are known to be causative to the improved phenotype. The previous study performed by Segretin *et al.* [25] identified six amino acid substitutions in the LRR domain that enhanced AVR3a^{EM} recognition: two of these were also found in our primary screen (L668P, K920E). The fact that we did not identify all the LRR mutations identified by Segretin *et al.* [25] and that they did not identify more mutations in the LRR region, demonstrates that neither screen was exhaustive.

The initial screen identified mutants with enhanced recognition of AVR3a^{EM} when expressed from the strong 35S promoter. Amino acid changes in LRRs #3 and #15, and thus within regions known to be under diversifying selection [13], were prevalent in the clones from the first round of screening, suggesting these regions might be of particular importance. This suggestion was supported by the fact that all the clones from the second round of shuffling contained an amino acid change, E620D, in LRR #3 and sometimes additional changes in LRR #15, while all of the clones from the third round contained the E620D change and one or two amino acid changes in LRR #15. Using DNA shuffling it is more difficult to bring together combinatorially beneficial mutations that are in close proximity. Furthermore, random mutagenesis as a source of diversity has limitations in that single nucleotide changes can only convert a codon for one amino acid to a limited set of codons for other amino acids rather than all twenty possible amino acids. To circumvent the former problem a more directed approach was adopted in which a library that contained all permutations of the amino acid changes thought important (one in LRR #2, two in LRR #3 and four in LRR #15) was produced for screening. The aptness of this approach was shown by the fact that clones were obtained with enhanced recognition responses to AVR3a^{EM} compared to the best performing clone from the third round of shuffling, Rd3-1, and that one of the possible 128 forms was prevalent. This form contained two amino acid changes in close proximity in each of LRRs #3 and #15; a combination that would have been difficult to obtain through DNA shuffling. Interestingly, an amino acid change at one of these positions, Q931, was also found in one of the clones recovered by Segretin *et al.* [25], though their "de-convolution" did not show their substitution at this position, proline instead of arginine, to improve AVR3a^{EM} recognition. As shown by Segretin *et al.* [25] for the R3a mutants with enhanced AVR3a^{EM} recognition, we did not note any reduction in the AVR3a^{KI} recognition responses of the clones we isolated.

Our studies show that the recognition of AVR3a^{EM} by the R3a* mutants we have isolated recapitulates the mechanistic processes of recognition of AVR3a^{KI} by the wild-type *R3a* gene. It has previously been reported [14] that the HR triggered by R3a recognition of AVR3a^{KI} is dependent on the ubiquitin ligase-associated protein SGT1 and HSP90. VIGS of *SGT1* and, to a lesser degree, of *HSP90* in our experiments inhibited the cell death responses induced by recognition of AVR3a^{EM} by our R3a* mutants. Similarly, it has been shown that wild-type R3a re-localises from the cytoplasm to late endosomal compartments when co-expressed with AVR3a^{KI}, but not when co-expressed with AVR3a^{EM} [15]. We found that the mutants from the three later rounds still re-localised to endosomal compartments when co-expressed with AVR3a^{KI} and, importantly, also re-localised to the same vesicles when co-expressed with AVR3a^{EM}. The earlier study by Engelhardt *et al.* [15] also showed that the effector AVR3a^{KI} itself relocalises from the cytoplasm to endosomes when co-expressed with wild-type R3a and is in close physical proximity to R3a, whereas AVR3a^{EM} remains distributed through the cytoplasm. Our BiFC experiments show that AVR3a^{KI} and the normally unrecognized form, AVR3a^{EM} both traffic from the cytoplasm to vesicles when co-expressed with the R3a* forms. This re-localisation of R3a and AVR3^{KI} was shown to be a prerequisite for the development of the HR [15]. Thus, the gain of AVR3a^{EM} recognition R3a* variants have gained many aspects of the mechanism of the wild-type R3a response to AVR3a^{KI}.

Although the R3a* variants produced in this study responded to AVR3a^{EM} and produced HR responses when the elicitor was transiently expressed via *Agrobacterium*, critically they only provided resistance to *P. infestans* isolates that express AVR3a^{KI} and not to isolates that express only AVR3a^{EM}. Both transient expression via *Agrobacterium* in *N. benthamiana* and stable transgenic expression in potato corroborated this finding. Failure to protect from the pathogen itself was also reported for the R3a mutants identified by Segretin *et al.* [25]. That this was the case for mutants with single amino acid changes is perhaps not surprising given the large differences from the wild-type R3a/AVR3a^{KI} response we observed when first and second round clones were expressed from the *R3a* promoter with AVR3a^{EM}. For some pathogen/*R* gene combinations, e.g. PVX and *Rx*, the resistance responses can be separated from the HR [46] though the induction of necrotic responses by transient expression of the elicitor protein has been used to identify Rx mutations that, when expressed transgenically in the model host *N. benthamiana*, provide resistance to the pathogen itself. However, for *P. infestans* it has been suggested that the strength of the HR correlates with resistance levels [47].

A recent, secondary mutation study of *Rx* provides some evidence that stepwise artificial evolution could be required to obtain an optimum combination between effector recognition and subsequent *R* gene activation and signal transduction [17]. In the *Rx* study, the resistance provided by one of the mutations in the LRR domain to PopMV was improved by random mutagenesis of the CC-NB-ARC1-ARC2 domains [17]. In addition to constitutively active mutants that by themselves gave necrotic responses, four mutants with enhanced responses to PopMV were isolated. For three of these mutants the improved phenotype was conferred by a single amino acid change, while for the fourth a pair of amino acid substitutions was required. The mutations, which affect activation sensitivity, were found to be located around the nucleotide-binding pocket of Rx. As mentioned previously, we have evidence that neither this screening nor the efforts from Segretin *et al.* [25] were exhaustive as both approaches yielded novel beneficial mutations. Whereas our study of *R3a* focused solely on the LRRs, in Segretin *et al.* [25] the entire *R3a* gene was subjected to mutagenesis. The latter study identified eight single amino acid changes that enhanced responses to AVR3a^{EM}. Out of these, six occurred in the LRR domain and one in the CC domain. This substitution enhanced the response to AVR3a^{EM} but also showed some auto-activation. The final substitution, found in the NB-ARC domain, sensitised the AVR3a^{EM} response and broadened specificity to include an elicitor from another *Phytophthora* species [25]. Interestingly, this change occurred in the nucleotide-binding pocket and is adjacent to one of the sensitizing mutations found in *Rx* [17]. Broadening resistance gene specificity merely by introducing sensitizing mutations without improving recognition may have detrimental consequences in the field. However, a natural precedent for this has been found in PM3 resistance protein alleles in which the substitution of two amino acids in the NB domain enhances the HR and broadens the spectrum of resistance to wheat powdery mildew isolates [48]. Thus, additional efforts to combine novel mutations in R3a domains responsible for AVR3a^{EM} recognition (LRR) and response (CC-NB-ARC1/ARC2) could further improve R3a* variants that already display gain of recognition and mechanistic re-localisation to ultimately yield genes that provide effective resistance in potato against isolates expressing AVR3a^{EM}. Considering the importance of AVR3a to *P. infestans*, such a resistance, combined with other, mechanistically distinct *R* genes could provide a step towards more durable late blight control.

Supporting Information

Figure S1 R3a* variants are not auto-activators. *N. benthamiana* leaves were infiltrated with *Agrobacterium* cultures designed to express R3a or the R3a* variants from the strong 35S promoter. Mixtures of cultures designed to co-express AVR3a^{KI} (KI) were used as positive controls for the induction of cell death. Leaves were examined under white-light and UV-B illumination. Photograph of representative leaf was taken five days after infiltration. In the absence of elicitor the R3a* variants, like R3a, do not produce visible cell death.

Figure S2 HR responses resulting from R3a* recognition of AVR3a^{EM} and AVR3a^{KI}, like those caused by wild-type R3a recognition of AVR3a^{KI}, are dependent on SGT1 and HSP90. SGT1- and HSP90-silenced plants were produced using TRV-based vectors. These plants and control plants inoculated with TRV:*eGFP* were infiltrated with different combinations of *Agrobacterium* cultures designed to express R3a, R3a* variants, AVR3a^{KI} (KI) or AVR3a^{EM} (EM). Photographs show representative HR responses induced by each of the different mixtures on control TRV:eGFP inoculated plants, SGT1-silenced plants and HSP90-silenced plants. The non-host bacterial pathogen *Erwinia amylovora* was used as a control for an SGT1- and HSP90-independnet HR response.

Figure S3 Western blot analysis showing integrity of YFP fusion proteins. Soluble protein extracts were prepared from *N. benthamiana* leaf tissue two days after infiltration with cultures designed to express YFP fusions to R3a, Rd2-1, Rd3-1 or Rd4-1. The blot was probed with anti-GFP antibodies as described by Engelhardt *et al.* (2012). Protein sizes are indicated

in kilodaltons (kD) and protein loading is shown by Ponceau S (PS) staining.

Figure S4 YFP fusions to R3a and the R3a* variants localize to the cytoplasm in the absence of AVR3a. *N. benthamiana* leaves were infiltrated with cultures designed to express YFP fusions to R3a, Rd2-1, Rd3-1 or Rd4-1. Leaf tissue was examined two days after infiltration under a confocal laser scanning microscope. Representative images are from five independent experiments. Scale bar = 20 μm.

Figure S5 In the presence of AVR3a^{EM} YFP fusions to R3a* variants, but not YFP-R3a, re-localize to vesicles labelled by the prevacuolar compartment marker PS1-CFP. *N. benthamiana* leaves were infiltrated with mixtures of cultures designed to express PS1-CFP, AVR3a^{EM} and YFP fusions to R3a, Rd2-1, Rd3-1 or Rd4-1. Tissue was examined two days after infiltration under a confocal laser scanning microscope. The left-hand panel shows YFP signal, the right-hand panel CFP signal and the central panel displays the merged signals. Representative images are from three independent experiments. Scale bar = 10 μm.

Figure S6 YC-AVR3a^{EM} reconstitutes YFP fluorescence with YN fusions to the R3a* variants at vesicles labelled by the prevacuolar compartment marker PS1-CFP. Generation of the YFP signal indicates that AVR3aEM and the R3a* variants are in close proximity at the vesicles. *N. benthamiana* leaves were infiltrated with mixtures of cultures designed to express PS1-CFP, YC-AVR3a^{EM} and YN fusions to Rd2-1, Rd3-1 or Rd4-1. Tissue was examined 2 d after infiltration under a confocal laser scanning microscope. Left-hand panel, YFP signal; right-hand panel, CFP signal; central panel, merged signals. Representative images from three experiments. Scale bar = 20 μm.

Acknowledgments

The authors acknowledge Dr Sanwen Huang for kindly providing sequences of wild type R3a and the native regulatory elements.

Author Contributions

Conceived and designed the experiments: SC PCB PRJB IH. Performed the experiments: SC LS PCB SE BH NC KM PSMVW XC. Analyzed the data: SC LS PCB CA NC PRJB IH. Wrote the paper: IH SC LS CA PRJB.

References

1. Jones JDG, Dangl JL (2006) The plant immune system. Nature 444: 323–329.
2. van der Biezen EA, Jones JDG (1998) Plant disease-resistance proteins and the gene-for-gene concept. Trends Biochem Sci 23: 454–456.
3. Heath MC (2000) Hypersensitive response-related death. Plant Mol Biol 44: 321–334.
4. Hein I, Gilroy EM, Armstrong MR, Birch PRJ (2009) The zig-zag-zig in oomycete–plant interactions. Mol Plant Pathol 10: 547–562.

5. Jupe F, Pritchard L, Etherington GJ, MacKenzie K, Cock PJA, et al. (2012) Identification and localisation of the NB-LRR gene family within the potato genome. BMC Genomics 13: 75.

6. Jupe F, Witek K, Verweij W, Śliwka J, Pritchard L, et al. (2013) Resistance gene enrichment sequencing (RenSeq) enables reannotation of the NB-LRR gene family from sequenced plant genomes and rapid mapping of resistance loci in segregating populations. Plant J 76: 530–544.

7. Birch PRJ, Boevink PC, Gilroy EM, Hein I, Pritchard L, et al. (2008) Oomycete RXLR effectors: delivery, functional redundancy and durable disease resistance. Curr Opin Plant Biol 11: 373–379.

8. Vleeshouwers VG, Raffaele S, Vossen JH, Champouret N, Oliva R, et al. (2011) Understanding and exploiting late blight resistance in the age of effectors. Annu Rev Phytopathol. 49:507–531.

9. Bos JIB, Armstrong MR, Gilroy EM, Boevink PC, Hein I, et al. (2010) Phytophthora infestans effector AVR3a is essential for virulence and manipulates plant immunity by stabilizing host E3 ligase CMPG1. Proc Natl Acad Sci USA 107: 9909–9914.

10. Vetukuri RR, Tian Z, Avrova AO, Savenkov EI, Dixelius C, et al. (2011) Silencing of the PiAvr3a effector-encoding gene from Phytophthora infestans by transcriptional fusion to a short interspersed element. Fungal Biol 115: 1225–1233.

11. Armstrong MR, Whisson SC, Pritchard L, Bos JIB, Venter E, et al. (2005) An ancestral oomycete locus contains late blight avirulence gene Avr3a, encoding a protein that is recognized in the host cytoplasm. Proc Natl Acad Sci USA 102: 7766–7771.

12. Cárdenas M, Grajales A, Sierra R, Rojas A, González-Almario A, et al. (2011) Genetic diversity of Phytophthora infestans in the Northern Andean region. BMC Genet 12: 23.

13. Huang S, van der Vossen EAG, Kuang H, Vleeshouwers VGAA, Zhang N, et al. (2005) Comparative genomics enabled the isolation of the R3a late blight resistance gene in potato. Plant J 42: 251–261.

14. Bos JIB, Kanneganti T-D, Young C, Cakir C, Huitema E, et al. (2006) The C-terminal half of Phytophthora infestans RXLR effector AVR3a is sufficient to trigger R3a-mediated hypersensitivity and suppress INF1-induced cell death in Nicotiana benthamiana. Plant J 48: 165–176.

15. Engelhardt S, Boevink PC, Armstrong MR, Ramos MB, Hein I, et al. (2012) Relocalization of late blight resistance protein R3a to endosomal compartments is associated with effector recognition and required for the immune response. Plant Cell 24: 5142–5158.

16. Farnham G, Baulcombe DC (2006) Artificial evolution extends the spectrum of viruses that are targeted by a disease-resistance gene from potato. Proc Natl Acad Sci USA 103: 18828–18833.

17. Harris CJ, Slootweg EJ, Goverse A, Baulcombe DC (2013) Stepwise artificial evolution of a plant disease resistance gene. Proc Natl Acad Sci USA, 110: 21189–21194.

18. Stemmer WP (1994) DNA shuffling by random fragmentation and reassembly: in vitro recombination for molecular evolution. Proc Natl Acad Sci USA 91: 10747–10751.

19. Bernal A, Pan Q, Pollack J, Rose L, Willets N, et al. (2005) Functional dissection of the Pto resistance gene using DNA shuffling. J Biol Chem 280: 23073–23083

20. Simpson CG, Lewandowska D, Liney M, Davidson DDavidson D, Chapman S, et al. (2014) Arabidopsis PTB1 and PTB2 proteins negatively regulate splicing of a mini-exon splicing reporter and affect alternative splicing of endogenous genes differentially. New Phytol 203: 424–436.

21. Leung DW, Chen E, Goeddel DV (1989) A method for random mutagenesis of a defined DNA segment using a modified polymerase chain reaction. Technique, 1: 11–15.

22. Lazo GR, Stein PA, Ludwig RA (1991) A DNA transformation-competent Arabidopsis genomic library in Agrobacterium. Biotechnology 9: 963–967.

23. van der Fits L, Deakin EA, Hoge JH, Memelink J (2000) The ternary transformation system: constitutive virG on a compatible plasmid dramatically increases Agrobacterium-mediated plant transformation. Plant Mol Biol 43: 495–502.

24. Duan H, Richael C, Rommens CM (2012) Overexpression of the wild potato eIF4E-1 variant Eva1 elicits Potato virus Y resistance in plants silenced for native eIF4E-1. Transgenic Res 21: 929–938.

25. Segretin ME, Pais M, Franceschetti M, Chaparro-Garcia A, Bos JIB, et al. (2014) Single amino acid mutations in the potato immune receptor R3a expand response to Phytophthora effectors. Mol Plant Microbe Interact 27: 624–637.

26. Liu Y, Burch-Smith T, Schiff M, Feng S, Dinesh-Kumar SP (2004) Molecular chaperone Hsp90 associates with resistance protein N and its signaling proteins SGT1 and Rar1 to modulate an innate immune response in plants. J Biol Chem 279: 2101–2108.

27. Azevedo C, Betsuyaku S, Peart J, Takahashi A, Noel L, et al. (2006) Role of SGT1 in resistance protein accumulation in plant immunity. EMBO J, 25, 2007–2016.

28. Gilroy EM, Hein I, van der Hoorn R, Boevink PC, Venter E, et al. (2007) Involvement of cathepsin B in the plant disease resistance hypersensitive response. Plant J 52: 1–13.

29. Saint-Jean B, Seveno-Carpentier E, Alcon C, Neuhaus JM, Paris N (2010) The cytosolic tail dipeptide Ile-Met of the pea receptor BP80 is required for recycling from the prevacuole and for endocytosis. Plant Cell, 22: 2825–2837.

30. Walter M, Chaban C, Schütze K, Batistic O, Weckermann K, et al. (2004) Visualization of protein interactions in living plant cells using bimolecular fluorescence complementation. Plant J 40: 428–438.

31. Saunders DG, Breen S, Win J, Schornack S, Hein I, et al. (2012) Host protein BSL1 associates with Phytophthora infestans RXLR effector AVR2 and the Solanum demissum Immune receptor R2 to mediate disease resistance. Plant Cell 24: 3420–3434.

32. Li G, Huang S, Guo X, Li Y, Yang Y, et al. (2011) Cloning and characterization of R3b; members of the R3 superfamily of late blight resistance genes show sequence and functional divergence. Mol Plant Microbe Interact 24: 1132–1142.

33. Foster SJ, Park TH, Pel M, Brigneti G, Sliwka J, et al. (2009) Rpi-vnt1.1, a Tm-2(2) homolog from Solanum venturii, confers resistance to potato late blight. Mol Plant Microbe Interact 22: 589–600.

34. Maffei ME, Arimura G, Mithöfer A (2012) Natural elicitors, effectors and modulators of plant responses. Nat Prod Rep 29: 1288–1303.

35. Newman MA, Sundelin T, Nielsen JT, Erbs GI (2013). MAMP (microbe-associated molecular pattern) triggered immunity in plants. Front Plant Sci 16;4: 139

36. Deslandes L, Rivas S (2012) Catch me if you can: bacterial effectors and plant targets. Trends Plant Sci. 17: 644–655

37. Raffaele S, Win J, Cano LM, Kamoun S (2010) Analyses of genome architecture and gene expression reveal novel candidate virulence factors in the secretome of Phytophthora infestans. BMC Genomics. 16: 637.

38. Haas BJ, Kamoun S, Zody MC, Jiang RHY, Handsaker RE, et al. (2009) Genome sequence and analysis of the Irish potato famine pathogen Phytophthora infestans. Nature 461: 393–398.

39. Rietman H, Bijsterbosch G, Cano LM, Lee HR, Vossen JH, et al. (2012) Qualitative and quantitative late blight resistance in the potato cultivar Sarpo Mira is determined by the perception of five distinct RXLR effectors. Mol Plant Microbe Interact 25: 910–919.

40. Van Poppel PM, Guo J, van de Vondervoort PJ, Jung MW, Birch PR, et al. (2008) The Phytophthora infestans avirulence gene Avr4 encodes an RXLR-dEER effector. Mol Plant Microbe Interact 21: 1460–1470.

41. Wang Q, Han C, Ferreira AO, Yu X, Ye W, et al. (2011) Transcriptional programming and functional interactions within the Phytophthora sojae RXLR effector repertoire. Plant Cell. 23: 2064–2086

42. Gilroy EM, Tayor RM, Hein I, Boevink P, Sadanandom A, et al. (2011) CMPG1-dependent cell death follows perception of diverse pathogen elicitors at the host plasma membrane and is suppressed by Phytophthora infestans RXLR effector AVR3a. New Phytol 190: 653–666.

43. Yoshida K, Schuenemann VJ, Cano LM, Pais M, Mishra B, et al. (2013) The rise and fall of the Phytophthora infestans lineage that triggered the Irish potato famine. eLife, 2: e00731.

44. McDowell JM, Simon SA (2006) Recent insights into R gene evolution. Mol Plant Pathol 7, 437–448.

45. Ori N1, Eshed Y, Paran I, Presting G, Aviv D, et al. (1997) The I2C family from the wilt disease resistance locus I2 belongs to the nucleotide binding, leucine-rich repeat superfamily of plant resistance genes. Plant Cell. 9: 521–532.

46. Bendahmane A, Kanyuka K, Baulcombe DC (1999) The Rx gene from potato controls separate virus resistance and cell death responses. Plant Cell, 11: 781–792.

47. Vleeshouwers VGAA, van Dooijeweert W, Govers F, Kamoun S, Colon LT (2000) The hypersensitive response is associated with host and nonhost resistance to Phytophthora infestans. Planta 210: 853–864.

48. Stirnweiss D, Milani SD, Jordan T, Keller B, Brunner S (2014) Substitutions of two amino acids in the nucleotide-binding site domain of a resistance protein enhance the hypersensitive response and enlarge the PM3F resistance spectrum in wheat. Mol Plant Microbe Interact 27: 265–276.

Permissions

All chapters in this book were first published in PLOS ONE, by The Public Library of Science; hereby published with permission under the Creative Commons Attribution License or equivalent. Every chapter published in this book has been scrutinized by our experts. Their significance has been extensively debated. The topics covered herein carry significant findings which will fuel the growth of the discipline. They may even be implemented as practical applications or may be referred to as a beginning point for another development.

The contributors of this book come from diverse backgrounds, making this book a truly international effort. This book will bring forth new frontiers with its revolutionizing research information and detailed analysis of the nascent developments around the world.

We would like to thank all the contributing authors for lending their expertise to make the book truly unique. They have played a crucial role in the development of this book. Without their invaluable contributions this book wouldn't have been possible. They have made vital efforts to compile up to date information on the varied aspects of this subject to make this book a valuable addition to the collection of many professionals and students.

This book was conceptualized with the vision of imparting up-to-date information and advanced data in this field. To ensure the same, a matchless editorial board was set up. Every individual on the board went through rigorous rounds of assessment to prove their worth. After which they invested a large part of their time researching and compiling the most relevant data for our readers.

The editorial board has been involved in producing this book since its inception. They have spent rigorous hours researching and exploring the diverse topics which have resulted in the successful publishing of this book. They have passed on their knowledge of decades through this book. To expedite this challenging task, the publisher supported the team at every step. A small team of assistant editors was also appointed to further simplify the editing procedure and attain best results for the readers.

Apart from the editorial board, the designing team has also invested a significant amount of their time in understanding the subject and creating the most relevant covers. They scrutinized every image to scout for the most suitable representation of the subject and create an appropriate cover for the book.

The publishing team has been an ardent support to the editorial, designing and production team. Their endless efforts to recruit the best for this project, has resulted in the accomplishment of this book. They are a veteran in the field of academics and their pool of knowledge is as vast as their experience in printing. Their expertise and guidance has proved useful at every step. Their uncompromising quality standards have made this book an exceptional effort. Their encouragement from time to time has been an inspiration for everyone.

The publisher and the editorial board hope that this book will prove to be a valuable piece of knowledge for researchers, students, practitioners and scholars across the globe.

List of Contributors

Dawei Yu
Embryo Biotechnology and Reproduction Laboratory, Institute of Animal Science, Chinese Academy of Agricultural Sciences, Beijing, China
State Key Laboratories of Agrobiotechnology, College of Biological Science, China Agricultural University, Beijing, China

Shoufeng Zhang and Jinxia Zhang
Institute of Military Veterinary, Academy of Military Medical Science, Changchun, China

Weihua Du, Zongxing Fan, Haisheng Hao, Yan Liu, Xueming Zhao, Tong Qin and Huabin Zhu
Embryo Biotechnology and Reproduction Laboratory, Institute of Animal Science, Chinese Academy of Agricultural Sciences, Beijing, China

Dongping Zhang, Li Chen, Bing Lv, Yun Chen, XuejiaoYan and Jiansheng Liang
Department of Biotechnology, College of Bioscience and Biotechnology, Yangzhou University, Jiangsu, China

Dahong Li
Department of Biological Engineering, Huanghuai University, Zhumadian City, Henan Province, China

Jingui Chen
Biosciences Division, Oak Ridge National Laboratory, Oak Ridge, Tennessee, United States of America

Dipak Kumar Sahoo and Indu Bhushan Maiti
KTRDC, College of Agriculture, Food and Environment, University of Kentucky, Lexington, Kentucky, United States of America

Nrisingha Dey
Department of Gene Function and Regulation, Institute of Life Sciences, Bhubaneswar, Odisha, India

Yulin Tang, Zhan Gao, Zhonghua Ou, Yajing Wang and Yizhi Zheng
Shenzhen Key Laboratory of Microbial and Gene Engineering, College of Life Sciences, Shenzhen University, Shenzhen, Guangdong, People's Republic of China

Yan Cao and Jianbin Qiu
The Key Laboratory for Marine Bioresource and Eco-environmental Science, College of Life Sciences, Shenzhen University, Shenzhen, Guangdong, People's Republic of China

Amal Arachiche, María de la Fuente and Marvin T. Nieman
Department of Pharmacology, Case Western Reserve University, Cleveland, Ohio, United States of America

Haiyang Nan and Sijia Lu
The Key of Soybean Molecular Design Breeding, Northeast Institute of Geography and Agroecology, Chinese Academy of Sciences, Nangang District, Harbin, China

Dong Cao, Lili Tang, Xiaohui Yuan, Baohui Li and Fanjiang Kong
The Key of Soybean Molecular Design Breeding, Northeast Institute of Geography and Agroecology, Chinese Academy of Sciences, Nangang District, Harbin, China
University of Chinese Academy of Sciences, Beijing, China

Dayong Zhang
Institute of Biotechnology, Jiangsu Academy of Agricultural Sciences, Nanjing, China

Ying Li
State Key Laboratory of Tree Genetics and Breeding, Northeast Forestry University, Harbin, China

Narendra Tuteja, Mst. Sufara Akhter Banu, Kazi Md. Kamrul Huda1, Sarvajeet Parul Jain, Xuan Hoi Pham and Renu Tuteja
International Centre for Genetic Engineering and Biotechnology, Aruna Asaf Ali Marg, New Delhi, India,

Singh Gill
Stress Physiology and Molecular Biology Lab, Centre for Biotechnology, MD University, Rohtak, India

Victor J. Pai, Bin Wang, Xiangyong Li and Jing X. Kang
Laboratory for Lipid Medicine and Technology (LLMT), Massachusetts General Hospital and Harvard Medical School, Boston, Massachusetts, United States of America,

Lin Wu
Cutaneous Biology Research Center, Massachusetts General Hospital and Harvard Medical School, Boston, Massachusetts, United States of America

Jinyi Liu, J. Hollis Rice Nana Chen and Tarek Hewezi
Department of Plant Sciences, University of Tennessee, Knoxville, Tennessee, United States of America,

Thomas J. Baum
Department of Plant Pathology and Microbiology, Iowa State University, Ames, Iowa, United States of America

Jirong Wu, Mingzheng Yu, Jianhong Xu, Juan Du, Fang Ji, Fei Dong and Jianrong Shi
Institute of Food Safety and Detection, Jiangsu Academy of Agricultural Sciences, Nanjing, China
Key Lab of Food Quality and Safety of Jiangsu Province—State Key Laboratory Breeding Base, Nanjing, China
Jiangsu Center for GMO evaluation and detection, Nanjing, China

Xinhai Li
Institute of Crop Sciences, Chinese Academy of Agricultural Sciences, Beijing, China

Pan Liao, Hui Wang, Mingfu Wang, An-Shan Hsiao and Mee-Len Chye
School of Biological Sciences, The University of Hong Kong, Hong Kong, China,

Thomas J. Bach
Centre National de la Recherche Scientifique, UPR 2357, Institut de Biologie Moléculaire des Plantes, Strasbourg, France

Katja U. Beiser, Anne Glaser, Isabell Scholl, Ralph Röth and Gudrun A. Rappold
Department of Human Molecular Genetics, Heidelberg University Hospital, Heidelberg, Germany

Kerstin Kleinschmidt and Wiltrud Richter
Division of Experimental Orthopaedics, Orthopaedic University Hospital, Heidelberg, Germany

Li Li and Norbert Gretz
Medical Research Center (ZMF), Medical Faculty Mannheim at Heidelberg University, Mannheim, Germany,

Gunhild Mechtersheimer
Institute of Pathology, Heidelberg University Hospital, Heidelberg, Germany,

Marcel Karperien
Department of Developmental Bioengineering, University of Twente, Enschede, The Netherlands,

Antonio Marchini
Department of Human Molecular Genetics, Heidelberg University Hospital, Heidelberg, Germany
German Cancer Research Center (DKFZ), Heidelberg, Germany

Relu Cocoş
Chair of Medical Genetics, "Carol Davila" University of Medicine and Pharmacy, Bucharest, Romania
Genome Life Research Centre, Bucharest, Romania

Alina Şendroiu
Family Medical Centre, Rucar, Romania

Sorina Schipor
National Institute of Endocrinology "C. I. Parhon", Bucharest, Romania

Laurenţiu Camil Bohîlţea
Chair of Medical Genetics, "Carol Davila" University of Medicine and Pharmacy, Bucharest, Romania
Sf. Pantelimon Clinical Emergency Hospital, Bucharest, Romania

Ionuţ Şendroiu
Family Medical Centre, Rucar, Romania

Florina Raicu
Chair of Medical Genetics, "Carol Davila" University of Medicine and Pharmacy, Bucharest, Romania
Francisc I. Rainer Anthropological Research Institute, Romanian Academy, Bucharest, Romania

Yongqiang Gao and Changshui Liu
CAS Key Laboratory of Biobased Materials, Qingdao Institute of Bioenergy and Bioprocess Technology, Chinese Academy of Sciences, Qingdao, China
University of Chinese Academy of Sciences, Beijing, China

Yamei Ding
Institute of Oceanology, Chinese Academy of Sciences, Qingdao, China

Chao Sun and Rubing Zhang
CAS Key Laboratory of Biobased Materials, Qingdao Institute of Bioenergy and Bioprocess Technology, Chinese Academy of Sciences, Qingdao, China

Mo Xian and Guang Zhao
CAS Key Laboratory of Biobased Materials, Qingdao Institute of Bioenergy and Bioprocess Technology, Chinese Academy of Sciences, Qingdao, China
Collaborative Innovation Center for Marine Biomass Fibers, Materials and Textiles of Shandong Province, Qingdao, China

Hyo-Hyoun Seo
Medicinal Nanomaterial Institute, BIO-FD&C Co. Ltd., Incheon, Korea

Sangkyu Park, Kyoungwhan Back and Oksoo Han
Department of Biotechnology, Chonnam National University, Gwangju, Korea

Soomin Park
Research Institute, National Agricultural Products Quality Management Service, Seoul, Korea

Byung-Jun Oh
Biological Control Center, Jeonnam Bioindustry Foundation, JeollaNamdo, Korea

Jeong-Il Kim and Young Soon Kim
Department of Biotechnology, Chonnam National University, Gwangju, Korea
Kumho Life Science Laboratory, Chonnam National University, Gwangju, Korea

Yan Liu, Xiaomei Zhu, Qiaolian Wen, Xi Wang and Hong Pan
Graduate School, Peking Union Medical College, Beijing, China
National Research Institute for Family Planning, Beijing, China

Fengyu Wang and Congmin Li
Henan Research Institute of Population and Family Planning, Key Laboratory of Population Defects Intervention Technology of Henan Province, Zhengzhou, China

Yuan Wu
Cardiac Surgery Department, Xiamen Heart Center, Organ Transplantation Institute of Xiamen University, Xiangan District, Xiamen, China

Sainan Tan
Key Laboratory of Genetics and Birth Health of Hunan Province, Family Planning Institute of Hunan Province, Chang sha, China

Jing Wang
Key Laboratory of Genetics and Birth Health of Hunan Province, Family Planning Institute of Hunan Province, Chang sha, China

Xu Ma
Graduate School, Peking Union Medical College, Beijing, China National Research Institute for Family Planning, Beijing, China
World Health Organization Collaborating Centre for Research in Human Reproduction, Beijing, China

Yan Zhao, Tao Xu, Chun-Ying Shen, Guang-Hui Xu, Shi-Xuan Chen, Mei-Jing Li, Xin Deng, Li-Li Wang, Yan Zhu and Wei-Tao Lv
Key Laboratory of Plant Resources, Institute of Botany, Chinese Academy of Sciences, Beijing, China

Li-Zhen Song
Key Laboratory of Plant Resources, Institute of Botany, Chinese Academy of Sciences, Beijing, China
Key Laboratory of Plant Molecular Physiology, Institute of Botany, Chinese Academy of Sciences, Beijing, China

Zhi-Zhong Gong
State Key Laboratory of Plant Physiology and Biochemistry, College of Biological Sciences, China Agricultural University, Beijing, China
Chun-Ming Liu
Key Laboratory of Plant Molecular Physiology, Institute of Botany, Chinese Academy of Sciences, Beijing, China

Nasir Ahmed Rajput, Meixiang Zhang, Yanyan Ru, Tingli Liu, Jing Xu, Li Liu, Daolong Dou and Joseph Juma Mafurah
Department of Plant Pathology, Nanjing Agricultural University, Nanjing, China

Tetsuya Chujo, Satoshi Ogawa, Yuka Masuda, Takafumi Shimizu, Satoshi Ogawa, Yuka Masuda, Takafumi Shimizu, Kazunori Okada and Hideaki Nojiri
Biotechnology Research Center, The University of Tokyo, Bunkyo-ku, Tokyo, Japan

Koji Miyamoto and Hisakazu Yamane
Biotechnology Research Center, The University of Tokyo, Bunkyo-ku, Tokyo, Japan
Department of Biosciences, Teikyo University, Utsunomiya, Tochigi, Japan

Mitsuko Kishi-Kaboshi, Akira Takahashi, Yoko Nishizawa and Eiichi Minami
Genetically Modified Organism Research Center, National Institute of Agrobiological Sciences, Tsukuba, Ibaraki, Japan

Yulong Guo, Yao Han, Jing Ma, Huiping Wang and Mingyang Li
Chongqing Engineering Research Center for Floriculture, Key Laboratory of Horticulture Science for Southern Mountainous Regions, Ministry of Education, College of Horticulture and Landscape Architecture, Southwest University, Chongqing, China

Xianchun Sang
College of Agronomy and Biotechnology, Southwest University, Chongqing, China

Sean Chapman, Brian Harrower and Kara McGeachy
Cell and Molecular Sciences, James Hutton Institute, Invergowrie-Dundee, United Kingdom

Laura J. Stevens, Pauline S. M. Van Weymers and Paul R. J. Birch
Cell and Molecular Sciences, James Hutton Institute, Invergowrie-Dundee, United Kingdom
Division of Plant Sciences, University of Dundee at James Hutton Institute, Invergowrie-Dundee, United Kingdom
Dundee Effector Consortium, Invergowrie-Dundee, United Kingdom

Petra C. Boevink, Xinwei Chen and Ingo Hein
Cell and Molecular Sciences, James Hutton Institute, Invergowrie-Dundee, United Kingdom
Dundee Effector Consortium, Invergowrie-Dundee, United Kingdom

Stefan Engelhardt
Division of Plant Sciences, University of Dundee at James Hutton Institute, Invergowrie-Dundee, United Kingdom
Dundee Effector Consortium, Invergowrie-Dundee, United Kingdom

Colin J. Alexander
Biomathematics and Statistics Scotland, Invergowrie-Dundee, United Kingdom

Nicolas Champouret
J.R. Simplot Company, Simplot Plant Sciences, Boise, Idaho, United States of America

Index